AF806648

Electrokinetics Across Disciplines and Continents

Alexandra B. Ribeiro • Eduardo P. Mateus
Nazaré Couto
Editors

Electrokinetics Across Disciplines and Continents

New Strategies for Sustainable Development

Springer

Editors
Alexandra B. Ribeiro
CENSE, Departamento de Ciências e Engenharia do Ambiente
Faculdade de Ciências e Tecnologia
Universidade Nova de Lisboa
Caparica, Portugal

Eduardo P. Mateus
CENSE, Departamento de Ciências e Engenharia do Ambiente
Faculdade de Ciências e Tecnologia
Universidade Nova de Lisboa
Caparica, Portugal

Nazaré Couto
CENSE, Departamento de Ciências e Engenharia do Ambiente
Faculdade de Ciências e Tecnologia
Universidade Nova de Lisboa
Caparica, Portugal

ISBN 978-3-319-20178-8 ISBN 978-3-319-20179-5 (eBook)
DOI 10.1007/978-3-319-20179-5

Library of Congress Control Number: 2015946866

Springer Cham Heidelberg New York Dordrecht London
© Springer International Publishing Switzerland 2016
This work is subject to copyright. All rights are reserved by the Publisher, whether the whole or part of the material is concerned, specifically the rights of translation, reprinting, reuse of illustrations, recitation, broadcasting, reproduction on microfilms or in any other physical way, and transmission or information storage and retrieval, electronic adaptation, computer software, or by similar or dissimilar methodology now known or hereafter developed.
The use of general descriptive names, registered names, trademarks, service marks, etc. in this publication does not imply, even in the absence of a specific statement, that such names are exempt from the relevant protective laws and regulations and therefore free for general use.
The publisher, the authors and the editors are safe to assume that the advice and information in this book are believed to be true and accurate at the date of publication. Neither the publisher nor the authors or the editors give a warranty, express or implied, with respect to the material contained herein or for any errors or omissions that may have been made.

Printed on acid-free paper

Springer International Publishing AG Switzerland is part of Springer Science+Business Media (www.springer.com)

Preface

The book *Electrokinetics Across Disciplines and Continents—New Strategies for Sustainable Development* aims to discuss and deepen the knowledge about electrokinetic (EK) process. The EK process could be used as an integrated approach for new strategies aiming at sustainable development and for supporting waste strategies worldwide. The conciliation of the EK process in the recovery of secondary resources, remediation, and conservation is a multidisciplinary, novel approach that opens new technical possibilities for waste minimization, through the upgradation of particulate waste products and recovery of secondary resources for industrial, agricultural, or social use. The EK process can also be coupled with phytoremediation and integrated with nanotechnology, enlarging the scope of its application. This was the basis and the motivation for this work.

The insights provided in this book are mainly based on a compilation of the works developed in the scope of ELECTROACROSS (*electrokinetics across disciplines and continents: an integrated approach to finding new strategies to sustainable development*), an FP7 People International Research Staff Exchange Scheme (IRSES) project.

The book is divided into five main parts: (I) Introduction and Overview of the Process; (II) Remediation of Contaminants and Recovery of Secondary Resources with Socio-Economical Value; (III) Conservation of Cultural Heritage and Use in Construction Material; (IV) Modeling of the Electrokinetic Process; and (V) Coupling Electrokinetic Process with Other Technologies to Enhance Performance and Sustainability (analytical, nano-, and phytotechnologies).

The book starts with an overview of EK soil remediation followed by influence of soil structure on EK dewatering process, EK enabled de-swelling of clay and soil stabilization, sustainable power generation from salinity gradient energy, and adsorption processes. The issue of phosphorus recovery in water and wastewater treatment plants by EK is discussed, together with remediation of copper mine tailings or as an alternative for soil and compost characterization. Life cycle assessment of EK remediation, electro-desalination of buildings damaged by salt weathering and incorporation of fly ashes as substitute for cement in mortar are also presented. A coupled reactive-transport model for EK remediation is discussed as well as the modeling of the transport of EK and zero valent iron nanoparticles and EK remediation of a mercury-polluted soil. The coupling of electrokinetics with phytotechnologies for arsenic removal or with nanoremediation for organic removal is also discussed, together with a broader range of topics regarding phytoremediation of pharmaceuticals and personal care products or inorganic compounds. The last part is devoted to analytical methodologies that allow detection and monitoring of contaminants in specific matrices.

We do hope you will find this book of interest, and we would like to thank all those who contributed to it.

Caparica, Portugal
30 March 2015

Alexandra B. Ribeiro
Eduardo P. Mateus
Nazaré Couto

Contents

Part I Introduction and Overview of the Process

1 **Electrokinetic Soil Remediation: An Overview** ... 3
Henrik K. Hansen, Lisbeth M. Ottosen, and Alexandra B. Ribeiro

2 **Soil Structure Influence on Electrokinetic Dewatering Process** ... 19
Vikas Gingine and Rafaela Cardoso

3 **Electrokinetically Enabled De-swelling of Clay** ... 43
Reena A. Shrestha, Angela P. Zhang, Eduardo P. Mateus, Alexandra B. Ribeiro, and Sibel Pamukcu

4 **Sustainable Power Generation from Salinity Gradient Energy by Reverse Electrodialysis** ... 57
Sylwin Pawlowski, João Crespo, and Svetlozar Velizarov

5 **The Kinetic Parameters Evaluation for the Adsorption Processes at "Liquid–Solid" Interface** ... 81
Svetlana Lyubchik, Elena Lygina, Andriy Lyubchyk, Sergiy Lyubchik, José M. Loureiro, Isabel M. Fonseca, Alexandra B. Ribeiro, Margarida M. Pinto, and Agnes M. Sá Figueiredo

Part II Remediation of Contaminants and Recovery of Secondary Resources with Socio-Economical Value

6 **Electrochemical Process for Phosphorus Recovery from Water Treatment Plants** ... 113
Nazaré Couto, Margarida Ribau Teixeira, Paula R. Guedes, Eduardo P. Mateus, and Alexandra B. Ribeiro

7 **Electrochemical Process for Phosphorus Recovery from Wastewater Treatment Plants** 129
Paula R. Guedes, Nazaré Couto, Eduardo P. Mateus, and Alexandra B. Ribeiro

8 **Electrokinetic Remediation of Copper Mine Tailings: Evaluating Different Alternatives for the Electric Field** 143
Henrik K. Hansen, Adrián Rojo, Claudia Gutiérrez, Pernille E. Jensen, and Lisbeth M. Ottosen

9 **Electrokinetics as an Alternative for Soil and Compost Characterization** 161
Alejandro Serna González, Lucas Blandón Naranjo, Jorge Andrés Hoyos, and Mario Víctor Vázquez

10 **Life Cycle Assessment of Soil and Groundwater Remediation: Groundwater Impacts of Electrokinetic Remediation** 173
Luís M. Nunes, Helena I. Gomes, Margarida Ribau Teixeira, Celia Dias-Ferreira, and Alexandra B. Ribeiro

Part III Conservation of Cultural Heritage and Use in Construction Material

11 **Electro-desalination of Buildings Suffering from Salt Weathering** 205
Lisbeth M. Ottosen and Henrik K. Hansen

12 **Incorporation of Different Fly Ashes from MSWI as Substitute for Cement in Mortar: An Overview of the Suitability of Electrodialytic Pre-treatment** 225
Cátia C. Magro, Paula R. Guedes, Gunvor M. Kirkelund, Pernille E. Jensen, Lisbeth M. Ottosen, and Alexandra B. Ribeiro

Part IV Modeling of the Electrokinetic Process

13 **A Coupled Reactive-Transport Model for Electrokinetic Remediation** 251
Juan Manuel Paz-García, María Villén-Guzmán, Ana García-Rubio, Stephen Hall, Matti Ristinmaa, and César Gómez-Lahoz

14 **Electrokinetics and Zero Valent Iron Nanoparticles: Experimental and Modeling of the Transport in Different Porous Media** 279
Helena I. Gomes, José M. Rodríguez-Maroto, Alexandra B. Ribeiro, Sibel Pamukcu, and Celia Dias-Ferreira

15 **Feasibility Study of the Electrokinetic Remediation of a Mercury-Polluted Soil** 295
Ana García-Rubio, María Villén-Guzmán, Francisco García-Herruzo, José M. Rodríguez-Maroto, Carlos Vereda-Alonso, César Gómez-Lahoz, and Juan Manuel Paz-García

Part V Coupling Electrokinetic Process with Other Technologies to Enhance Performance and Sustainability

16 **Phytoremediation Coupled to Electrochemical Process for Arsenic Removal from Soil** 313
Paula R. Guedes, Nazaré Couto, Alexandra B. Ribeiro, and Dong-Mei Zhou

17 **Nanoremediation Coupled to Electrokinetics for PCB Removal from Soil** 331
Helena I. Gomes, Guangping Fan, Lisbeth M. Ottosen, Celia Dias-Ferreira, and Alexandra B. Ribeiro

18 **Removal of Pharmaceutical and Personal Care Products in Aquatic Plant-Based Systems** 351
Ana R. Ferreira, Nazaré Couto, Paula R. Guedes, Eduardo P. Mateus, and Alexandra B. Ribeiro

19 **Phytoremediation of Inorganic Compounds** 373
Bruno Barbosa, Jorge Costa, Sara Boléo, Maria Paula Duarte, and Ana Luisa Fernando

20 **Sensing of Component Traces in Complex Systems** 401
Maria Raposo, Paulo A. Ribeiro, Nezha El Bari, and Benachir Bouchikhi

21 **Analysis of Endocrine Disrupting Chemicals in Food Samples** 427
Miriany A. Moreira, Leiliane C. André, Marco D.R. Gomes da Silva, and Zenilda L. Cardeal

22 **Multidimensional Chromatographic Techniques for Monitoring and Characterization of Environmental Samples** 439
Eduardo P. Mateus, Marco D.R. Gomes da Silva, Alexandra B. Ribeiro, and Philip Marriott

Index 455

Contributors

Leiliane C. André Departamento de Análises Clínicas e Toxicológicas, Faculdade de Farmácia (FaFar), Universidade Federal de Minas Gerais, Belo Horizonte, MG, Brazil

Bruno Barbosa MEtRiCS, Departamento de Ciências e Tecnologia da Biomassa, Faculdade de Ciências e Tecnologia, Universidade Nova de Lisboa, Caparica, Portugal

Sara Boléo MEtRiCS, Departamento de Ciências e Tecnologia da Biomassa, Faculdade de Ciências e Tecnologia, Universidade Nova de Lisboa, Caparica, Portugal

Benachir Bouchikhi Sensor Electronic & Instrumentation Group, Department of Physics, Faculty of Sciences, Moulay Ismaïl University, Meknes, Morocco

Lucas Blandón Naranjo Interdisciplinary Group of Molecular Studies (GIEM), Chemistry Institute, Faculty of Exact and Natural Sciences, University of Antioquia, Medellín, Colombia

Zenilda L. Cardeal Departamento de Química, Instituto de Ciências Exatas (ICEx) Universidade Federal de Minas Gerais, Belo Horizonte, MG, Brazil

Rafaela Cardoso ICIST, Instituto Superior Técnico, University of Lisbon, Lisbon, Portugal

Jorge Costa MEtRiCS, Departamento de Ciências e Tecnologia da Biomassa, Faculdade de Ciências e Tecnologia, Universidade Nova de Lisboa, Caparica, Portugal

Nazaré Couto CENSE, Departamento de Ciências e Engenharia do Ambiente, Faculdade de Ciências e Tecnologia, Universidade Nova de Lisboa, Caparica, Portugal

Key Laboratory of Soil Environment and Pollution Remediation, Institute of Soil Science, Chinese Academy of Sciences, Nanjing, China

João Crespo LAQV-REQUIMTE, Departamento de Química, Faculdade de Ciências e Tecnologia, Universidade Nova de Lisboa, Caparica, Portugal

Nezha El Bari Biotechnology Agroalimentary and Biomedical Analysis Group, Department of Biology, Faculty of Sciences, Moulay Ismaïl University, Meknes, Morocco

Celia Dias-Ferreira CERNAS—Research Center for Natural Resources, Environment and Society, Escola Superior Agraria de Coimbra, Instituto Politecnico de Coimbra, Coimbra, Portugal

Maria Paula Duarte MEtRiCS, Departamento de Ciências e Tecnologia da Biomassa, Faculdade de Ciências e Tecnologia, Universidade Nova de Lisboa, Caparica, Portugal

Guangping Fan Key Laboratory of Soil Environment and Pollution Remediation, Institute of Soil Science, Chinese Academy of Sciences (ISSCAS), Nanjing, China

Ana Luisa Fernando MEtRiCS, Departamento de Ciências e Tecnologia da Biomassa, Faculdade de Ciências e Tecnologia, Universidade Nova de Lisboa, Caparica, Portugal

Ana R. Ferreira CENSE, Departamento de Ciências e Engenharia do Ambiente, Faculdade de Ciências e Tecnologia, Universidade Nova de Lisboa, Caparica, Portugal

Agnes M. Sá Figueiredo Instituto de Microbiologia Paulo de Góes, Universidade Federal do Rio de Janeiro, Rio de Janeiro, Brazil

Isabel M. Fonseca REQUIMTE, Departamento de Química, Faculdade de Ciências e Tecnologia, Universidade Nova de Lisboa, Caparica, Portugal

Francisco García-Herruzo Department of Chemical Engineering, University of Málaga, Málaga, Spain

Ana García-Rubio Department of Chemical Engineering, University of Málaga, Málaga, Spain

Vikas Gingine ICIST, Instituto Superior Técnico, University of Lisbon, Lisbon, Portugal

Helena I. Gomes CENSE, Departamento de Ciências e Engenharia do Ambiente, Faculdade de Ciências e Tecnologia, Universidade Nova de Lisboa, Caparica, Portugal

CERNAS—Research Center for Natural Resources, Environment and Society, Escola Superior Agraria de Coimbra, Instituto Politecnico de Coimbra, Coimbra, Portugal

Department of Civil Engineering, Technical University of Denmark, Lyngby, Denmark

Marco D.R. Gomes da Silva LAQV, REQUIMTE, Departamento de Química, Faculdade de Ciências e Tecnologia, Universidade Nova de Lisboa, Caparica, Portugal

César Gómez-Lahoz Department of Chemical Engineering, University of Málaga, Málaga, Spain

Paula R. Guedes CENSE, Departamento de Ciências e Engenharia do Ambiente, Faculdade de Ciências e Tecnologia, Universidade Nova de Lisboa, Caparica, Portugal

Key Laboratory of Soil Environment and Pollution Remediation, Institute of Soil Science, Chinese Academy of Sciences, Nanjing, China

Claudia Gutiérrez Departamento de Ingeniería Química y Ambiental, Universidad Técnica Federico Santa Maria, Valparaíso, Chile

Stephen Hall Division of Solid Mechanics, University of Lund, Lund, Sweden

Henrik K. Hansen Departamento de Ingeniería Química y Ambiental, Universidad Técnica Federico Santa María, Valparaíso, Chile

Jorge Andrés Hoyos Interdisciplinary Group of Molecular Studies (GIEM), Chemistry Institute, Faculty of Exact and Natural Sciences, University of Antioquia, Medellín, Colombia

Pernille E. Jensen Department of Civil Engineering, Technical University of Denmark, Lyngby, Denmark

Gunvor M. Kirkelund Department of Civil Engineering, Technical University of Denmark, Lyngby, Denmark

José M. Loureiro Faculdade de Engenharia, Universidade do Porto, Porto, Portugal

Elena Lygina REQUIMTE, Departamento de Química, Faculdade de Ciências e Tecnologia, Universidade Nova de Lisboa, Caparica, Portugal

Sergiy Lyubchik CQE, Departamento de Engenharia Química, Instituto Superior Técnico de Lisboa, Lisbon, Portugal

Svetlana Lyubchik REQUIMTE, Departamento de Química, Faculdade de Ciências e Tecnologia, Universidade Nova de Lisboa, Caparica, Portugal

Andriy Lyubchyk CENIMAT, Departamento de Engenharia de Materiais e Cerâmica, Faculdade de Ciências e Tecnologia, Universidade Nova de Lisboa, Caparica, Portugal

Cátia C. Magro CENSE, Departamento de Ciências e Engenharia do Ambiente, Faculdade de Ciências e Tecnologia, Universidade Nova de Lisboa, Caparica, Portugal

Department of Civil Engineering, Technical University of Denmark, Lyngby, Denmark

Philip Marriott Australian Centre for Research on Separation Science, School of Chemistry, Monash University, Clayton, VIC, Australia

Eduardo P. Mateus CENSE, Departamento de Ciências e Engenharia do Ambiente, Faculdade de Ciências e Tecnologia, Universidade Nova de Lisboa, Caparica, Portugal

Miriany A. Moreira Departamento de Química, Instituto de Ciências Exatas (ICEx) Universidade Federal de Minas Gerais, Belo Horizonte, MG, Brazil

Luís M. Nunes CERIS—Civil Engineering Research and Innovation for Sustainability, Faculdade de Ciências e Tecnologia, Universidade do Algarve, Faro, Portugal

Lisbeth M. Ottosen Department of Civil Engineering, Technical University of Denmark, Lyngby, Denmark

Sibel Pamukcu Department of Civil and Environmental Engineering, Lehigh University, Bethlehem, PA, USA

Sylwin Pawlowski LAQV-REQUIMTE, Departamento de Química, Faculdade de Ciências e Tecnologia, Universidade Nova de Lisboa, Caparica, Portugal

Juan Manuel Paz-García Division of Solid Mechanics, University of Lund, Lund, Sweden

Margarida M. Pinto ISQ-R&D Department, Instituto de Soldadura e Qualidade, Porto Salvo, Portugal

Maria Raposo CEFITEC, Departamento de Física, Faculdade de Ciências e Tecnologia, Universidade Nova de Lisboa, Caparica, Portugal

José M. Rodríguez-Maroto Department of Chemical Engineering, University of Málaga, Málaga, Spain

Alexandra B. Ribeiro CENSE, Departamento de Ciências e Engenharia do Ambiente, Faculdade de Ciências e Tecnologia, Universidade Nova de Lisboa, Caparica, Portugal

Paulo A. Ribeiro CEFITEC, Departamento de Física, Faculdade de Ciências e Tecnologia, Universidade Nova de Lisboa, Caparica, Portugal

Matti Ristinmaa Division of Solid Mechanics, University of Lund, Lund, Sweden

Adrián Rojo Departamento de Ingeniería Química y Ambiental, Universidad Técnica Federico Santa Maria, Valparaíso, Chile

Alejandro Serna González Interdisciplinary Group of Molecular Studies (GIEM), Chemistry Institute, Faculty of Exact and Natural Sciences, University of Antioquia, Medellín, Colombia

Reena A. Shrestha Department of Civil and Environmental Engineering, Lehigh University, Bethlehem, PA, USA

Department of Petroleum Engineering, The Petroleum Institute, Abhu Dhabi, UAE

Margarida Ribau Teixeira CENSE, Faculdade de Ciências e Tecnologia, Universidade do Algarve, Faro, Portugal

Mario Víctor Vázquez Interdisciplinary Group of Molecular Studies (GIEM), Chemistry Institute, Faculty of Exact and Natural Sciences, University of Antioquia, Medellín, Colombia

Svetlozar Velizarov LAQV-REQUIMTE, Departamento de Química, Faculdade de Ciências e Tecnologia, Universidade Nova de Lisboa, Caparica, Portugal

Carlos Vereda-Alonso Department of Chemical Engineering, University of Málaga, Málaga, Spain

María Villén-Guzmán Department of Chemical Engineering, University of Málaga, Málaga, Spain

Angela P. Zhang Department of Civil and Environmental Engineering, Lehigh University, Bethlehem, PA, USA

Dong-Mei Zhou Key Laboratory of Soil Environment and Pollution Remediation, Institute of Soil Science, Chinese Academy of Sciences, Nanjing, China

Part I
Introduction and Overview of the Process

Chapter 1
Electrokinetic Soil Remediation: An Overview

Henrik K. Hansen, Lisbeth M. Ottosen, and Alexandra B. Ribeiro

1.1 Introduction

When F. F. Reuss more than 200 years ago, in 1809, discovered that water can be transported within a porous material by applying an electric DC field (Reuss 1809), he had probably not thought that this finding would inspire thousands of researchers to apply this phenomenon on a huge amount of porous materials. Several of these materials were found to have the same basic properties as the compacted clay that Reuss tested, and therefore the same effect has been obtained. The phenomenon that Reuss discovered was later defined as *electroosmosis*.

In the 1930–1940s, L. Casagrande tested the principle of electroosmosis on real soil in the field for the first time. He consolidated soft clays by means of an electric current (Casagrande 1948). This is many times considered as "the starting point" for electroosmosis as an engineering tool. During the last 50–60 years, the electrokinetic (EK) phenomena, that occur when applying electric DC fields, have been tried to be used in the field in solving different emerging and practical problems. The particularity of the water transport in a DC field has led various engineers to test this idea on problems where conventional methods have shown their limitations and were inefficient for the purpose.

H.K. Hansen (✉)
Departamento de Ingeniería Química y Ambiental, Universidad Técnica Federico Santa Maria, Avenida España 1680, Valparaíso, Chile
e-mail: henrik.hansen@usm.cl

L.M. Ottosen
Department of Civil Engineering, B118, Technical University of Denmark, 2800 Kgs., Lyngby, Denmark

A.B. Ribeiro
CENSE, Departamento de Ciências e Engenharia do Ambiente, Faculdade de Ciências e Tecnologia, Universidade Nova de Lisboa, Caparica 2829-516, Portugal

© Springer International Publishing Switzerland 2016

A.B. Ribeiro et al. (eds.), *Electrokinetics Across Disciplines and Continents*,
DOI 10.1007/978-3-319-20179-5_1

Table 1.1 Main pilot- and full-scale applications of EK (Page and Page 2002; Virkutyte et al. 2002; Lageman et al. 2005; Ottosen et al. 2008; Yeung 2011; Alshawabkeh 2009; Kim et al. 2012; Gomes 2014; Symes 2012; USEPA 1997)

Year	Field-/pilot-scale remediation
1936	Application to remove excess salts from alkali soils in India
1939	Application to reverse the seepage flow direction and stabilize a long railroad cut at Salzgitter, Germany
1976	Desalination of concrete, Federal Highway Association, USA
1987	EK pilot project at former paint factory by Groningen, the Netherlands conducted by Geokinetics International, Inc.
1991	EK remediation of soil contaminated with chlorinated solvents at Anaheim, USA, conducted by Environmental & Technology Services
1992	Cadmium and other metals removal by EK at Woensdrecht, the Netherlands, at a former Dutch Royal Air Force base (Geokinetics International, Inc.). This project is considered to be the largest electrokinetic project completed worldwide (3600 cubic yard area)
1993	Electro-bioreclamation pilot project at former industrial site with diesel fuel and aromatics at Vorden, the Netherlands
1994	Injection of chemical conditioners, Electrokinetics, Inc., US Army Waterways Experiment Station, Vickburg, MS, USA
1994	EK was used in situ in the Old TNX Basin at the Savannah River Site in South Carolina, USA, to remediate mercuric nitrate contamination in unsaturated soil
1995	In situ remediation of Uranium contaminated soils, Oak Ridge, K25 Facility, Oak Ridge, TN, USA
1995	Pilot project Lasagna, Paducah Site, contaminated with TCE, Kentucky, USA
1995–1996	Removal of lead from soils in situ at Firing Range 24A located in Fort Polk, LA, USA
1997	Electrokinetic demonstration at the unlined chromium acid pit, Sandia National Laboratory, USA
1998	Field-scale demonstration of chromium and copper remediation, Point Mugu, CA, USA
2005	Pilot-scale electrochemical cleanup of lead contaminated soil in a firing range, USA
2006–2008	Quicrez Industrial Site in Fond du Lac, Wisconsin, USA. Soil contaminated with TCE
2010	EK treatment has been used to successfully stabilize a Victorian railway embankment in London, UK. The embankment was 9 m high with side slopes of 22°
2011	Pilot-scale application in a rice field near a zinc refinery plant located at Jang Hang, South Korea
2011–2012	Electroosmotic stabilization of a 7 m high failing embankment on the A21 in Kent, UK

In Table 1.1, some of these applications in full- or pilot scale are summarized, and this again shows the variety of applications and positive results when applied electric fields to porous media (Page and Page 2002; Virkutyte et al. 2002; Lageman et al. 2005; Ottosen et al. 2008; Yeung 2011; Alshawabkeh 2009; Kim et al. 2012; Gomes 2014; Symes 2012; USEPA 1997). These go from soil consolidation, soil decontamination, soil dewatering, and densification to salt removal and electroosmotic stabilization. The company Geokinetics International, Inc., conducted around 13 full- and pilot-scale EK projects between 1987 and 1998 (USEPA 1997).

Small particles were also found to move in the electric field and always in opposite direction than the electroosmotic flow. This phenomenon of particle movement is named electrophoresis. In fact, the electrophoresis has also shown great uses in our daily life such as in food analysis, separation and analysis of biological samples (e.g., DNA and proteins), separation of nanoparticles and environmental samples (e.g., pesticide and other pollutant analysis), and as a deposition technique in ceramics processing (Dolnik 2008; García-Cañas and Cifuentes 2008; Boccaccini and Zhitomirsky 2002; Besra and Liu 2007).

During the late 1980s, a lot of focus of the leading researchers and practitioners turned onto heavy metal removal from contaminated soil, sediments, and sludge (Lageman et al. 2005; Yeung 2011). Again they applied an electric DC field and discovered that contaminants were transported towards the electrodes. In the case of heavy metal contaminated soil, electroosmosis could not contribute as the only main reason for the heavy metal movement in the solid materials, since the metals in most cases were present in soil solution as ions during remediation. These ions were then migrating—mostly referred to as by electromigration—towards the electrode of opposite their electrical charge in the DC field. The process is called several names in the literature (Gomes 2014) such as electrokinetics, electroreclamation, electroremediation, and electroosmosis remediation—but in this chapter, it will be referred to as electrokinetic remediation or abbreviated as EKR.

In comparison to other competing soil remediation technologies available at that time, EKR showed several potential advantages. While the success of the technology was sensitive to many physicochemical variables, such as cation exchange capacity, surface charge and geochemistry, its major advantages included (Alshawabkeh 2009):

- It could be implemented in situ with minimal soil disruption.
- It was well suited for fine-grained, heterogeneous media, where other techniques were ineffective.
- The technology minimized the post-treatment volume of waste material.

Moreover, companies specialized in electrokinetic remediation were also established during the period. Prominent commercial establishments working on different applications of the process include Holland Environment BV of Doorn, the Netherlands; Electro-Petroleum, Inc., of Wayne, PA, USA; Geokinetics International, Inc., Palo Alto, CA, USA; and Electrokinetics, Inc., of Baton Rouge, LA, USA (Yeung 2011).

Some of the reasons that are mentioned to explain why EKR today still is considered as a treatment method under development and not fully developed are related to differences between practices in the application in the field and practices in laboratory investigation. Furthermore, it is clear that the behavior of the solid material is different in the field than in the laboratory, and scaling up of the process is still challenging.

1.2 EKR: Who Is Working on It?

Today, there is still a discussion going on about who was the first in applying electric fields for the specific aim of soil remediation. Several pioneers developed the method independently of each other in the end of the 1980s. The most active and consistent from that time were the following who really contributed to the development and distribution of the method to the public.

The group of Yalcin Acar was a true pioneer in the field, and the scientific paper by Acar, Y. B., and Alshawabkeh, A. "Principles of Electrokinetic Remediation" published in Environmental Science and Technology (1993) is one of the earliest complete publications about electrokinetic remediation and is still considered a milestone in EKR. The group holds various patents on the methodology.

In the 1990s, Ronald F. Probstein introduced one of the concepts of electrokinetic soil remediation. The basic procedure was patented and licensed to an industrial firm for further development, and today the subject has become one that is widely studied and applied worldwide. The idea was first published by R. F. Probstein and R. E. Hicks in "Removal of contaminants from soils by electric fields," in Science (1993).

At the same time, Renout Lageman and his team were developing the concept and using it in the field for real cleanup purposes in the Netherlands. The concept and successful results (Electroreclamation: applications in the Netherlands) were published in Environmental Science and Technology—another paper today considered as a milestone of EKR (Lageman and Pool 1993).

Later, during the 1990, research groups were created all over the world. Even if several of the early EKR groups are not active in the field anymore and have switched to other topics of interest, it is important to mention the groups that are still active in applying EKR after decade-long research and development investment in the process—maybe not only to soils but to a diversity of different solid waste products or as a tool to solve an emerging environmental problem. These groups have always been active to promote their results publically.

At Lehigh University, Bethlehem, PA, USA, the group headed by Sibel Pamukcu has been conducting leading research on electrochemical processing of soil and groundwater. Especially, new focus has been based on the application of direct current electric field for enhanced recovery of oil, and for in situ destruction of contaminants through enhanced redox in clay-rich soils and rock formations.

At Northeastern University, Boston, USA, the group headed by Akram Alshawabkeh has been involved with EKR since the very early days of the application of the method, and therefore plays a significant and important role in the development of the process—especially on field application and theoretical understand of the ongoing processes during the treatment.

At University of Illinois, Chicago, USA, the group headed by Krishna R. Reddy has participated in the field since the 1990s, plays a role as one of the scientific front leaders of EKR in the development of the process in geological and environmental engineering, and has an extended experience in application of the method in the field.

At the Technical University of Denmark, the research group headed by Lisbeth M. Ottosen has for decades been developing the method to deal not only with soil contamination but a variety of solid waste normally deposited in safe landfills as hazardous waste. The group has several patents on their developments and is also applying the method to salt extraction from building materials, phosphorus recovery from wastewater treatment sludge and sludge ash, municipal solid waste incineration ashes, soil remediation, and monument restoration.

At Universidad de Malaga, Spain, the research group headed by José Rodriguéz-Maroto has been investigating the different processes occurring during EKR and been modeling the process—not only for soil but different waste materials. The understanding of the complex characteristics of the contaminated matrix has been their main target resulting in important findings for the academic society and field applications.

At Universidade Nova de Lisboa, Portugal, the group headed by Alexandra B. Ribeiro has, since the late 1990s, focused on specific environmental topics with urgent importance within the European Union such as chromated copper arsenate-treated and creosote-treated timber waste, herbicide removal, pharmaceutical and personal care products and other emerging organic contaminants from soil and wastewater, sewage sludge and phosphorus removal, nanofiltration concentrates—phosphorus recovery and microcystins degradation, in line with the development of new analytical technologies.

At Chonbuk National University in Jeonju, South Korea, the group headed by Kitae Baek started early 2000 and has by now settled as a leading EKR research group in Asia. Several field- and pilot-scale studies by this group have demonstrated its capacity to perform an adequate scaling up of the process.

1.3 Scientific Advances in EKR

In general, the pioneers of electrokinetic remediation from the end of 1980 were mostly from the USA. An important milestone was the Workshop on Electrokinetic Treatment and Its Applications in Environmental-Geotechnical Engineering for Hazardous Waste Site Remediation held in Seattle, WA, USA in 1986, which in the USA is considered as a starting point for EK (Yeung 2011). In this workshop, the state-of-understanding of the fundamentals and mechanisms of electrokinetics in migrating fluid and chemicals in soil and potential applications of electrokinetics for remediation of hazardous waste sites were summarized in by leading USA researchers and university professors such as James K. Mitchell of the University of California at Berkeley, Donald H. Gray of the University of Michigan at Ann Arbor and Harold W. Olsen of the U.S. Geological Survey, Ronald F. Probstein and Patricia C. Renaud of the Massachusetts Institute of Technology, John F. Ferguson of the University of Washington, and Burton A. Segall of the University of Massachusetts at Lowell.

Journal of Hazardous Materials, in its volume 55 (issues 1–3) of 1997, dedicated these issues to the Yalcin Acar—one of the pioneers the EKR field that died tragically in a car accident. In this volume, named Electrochemical Decontamination of Soil and Water, a series of EKR-related manuscripts were gathered from experts in the field, mostly from the North America continent. Therefore, this special volume can be considered as an important early scientific contribution to the understanding and distribution of fundamental findings using EKR for polluted soil and ground water.

In 1997, École Nationale Supérieure des Mines d'Albi-Carmaux in the southern France, invited researchers worldwide to the first known European symposium on the specific topic of electrokinetic remediation. Here for the first time, leading EKR researchers and practitioners could present and discuss findings and problems with the technique. The success and need of this event was so evident that in the following decades a series of symposia have been arranged to follow up—as the later defined EREM symposia. In most cases, the results from the conferences have been published in scientific journals open for the general public. The latest in the series, EREM 2014 was hosted in Malaga, Spain in 2014, confirmed the world spread interest of electrokinetic processes, counting with participants from all continents of the world.

Several scientific textbooks and handbooks on soil contamination and remediation, environmental restoration methods, and environmental engineering were published early 2000, where chapters or parts were dedicated to EKR. As examples one can mention: "Remediation Engineering of Contaminated Soils" edited by Wise et al. (2000), "Environmental Restoration of Metals-Contaminated Soil" edited by Iskanda (2000), "Geoenvironmental Engineering: Site Remediation, Waste Containment, and Emerging Waste Management Technologies" by Sharma and Reddy (2004), "Trace Elements in the Environment. Biogeochemistry, Biotechnology, and Bioremediation" edited by Prasad et al. (2006).

In 2009, K. R. Reddy and C. Cameselle edited what is known to be the first scientific book entirely addressed to EKR with the title "Electrochemical Remediation Technologies for Polluted Soils, Sediments and Groundwater"—covering a wide range of aspects during the use of electric fields applied to porous contaminated media to remove contaminants such as heavy metals, radionuclides, herbicides, polycyclic aromatic hydrocarbons, chlorinated organic compounds, and mixed contaminants. In this book, a variety of the world's leading scientists, engineers, and decisions makers within the field of EKR have contributed with their knowledge and expertise. With time, this book could be expected to be one of many of the necessary building stones in the further development of this method and technology.

In 2011, a consortium of a number of important research institutions introduced the project "Electrokinetics across disciplines and continents: an integrated approach to finding new strategies to sustainable development (ELECTROACROSS)." This EU-financed project coordinated by Dr. A. B. Ribeiro from the UNL group has the main objective to conciliate of the electrokinetic transport processes in the recovery of secondary resources, remediation, and conservation as a multidisciplinary novel approach that opens new technical possibilities for waste minimization, through upgrading of particulate waste products and the recovery of secondary resources for industrial, agricultural, or social use. This objective is achieved mainly through knowledge transfer activities, among a network of European and other continents centers of excellence, and ELECTROACROSS may be one way to help with the difficult integration of different scientific and practical fields to develop the EK process for the better of the environment.

1.4 EKR: Theoretical Aspects

In this chapter, the basic transport phenomena and reactions in soil during EKR will be summarized. It is not the objective to perform a thorough theoretical analysis—this has been done elsewhere in the literature and therefore one can refer to those works (Ottosen et al. 2008; Yeung 2011; Alshawabkeh 2009; Wise et al. 2000; Iskanda 2000; Sharma and Reddy 2004; Prasad et al. 2006; Reddy and Cameselle 2009). In this book chapter, only a basic introduction is given.

EKR uses an electric DC field with potential gradients around 1 V/cm between working electrodes—creating current densities of the order of milliamps per square centimeter applied to the cross-sectional area of soil mass between the electrodes (Prasad et al. 2006; Reddy and Cameselle 2009). The main interest of EK soil remediation in environmental cleanup operations lies in an attempt to concentrate and confine contaminants close to an electrode and remove them if possible. Inorganic contaminants will be transported as ions with electromigration, and organic contaminants and uncharged species of inorganic contaminants will be transported be electroosmosis towards the electrodes.

The principle in a classic EKR setup can be seen in Fig. 1.1. The contaminated soil—or other waste—is typically placed in the middle between electrodes separated physically from the soil in different manners ((I) and (II)). The cathode and anode are then connected to an external DC power supply. Ions present in the soil solution would then carry the current between the anode and cathode. When the contaminants are present in ionic form (e.g., Cr^{3+}, Pb^{2+}, or Cu^{2+} in Fig. 1.1), then

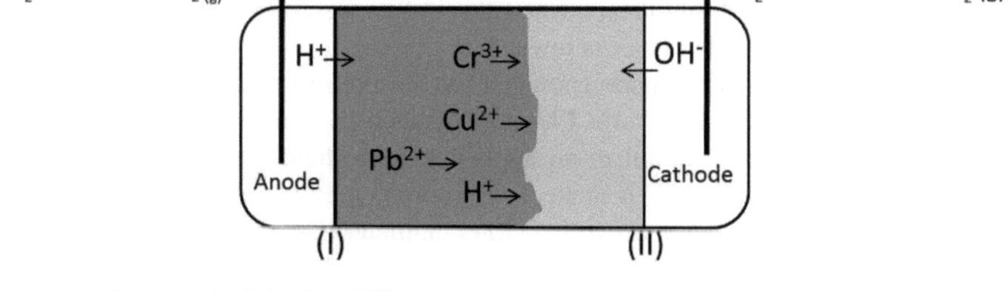

Fig. 1.1 General principle of an EKR setup

the electric field would transport them to the electrode of opposite charge, and they can be recovered or removed from the electrolyte solutions. The electrode reactions, which convert the electric DC field applied with the external electric circuit to ionic transport in the porous materials, are of course of crucial importance for the EKR process.

In most cases, the electrode processes are electrolysis of water—at the cathode resulting in hydrogen gas and OH^- generation and at the anode oxygen and H^+ production. Other cathode reactions used or observed in EKR are reduction of metal cations—such as $Cu^{2+} + 2e^- \rightarrow Cu$, or $Fe^{3+} + e^- \rightarrow Fe^{2+}$. Competing anode reactions could be or $Fe^{2+} \rightarrow Fe^{3+} + e^-$, or even chlorine production when treating very saline soils or waste materials ($2Cl^- \rightarrow 2e^- + Cl_2$). In the case of water electrolysis, the electrode reactions would generate H^+ and OH^- (see Fig. 1.1), which would penetrate the soil from each side. Several times (Alshawabkeh 2009; Yeung 2011) it has been found that this created an acid front moving from the anode towards the cathode—dissolving important amounts of contaminants from the soil. On the other hand, the alkaline front generated from the cathode towards the anode would precipitate the ionic contaminants such as heavy metals as hydroxides within the soil matrix. The result was that the electrical resistance increased drastically and the remediation process nearly stopped. In order to avoid the unwanted alkaline front to form, different solutions have been suggested such as pH control in the cathode compartment (Ho Lee and Yang 2000) or use of ion exchange membranes (Hansen et al. 1999).

When a voltage is applied across a fine-porous material, the electromigration results in a water movement towards the positive or negative electrode dependent on the overall surface charge of the porous material. Both counter- and co-ions will move towards the electrode of opposite charge. Since the counter-ions are in excess to the co-ions in the electric double layer, a net flow of ions across the electrode of opposite sign compared to the surfaces of the porous material will occur, and water molecules are pushed or dragged towards the electrode together with the counter-ions. The amount of water moved per ion is very large compared to the hydration numbers and electroosmosis is more than transport of hydration water (Ottosen et al. 2008).

Under normal operating conditions, soil pores have negatively charged surfaces, and there will then be a net flow of cations towards the cathode compared to the flow of anions. This causes then the water to be dragged along towards the cathode. While electromigration occurs in all moist, porous materials in an applied electric field, electroosmosis is only significant in materials with fine pores. Further, electroosmosis is most significant when the ion concentration in the pore water outside the electric double layer is low.

Another electrokinetic phenomenon that has been observed during electrokinetic soil remediation is electrophoresis. Electrophoresis is the transport of small charged particles in a stationary liquid in an applied electric field (Ottosen et al. 2008). Traditionally, electrophoresis is neglected in porous media, though colloidal particles may be transported in pores with a larger diameter than the particle. That small particles can move in capillaries is known from the technique of capillary electrophoresis, which is a method designed to separate species based on their size to charge ratio. In a porous material, the pores are seldom straight and electrophoretic transport will not happen over long distances due to the tortuosity of the pores.

Most porous materials (e.g., ion-exchange membranes, clays, concretes) carry a surface charge and have in common the presence of pores with unbalanced charges and corresponding unbound ions (Marry et al. 2003). The charged surfaces are counterbalanced by ions of opposite sign in the diffuse electric double layer. Ions in the solution with the same sign as the charged surface are co-ions and they are represented to a much lesser extent in the electric double layers than the counter-ions. Charge balance is always maintained throughout the system at all times, the porous media and electrolyte system must be electrically neutral. Charges cannot be added to, formed in, or removed from the system without addition, formation or removal of an equal number of the opposite charge. The electric current can be carried by ions in the electric double layer and by the ions dissolved in the brine, with rather different characteristics (Pengra and Wong 1996). Unlike in solutions, the ions in a porous material are not able to move by electromigration directly to the opposite pole by the shortest route. Instead, they have to find their way along the tortuous pores and around the particles or air filled voids that block the direct path. Moreover, the ions can be transported only in continuous pores, but not in closed ones and ions are only transported in the liquid phase. The electromigration rate of ions in the porous media depends on the pore volume, geometry, and the water content. Charged porous media are filled at least by counterions and often water and occasionally co-ions. The dynamic of counter-ions, water molecules, and co-ions depend strongly on the water content of the medium. For very compacted media, the water content is low and the ions are slowed down. For less compacted media, the dynamics of inserted water and counter-ions tend towards that in bulk solution.

The last transport phenomena during EKR that are not considered as important as the ones mentioned above are diffusion and advection. Diffusion is the movement of species under a chemical concentration gradient. Under normal EKR operation conditions, chemical gradients are typically not so high but in the narrow zone where the acid and basic fronts are to meet is an example, where diffusion becomes important. Advection by hydraulic gradients is not an important contribution to global transport (Prasad et al. 2006). However, one of the applications of EKR treatment is to act as a reactive barrier to avoid the advance of contaminants into groundwater, and here hydraulic gradients convert in a driving force to the movement of water.

Next to transport processes, several other processes occur within the porous material during application of the electric field: ion exchange, complexation, desorption/adsorption, dissolution/precipitation, degradation, pH changes, redox changes, phase changes, and structural changes (Ottosen et al. 2008; Yeung 2011; Prasad et al. 2006; Guedes et al. 2014). Many of these are related, e.g., when the pH in the soil during EKR drops due to the arrival of the acid front, metals precipitated as carbonates or oxides can dissolve and move in the electric field. Again, ion exchange and complexation are strongly affected by the pH, and the adsorption capacity of humus and clays—typically present in natural soil—depends on pH too. Redox changes occur when oxidants or reducing agents are transported with the electric field, and these can also affect dissolution or precipitation of different species.

1.5 Practical Application Versus Academical Research

In early days of EKR (late 1980s–beginning 1990s), the practical application of EKR in the field dominated scientific research on the topic. Several field projects were carried out—especially in the USA (Superfund program) as indicated in Table 1.1. The competition and potential market for EKR in those days were interesting, and several patents on different concepts, setups, detailed equipment, and EKR remediation conditions were accepted. This can be seen from Fig. 1.2, where the number of US patents (obtained from The United States Patent and Trademark Office (USPTO)) including the words "electrokinetic soil remediation" is given for different years. As can be seen, the number of patents is high in the 1990s and early 2000 but is decreasing further into the new millennium, and around 2010 the numbers is less than 50 % of the numbers during the 1990s.

On the other hand, the scientific interest of EK is continuously increasing in the latest years. Making a similar search as with patents, one can search the Thomson Reuters Web of Science (Thomson Reuters Web of Science) to find the number of publications including the words "electrokinetic soil remediation" in the abstract, keywords, or topic. With this information, Fig. 1.3 is obtained, showing the number of journal publications for different years. The tendency shows an increasing number with time—in contrast to the dropping number of patents. Figure 1.4 shows where the publications are generated. This figure confirms the tendency that North America researchers were very active during the 1990s; but in recent years, other continents are becoming leaders in research on the electrokinetic soil remediation. Looking into details of the manuscripts, especially since 2009, the number of publications from Asia is passing the number of publications from the rest of the world combined.

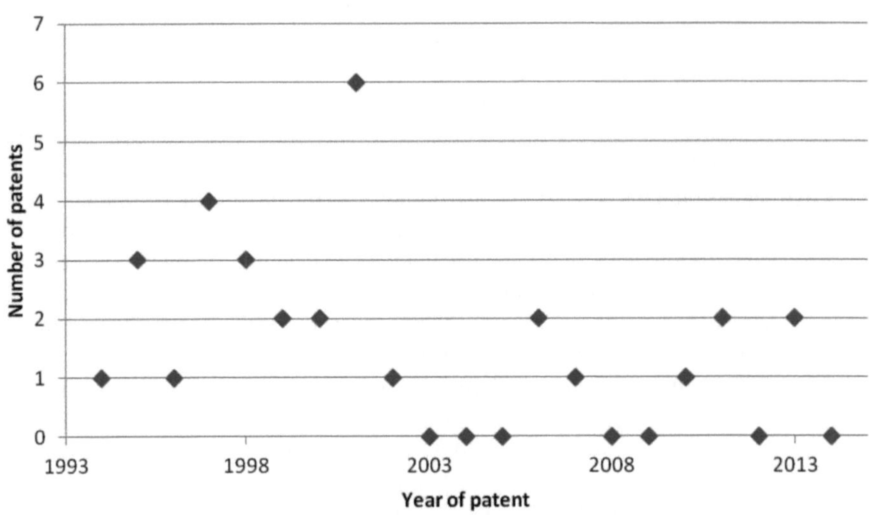

Fig. 1.2 Number of US patents on "electrokinetic soil remediation" from 1993 to 24th of September 2014 [Based on data published in (USPTO)]

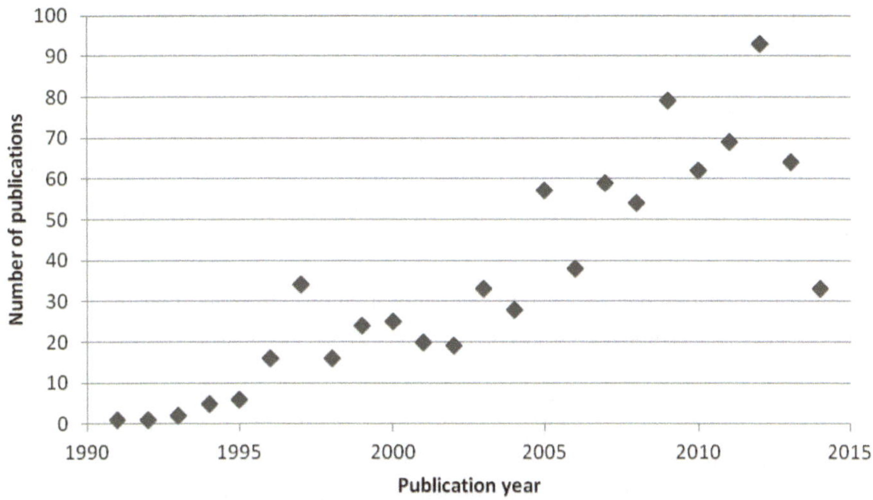

Fig. 1.3 Number of scientific publications on "electrokinetic soil remediation" registered by Thomson Reuters Web of Science from 1991 until 24th September 2014 (based on data published in Thomson Reuters Web of Science)

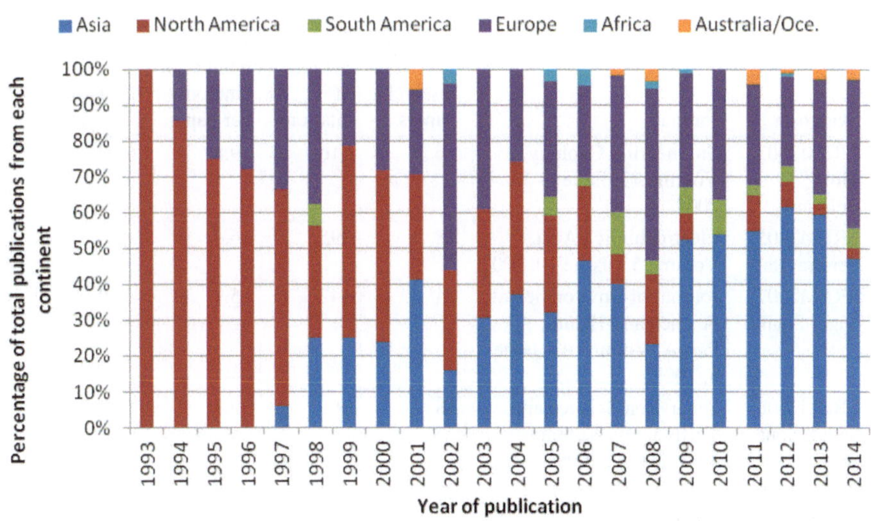

Fig. 1.4 Origin of publications on "electrokinetic soil remediation" registered by Thomson Reuters Web of Science from 1993 until 24th September 2014. Continent of corresponding author. The percentage of the total number of publications from each continent is shown (based on data published in (Thomson Reuters Web of Science))

So this leaves some questions behind: Is EKR becoming an exclusively scientific treatment method? Or are the perfect practical setups already developed? Is Asia the future of EKR? Does this steep increase in the number of scientific publications really mean that the publications have an impact?

At least the last question can partly be answered by a simply study on how many times EKR publications are cited. Taking the special issues dedicated to the earlier mentioned EREM symposia and using Thomson Reuters Web of Science to find out how many times the manuscripts are cited, Table 1.2 can be constructed showing year of publication (EREM), number of publications, number of total citations, and number of citations per year. From this table it can be deducted that the scientific papers do have an impact on the scientific level. An average citation ratio of around twice a year is considerably high compared to typical average values given by Thomson Reuters (Thomson Reuters' Essential Science Indicators database), e.g., publications within the scientific area of "engineering" with around one citation per year (Thomson Reuters Essential Science Indicators). Here, one should mention that some papers in the special editions are cited much more than others and therefore have a higher weight in the citation ratio. Furthermore, the numbers for last couple of EREM special issues should be taken with care since they were only published a few years ago.

Table 1.2 EREM symposia special editions

Symposia	Journal	Number of research papers	Total number of citations	Total number of citations per paper	Citations/ (paper·year)[a]
EREM2003, Mol, Belgium	Engineering Geology, Volume 77, Issue 3–4, 2005	16	316	19.8	2.2
EREM2005, Ferrara, Italy	Electrochimica Acta, Volume 52, Issue 10, 2007	18	295	16.4	2.3
EREM2007, Vigo, Spain	Journal of Environmental Science and Health, Part A, Volume 43, Issue 8, 2008	24	142	5.9	1.0
EREM2008, Seoul, South Korea	Separation Science and Technology, Volume 44, Issue 10, 2009	18	163	9.1	1.8
EREM2009, Lisbon, Portugal	Journal of Applied Electrochemistry, Volume 40, Issue 6, 2010	27	123	4.6	1.1
EREM2010, Kaohsiung, Taiwan	Separation and Purification Technology, Volume 79, Issue 2, 2011	24	196	8.2	2.7
EREM2011, Utrecht, The Netherlands	Electrochimica Acta, Volume 86, 2012	29	59	2.0	1.0

Number of manuscripts and number of total and yearly citations. (Data obtained from Thomson Reuters Web of Science, September 24th, 2014.)

[a]Total number of citations per year divided by (2014—publication year)

1.6 EKR Tendencies

Another reason for the increasing number of scientific publications is due to the fact that researchers are somehow turning away from the classic EKR setup and soil as the waste product. Several new EKR concepts are under development and several new waste materials are tested.

Some general rules or conditions have to be fulfilled to use EK in the classical manner meaning applying a DC field to a wet solid matrix with defined porous characteristics can be listed (Virkutyte et al. 2002; Ottosen et al. 2008). These are mainly:

- Fine porous materials.
- Low hydraulic conductivity.
- Water-soluble contaminants if there are any poorly soluble contaminants, it may be essential to add solubility-enhancing reagents.
- Relatively low concentrations of ionic materials in the water.

This has led to research and development of the EK process applied to waste products and environmental problems such as (Kirkelund et al. 2013; Ferreira et al. 2005; Pazos et al. 2010; Jakobsen et al. 2004; Kim et al. 2011; Hansen et al. 2005; Gomes et al. 2014; Pedersen 2014):

1. Fly ash and gas cleaning residues from municipal solid waste incineration
2. Fly ash from biomass combustion
3. Sewage sludge
4. Fly ash from sewage sludge incineration
5. Mine tailings
6. Harbor sediments
7. Marine dredging
8. PCB or TBT contaminated soil

Lately, enhancements of the EK process have been suggested such as (Kirkelund et al. 2013; Cang et al. 2013; Velizarova et al. 2004; Ribeiro et al. 2000; Ottosen et al. 2005; Nystroem et al. 2006; Saichek and Reddy 2003; Rojo et al. 2012; Hansen and Rojo 2007; Gomes et al. 2014):

1. Use of pH control.
2. Treatment of waste products in slurries/suspensions with continuous agitation.
3. Adding or complexing agents (in the case of heavy metals).
4. Use of desorbing agents.
5. Adding of surfactants (in the case of organic contaminants).
6. Adding of oxidants/reducing agents to change oxidation state of contaminants and increase their mobility.
7. Use of pulsed and/or alternating current.

Also recently, EK has been combined or coupled with several other remediation technologies such as nanofiltration, nano particles, soil washing, bioremediation, chemical oxidation or reduction-based methods, permeable reactive barriers, and phytoremediation in order to create a more widespread remediation solution for complex problems (Couto et al. 2013, 2015; Gomes et al. 2012).

Even if this EKR overview is addressed to the environmental issues of the electrokinetic phenomena, it is worth mentioning the potential for EKR in related areas—such as civil engineering. Some of the processes where electromigration is utilized in civil engineering (Ottosen et al. 2008) are:

- *Removal of chloride from concrete* (*desalination*). Chloride ions are removed from otherwise sound concrete to stop or to hinder reinforcement corrosion.
- *Re-alkalization of carbonated concrete*. Re-alkalization is used for carbonated reinforced concrete structures and the purpose is the re-establishment of alkalinity around the reinforcement and in the cover zone to protect the reinforcement against corrosion.
- *Crack closure in concrete*. Concrete with cracks is vulnerable to penetration of water and chlorides. In this method, filling of cracks occurs with compounds as, e.g., $CaCO_3$ and $Mg(OH)_2$, where the necessary ions are supplied by electromigration. The method is primarily used for marine structures.
- *Injection of organic corrosion inhibitors into concrete*. Increase in penetration rate of corrosion inhibitors (as amino compounds that undergo protonation to form cationic species in solution) with electromigration.
- *Re-impregnation of wood in structures*. Wood in constructions attacked by decay is either replaced with new wood or a wood preservative (most often boron compounds) can be sprayed on the surface. In the latter case, the diffusion of boron into the wood is slow but the transport rate can be increased if the main transport mechanism is electromigration instead of diffusion.
- *Removal of salts from brick masonry*. Salts are removed from masonry that is suffering from salt weathering (a process that is caused by high salt concentrations) in an applied electric field.

The referred EKR in related areas are just some, being difficult, if not impossible, to conclude "where do we go now?" once EKR applications continue to emerge.

References

Acar YB, Alshawabkeh A (1993) Principles of electrokinetic remediation. Environ Sci Technol 27(13):2638–2647

Alshawabkeh AN (2009) Electrokinetic soil remediation: challenges and opportunities. Sep Sci Technol 44:2171–2187

Besra L, Liu M (2007) A review on fundamentals and applications of electrophoretic deposition (EPD). Prog Mater Sci 52(1):1–61

Boccaccini AR, Zhitomirsky I (2002) Application of electrophoretic and electrolytic deposition techniques in ceramics processing. Curr Opin Solid St M 6(3):251–260

Cang L, Fan G-P, Zhou D-M, Wang Q-Y (2013) Enhanced-electro kinetic remediation of copper–pyrene co-contaminated soil with different oxidants and pH control. Chemosphere 90(8):2326–2331

Casagrande L (1948) Electro-osmosis in soil. Geotechnique 1:1959

Couto N, Guedes P, Mateus EP, Santos C, Ribau Teixeira M, Nunes LM, Hansen HK, Gutierrez C, Ottosen LM, Ribeiro AB (2013) Phosphorus recovery from a water reservoir—potential of nanofiltration coupled to electrodialytic process. Waste Biomass Valoriz 4(3):675–681. doi:10.1007/s12649-012-9194-7

Couto N, Guedes P, Zhou D-M, Ribeiro AB (2015) Integrated perspectives of a greenhouse study to upgrade an antimony and arsenic mine soil—potential of enhanced phytotechnologies. Chem Eng J 262: 563–570. http://dx.doi.org/10.1016/j.cej.2014.09.021

Dolnik V (2008) Capillary electrophoresis of proteins 2005–2007. Electrophoresis 29:143–156

ELECTROACROSS—Electrokinetics across disciplines and continents: an integrated approach to finding new strategies for sustainable development (European Union financed project—FP7-PEOPLE-2010-IRSES-269289). http://sites.fct.unl.pt/electroacross/

Ferreira C, Jensen P, Ottosen L, Ribeiro A (2005) Removal of selected heavy metals from MSW fly ash by the electrodialytic process. Eng Geol 77(3–4):339–347

García-Cañas V, Cifuentes A (2008) Recent advances in the application of capillary electromigration methods for food analysis. Electrophoresis 29:294–309

Gomes HI, Dias-Ferreira C, Ribeiro AB (2012) Electrokinetic remediation of organochlorines in soil: enhancement techniques and integration with other remediation technologies. Chemosphere 87(10):1077–1090

Gomes HICR (2014) Coupling electrokinetics and iron nanoparticles for the remediation of contaminated soils. PhD thesis, Faculdade de Ciencias e Tecnologia, Universidade Nova de Lisboa, Portugal

Gomes HI, Dias-Ferreira C, Ottosen LM, Ribeiro AB (2014) Electrodialytic remediation of PCB contaminated soil with iron nanoparticles and two different surfactants. J Coll Interf Sci 433: 189–195. http://dx.doi.org/10.1016/j.jcis.2014.07.022

Guedes P, Mateus EP, Couto N, Rodríguez Y, Ribeiro AB (2014) Electrokinetic remediation of six emerging organic contaminants from soil. Chemosphere 117: 124–131. http://dx.doi.org/10.1016/j.chemosphere.2014.06.017

Hansen HK, Ottosen LM, Hansen L, Kliem BK, Villumsen A, Bech-Nielsen G (1999) Electrodialytic remediation of soil polluted with heavy metals: Key parameters for optimization of the process. Chem Eng Res Des 77(3):218–222

Hansen HK, Rojo A, Ottosen LM (2005) Electrodialytic remediation of copper mine tailings. J Hazard Mater 117(2–3):179–183

Hansen HK, Rojo A (2007) Testing pulsed electric fields in electroremediation of copper mine tailings. Electrochim Acta 52(10):3399–3405

Ho Lee H-H, Yang J-W (2000) A new method to control electrolytes pH by circulation system in electrokinetic soil remediation. J Hazard Mater 77(1–3):227–240

Iskanda IK (ed) (2000) Environmental restoration of metals—contaminated soil. CRC, London

Jakobsen MR, Fritt-Rasmussen J, Nielsen S, Ottosen LM (2004) Electrodialytic removal of cadmium from wastewater sludge. J Hazard Mater 106(2–3):127–132

Kim K-J, Kim D-H, Yoo J-C, Baek K (2011) Electrokinetic extraction of heavy metals from dredged marine sediment. Sep Purif Technol 79(2):164–169

Kim WS, Park GY, Kim DH, Jung HB, Ko SH, Baek K (2012) *In situ* field scale electrokinetic remediation of multi-metals contaminated paddy soil: influence of electrode configuration. Electrochim Acta 86:89–95

Kirkelund GM, Damoe AJ, Ottosen LM (2013) Electrodialytic removal of Cd from biomass combustion fly ash suspensions. J Hazard Mater 250–251:212–219

Lageman R, Pool W (1993) Electro reclamation: applications in the Netherlands. Environ Sci Technol 27:2648–2650

Lageman R, Clarke RL, Pool W (2005) Electro-reclamation, a versatile soil remediation solution. Eng Geol 77:191–201

Marry V, Dufreche JF, Jardat M, Meriguet G, Turq P, Grun F (2003) Dynamics and transport in charged porous media. Coll Surface A 222:147–153
Nystroem GM, Pedersen AJ, Ottosen LM, Villumsen A (2006) The use of desorbing agents in electrodialytic remediation of harbor sediment. Sci Total Environ 357:25–37
Ottosen LM, Pedersen AJ, Ribeiro AB, Hansen HK (2005) Case study on the strategy and application of enhancement solutions to improve remediation of soils contaminated with Cu, Pb and Zn by means of electrodialysis. Eng Geol 77:317–329
Ottosen LM, Christensen IV, Rorig-Dalgård I, Jensen PE, Hansen HK (2008) Utilization of electromigration in civil and environmental engineering—processes, transport rates and matrix changes. J Environ Sci Health A 43:795–809
Page MM, Page CL (2002) Electro remediation of contaminated soils. J Environ Eng 128:208–219
Pazos M, Kirkelund GM, Ottosen LM (2010) Electrodialytic treatment for metal removal from sewage sludge ash from fluidized bed combustion. J Hazard Mater 176:1073–1078
Pengra DB, Wong PZ (1996) Electrokinetic phenomena in porous media. Mat Res Soc Symp Proc 407:3–14
Pedersen KB (2014) Applying multivariate analysis to developing electrodialytic remediation of harbour sediments from arctic locations. PhD dissertation, Faculty of Sciences and Technology, The Arctic University of Norway, Norway, 293p
Prasad MNV, Sajwan KS, Naidu R (eds) (2006) Trace elements in the environment. Biogeochemistry, biotechnology, and bioremediation. CRC, Boca Raton
Probstein RF, Hicks RE (1993) Removal of contaminants from soils by electric fields. Science 260:498–503
Reddy KR, Cameselle C (eds) (2009) Electrochemical remediation technologies for polluted soils, sediments and groundwater. Wiley, Hoboken
Reuss FF (1809) Sur un nouve leffet de l'électricité galvanique. Mem Soc Imp Naturalists Moscow 2:327–336
Ribeiro AB, Mateus EP, Ottosen LM, Bech-Nielsen G (2000) Electrodialytic removal of Cu, Cr and As from chromated copper arsenate-treated timber waste. Environ Sci Technol 34 (5):784–788. doi:10.1021/es990442e
Rojo A, Hansen HK, Cubillos M (2012) Electrokinetic remediation using pulsed sinusoidal electric field. Electrochim Acta 86:124–129
Saichek RE, Reddy KR (2003) Effects of system variables on surfactant enhanced electrokinetic removal of polycyclic aromatic hydrocarbons from clayey soils. Environ Technol 24:503–515
Sharma HD, Reddy KR (eds) (2004) Geoenvironmental engineering: site remediation, waste containment, and emerging waste management technologies. Wiley, Hoboken
Symes C (2012) A little nail treatment. Ground Engineering, Feb. 2012: 18–20
Thomson Reuters Web of Science, http://thomsonreuters.com/thomson-reuters-web-of-science/. Accessed 24 Sept 2014
Thomson Reuters Essential Science Indicators database, http://thomsonreuters.com/thomson-reuters-web-of-science/. Accessed 24 Sept 2014
US Environmental Protection Agency (1997) Report EPA 402-R-97-006: resource guide for electrokinetics laboratory and field processes applicable to radioactive and hazardous mixed wastes in soil and groundwater from 1992 to 1997.
United States Patent and Trademark Office (USPTO). http://www.uspto.gov/. Accessed 24 Sept 2014
Velizarova E, Ribeiro AB, Mateus EP, Ottosen LO (2004) Effect of different extracting solutions on electrodialytic remediation of CCA-treated wood waste. Part 1. Behaviour of Cu and Cr. J Hazard Mater 107(3):103–113
Virkutyte J, Sillanpaa M, Latostenmaa P (2002) Electrokinetic soil remediation—critical overview. Sci Total Environ 289:97–121
Wise DL, Trantolo DJ, Cichon EJ, Inyang HI, Scottmeister U (eds) (2000) Remediation engineering of contaminated soils. Marcel Dekker, New York
Yeung AT (2011) Milestone developments, myths, and future directions of electrokinetic remediation. Sep Purif Technol 79:124–132

Chapter 2
Soil Structure Influence on Electrokinetic Dewatering Process

Vikas Gingine and Rafaela Cardoso

2.1 Introduction

The control of water content or water movement in soils is fundamental not only to ensure some desired characteristics but also for different applications, such as consolidation, dewatering, conditioning, and decontamination. In case of consolidation, water is removed mainly in soft clayey soils to improve their stiffness and strength. Dewatering is done mainly in mining sludge to reduce water content and therefore the storage volume is reduced and disposal can be done in landfill because it is no longer in the liquid phase. Conditioning consists in controlling the water amount and composition in the soil, and is done to promote certain chemical and bacterial activities (biocement or bioremediation) because oxygen availability increases and heat may be generated. In case of decontamination, water movement is done by flushing water through the soil aiming to mobilize certain contaminants or to inject chemical treatment.

Electrokinetic treatment (EKT) promotes water movement in soils under electrical gradients and therefore it can be used in these applications. There are several examples in the literature concerning the use of EKT in Geotechnical and Environmental engineering practice. For example, EKT was used for strengthening of foundations like skirted foundation and caisson anchors embedded in marine clay and offshore calcareous sand (Micic et al. 2001; Rittirong 2008). Researchers have been working to prove the efficiency of the EKT remediation in in situ remediation of low-permeability and heterogeneous soils that have been contaminated by ionic species, organics, heavy metals, radionuclides, or a combination of these contaminants (Acar et al. 1993; Reddy 2010; Gabrieli and Alshawabkeh 2010). This technique was investigated also for precise permeability estimations of low

V. Gingine (✉) • R. Cardoso
ICIST, Instituto Superior Técnico, University of Lisbon,
Av Rovisco Pais, 1, Lisbon 1049-001, Portugal
e-mail: vikas.gingine@ist.utl.pt

© Springer International Publishing Switzerland 2016

A.B. Ribeiro et al. (eds.), *Electrokinetics Across Disciplines and Continents*,
DOI 10.1007/978-3-319-20179-5_2

Table 2.1 Main soil properties for EKT suitability (adapted from Jones et al. (2006))

Property	Range of values
Plasticity index	5–30 %
In situ water content	0.6–1.0 liquid limit
Percentage of material with diameter $D < 0.075$ mm	>50 %
Organic content	<10
Hydraulic conductivity	$<10^{-8}$ m/s
Electrical conductivity	0.05–0.005 S/m
Undrained shear strength	<55 kPa
Oedometric modulus	0.3–1.5 MPa
Coefficient of consolidation	$<1.5 \times 10^{-6}$ m^2/s

permeability rocks without applying any fluid pressure gradient (Pengra et al. 1999). Electrokinetic geosynthetics (EKG) were developed for combining the functions of geosynthetics, such as drainage, reinforcement, and filtration, with electroosmosis to perform in situ dewatering of sewage sludge lagoons (Walker and Glendinning 2002) or for belt press dewatering of mining sludge (Lamont-Black et al. 2006).

Concerning economic benefits, the use of EKG permits wastes to be solidified prior to disposal. In the case of lagooned wastes, this means that these features can be remediated at a viable cost with the additional benefit of releasing brown field land for development. In the mining industry, the use of electrokinetic dewatering of tailings produces water recovery and a major reduction in the carbon footprint of mining operations, both of which are growing priorities. In terms of waste reduction and efficiency, full-scale field trials have shown that the invention of geosynthetics incorporating EKT has meant very efficient utilization of electrokinetic functions and cost reductions (Jones et al. 2006).

EKT has some advantages over the cases when water movement is caused by hydraulic gradients, namely when the coefficient of hydraulic conductivity of the soils is very low and the chemical nature of the percolating fluid and its interaction with the percolated medium is relevant. Jones et al. (2006) synthesized the main proprieties of soils which would help determine the usefulness of EKT (Table 2.1). Basically, they refer to fine soils and are as follows: Atterberg limits, in situ water content, percentage of fines (material with diameter $D < 0.075$ mm) and organic content, saturated hydraulic permeability and electrical conductivity, undrained shear strength, oedometer modulus or confined stiffness, and coefficient of consolidation. However, it is most common to consider that EKT can be more efficient in soils where their coefficient of hydraulic conductivity, k_h, is equal or smaller than the coefficient of electroosmotic permeability, k_e, as shown in Table 2.2 (materials organized from more to less suitable for using EKT). For this reason, clayey soils are the kind of materials for which the use of EKT is advantageous.

There are several studies focused on the electrical properties of clays. Indeed, due to the small size of clay particles, their behavior is ruled by the interactive electrochemical forces rather than the gravity forces. Their natural negative

Table 2.2 Typical values for the coefficients of hydraulic and electroosmotic permeability (adapted from Mitchell and Soga (2005))

Material	Water content (%)	$k_e \times 10^{-9}$ (m^2/V/s)	$k_h \times 10^{-9}$ (m/s)	k_h/k_e
London clay	52.3	5.8	0.1	0.02
Boston blue clay	50.8	5.1	0.1	0.02
Silty clay	32.0	3.0–6.0	0.12–0.65	0.02–0.22
Quick clay	31.0	2.0–2.5	0.2	0.08–0.10
Bootlegger clove clay	30.0	2.4–5.0	0.2	0.04–0.08
Kaolin	67.7	5.7	1.0	0.18
Rock flour	27.2	4.5	1.0	0.22
Clayey silt	31.7	5.0	10.0	2.00
Mica powder	49.7	6.9	100.0	14.5
Fine sand	26.0	4.1	1000.0	244
Quartz powder	26.5	4.3	1000.0	233

electrical charge at the particles surface, forming a double layer, explains water sensitivity and osmotic effects. A proof of this is the fact that EKT with/without chemicals introduces a significant change in the consistency limits of some expansive clayey soils (Jayasekera and Hall 2007; Abdullah and Al-Abadi 2010; Gingine et al. 2013). Moreover, the existence of electrochemical forces between the particles interacting with water molecules generates attraction/repulsion forces, therefore explaining the different structures (arrangement of particles) found when clayey materials are compacted with different energies and with different water contents (Lambe 1958; Sivakumar and Wheeler 2000).

Soil structure is determinant for the mechanical properties of soils, stiffness, and strength and can be quantified by means of voids ratio. Voids ratio decreases when there is a mechanical load in soils through the dissipation of pore pressures and the consequent increment of effective stresses. The coefficient of hydraulic conductivity also depends on voids ratio, as well as volume changes measured on wetting. As changes in water content also cause deformations in soils, such as shrinkage on drying, or swelling or collapse on wetting, hence hydraulic and mechanical behavior are coupled. When the mineralogical properties of soils or the composition of the percolating fluid influence the amplitude of the deformations, there is a chemical interaction and therefore coupled chemo-hydro-mechanical behavior needs to be considered. If water flow is achieved by applying an electrical gradient, then a coupled electro-hydro-mechanical analysis must be performed. Despite the complex phenomena like heat generated by the electrodes, evolution of gases, precipitation of the metal (electrode) oxides, and other chemical changes, different theoretical models have been developed considering various parameters and electrical properties of the soil (Alshawabkeh and Acar 1996; Ribeiro and Mexia 1997). The basic phenomena of electroosmotic consolidation and decontamination are so far well understood but the unpredictable reactions make the theory more complex. The analysis of such models is out of the scope of this chapter.

In this chapter, the equations used to describe EK processes in the soil from a geotechnical point of view are reviewed emphasizing the electro-hydraulic coupled behavior and how it is affected by soil structure. The results from tests performed on samples of white kaolin compacted with known structure are presented as an example. The main purpose of the experimental study was to study the effect of soil structure in its hydraulic, electrical, and electroosmotic conductivities. The results showing changes in these conductivities are interpreted considering the existence of two types of voids, the microvoids of the clay aggregates and the macrovoids between these aggregates, and identifying the dominant pore size for each phenomenon.

2.2 Some Theoretical Concepts

Some basic concepts of EKT are explained below and focused on the use of this technique for soil dewatering. Some concepts of soil structure generated by compaction processes are reviewed emphasizing on how this physical property influences the hydraulic and electrical properties.

2.2.1 Electrokinetic Processes in Soils

Electrokinetics refers to the relationship between electrical potential and the movement of water and charged particles. Under a DC voltage water flows by electroosmosis from the anode (+) to the cathode (−) as long as the voltage gradient is kept constant (Fig. 2.1). Electrophoresis (movement of charged colloids) and electromigration (movement of ions) also occur, along with water electrolysis, heat generation, and REDOX reactions; however, they are not as relevant as electroosmosis for saturated soils.

Electroosmotic efficiency is defined as the quantity of water moved per unit of electricity. EK process is more efficient for porous-saturated materials, i.e., when the pores are completely filled with fluid, generally water. The soils for which electrokinetics is more efficient are those having clay minerals with low cation exchange capacity (CEC), low valency exchange cations, high surface charge density, and high surface area. These are the characteristics of clayey soils; however, electroosmotic flow has been observed in other materials, for example in quartz powder, rock flour, and several types of sludge (ochre and alum, humic, anaerobic digested, surplus activated and primary) (Jones et al. 2004). In all cases, it is assumed the solid particles in suspension are treated as colloids because of their smaller size (less than 2 μm) and electrostatic forces prevail instead of gravitational forces.

As well known and presented schematically in Fig. 2.1, water forms a boundary layer in the surface of the charged particles, named as double layer because it has an

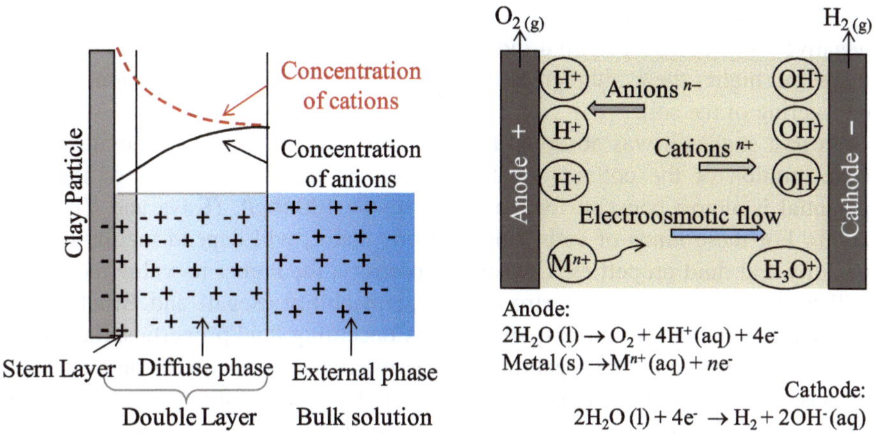

Fig. 2.1 Double layer in clays and electrokinetic processes

inner immobile zone (Stern layer) and an outer mobile zone. The electrical potential at the junction between these layers is the zeta potential. It is generally expressed in mV and is determined by the phenomenon of electrophoresis, which is the movement of the charged colloids under a given electric field. Electrostatic forces at the surface of the Stern layer drag positively charged cations with their surrounding water molecules towards the negatively charged cathode.

Different theories, like the Schmid theory, the Spiegler friction model, the ion hydration model, and the Gray–Mitchell approach, have been proposed to quantify the coefficient of electroosmotic conductivity on the basis of different assumptions of ion distribution in the pore fluid (Mitchell and Soga 2005). However, the Helmholtz-Smoluchowski theory is most commonly used and hence is described below.

Accordingly with Helmholtz-Smoluchowski theory, the rate of this water flow is controlled by the balance between the electrical forces causing water movement and frictional forces retarding water movement. In this case, zeta potential ζ is defined by

$$\zeta = \frac{\eta u}{\varepsilon_T \varepsilon_o E} \tag{2.1}$$

where u is the velocity of the particle, E is the electrical field, η is the fluid viscosity, ε_T is the relative permittivity of the pore fluid, and ε_o is the permittivity of free space. This potential can be used to compute the coefficient of electroosmotic permeability k_e

$$k_\mathrm{e} = -\frac{\varepsilon_\mathrm{w} n \zeta}{\mu \tau} \tag{2.2}$$

where ε_w is the permittivity of pore water ($\varepsilon_\mathrm{w} = 80$ F/m for pure water), τ is tortuosity ($\tau = 0.25$ is a typical value), n is the porosity of medium, and μ is the dynamic viscosity of water ($\mu = 10^{-3}$ N s/m^2) (Kaya and Yukselen 2005). The negative value of zeta potential indicates positive rate of electroosmotic flow and, therefore, higher the k_e, higher is the rate of electrokinetic consolidation, and the dewatering of the soil.

In past studies it was observed that, for soft soils, the effect of the amount of concentration of the colloids in the solution on the measurement of the zeta potential is almost constant for the values above 100 mg/L (Kaya and Yukselen 2003). For these kinds of soils, this confirms that k_e will depend mainly on soil porosity and fluid properties. Concerning porosity, the higher its value the higher will be the coefficient of electroosmotic permeability (Segall and Bruell 1992; Gabrieli et al. 2008) as mobility increases. Concerning fluid properties, the coefficient of electroosmotic permeability k_e is proportional to electrical conductivity σ because electrical conductivity (the inverse of electrical resistivity) is related with the flow of electrical current through the medium, which is easier if the fluid polarizes in response to an electric field. However, the presence of high concentration of ions may generate ions diffusion and contra flows (Fig. 2.1), and for this reason, EKT is more efficient for fluids with low salinity and high pH.

For low salt concentrations, the cations are affected by the double layers of the minerals and their dominant paths are located in the grain–water interface, whereas the anions always move in the pore space and are affected by the tortuosity of medium, while for high salt concentrations, both cations and anions are moving in the pore space and conductivity of cations equals that of anions. As a consequence, zeta potential decreases and therefore k_e also decreases, reducing EKT efficiency. This is particularly important during EKT because pH tends to change. In Fig. 2.1, it is also shown the formation of an acidic zone near the anode and a basic one near the cathode due to the release of hydrogen and hydroxyl ions in the pore water (electrolysis). In this case, the treatment efficiency may be reduced along time. Because the zeta potential value is influenced by the electrolytic composition, the value obtained after testing contaminated soils plays an important role in designing the treatment of soil contaminated with different metals and chemicals.

2.2.2 Dewatering and Hydrodynamic Consolidation of Soils

Water flows in response to differences in pressure, which is named as water head. The rate of water flow depends on the coefficient of hydraulic permeability, however as mentioned earlier, in fine soils such as clays this value is very small and for this reason there was the idea of inducing flow through changes in electrical potential. The literature on EKT goes back to 1940s, when authors like Casagrande started experimenting on electrokinetic properties of saturated clays. Apart from an effort of stabilizing the soils, the studies also focused on the importance of drainage conditions at the electrodes and pore pressures developed throughout the soil sample and the resulting consolidation (Esrig 1968; Arnold 1973).

Drainage conditions are important because water pressure generated by electroosmotic flow at the cathode will generate a counteracted hydraulic flow. Anode must be closed in this case to stop this hydraulic flow, and this is the configuration adopted for dewatering or hydrodynamic consolidation. If open conditions are ensured at the anode and cathode, i.e., there is no pressure gradient across the boundary, then water flow is caused only by EK. Such drainage configuration is adopted to flush out the contaminants out from the soil using clean water.

The theory of electroosmotic consolidation and the analytical solution for one-dimensional flow were presented in 1960s (Esrig 1968; Mitchell and Soga 2005). The electroosmotic volume flow rate can be described by (2.3)

$$Q_{\mathrm{e}} = k_{\mathrm{e}} \nabla E A \tag{2.3}$$

where Q_{e} is the electroosmotic volume flow (m^3/s), k_{e} is the coefficient of electroosmotic permeability ($m^2/V/s$) already presented, ∇E is the gradient of the direct-current electrical field applied (V/m), and A is the total cross-sectional area perpendicular to the direction of flow (m^2). For unidirectional flow along the position x, the electroosmotic volume flow can be described by (2.4), where k_{e} is measured when water percolates the soil under constant voltage gradient $\mathrm{d}V/\mathrm{d}x$

$$Q_{\mathrm{e}} = k_{\mathrm{e}} \frac{\partial V}{\partial x} A \tag{2.4}$$

This equation is analogous to the well-known Darcy's law used to compute the volume flow of water Q_{h} driven by a hydraulic gradient in saturated soils

$$Q_{\mathrm{h}} = k_{\mathrm{h}} i A = k_{\mathrm{h}} \frac{\partial H}{\partial x} A \tag{2.5}$$

where k_{h} is the coefficient of hydraulic conductivity, or saturated permeability (m/s), and i is the hydraulic gradient given by the change in water head ∂H along the direction of flow ∂x caused by changes in pore water pressure. In case of hydrodynamic consolidation, changes in water pressure are caused by stress increment which is entirely transmitted to the continuous fluid phase. This is because water is much stiffer than the solid soil skeleton made of particles with a given arrangement, or structure, accordingly with Terzaghi's theory. In this case, effective stresses increase while pore water pressure reduces along time in asymptotic manner, and therefore soil reduces volume as a result for the arrangement of particles. In this case, the hydraulic permeability reduces along time, and this is a hydro-mechanical coupled effect.

These equations do not consider any coupling between electrical and hydraulic phenomena. If the drainage is kept open at the anode, the electric energy will act as a gradient just to cause the electromigration and electroosmosis, without any consolidation and therefore (2.3) is valid. However, hydrodynamic consolidation occurs during the electrokinetic treatment if the anode side drainage is kept closed because there are changes in interstitial pressure and for this reason effective stresses also change and therefore the soil experiences volume changes. In this case, water velocity v can be described by (2.6), where u is the increment of interstitial pressure due to the electrical gradient applied and k_e and k_h were already explained. It is worth to note that the hydraulic conductivity considered is measured when water percolates due to differences in water potential only, and not driven by external stress increment.

$$v = \frac{Q}{A} = \frac{k_h}{k_e}\frac{\partial u}{\partial x} + k_e\frac{\partial V}{\partial x} \tag{2.6}$$

Concerning (2.6) it is worth to note that the relationship between k_h and k_e determines the relative importance of hydraulic flow versus electroosmotic flow. For electroosmotic flow to be more important that the hydraulic one this relationship must be the smallest as possible, as previously discussed and presented in Table 2.2.

Both conductivities change as the treatment proceeds due to the fact that the soil properties and electric field changes causing nonuniformity in the water content and voids ratio (Mitchell and Soga 2005). Dislike the hydraulic permeability, and as already mentioned, electroosmotic permeability is time dependent and is considerably influenced over the period of treatment. It is independent of grain size but, as hydraulic permeability, it depends on soil voids ratio, or porosity, as its definition given by (2.2). k_e dependence on soil porosity indicates that it will decrease as the soil reduces volume, and for this reason water flow will reduce along EK treatment if soil volume is allowed to change. Thus, drainage conditions are important.

Since the soil properties change throughout the EKT, the study of interrelationships between the mechanical, hydraulic, and electrical properties is necessary. In addition, the unpredictable chemical reactions make the theory more complex and the equations to quantify the effects. Their description is out of the scope of this chapter.

2.2.3 Soil Structure and Its Effects on Hydraulic (and Electrical) Behavior

Compacted soils are preferred to natural soils because their structure can be easily reproduced and different structures can be generated depending from the compaction process adopted. Being this a mechanical process (a given energy is applied for a fixed amount of water), the most significant changes in clayey soils are mainly in the size of the clay aggregates and in their arrangement. In general, two different pore sizes can be distinguished in clayey soils: the macropores and the micropores. The first are the large pores between clay aggregates and the second are the small pores of the clay aggregates. Both kinds of pores are accounted in soil porosity and in voids ratio. The arrangement of the aggregates corresponds to the macrostructure and the arrangement of the clay minerals forming the aggregates is the microstructure, and for this reason compacted clayey soils are usually named double structured materials. Figure 2.2a is a scheme presenting the two different types of pores and structures.

The quantity of water used in the compaction rules soil structure. For the same compaction effort, optimum point corresponds to maximum dry volumetric weight or minimum voids ratio, being soil structure flocculated or dispersive below and above that value, respectively (Lambe 1958). The different structural arrangements can be explained by electrochemical interaction between water and clay minerals prevailing in the edges of the clay aggregates, in a mechanism similar to that explaining the thickness of the double layer.

In most of the practical applications, the size of the micropores remains practically unchanged independently from compaction process because it is controlled by the amount of water in the aggregates, and it is assumed that they are fully saturated (Romero et al. 1999; Monroy et al. 2007; Thom et al. 2007; Cardoso et al. 2011). When structure is flocculated, in the drying branch, the macropores are very large where as they are smaller when structure is dispersive in the wetting branch (Fig. 2.2a). For the same voids ratio, it is possible to have dispersive or flocculated structures, which means that the volume of voids is the same as it is the sum of the volume of both micro- and macropores.

For the same voids ratio, if the volume of micropores remains practically unchanged, it is expected large macropores in flocculated structures, and small macropores in dispersive structures. Hydraulic conductivity therefore is different because fluid percolation is made mainly through the macropores as they are the largest channels in the porous medium.

The size of macropores change in case on stress changes, due to hydrodynamic consolidation, or if soil experiences significant volume changes caused by wetting or drying, due to expansiveness phenomena. Figure 2.2b illustrates both mechanisms. When unsaturated clayey soils are wetted with water, the size of the micropores increases and therefore aggregates increase volume. These microstructural volume changes are caused by chemical interaction between water and the clay minerals or osmosis, already discussed. This volume increment has different consequences on global volumetric behavior depending from the arrangement of the aggregates and their confinement and is simulated in constitutive models for unsaturated soils using interaction functions between the two structural levels (Gens and Alonso 1992; Cardoso et al. 2013). In fact, clays may exhibit overall volume increment (swelling), which generally corresponds to the increment of the size of the macropores, or to volume decrease (collapse), which corresponds to the decrease of the size of the macropores.

Electroosmotic flow depends on the electrical conductivity of the medium and for that the presence of water is fundamental. It occurs through the macropores because electrical current flows through the continuous fluid phase, but the water in the micropores has an important contribution and for this reason the influence of both kinds of pores must be considered when investigating EKT.

The comparison between hydraulic conductivity and electroosmotic conductivity for soils compacted with different and known structures may help clarifying the

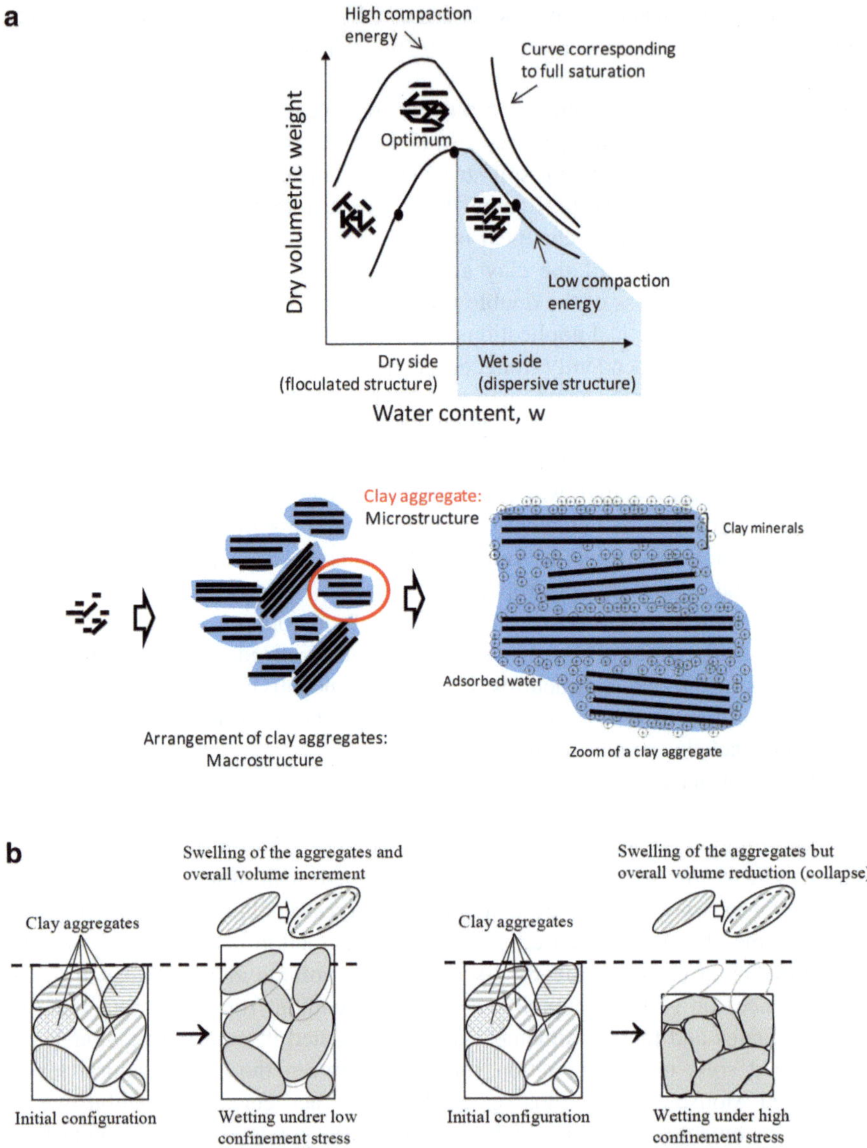

Fig. 2.2 Structure of compacted clayey soils: (**a**) structure induced by compaction; (**b**) volume change mechanisms

relevance of both types of pores in the electrical current flow in soils because hydraulic conductivity is controlled mainly by the size of macropores therefore the differences in tendencies may be attributed to the micropores. The dependence from soil structure of both parameters will be investigated deeper in the research presented, in which commercial kaolin is compacted with different structures and the hydraulic, electrical, and electroosmotic conductivities will be measured.

It is believed that the main applications of EKT in the future are in the domain of environmental engineering, where contaminants are present in pore fluid affecting its electrical conductivity and also the size of micropores. For this reason, a better knowledge of the coupled chemo-electro-hydraulic behavior of clays linked to their mechanical behavior (volume changes and consequently stiffness and strength changes) caused by the interaction between the two structural levels may help defining better models also in this domain.

2.3 Experimental Study

The evaluation of the electrokinetic properties of the materials is necessary to identify the viability of EKT on the type of soil and to calibrate some models. These properties play an important role in planning the execution of the treatment in field, either for electroosmotic consolidation, dewatering, or decontamination of the site.

Application of electric field affects the physical, chemical, and electrokinetic properties and thus changes structure of the soil. The results of a laboratorial study performed on compacted Kaolin are presented, in which hydraulic, electrical, and electroosmotic permeability were measured considering the effect of voids ratio and molding water content in specimens preparation.

2.3.1 Kaolin Investigated

Commercially available white kaolin clay was adopted in this experimental research aiming to simplify data treatment because it is a Portuguese non-expansive clayey soil, therefore no relevant structural changes are expected on full saturation and for this reason the study on how structure affects soils electro-hydro-mechanical properties can be more easily done. The material was supplied in powder, with 68 % in mass of grains with silt size (between 0.075 and 0.002 mm) and 31 % with clay size (less than 0.002 mm) and solid volumetric weight of 26.1 kN/m^3. Liquid limit is 52 % and plasticity index is 22 %; therefore, the material classifies as highly plastic silt (MH) accordingly with the Unified Soil Classification System.

Pore size distribution was measured using Mercury Intrusion Porosimetry, MIP (Autopore IV 9500), in destructured samples of similar kaolin (i.e., samples mixed with water content 1.5 times higher than the liquid limit, as suggested by Burland (1990)), in which it can be assumed the inexistence of macrovoids. This is expected to be the dimension of the micropores of the soil studied. An almost uniform pore size distribution was found, with a peak at 310 nm diameter as shown in Fig. 2.3.

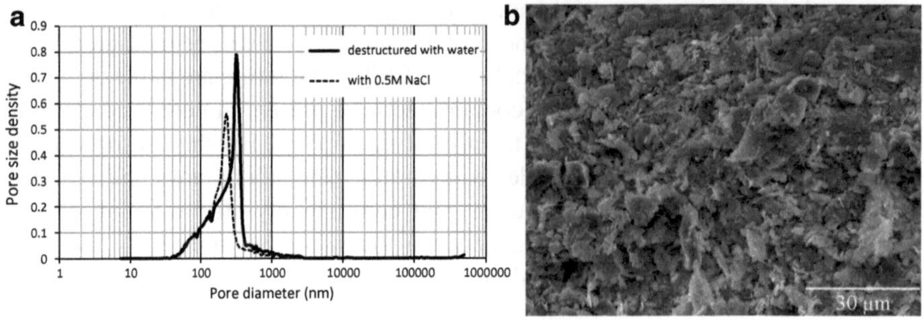

Fig. 2.3 Destructured kaolin: (**a**) pore size distributions with distilled water and with 0.5 M NaCl; (**b**) SEM photograph for the destructured case with distilled water

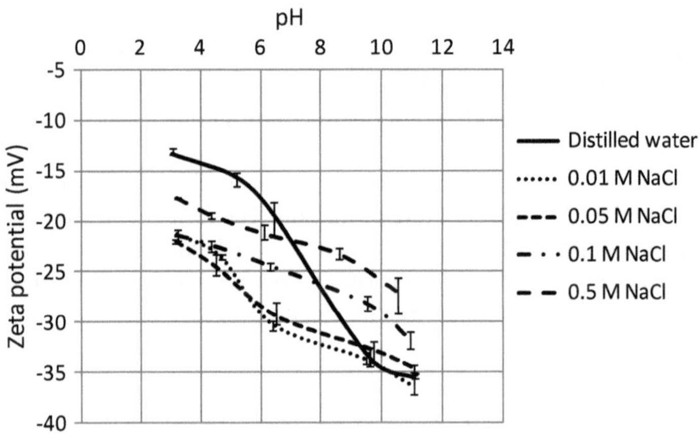

Fig. 2.4 Zeta potential curves and the effect of different concentrations of NaCl

This figure also presents a photograph from Scanning Electron Microscopy, SEM, where the homogeneous disposition of the minerals confirms the MIP result.

The hydraulic conductivity and electroosmotic conductivity of destructured samples with voids ratio $e = 1.5$ and water content $w = 65\ \%$ were measured (Cardoso and Santos 2013) and the values of $k_h = 5 \times 10^{-10}$ m/s and $k_e = 2 \times 10^{-9}$ m/s (3.2 V/cm, DC) were found, both within the ranges presented in Table 2.1 for similar materials. The electrical conductivity of the water extracted from the soil was $\sigma = 2.7$ S/m, which identifies the presence of dissolved salts from the soil because distilled water was used in samples preparation.

Zeta potential was measured for increasing values of pH by chemical conditioning of the anolyte and catholyte solutions, done respectively by 0.1 M solutions of HCl and NaOH. The zeta meter 3.0 used allowed applying 20–300 V. Data is presented in Fig. 2.4, where it can be seen negative zeta potential values for this clay. This information, plus the values of hydraulic conductivity and electrical resistivity

Table 2.3 Changes in consistency limits for different concentrations of NaCl

Fluid	Liquid limit w_L (%)	Plasticity limit w_P (%)	Plasticity index IP (%)
Distilled water	52	30	22
NaCl 0.01 M	52	30	22
NaCl 0.05 M	50	32	22
NaCl 0.10 M	50	29	21
NaCl 0.50 M	48	25	23

already mentioned, show that that EKT can be an adequate dewatering technique for this clayey silt.

The effect of the presence of salts in the pore fluid on clay water sensitivity was also studied by mixing the material with solutions prepared with different concentrations of NaCl (0.01, 0.05, 0.1, and 0.5 M). The consistency limits decrease as shown in Table 2.3; therefore, the molding capacity of the soil, or its plasticity, reduces with increasing NaCl concentration.

This expected result is consistent with the zeta potential curves shown in Fig. 2.4 for the same salt concentrations. In fact, as salt concentration increases zeta potential decreases for the lower pH and increases for the higher, resulting in more flat curves; however, the final zeta potential is less negative than the one measured using distilled water.

The ions supplied to the solution contribute to reduce the thickness of the double layer, and therefore clay particles reduce mobility. When in the reconstituted material, the repulsion between clay minerals reduces, and the size of the pores of the clay aggregates decrease. This result may be confirmed in MIP tests performed in samples percolated with a solution prepared with 0.5 M of NaCl also shown in Fig. 2.2, where it can be seen the reduction of pores' size because the peak displaces to 220 nm diameter.

2.3.2 *Samples Preparation and Structure Effect on Electrical and Hydraulic Conductivity*

The specimens investigated, in a total of nine, were compacted for three different void ratios, 0.90, 1.05, and 1.20 (±0.005), each void ratio with the molding water contents of 20, 25, and 30 % (±1.0 %). Table 2.4 presents the different characteristics of the specimens, including the water content corresponding to their full saturation (all voids filled with water).

The nine points were selected in order to study the effect on electro-hydraulic behavior of voids ratio and, for the same voids ratio, to create flocculated and dispersed structures (Fig. 2.5). Low dry unit weight was assumed to ensure that the effects would be visible. Figure 2.5 presents their position in relation to the compaction curve obtained using 50 % of standard Proctor energy.

Table 2.4 Characteristics of the compacted specimens

Void ratio, *e*	Porosity, *n* (%)	Dry volumetric weight (kN/m^3)	Saturated water content (%)
0.90	55	13.8	34.5
1.05	51	12.8	42.0
1.20	47	11.9	46.0

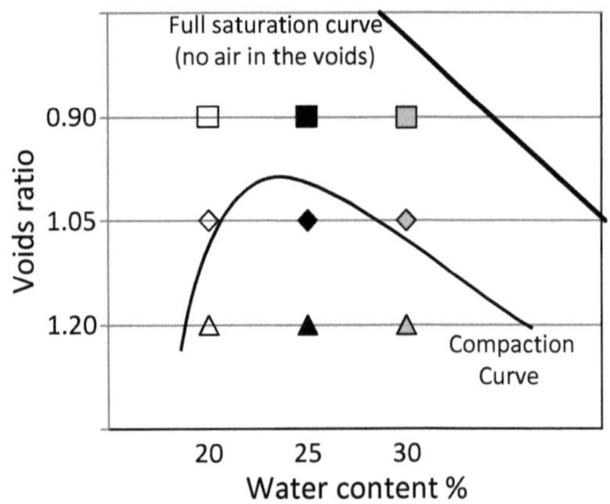

Fig. 2.5 Details of the compacted specimens

Different electrical conductivities were measured for the specimens with the molding water content (in Fig. 2.6a and Table 2.5). Electrical conductivity σ was measured adopting Wenner method described in BS1317 (BSI, 1990), in cylindrical samples prepared for this purpose and using steel nails as electrodes. For the same voids ratio, the electrical conductivity increases with molding water content simply because there is large amount of water present in the voids. However, this increment is more notorious for the specimen prepared with $e = 1.02$, and this can be explained by the transition from flocculated to dispersive structures (the dry to the wet side of the compaction curve), and by the fact that the soil is not saturated in all cases. The existence of air bubbles inside the samples reduces electrical conductivity because air is an electrical isolator. In flocculated structures, because they are in the dry side, the size of the bubbles is larger and therefore σ decreases. The results measured for the other voids ratio studied are consistent with the above. Accordingly with Fig. 2.6a, the electrical conductivity is slightly smaller for the specimens compacted in the dry side or with small voids ratio than for the specimens compacted in the wet side or with large voids ratio.

Electrical conductivity increases for all specimens to values closer to 0.036 S/m when they are fully saturated, also in Fig. 2.6a. This result confirms that the electrical conductivity is more sensitive to the amount of water present in the soil than to its structure, and for this reason the measurements in electrical conductivity may be used only as an indicator of a positive response of soil to EKT, as previously discussed (see Table 2.1). It is worth to note that the saturated electrical

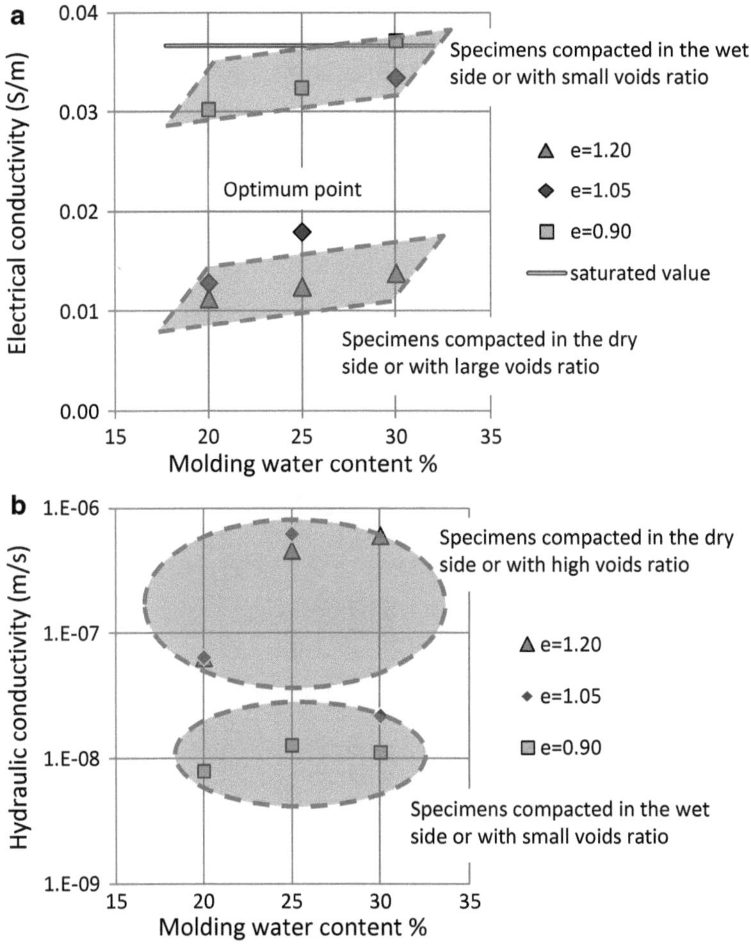

Fig. 2.6 Characterization of the compacted specimens: (**a**) electrical conductivity for the molding water content; (**b**) saturated hydraulic conductivity

conductivity of the soil is smaller than the value found for the pore fluid, which confirms that the solids work as electrical insulators.

Structural effects are more evident when the hydraulic conductivities measured in saturated specimens are compared, even if the order of magnitude of the values expected for all specimens is within the ranges of values measured in similar kinds of clays (between 10^{-7} and 10^{-10} m/s). Hydraulic conductivity was measured in the triaxial chamber in cylindrical specimens (7 cm diameter and 14 cm height), after full saturation following ASTM D4647 (D4767-11 2011). The results are presented in Fig. 2.6b and in Table 2.5, where it can be seen that the k_h of the samples compacted in the dry side of the compaction curve or with high voids ratio is at least one order of magnitude larger than that of the samples compacted in the wet side or

Table 2.5 Synthesis of the values measured for all specimens

Voids ratio	0.90	0.90	0.90	1.05	1.05	1.05	1.20	1.20	1.20
w (%)	20	25	30	20	25	30	20	25	30
σ (S/m)[a]	0.0303	0.0325	0.0372	0.0129	0.0180	0.0335	0.0113	0.0125	0.0138
k_h (m/s)	8.0E-09	1.3E-08	1.1E-08	6.5E-08	6.3E-07	2.2E-08	6.5E-08	4.7E-07	6.1E-07
k_e (m^2/V/s)	1.3E-09	9.3E-10	9.5E-10	1.4E-09	1.2E-09	9.0E-10	1.3E-09	1.9E-09	1.7E-09
k_h/k_e	6	14	12	48	544	25	50	247	364

[a]For the molding water content

with small voids ratio. This result is explained by structural arrangement from compaction process, as previously discussed, because water flowing is easier through the large macropores when the structure is flocculated or voids ratio is large. For this reason, even if porosity affects this coefficient, it is dominated mainly by the size of the macropores.

To conclude, the results found show that k_h is larger for soils with large macropores, while σ is independent from structure as long as the soil is fully saturated. For unsaturated soils, σ is more for the lesser voids ratio, where the smallest is the size of the macropores or the largest is their degree of saturation, because it means that the size of the air bubbles are also small.

2.3.3 *Experimental Setup for EKT and Results*

The electroosmotic permeability test apparatus was modified from an oedometer apparatus, as described by Cardoso and Santos (2013). The schematic diagram and photograph of the experimental setup developed are shown in Fig. 2.7, where it can be seen the oedometer cell, the loading system, the transducer LVDT for measuring vertical displacements, the current/voltage source, and the system for water collection. The oedometer cell has a special provision for the inlet and outlet of the water as well as for the electrical wires for the electrodes, 2 mm thick silver metal strips with 2 cm side. The electrodes are placed in between the soil and the porous stone. Pure silver metal was used because of its excellent electrical conductivity; however, it suffers oxidation during EKT. A pressure of 12 kPa was applied to ensure proper contact between the electrodes and the soil. The anode is the top and the cathode is the bottom of the sample. The DC voltage/current source EX354RD, with two outputs (each with capacity 0–35 V and 0–4 A) was used. The tests were performed keeping voltage constant and the samples submerged in distilled water but electrically isolated in the contacts using tape.

The samples tested were compacted inside the oedometer ring (5 cm diameter and 2 cm height), as those investigated for measuring the hydraulic and electrical conductivities previously shown in Fig. 2.4. Total test duration was 6 h for each sample after its complete saturation and was divided into three stages of 2 h each. Increasing voltage gradients were applied in each stage 2.5 V/cm, 4 V/cm, and 6 V/cm, respectively, and kept constant in each stage. The selection of these voltages was based on trials performed to achieve a current difference near 10, 20, and 30 mA, respectively, in each test stage, which are considered efficient for clayey soils (Jones et al. 2006). The tests were conducted for 6 h maximum to prevent significant reduction of k_e along time and avoid any other complex physical and chemical changes in the soil, as their main aim was to determine the electroosmotic permeability k_e. In spite of the short duration of the test corrosion of the silver electrodes and evident signs of contamination of the clay with silver oxide occurred, as shown in Fig. 2.8.

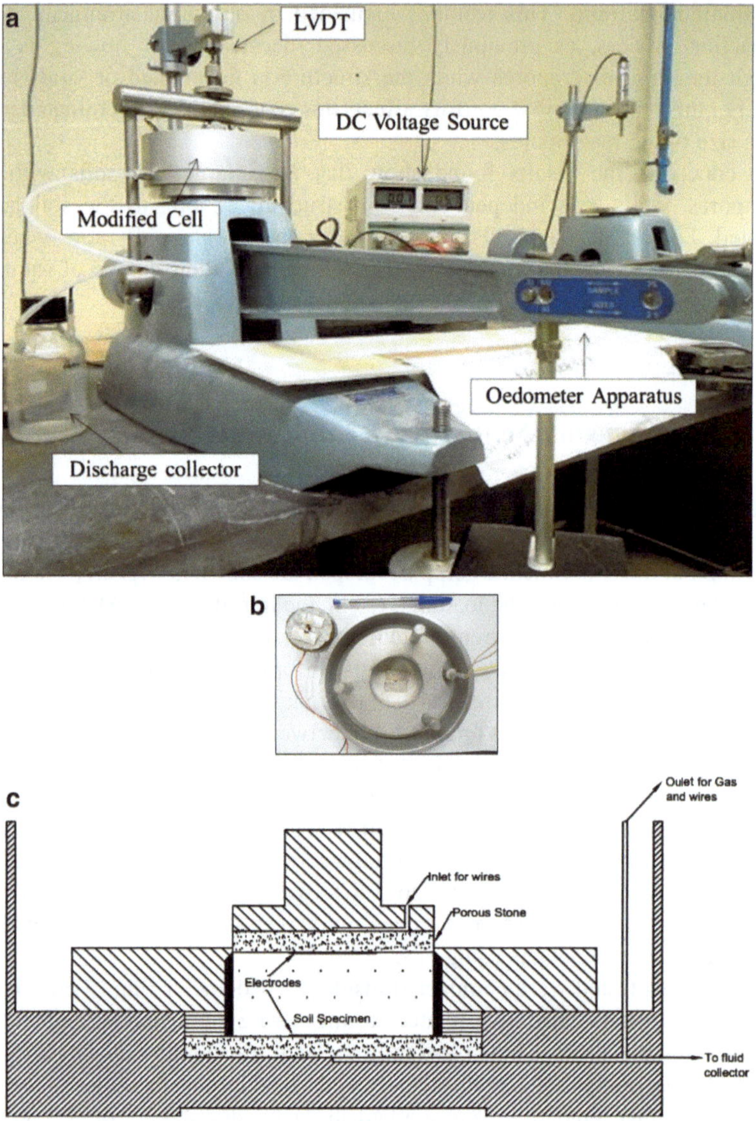

Fig. 2.7 Experimental setup: (**a**) overview; (**b**) top view of the oedometer; (**c**) scheme of the oedometric cell

To avoid consolidation of the soil by the electroosmotic treatment, the drainage conditions were kept open at the anode and cathode. No deformations were registered for the soil confirming that here was no consolidation. The pH value of the water collected at the cathode was also noted at the end of the tests. Its value increased from pH near 7 before the test to pH between 8.8 and 10.2. This confirms EK flow and is consistent with the equations previously shown in Fig. 2.1.

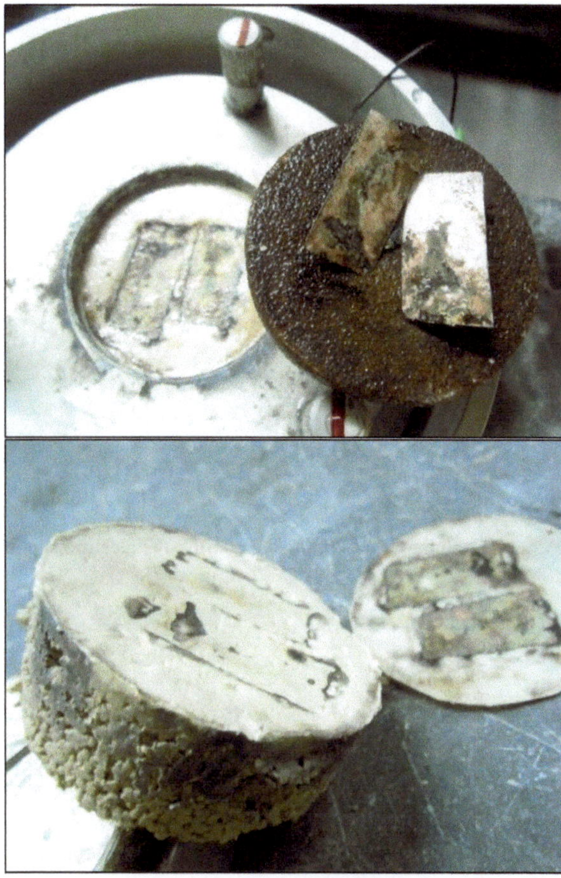

Fig. 2.8 Corrosion of the silver electrodes and evident signs of contamination of the clay with silver oxide (sample with $e = 1.20$ and $w = 25$ %)

The volume of water collected (at the cathode exit) was recorded throughout the test. Figure 2.9 presents the ranges of volumes measured for each void ratio tested, being the upper limit the curve collected for the sample prepared with $w = 30$ % molding water content and the lower for the sample prepared with $w = 20$ %. It can be seen that the slopes of the curves increase with increasing voltage gradient computed for each 2 h of test, and that the larger volumes were collected in the samples with larger voids ratio, which are those with larger porosity and also those with larger macrovoids.

The coefficient of electroosmotic conductivity for each specimen was computed using (2.4) with the data from Fig. 2.9. The values found along time are presented in Fig. 2.10, where it can be seen they are around 10^{-9} $m^2/V/s$ independently from the sample studied. They are not constant because of experimental error and also because the conductivities change due to the fact that the soil properties and electric field also change (Mitchell and Soga 2005).

If the average values along time (Table 2.5) are considered for comparison purposes, as shown in Fig. 2.11 the electroosmotic conductivities computed for

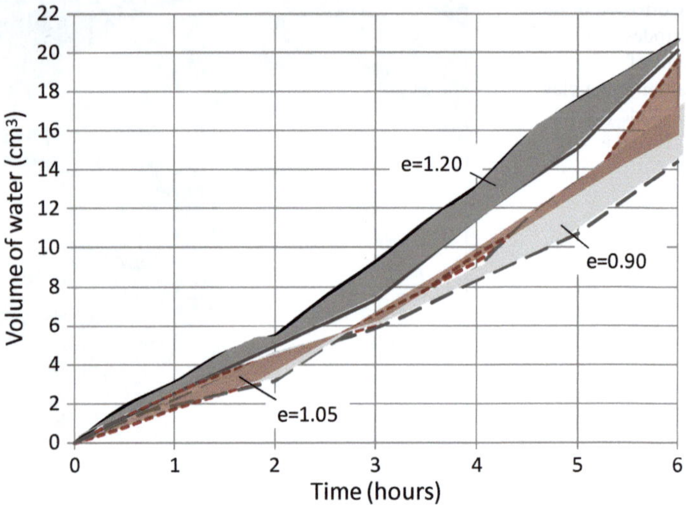

Fig. 2.9 Volume of water collected during the electroosmotic test

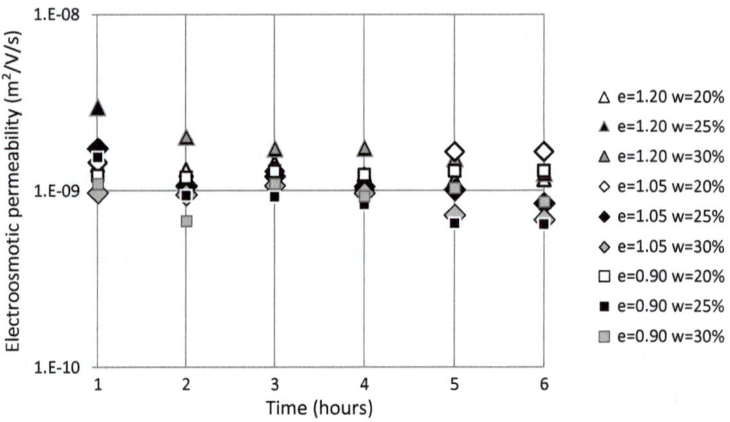

Fig. 2.10 Electroosmotic permeability computed along time

the samples with $e = 1.20$ ($n = 55$ %, Table 2.4) are slightly larger than those found for the samples with $e = 0.90$ ($n = 47$ %), which confirms that this coefficient depends on porosity. Higher differences would be expected if the samples investigated would have more contrasting porosities.

The differences found in the values of k_e for the samples investigated are almost irrelevant, in particular when compared with the differences found for the saturated hydraulic permeability (Fig. 2.6b and in Table 2.5) because the values measured for k_h differed more than one order of magnitude depending on soil structure. While the hydraulic permeability is strongly dependent on the size of the macropores because

Fig. 2.11 Electroosmotic conductivity (average values measured in the entire test duration)

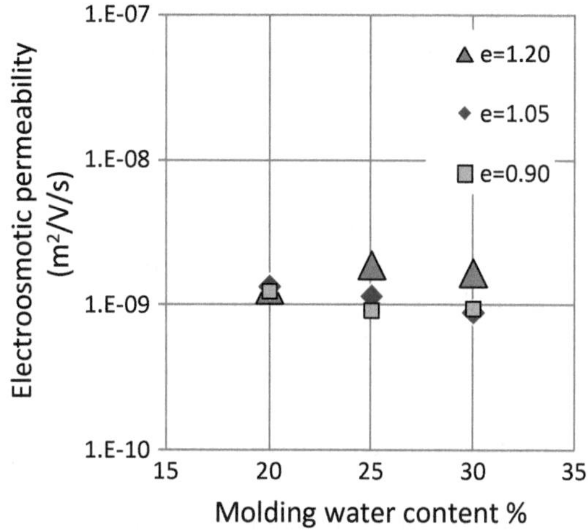

they are the main channels for water flow, the electroosmotic permeability depends on global porosity and therefore the electric flow is done through all types of pores filled with water. Assuming that the microporosity is almost the same for all specimens independently from their preparation, the similar k_e measured may indicate that this property may be controlled mainly by the size of the micropores, where water flows driven by electroosmotic effects. Further tests must be done to confirm this result using different clays with similar zeta potential so that different micropore sizes could be generated through mechanical compaction.

To conclude, the relationship found of k_h and k_e is presented in Table 2.5 for all samples. Their values are larger than the expected in some cases previously presented in Table 2.1 because of the different voids ratio of the specimens and also because of experimental error. The lesser values were found for the samples prepared with the lowest voids ratio, which proves that even small electric gradients can cause similar flow induced by the higher pressures in a dense compacted soil. This confirms that EKT dewatering treatment is more efficient for denser materials because it is when water percolation is more difficult to achieve by applying hydraulic gradients.

2.4 Conclusions

The theoretical study and practical applications of EKT were presented in this chapter according to a Geotechnical engineer's interest. EKT helps to increase the rate of hydro-electro consolidation of a fine-grained soil. The complexity in the EKT is severe and hence a chemo-electro-hydraulic-mechanical coupling was performed and studied. Taken into consideration the voids ratio and the molding water contents, the structure of a fine-grained soil, especially the arrangement of clay aggregates plays an important role in EKT. Various hydraulic and electrical properties were measured for the Kaolin soil and significant relations were observed between them, which were related to EKT using different theories. The effect of flocculent and dispersed structure on the electro-hydro properties of a clayey soil can be well explained with the help of the micropores and the macropores of the soil.

Experimental data shows that the hydraulic and electrical conductivity of this soil depends mainly on the size of the macropores, while the electroosmotic permeability depends on global porosity due to electro-chemical interaction between the water and the clay mineral, and for this reason the water present in the micropores is relevant and affects this parameter. The study must be repeated using different clays to confirm the importance of the micropore's size for this parameter.

EKT is believed to have a significant impact in the domain of Environmental Engineering. Even a smaller electric gradient is able to create a same flow as that induced by a high hydraulic gradient. From a hydraulic point of view, it is not expected that the presence of contaminant will affect deeply the hydraulic properties of the soils; however, due to their chemical interaction with clay minerals, they may change micropores and therefore may interfere with mechanical behavior, and certainly will affect soil's electrical and electroosmotic conductivities. The study of interrelationships between the mechanical, hydraulic, and electrical properties of low pervious soils such as the one studied and of clayey soils in general is necessary for the definition of better models based on their coupled chemo-electro-hydro-mechanical behavior, which will help in designing efficient and economically advantageous EKT solutions.

Acknowledgements The authors acknowledge Professor Susana Freitas from INESC-MN for providing the equipment for measuring the electrical conductivity, Doctor Remígio Machado from the Department of Chemical Engineering of IST for helping using the zeta meter and FCT for funding the experimental tests presented (EXPEL/ECM-GEO/0109/2013). MOTA Ceramic Solutions for providing the soil for the research.

References

Abdullah W, Al-Abadi A (2010) Cationic–electrokinetic improvement of an expansive soil. Appl Clay Sci 47:343–350

Acar YB, Alshawabkeh AN, Gales R (1993) Fundamentals of extracting species from soils by electrokinetics. Waste Manage 13:141–151

Alshawabkeh AN, Acar YB (1996) Electrokinetic remediation II: theoretical model. J Geotech Eng 122:186–196

Arnold M (1973) Laboratory determination of the coefficient of electro-osmotic permeability of a soil. Géotechnique 23:581–588

Burland JB (1990) On the compressibility and shear strength of natural clays. Géotechnique 40:329–378

Cardoso R, Reis A, Maranha das Neves E, Caldeira L (2011) Hydro-mechanical behavior of a clayey soil compacted on the dry and wet side of optimum. In: Proceedings of VII Brazilian symposium on unsaturated soil, 29–31 August, Pirenópolis

Cardoso R, Santos JN (2013) An experimental study on the consolidation of soft clayey soils using electrochemical methods. In: Proceedings of 18th international conference on soil mechanics and geotechnical engineering, Paris

Cardoso R, Alonso EE, Maranha das Neves E (2013) A constitutive model for compacted expansive and bonded marls. Géotechnique 63:1116–1130

D4767-11 (2011) Standard test method for consolidated undrained triaxial compression test for cohesive soils, ASTM, West Conshohocken

Esrig M (1968) Pore pressures, consolidation and electrokinetics. J Soil Mech Found Div ASCE 94 (SM4):899–921

Gabrieli L, Alshawabkeh AN (2010) Influence of boundary conditions on transient excess pore pressure during electrokinetic applications in soil. J Appl Electrochem 40:1113–1121

Gabrieli L, Jommi C, Musso G, Romero E (2008) Influence of electroosmotic treatment on the hydro-mechanical behaviour of clayey silts: preliminary experimental results. J Appl Electrochem 38:1043–1051

Gens A, Alonso E (1992) A framework for the behaviour of unsaturated expansive clays. Can Geotech J 29:1013–1032

Gingine V, Shah R, Koteshwar R, Harikrishna P (2013) A review on study of electrokinetic stabilization of expansive soil. Int J Earth Sci Eng 6:176–181

Jayasekera S, Hall S (2007) Modification of the properties of salt affected soils using electrochemical treatments. Geotech Geol Eng 25:1–10

Jones JFPC, Lamont-Black J, Glendinning S, Bergado D, Eng T, Fourie A, Yang-Feng Z (2004) Recent research and applications in the use of electro-kinetic geosynthetics. In: Proceedings of EuroGeo4

Jones JFPC, Glendinning S, Huntley DT, Lamont-Black J (2006) Soil consolidation and strengthening using electrokinetic geosynthetics—concepts and analysis. In: Kuwano J, Kosedi J (eds) Geosynthetics. Millpress, Rotterdam

Kaya A, Yukselen Y (2003) Zeta potential of kaolinite in the presence of alkali, alkaline earth and hydrolyzable metal ions. Water Air Soil Poll 145:155–168

Kaya A, Yukselen Y (2005) Zeta potential of soils with surfactants and its relevance to electrokinetic remediation. J Hazard Mater 120:119–126

Lambe TW (1958) The structure of compacted clay. J Soil Mech Found Div ASCE 84(SM2):1654

Lamont-Black J, Huntley DT, Glendinning S, Jones CJFP (2006) Case history: the use of electrokinetic geosynthetics (EKG) in belt press dewatering. In: 8th international conference on geosynthetics, Yokohama

Micic S, Shang JQ, Lo KY (2001) Electrokinetic strengthening of marine clay adjacent to offshore foundations. In: Proceedings of XIII international offshore polar engineering conference, Norway, pp 694–701

Mitchell J, Soga K (2005) Fundamentals of soil behavior, 3rd edn. Wiley, New York

Monroy R, Zdravkovic L, Ridley A (2007) Fabric changes in compacted London clay due to variations in applied stress and suction. In: Proceedings of second international conference, mechanics of unsaturated soils, Springer, Germany, pp 41–48

Pengra DB, Li SX, Wong P-Z (1999) Determination of rock properties by low frequency AC electrokinetics. J Geophys Res 104(B12):29.485–29.508

Reddy K (2010) Technical challenges to in-situ remediation of polluted sites. Geotech Geol Eng 28:211–222

Romero E, Gens A, Lloret A (1999) Water permeability, water retention curve and microstructure of unsaturated compacted Boom clay. Eng Geol 54:117–127

Ribeiro AB, Mexia JT (1997) A dynamic model for the electrokinetic removal of copper from a polluted soil. J Hazard Mater 56:257–271

Rittirong A (2008) Effects of electrode configuration on electrokinetic stabilization for Caisson Anchors in calcareous sand. Geotech Geol Eng 134:352–365

Segall BA, Bruell CJ (1992) Electroosmotic contaminant removal processes. J Environ Eng 118:84–100

Sivakumar V, Wheeler S (2000) Influence of compaction procedure on the mechanical behaviour of an unsaturated compacted clay. Part 1: wetting and isotropic compression. Géotechnique 50:359–368

Thom R, Sivakumar R, Sivakumar V, Murray V, Mackinnon P (2007) Pore size distribution of unsaturated compacted kaolin: the initial states and final states following saturation. Géotechnique 57:469–479

Walker J, Glendinning S (2002) In situ dewatering of lagooned sewage sludge using electrokinetic geosynthetics (EKG). In: Proceedings of 7th European biosolids and organic residuals conference, Wakefield

Chapter 3
Electrokinetically Enabled De-swelling of Clay

Reena A. Shrestha, Angela P. Zhang, Eduardo P. Mateus, Alexandra B. Ribeiro, and Sibel Pamukcu

3.1 Introduction and Background

Clay minerals, such as smectites and mixed-layer illites, can expand in volume up to 20 times their original volume through inner crystalline adsorption of layers of water and by the hydration of the exchangeable cations on the clay surfaces. Hydration-swelling of clays is an important concern in geo-engineering when swelling clay deposits are encountered under or near constructed facilities, such as foundations, tunnels, embankments, boreholes, and wells. Upon swelling these clays tend to lose shear strength and can undergo large volume changes as they expand and shrink. Swelling soils exhibit alternating cycles of swelling and shrinkage, often causing substantial damage in civil infrastructure (Chen 1988; Nelson and Miller 1992). Furthermore, upon swelling they become retardant of flow and/or a liquid barrier due to substantial decrease in hydraulic conductivity. Studies on the loss of permeability of clay-containing sandstone and shale rock in response to swelling have been conducted over many decades, particularly for petroleum industry concerns of wellbore construction and operations. Researchers agree that the water sensitivity or loss of permeability of sandstones can be attributed to combination of effects including clay swelling in the rock pores, colloid and/or

R.A. Shrestha
Department of Civil and Environmental Engineering, Lehigh University, Bethlehem, PA 18015, USA

Department of Petroleum Engineering, The Petroleum Institute, Abhu Dhabi, UAE

A.P. Zhang • S. Pamukcu (✉)
Department of Civil and Environmental Engineering, Lehigh University, Bethlehem, PA 18015, USA
e-mail: sp01@lehigh.edu

E.P. Mateus • A.B. Ribeiro
CENSE, Departamento de Ciências e Engenharia do Ambiente, Faculdade de Ciências e Tecnologia, Universidade Nova de Lisboa, Caparica 2829-516, Portugal

© Springer International Publishing Switzerland 2016
A.B. Ribeiro et al. (eds.), *Electrokinetics Across Disciplines and Continents*,
DOI 10.1007/978-3-319-20179-5_3

salt plugging. In shale rock, swelling of the clay constituents is the major contributor to drilling problems in the wellbore management (Mohan and Fogler 1997). Similarly, drilling-mud filtrates can cause oil permeability to decrease, which can interfere with oil well operations. The injection pressures and the time required for water flooding can increase if clay blocking occurs due to clay dispersion and/or swelling, which can give rise to operational and economic problems in water reservoirs and oil recovery processes (Zhou and Law 1998; Wilson et al. 2014).

Chemical stabilization of high plasticity clay soils using calcium-based stabilizers, such as lime and cement has been practiced routinely over many decades to reduce their volume change potential. Its overall benefits include an increase in soil strength, stiffness and durability, and a reduction in soil plasticity and swell/shrink potential (Winterkorn and Pamukcu 1991). The calcium in the stabilizing agents exchanges with sodium (i.e., swelling potential maker) on clay surface and affects an increase in the permeability and strength of the clay matrix by reducing the dispersion tendency of the particles. This has been well known and practiced in petroleum production and well stability operations over many years. For instance, when formation clays are exposed to calcium chloride solution and then exposed to weaker solutions of calcium chloride or distilled water, considerably less permeability damage results than if the clays had been exposed to sodium chloride solutions first (Jones 1964). Among the physical soil stabilization methods there are only a few options for fine grained soils which include freezing, hydro fracture grouting, and electroosmotic dewatering. Electroosmotic dewatering has been shown a viable stabilization for high water content clays (Adamson et al. 1966; Alshawabkeh and Sheahan 2003; Tajudin 2012; Estabragh et al. 2014; Arutselvam 2014).

The $CaCl_2$ is a common salt with properties that make it particularly suitable for many soil stabilization applications (Jensen 2003). The calcium ions improve the soil strength by three mechanisms, ion replacement, mineralization, and precipitation of species in the pore fluid (Keykha et al. 2014). The $CaCl_2$ is a hygroscopic material which can attract and absorb moisture from its environment, it is highly soluble, and Ca^{2+} ions can replace the Na^{+} ions within the electrical double layers of clay particles. Replacement of a monovalent ion with a divalent ion the DDL promotes higher charge density, hence a reduction in the thickness of double layer. Smaller DDL lessens the dispersion tendency of the particle and allows for larger pore space for water conduction. Injection of $CaCl_2$ solution into tight clay soils which may be already hydrated and severely low in their ability to conduct liquids under a hydraulic gradient can be accomplished adequately by electroosmosis and electrokinetics. The Ca^{2+} ions migrate into soils through the processes of electromigration and electroosmotic advection.

This study reports the findings of a series of laboratory reactor and batch tests conducted to determine the efficacy of electrokinetic injection of $CaCl_2$ solution into fully hydrated bentonite (Na-smectite) and sand mixtures to reduce their electrokinetic potential and increase hydraulic conductivity, hence reverse the swelling condition.

3.2 Experimental Program

3.2.1 The EK Reactor and Test Protocols

Figure 3.1 shows the EK reactor setup used in the flow tests. The simple reactor design constituted of PVC tube cells ($L = 7.6$ cm; $D = 2.6$ cm) equipped with circular mesh titanium working electrodes at each end. Cotton cloth filters were inserted at soil–liquid interfaces. Soil sample was composed of mixtures of sand and bentonite with varying bentonite mass fractions of 25, 50, and 100 %. Table 3.1 presents the properties of the bentonite clay. The sand was fine size, clean silica sand. About 40 g of each mixture was packed in the tubular cells. Standing burettes connected to each end of a cell were used to manage the EO flow. The water levels in the burettes were adjusted to equilibrium periodically to avoid build-up of hydraulic pressures against the EO flow, or suction in favor of EO. Fully hydrated soil samples were permeated electrokinetically with water, and solutions of $CaCl_2$ at three different concentrations of Ca^{2+} ion (0.01, 0.05, and 0.1 M) in the direction of electroosmotic flow. All the tests were conducted for 48 h under constant electrical potential of 4 V/cm. Flow, current, pH, were measured periodically for the duration of the tests. Table 3.2 shows the list of tests run and the composition of the anolyte, catholyte, and the soil specimen for each test.

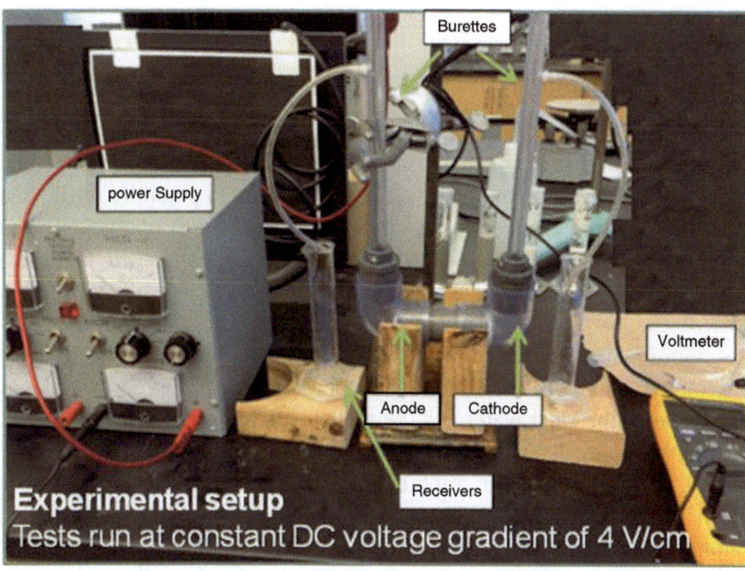

Fig. 3.1 Experimental setup for measuring EO flow in hydrated clay specimens

Table 3.1 Physicochemical properties of bentonite clay used in this study

Parameters	Value
pH	9.46
Conductivity	588 μS
pH @zpc	7
Dry bulk mass density	0.96 g/cm^3
Zeta potential	−29.5 mV

Table 3.2 Electrokinetic experiments—set

Set	Catholyte	Anolyte	Soil composition (Bentonite %)
1	Pure water	Pure water	100
2	Pure water	Pure water	50
3	Pure water	Pure water	25
4	Pure water	0.01 M $CaCl_2$	100
5	Pure water	0.01 M $CaCl_2$	50
6	Pure water	0.01 M $CaCl_2$	25
7	Pure water	0.05 M $CaCl_2$	100
8	Pure water	0.05 M $CaCl_2$	50
9	Pure water	0.05 M $CaCl_2$	25
10	Pure water	0.1 M $CaCl_2$	100
11	Pure water	0.1 M $CaCl_2$	50
12	Pure water	0.1 M $CaCl_2$	25

3.2.2 Swelling Behavior of Bentonite Test Sample

Batch tests were performed to determine the pre-test pH, mass density, electrical conductivity, pH at zero point charge (ZPC), and the zeta potential of the bentonite clay (Table 3.1). Pure bentonite can swell up to 300 times of the initial volume (Mollins et al. 1996). Bench scale column tests were conducted to observe the differences in swelling tendencies of bentonite in water (distilled water) and in solutions of $CaCl_2$ of three different concentrations (0.01, 0.05, and 0.1 M). For each test 5 g of bentonite was added into 20 mL of liquid in a standing glass column and the volume change of the bentonite suspension was observed by following the interface between the clear water and the suspension to drop over time for 48 h. The columns' tops were sealed with stretch wrap to limit evaporation.

Figure 3.2 presents the photographic images of the various test columns obtained over 48 h period of time. As observed, while the bentonite in water remain in dispersion, the others were suppressed to smaller volumes as settlement ensued. The Ca^{2+} is likely to exchange on the clay surface and reduce dispersion so gravity settlement and compression can occur. The transient change in the measured volume of clay that remains in suspension below the water clay interface is plotted

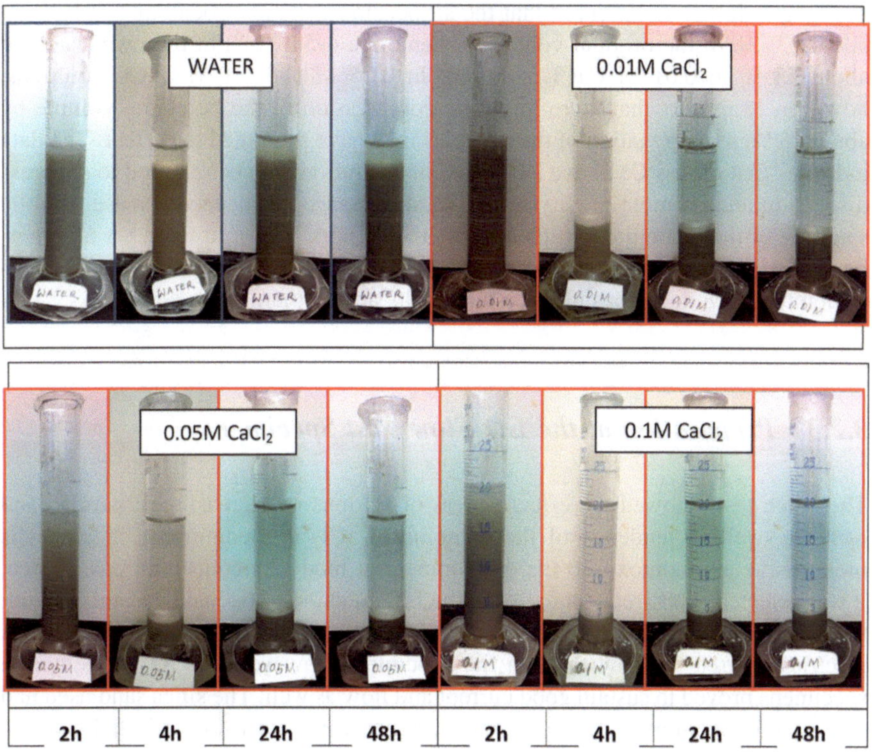

Fig. 3.2 Column tests to determine the effect of $CaCl_2$ concentration on clay swelling

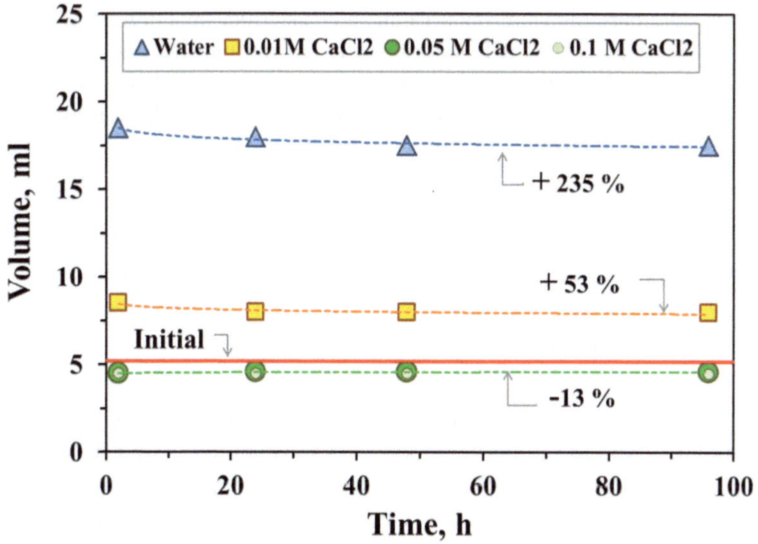

Fig. 3.3 Volume expansion of clay in free swelling tests in different concentration solutions of $CaCl_2$ and distilled water

in Fig. 3.3. It can be observed that the suspension volume of bentonite remains at about 235 % of the original volume of bentonite (~5.2 cm^3) in pure water and at about 53 % in 0.01 M $CaCl_2$ solution. In 0.05 M and 0.1 M $CaCl_2$ solutions, however, it appears that there might be compression of the bentonite volume by about 13 %, as aggregation of the particles may have promoted the effect. The data indicates that excess Ca above a critical concentration may be required to suppress dispersion and promote gravity settling and compression, hence decrease osmotic swelling of the bentonite particle (Norrish and Quirk 1954). In these set of experiments, 0.05 M Ca^{2+} appears to be sufficiently critical and increasing that concentration to 0.1 M does not influence the results any further.

3.2.3 Preparation of the EO Flow Test Specimens

The batch tests showed the effectiveness of Ca ions in decreasing dispersion and osmotic swelling tendency of the bentonite. $CaCl_2$ is a common salt, which is inexpensive and nontoxic to the environment at moderate concentrations. Hence, it was selected as the reagent to use in the EK injection tests. Sand was mixed with the bentonite clay initially to provide adequate electrolyte-filled pore space for electrical current to flow through the specimen. However, the 100 % bentonite specimens proved to sustain good current and flow as well. The silica sand used had a substantial amount of calcium content, hence contributed to the initial calcium content of the wet specimens as will be discussed later. During the test, the inflow through the anode and the outflow through the cathode were recorded, while keeping the hydraulic heads at equilibrium at the standing burettes. The equality of the quantity of flow into and out of the two burettes ensured steady, saturated flow. Since the burettes were open to air, any electrode reaction gases generated were free to escape. The $CaCl_2$ solution was fed through the anode burette. The calcium content of both the inflow and outflow were checked periodically by sampling. This ensured detection of Ca breakthrough when the Ca concentration levels at the cathode reached to that of the anode.

3.3 Results and Discussion of Electrokinetic Experiments

3.3.1 Electroosmotic Flow

Figure 3.4 presents the results of the electroosmotic tests with Ca^{2+} ion injection. The cumulative volume of outflow increased substantially with Ca^{2+} as opposed to water only, in all tests. This increase was most pronounced with 100 % bentonite specimen. The higher the clay content the higher the electroosmotic flow quantity was, as measured. The flow rate was almost constant beyond the first

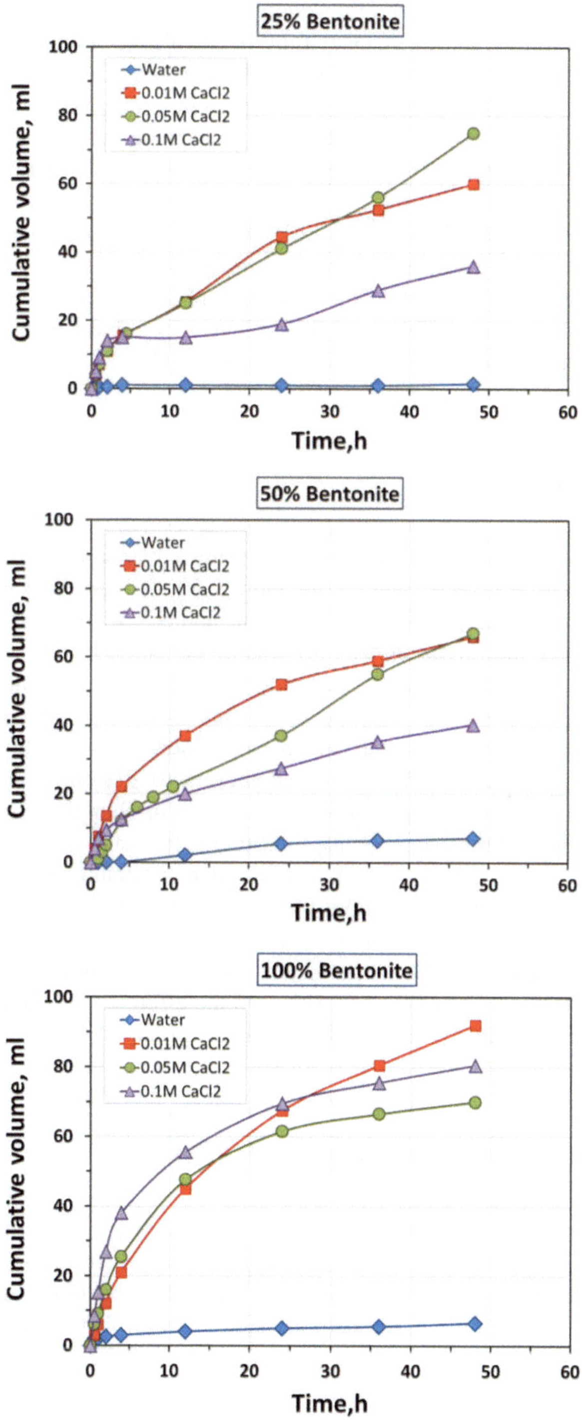

Fig. 3.4 Transient variation of the cumulative volume of electroosmotic flow in different bentonite content test specimens using different concentration solutions of $CaCl_2$

5 h of test run in the low bentonite content specimen, whereas it was clearly nonlinear with a decreasing rate over time in the higher bentonite content specimens. This data indicates reaching gradually a steady state of flow after which little or no changes in the pore space that influence the electroosmotic conductivity are to be expected. This equilibrium took place faster in highest content bentonite (100 % bentonite) specimen at about 24 h, then for the 50 % bentonite specimen at about 36 h.

In low bentonite content specimen (25 %), the low electrolyte concentration (0.01 M) appeared to cause faster increase in quantity of flow, whereas in high bentonite content specimen (100 %) the electrolyte concentration did not appear to affect the flow over time. This may be explained by the differences in post-test water content and pH distribution, as discussed below.

3.3.2 Water Content

In Fig. 3.5, the post-test distribution of the water contents of three specimens is presented. The initial water contents for the 25, 50, and 100 % bentonite content specimens were measured as 42 %, 55.5 %, and 77 %, respectively. The low bentonite content specimens showed no notable deviation from their initial water content distribution after the test. A general trend of decrease of water at anode and increase of water at cathode is observed in all specimens, which is typical of electroosmotic experiments conducted in closed containers (Reddy et al. 2002). Larger water content changes occurred in higher bentonite content specimens. This is attributed to larger quantities of water permeation through these specimens. The most notable result of the water content measurements came with higher electrolyte concentration runs (0.05 M, 0.1 M). Potentially due to calcium hydrate formation at increased $CaCl_2$ contents hence more water take up for the ensuing reactions, the quantity of flow decreased in these test runs. Formation of precipitates in the pore space can decrease the quantity of flow, as was observed with the higher electrolyte concentration test results in Fig. 3.5. Although the effect was not as pronounced in 100 % bentonite specimen, there was an decrease in rate of flow with time.

3.3.3 Zeta Potential and pH Distribution

The Fig. 3.6 shows the distribution of pH and zeta (ζ) potential in post-test specimens. pH measurements show the characteristic distribution of low values at anode and high values at cathode for the 25 and 50 % bentonite specimens. It is plausible that the high pH at the cathode side would affect the calcium hydrate formation at these locations. Yet, at higher bentonite concentration, the pH

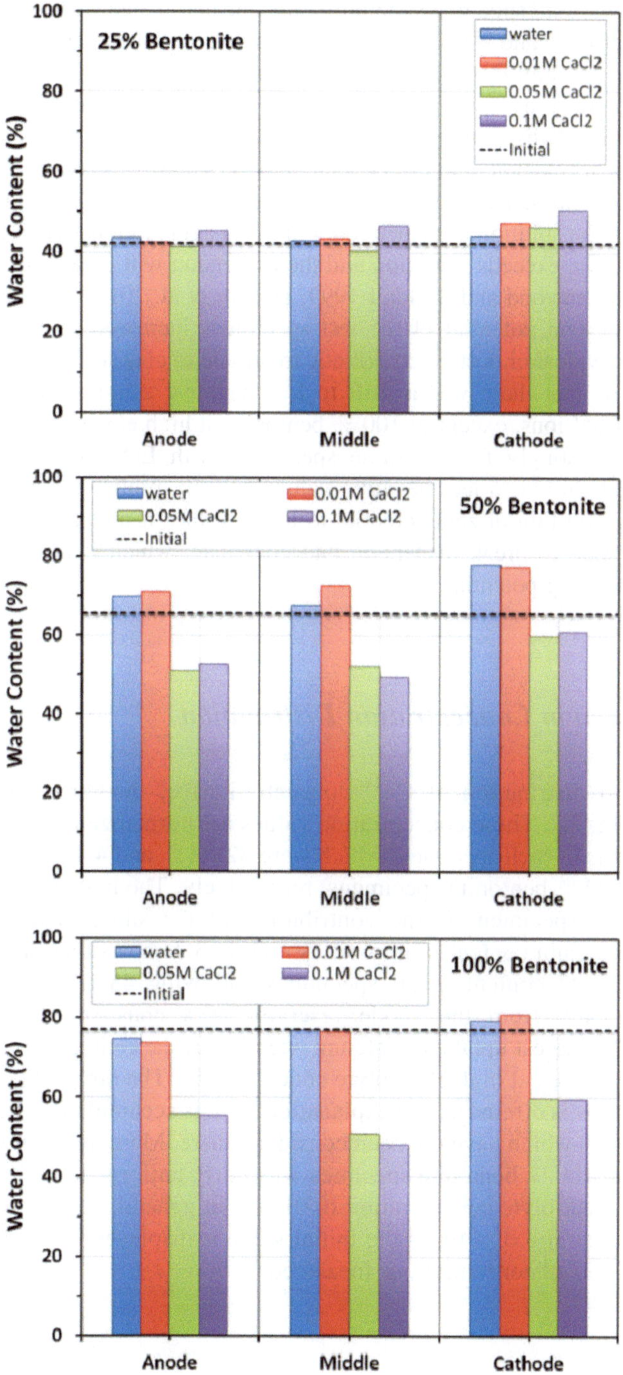

Fig. 3.5 Post EK test distribution of water content across the three test specimens with different $CaCl_2$ solution concentrations permeated by EO

remained low for all specimens including the one permeated with water only. This indicates that Ca^{2+} and Na^{1+} exchange has not taken place as rigorously to affect the pH of the system. Hence, the uptake of Ca^{2+} is slower in the 100 % bentonite system than the other two specimens of lower bentonite content. Indeed, the results of calcium distribution below confirm this conclusion.

The value of zeta potential can be related to the stability of colloidal dispersions. For molecules and particles that are small enough, a high zeta potential will confer stability, i.e., the solution or dispersion will resist aggregation. When the potential is low, attraction exceeds repulsion and the dispersion will break and the particles coagulate (Greenwood and Kendall 1999; Hanaor et al. 2012). From Fig. 3.6 it is observed that zeta potential of the test specimens increased from anode end to cathode end, typical of post-electrokinetic treatment of clay soils. The most notable observation is that there was a shift to less negative zeta-potential values with addition of Ca^{2+} ions, except in 100 % bentonite at high electrolyte concentration (0.1 M). Interestingly for the same specimen, with EO water conduction the zeta potential shifted to less negative values as well, probably controlled by pH changes. The shifting of zeta potential to lower values is a clear indication of the particles to tend to break dispersion and coagulate, which is also a measure of decreased swelling potential.

3.3.4 Calcium Concentration Distribution

The post-test distribution of Ca^{2+} concentrations of the three specimens are plotted in Fig. 3.7. These concentration values are normalized to the initial Ca^{2+} concentrations in each specimen, which were 32, 11.5, and 4.4 ppm for the 25 %, 50 %, and 100 % bentonite specimens, respectively. The higher Ca^{2+} content in low bentonite specimens is the contribution of the silica sand which had a substantial calcium content in impurities. Figure 3.7 shows that there is notable increase in Ca^{2+} content in all specimens following 48 h of Ca^{2+} injection electrokinetically from the anode reservoir at a constant concentration of 275 ppm. The largest uptake of calcium occurred at the cathode end of the 50 % bentonite specimen at high electrolyte concentration. The high pH (~11) and low quantity of flow is offered as one explanation for this accumulation, but there may be other factors which have not been considered here. More importantly, the Ca^{2+} uptake in the 100 % bentonite specimen was fairly uniform across the specimen for the low electrolyte concentration (0.01 M). In general, an average uptake of Ca^{2+} at a mass ratio of 4 times the initial value is displayed in most specimens, despite few spiked concentrations measured.

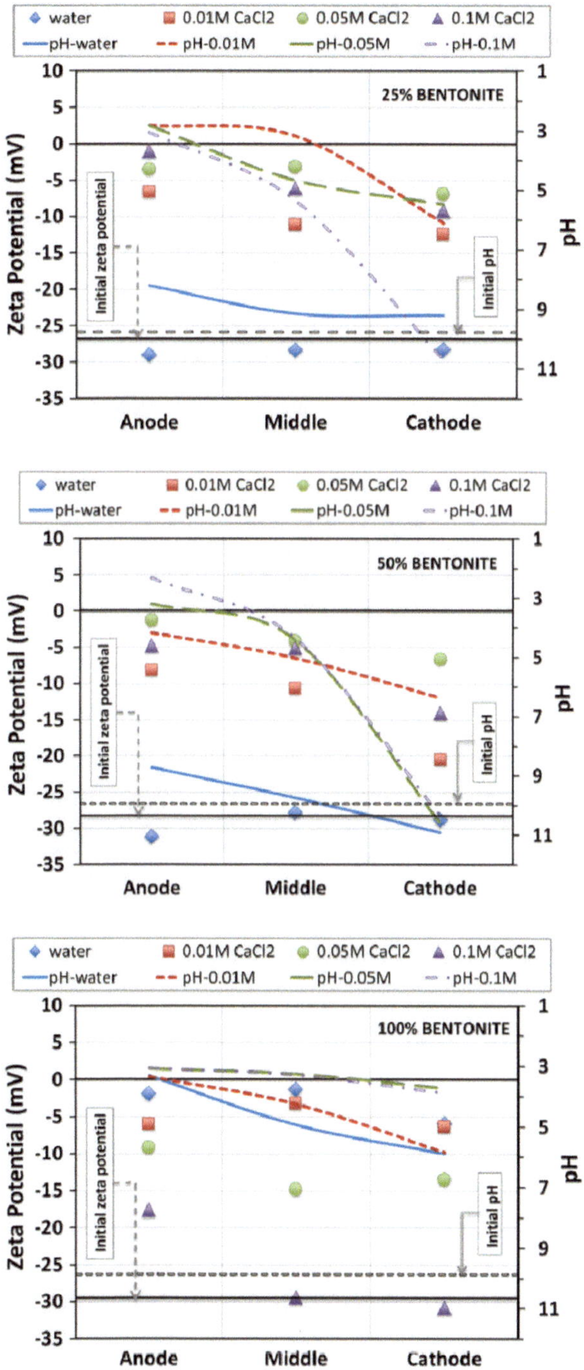

Fig. 3.6 Post EK test zeta potential and pH distribution across the three test specimens with different $CaCl_2$ solution concentrations permeated by EO

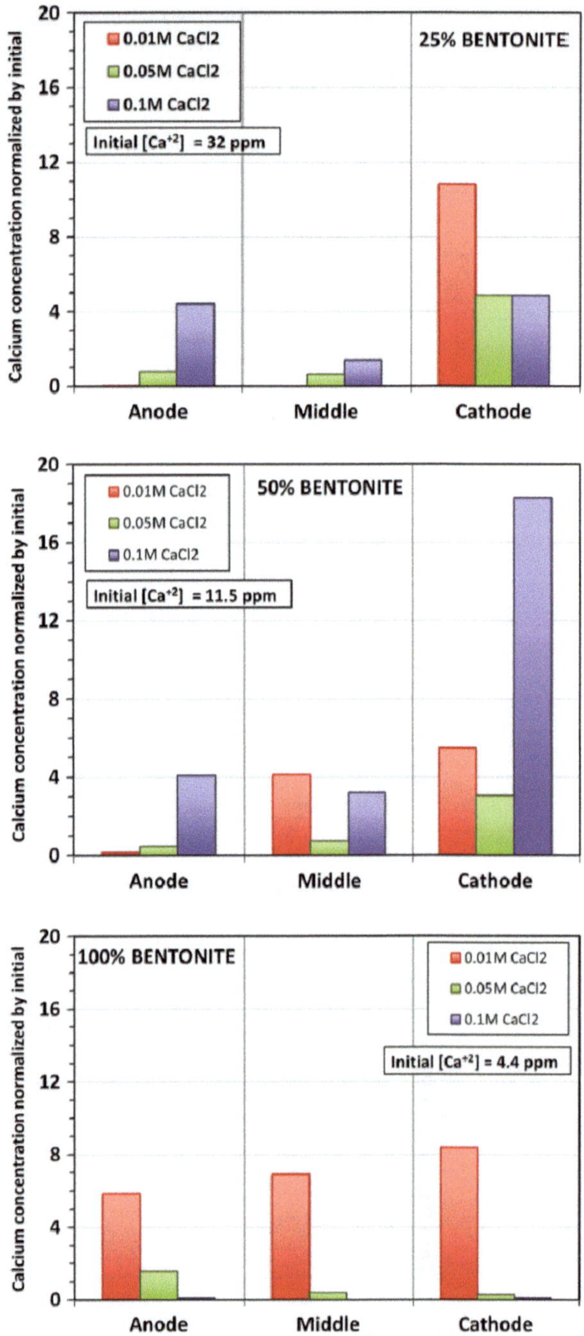

Fig. 3.7 Post EK test distribution of normalized Ca^{2+} across the three test specimens with different $CaCl_2$ solution concentrations permeated by EO

3.4 Conclusions

In conclusion, the electrokinetic drift of calcium ions into swelling clay stabilizes the formation and increases the volume rate of flow in laboratory tests. The decrease in zeta potential to less negative values indicate decrease in cation exchange capacity, reduction in surface charge deficiency and shrinking of DDL thickness for swelling clays. Beneficial shift of zeta potential appears to take hold even at higher pH vales (>7) for low Ca^{2+} concentration injections and for less than 100 % bentonite content in formation. Finally, a steady gain of Ca^{2+} is demonstrated, partially due to ion exchange and partially of due to formation of calcium hydrates at higher pH values and injection concentrations.

References

Adamson LG, Chilingar GV, Beeson CM (1966) Electrokinetic dewatering, consolidation and stabilization of soils. Eng Geol 1:291–304

Alshawabkeh AN, Sheahan TC (2003) Soft soil stabilization by ionic injection under electric fields. Proc Inst Civ Eng Ground Improv 7(4):177–185

Arutselvam A (2014) Soft clay stabilization using electro kinetic process. Int J Eng Technol Sci—IJETS, ISSN(P), 1 (II) 2349–3968, ISSN (O): 2349–3976

Chen FH (1988) Foundations on expansive soils. Elsevier, Amsterdam

Estabragh AR, Naseha M, Javadi AA (2014) Improvement of clay soil by electro-osmosis technique. Appl Clay Sci 95:32–36

Greenwood R, Kendall K (1999) Selection of suitable dispersants for aqueous suspensions of zirconia and titania powders using acoustophoresis. J Eur Ceram Soc 19(4):479–488

Hanaor DAH, Michelazzi M, Leonelli C, Sorrell CC (2012) The effects of carboxylic acids on the aqueous dispersion and electrophoretic deposition of ZrO_2. J Eur Ceram Soc 32(1):235–244. doi:10.1016/j.jeurceramsoc.2011.08.015

Jensen W (2003) Geotechnical policy and procedures manual. Research project SPR-1 (03) P552, prepared for the Nebraska Department of Roads (NDOR); Available on NDOR's website at http://www.nebraskatransportation.org/

Jones FO Jr (1964) Influence of chemical composition of water on clay blocking of permeability. J Pet Technol 16(4):441–446

Keykha HA, Huat BBK, Asadi A (2014) Electrokinetic stabilization of soft soil using carbonate-producing bacteria. Geotech Geol Eng 32:739–747

Mohan KK, Fogler HS (1997) Colloidally induced smectitic fines migration—existence of micro-quakes. AIChE J 43(3):565–576

Mollins LH, Stewart DI, Cousens TW (1996) Predicting the properties of bentonite sand mixtures. Clay Miner 31:243–252

Nelson JD, Miller DJ (1992) Expensive soils: problems and practice in foundation and pavement engineering. Wiley, New York

Norrish K, Quirk JP (1954) Crystalline swelling of montmorillonite—use of electrolytes to control swelling. Nature 173:255–256

Reddy KR, Saichek RE, Maturi K, Ala P (2002) Effects of soil moisture and heavy metal concentrations on electrokinetic remediation. Indian Geotech J 32(2):258–288

Tajudin SAA (2012) Electrokinetic stabilisation of soft clay. Doctor of Philosophy Thesis, School of Civil Engineering College of Engineering and Physical Sciences, The University of Birmingham, Birmingham

Wilson MJ, Wilson L, Patey I (2014) The influence of individual clay minerals of formation damage of reservoir standstones: a critical review with same insights. Clay Miner 49:147–164
Winterkorn HF, Pamukcu S (1991) Soil stabilization and grouting. In: Fang HY (ed) Foundation engineering handbook. Springer, Berlin, Chapter 9. XVIII, ISBN 978-0-412-98891-2
Zhou ZJ, Law HS (1998) Swelling clays in hydrocarbon reservoir; the bad, less bad and the useful. In: 7th Unitar international conference on heavy crude and tar sands. No. 1998.0570

Chapter 4
Sustainable Power Generation from Salinity Gradient Energy by Reverse Electrodialysis

Sylwin Pawlowski, João Crespo, and Svetlozar Velizarov

4.1 Background

Sustainable development cannot be achieved without a sustainable power generation. Energy represents from 1 to 10 % of industrial cost production (excluding personnel costs) (European Union 2011) and has been responsible in 2011, for almost 80 % of the greenhouse gas emissions in EU-28 (Eurostat 2014). Moreover, the power generation is pointed out as a key factor which may solve other worldwide problems, such as education or health care as well as water and food availability (Smalley 2005). For example, cooling for electricity production dominates the water use in industry (Forster 2014). The EU 2050 roadmap assumes that the power sector has the biggest potential for cutting greenhouse gas emissions, as electricity generated by renewable sources could partially replace fossil fuels used in transport and for heating (http://ec.europa.eu/clima/policies/roadmap/perspective/index_en.htm). In 2011, the World gross electricity generation was approximately 2.5 TW, but only 20.6 % was generated from renewable sources (Eurostat 2014). Therefore, decarbonizing our energy sources, which must be affordable, accessible, and sustainable, is essential (Chu and Majumdar 2012).

4.2 Salinity Gradient Energy

When water streams with different salt concentrations mix, energy is released due to a chemical potential difference between the two water bodies. The process is spontaneous, irreversible and leads to an increase in the system entropy, what gives

S. Pawlowski • J. Crespo • S. Velizarov (✉)
LAQV-REQUIMTE, Departamento de Química, Faculdade de Ciências e Tecnologia, Universidade Nova de Lisboa, Caparica 2829-516, Portugal
e-mail: s.velizarov@fct.unl.pt

© Springer International Publishing Switzerland 2016

A.B. Ribeiro et al. (eds.), *Electrokinetics Across Disciplines and Continents*,
DOI 10.1007/978-3-319-20179-5_4

an opportunity to capture a renewable energy (Kuleszo et al. 2010; Murray et al. 2013; Post et al. 2008; Ramon et al. 2011; Vermaas et al. 2013a). During the mixing process no CO_2 or other greenhouse gases, which may interfere with the global climate are released, no salts are "consumed" and virtually (unless the water must be pre-treated and transported (Post et al. 2008) this "fuel" has no cost (Jones and Finley 2003; Merz et al. 2012; Murray et al. 2013). Salinity gradient energy (SGE) can therefore be considered as a completely renewable and sustainable source of energy.

4.2.1 SGE Thermodynamic/Available Work

The theoretically available energy (i.e., exergy) can be estimated in terms of the Gibbs free energy (Murray et al. 2013; Post et al. 2008):

$$\Delta G_{\mathrm{mix}} = G_{\mathrm{b}} - (G_{\mathrm{c}} + G_{\mathrm{d}}) \tag{4.1}$$

where ΔG_{mix} is the change in Gibbs energy upon solutions mixing, G_{b}, G_{c}, and G_{d} are the Gibbs energies of the brackish, concentrated, and dilute solutions, respectively. The Gibbs energy of an ideal solution is equal to:

$$G = \sum_i \mu_i \cdot n_i \tag{4.2}$$

where μ is the chemical potential and n is the number of moles of a component i in the solution so the amount of released energy corresponds to the chemical potential difference between the brackish solution and the initial dilute and concentrated solutions. For ideal dilute solutions (i.e., with no change in the enthalpy, $\Delta H_{\mathrm{mix}} = 0$), (4.1) can be also rewritten in terms of the entropy change (ΔS) (Alvarez-Silva and Osorio 2015; Kuleszo et al. 2010; Post et al. 2008; Vermaas et al. 2013a):

$$\frac{\Delta G_{\mathrm{mix}}}{n_{\mathrm{b}}} = -T \cdot [\Delta S_{\mathrm{b}} - f \cdot \Delta S_{\mathrm{c}} - (1 - f) \cdot \Delta S_{\mathrm{d}}] \tag{4.3}$$

with

$$\Delta S = -R \cdot \sum_i x_i \cdot \ln(x_i) \tag{4.4}$$

where n_{b} is brackish water n° of moles, T is temperature, f is the fraction of seawater, and R is the universal gas constant. The power which can be generated by harvesting the released energy ($P_{\Delta G_{\mathrm{mix}}}$) depends therefore on the salinity difference between the two initial streams as well as on the solutions flow rate (Q) and molar volume (V_{mol}) (Vermaas et al. 2013a):

$$P_{\Delta G_{\mathrm{mix}}} = -T \cdot [\Delta S_{\mathrm{b}} - f \cdot \Delta S_{\mathrm{c}} - (1-f) \cdot \Delta S_{\mathrm{d}}] \cdot \frac{Q_{\mathrm{b}}}{V_{\mathrm{mol,b}}} \tag{4.5}$$

4.2.2 River–Sea Interface

A naturally occurring interface between water streams of differing salinities exists when, e.g., a river flows into a sea. The chemical potential difference between sea and river water is equivalent to 175–270 m of hydraulic head (Emami et al. 2013; Jones and Finley 2003; Kempener and Neumann 2014; Merz et al. 2012; Murray et al. 2013; Ramon et al. 2011; Vermaas et al. 2013a), what has inspired the name "silent waterfalls" for rivers mouth (Ramon et al. 2011). Mixing of equal volumes (1 m^3) of "river" (~0.01 M of NaCl) and "sea" (~0.50 M of NaCl) waters, release approximately 1.4 MJ of energy at 293 K (Kempener and Neumann 2014; Post et al. 2007; Vermaas et al. 2013a). If we consider, instead of 1 m^3, an infinite amount of sea water, 2.3 MJ would be released (Post et al. 2008). Moreover, unlike wind or solar energies, the SGE source is non-periodic due to the Earth's continuous evaporation/precipitation hydrological cycle (Merz et al. 2012; Murray et al. 2013) as schematically illustrated in Fig. 4.1.

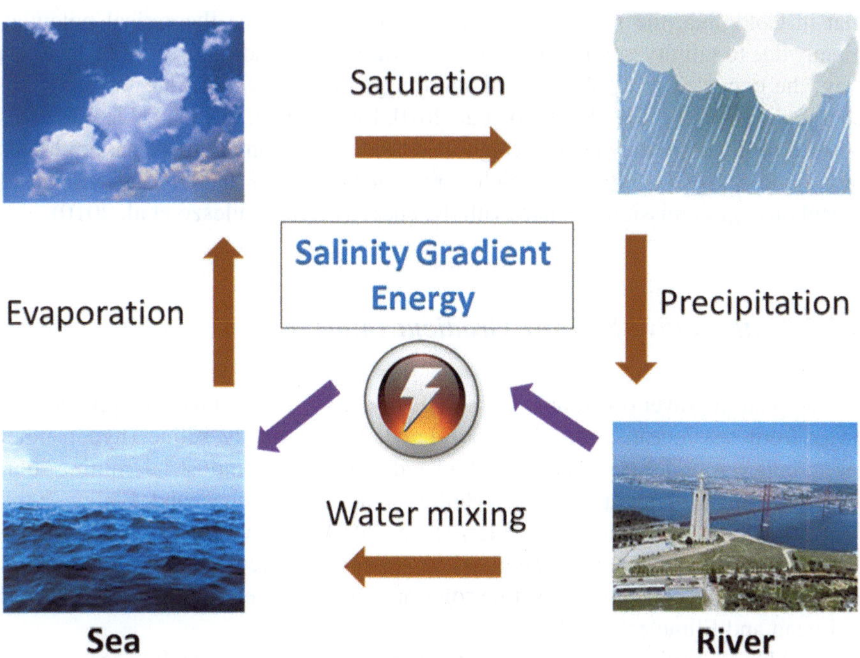

Fig. 4.1 Schematic illustration of Earth's water cycle, with identification of a possible source of salinity gradient energy. (The photo of the "river" is the Tagus River under the 25th of April Bridge in Lisbon.)

Table 4.1 Theoretical and technical potential of salinity gradient energy due to rivers discharge throughout the World (adapted from Kuleszo et al. (2010))

Region	No. of rivers	Discharge (m^3/s)	Theoretical potential (GW)	Technical potential (GW)
Africa	391	170,294	311	190
Asia	1243	236,769	374	206
Europe	779	74,569	94	56
North America	1878	191,434	321	189
Oceania	791	165,566	308	195
South America	390	320,078	316	148
World	5472	1,158,709	1724	983

The worldwide potential of all rivers discharge into seas/oceans has been estimated to be within the range of 1.4–2.6 TW (Emami et al. 2013; Kuleszo et al. 2010; Ramon et al. 2011). However, a feasible electricity amount which can be generated depends, besides river and sea salinities, on temperature and flow rate, streams salt composition, salinity gradient steepness, and environmental impacts (Kempener and Neumann 2014; Kuleszo et al. 2010; Ramon et al. 2011).

For example, rivers with very big flow rates, such as Amazon and Congo, have a diluting effect on the Atlantic Ocean salinity, what decreases the salinity gradient in their mouths (Huckerby et al. 2012; Kuleszo et al. 2010). Another study has shown that in Colombia, the site-specific potential is 31 % of the theoretical potential, mainly due to salinity gradient variations (Alvarez-Silva and Osorio 2015). In such a way, the expected realizable global salinity gradient potential, due to rivers discharge, is close to 1 TW (Kuleszo et al. 2010; Ramon et al. 2011) (Table 4.1). This power, if fully used, may lead to a significant reduction of greenhouse gas emissions by around 8 Gt CO_2-eq/year, which corresponds to 24 % of the total amount of greenhouse gas emissions related with the energy sector (Kuleszo et al. 2010).

4.2.3 Alternative Salinity Gradient Interfaces

The amount of power obtainable from SGE can be increased if hybrid applications, away from seas/oceans, are also considered. Inland highly saline (hypersaline) lakes, underground salt reserves/aquifers and wastewaters (industrial, mining solutions, desalination plants brines) (Helfer et al. 2013; Jones and Finley 2003; Kempener and Neumann 2014; Merz et al. 2012; Neuman et al. 2012) can be alternative sources of concentrated saline streams. Figure 4.2 shows the osmotic pressure of some concentrated saline solutions based on literature data summarized in Logan and Elimelech (2012):

Worldwide municipal wastewaters plants effluents, when drawn into seas, offer a possible potential of 18.5 GW (Logan and Elimelech 2012; Ramon et al. 2011).

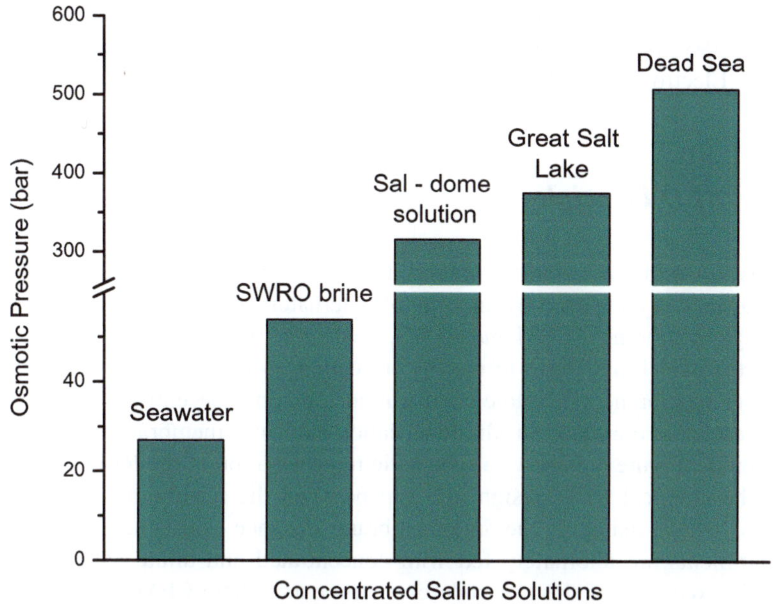

Fig. 4.2 The osmotic pressure of possible five sources of concentrated saline streams (*SWRO brine* brine from Sea Water Reverse Osmosis)

Mixture of equal volumes (1 m^3) of brine (~5 M of NaCl) and river (~0.01 M of NaCl) waters can release around 16 MJ of energy (Kempener and Neumann 2014; Post et al. 2007), besides providing inherent ecological benefits of diluting saline wastewater streams before disposal. The mentioned sources, despite of offering higher saline concentration difference, have limited overall potential compared to river–sea systems because of their smaller volumes, but still may found rather interesting site-specific applications, e.g., as energy recovery systems. Highly saline waters may be also stored in ponds for being used when necessary (Helfer et al. 2013; Kempener and Neumann 2014). Another opportunity of creating a salinity gradient interface could pass through utilization of thermolytic solutions (e.g., containing ammonium bicarbonate) in the osmotic heat engines used to capture energy from waste heat (Logan and Elimelech 2012; Luo et al. 2012; Neuman et al. 2012).

4.3 Reverse Electrodialysis

The SGE can be transformed into electric power only when appropriately harvested, as in a random mixing, the system quickly reaches chemical equilibrium, without giving the time to capture the released energy (Merz et al. 2012). The first concept of extracting electrical energy from such systems has been developed by Pattle in 1954 (Pattle 1954) and is very similar to the process currently known as Reverse Electrodialysis (RED).

4.3.1 RED Principle

RED is a membrane-based technology for controlled mixing of saline solutions. A RED stack is composed by an array of, alternately arranged, cation- and anion-exchange membranes (CEM and AEM), stacked between two electrodes. A basic unit of the stack is a cell pair formed by one cation-exchange membrane, one anion-exchange membrane, one dilute compartment, and one concentrated compartment. In order to close the stack, an additional cation-exchange membrane is added facing the cathode. Figure 4.3 shows a schematic representation of the RED principle.

In the classical RED design, the gap between the membranes is assured by spacers, which maintain the intermembrane distance constant. The so-formed compartments are alternately fed with concentrated and dilute saline solutions, with the exception of the electrode compartments, where CEM separate a dilute or a concentrated saline stream from a solution containing redox pair which is recirculated between the two terminal electrode compartments.

Since the membranes are semipermeable barriers, which, in ideal conditions, only allow the passage of anions (through AEM) or cations (through CEM), while blocking transport of co-ions and water, only the respective counter-ions

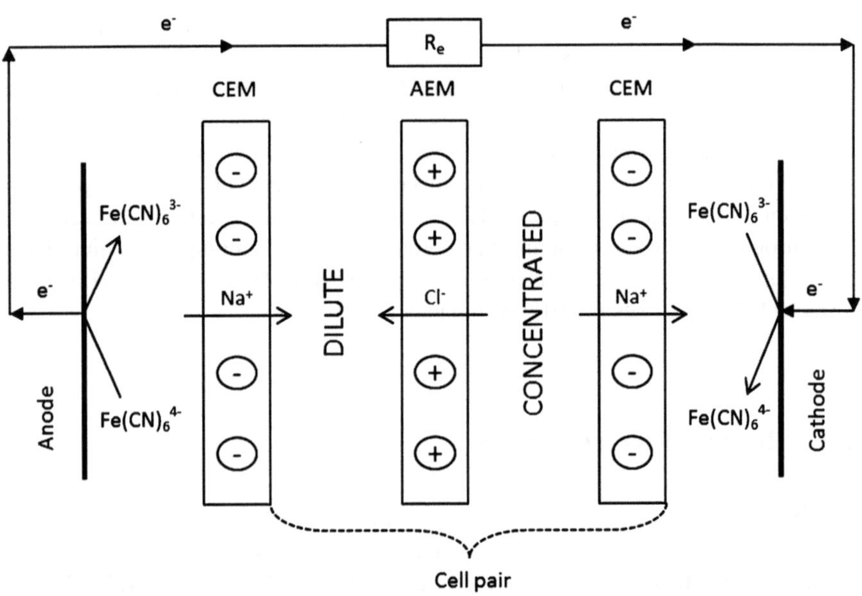

Fig. 4.3 Schematic representation of the RED principle (*CEM* cation-exchange membrane, *AEM* anion-exchange membrane, R_e external load resistance)

should migrate from a concentrated to a dilute saline stream. Due to the alternate arrangement of AEM and CEM, the cations and anions migrate in opposite directions. This movement leads to an establishment of an electric potential difference between the two electrodes. Thus, the chemical SGE is converted, by mean of an appropriate redox couple, into electrical energy, i.e., an electron flux from the anode to the cathode through an external circuit, resulting from the need of maintaining electro-neutrality in the solution recirculated in the electrode compartments.

4.3.2 RED Metrics

The main parameter determining a RED stack performance is the provided net power density ($P_{\text{net density}}$), which consists in a trade-off between the gross power density ($P_{\text{gross density}}$) obtained in the stack and the power spent for pumping the solutions (P_{pumping}):

$$P_{\text{net density}} = P_{\text{gross density}} - \frac{P_{\text{pumping}}}{2 \cdot N \cdot A_{\text{m}}} \tag{4.6}$$

where A_{m} is the area of one membrane and N is the number of cell pairs (Vermaas et al. 2011a, 2012). For the same hydrodynamic conditions in dilute and concentrated compartments, the power spent for pumping can be obtained as follows (Pawlowski et al. 2014a):

$$P_{\text{pumping}} = 2 \cdot (\Delta p_{\text{t}} \cdot Q) \tag{4.7}$$

where Q is feed solutions flow rate and Δp_{t} is the total pressure drop between the stack inlet and outlet. The factor 2 appears because there are two saline solutions which are pumped, the diluted and the concentrated one. The total pressure drop is the sum of partial pressure drops along the solutions pathway inside the RED stack. Most often the main partial pressure drops are those localized in the stack channels and in its branches (Pawlowski et al. 2014a). The power required for pumping the redox couple solution recirculated in the electrode compartments, especially for stacks with a big number of pairs, can be neglected (Strathmann 2010).

The gross power generated by a RED stack depends of the potential difference established between the electrodes (*OCV*—open circuit voltage), the stack internal resistance (R_i), and the external load resistance. The maximal value of gross power density is achieved when the external load resistance has the same value as the internal resistance (Veerman et al. 2008, 2009, 2010a). For such a case:

$$P_{\text{gross density}} = \frac{\text{OCV}^2}{4 \cdot R_i} \tag{4.8}$$

The open circuit voltage is the driving force of the RED process and represents the sum of potential differences over each membrane, which mainly depends on the activities $(\gamma \cdot C)$ ratio between the concentrated (c) and the diluted (d) saline solutions (Dlugolecki et al. 2009; Vermaas et al. 2012):

$$\mathrm{OCV} = 2N \cdot \alpha \cdot \frac{R \cdot T}{F} \cdot \ln\left(\frac{\gamma_\mathrm{c} \cdot C_\mathrm{c}}{\gamma_\mathrm{d} \cdot C_\mathrm{d}}\right) \tag{4.9}$$

In (4.9), α is the membrane permselectivity, R is the universal gas constant, T is the absolute temperature, F is the Faraday constant, γ is the activity coefficient, and C is the saline solution concentration on membrane surface.

The stack internal resistance (R_i) is the sum of *CEM* and *AEM* resistances, solutions resistances in the compartments, where h is the compartment thickness (intermembrane distance) and κ is the solution conductivity, diffusion boundary layers (DBLs) resistance (R_DBL), and electrode compartments resistance (R_el) (Brauns 2009; Vermaas et al. 2011a, b, 2012, 2014a, b):

$$R_i = N \cdot \left(R_\mathrm{CEM} + R_\mathrm{AEM} + \frac{h_\mathrm{c}}{\kappa_\mathrm{c}} + \frac{h_\mathrm{d}}{\kappa_\mathrm{d}} + R_\mathrm{DBL}\right) + R_\mathrm{el} \tag{4.10}$$

The DBL resistance is considered as a non-ohmic resistance, while all other resistances are of ohmic nature (Vermaas et al. 2011a, 2012, 2014a). In a large-scale stack, the resistance of the electrodes compartments is insignificant, compared to the other resistances (Vermaas et al. 2011a, b), and could be neglected. On the other hand, formation of DBLs must be carefully taken into consideration, since it leads to a decrease in potential difference over the membranes due to concentration polarization, which is especially important in the diluted compartments and at low flow rates (Kozmai et al. 2010; Pawlowski et al. 2014b).

4.3.3 *Electrokinetics in RED*

4.3.3.1 Ion-Exchange Membranes

An efficient ion transport in electromembrane processes, such as RED, firstly depends on the ion-exchange membranes performance. Besides, the price of the membranes is the main cost in terms of Levelized Cost of Electricity (LCOE) generated by RED (Daniilidis et al. 2014a). Ion-exchange membranes are capable of selective transport of the respective counter-ions from one phase to another due to the high concentration of oppositely charged functional groups fixed to the polymer matrix. Most common negatively charged fixed groups for CEM are sulfonic acid groups ($-SO_3$), while as positively charged groups of AEM, quaternary amine groups ($-N^+(CH_3)_3$) are usually chosen (Dlugolecki et al. 2008; Galama et al. 2014a; Strathmann et al. 2013).

The main properties of some commercially available ion-exchange membranes (Dlugolecki et al. 2008, 2010a; Guler et al. 2012, 2013, 2014a; Slagt 2012; Vermaas et al. 2011b), which can be used in RED are summarized in Table 4.2.

Table 4.2 Main properties of some commercially available ion-exchange membranes

Membrane	Area resistance (Ω cm^2)	Perm-selectivity (%)	Wet thickness (μm)	Ion-exchange capacity (meq/g)	Swelling degree (%)
Cation-exchange membranes					
Fuji (Slagt 2012)	>2.0	95.0	140		
Fumasep (Dlugolecki et al. 2008, 2010a; Guler et al. 2013)	2.14	89.5	113	1.14	29
Fumasep FKE (Dlugolecki et al. 2008)	2.46	98.6	34	1.36	12
Fumasep FKS (Guler et al. 2013)	1.50	94.2	40	1.54	13.5
Neosepta CM-1 (Dlugolecki et al. 2008)	1.67	97.2	133	2.30	20
Neosepta CMX (Dlugolecki et al. 2008, 2010a; Guler et al. 2012, 2013)	2.91–3.00	95.0–99.0	158–164	1.62	18
Qianqiu CEM (Guler et al. 2013)	1.97	82.0	205	1.21	33
Ralex CMH-PES (Dlugolecki et al. 2008; Guler et al. 2013; Vermaas et al. 2011b)	11.33–7.0	94.7–97.0	764–650	2.34	31
Selemion CMV (Dlugolecki et al. 2008; Guler et al. 2013)	2.29	98.8	101	2.01	20
Anion-exchange membranes					
Fuji (Guler et al. 2014a; Slagt 2012)	0.93	89.0–91.0	123–140		
Fumasep FAD (Dlugolecki et al. 2008, 2010a; Guler et al. 2013)	0.89	86.0	74	0.13–1.42	34
Fumasep FAS (Guler et al. 2013)	1.03	89.4	33	1.12	8
Neosepta AFN (Dlugolecki et al. 2008)	0.70	88.9	163	3.02	43
Neosepta AM-1 (Dlugolecki et al. 2008)	1.84	91.8	126	1.77	19

(continued)

Table 4.2 (continued)

Membrane	Area resistance (Ω cm^2)	Perm-selectivity (%)	Wet thickness (μm)	Ion-exchange capacity (meq/g)	Swelling degree (%)
Neosepta AMX (Dlugolecki et al. 2008, 2010a; Guler et al. 2012, 2013)	2.35	90.0–90.7	134	1.25–1.40	16–26
Qianqiu AEM (Guler et al. 2013)	2.85	86.3	294	1.33	35
Ralex AMH-PES (Dlugolecki et al. 2008; Guler et al. 2013; Vermaas et al. 2011b)	7.3–7.66	89.0–89.3	670–714	1.97	56
Selemion AMV (Guler et al. 2013)	3.15	87.3	124	1.78	17

The resistance values in Table 4.2 have been measured in 0.5 M NaCl solutions at 25 °C, while the permselectivity has been obtained from membrane potential measured across the membranes separating 0.5 M and 0.1 M NaCl solutions. In RED application, these values will depend on the actual saline solutions used, as the electrical resistance of an ion-exchange membrane strongly depends on the solution concentration, especially for salt concentrations <0.1 M NaCl, in which range the electrical resistance of the membrane strongly increases with the decrease in salt concentration in the solution (Dlugolecki et al. 2010a, b; Galama et al. 2014a; Geise et al. 2014a). The membrane resistance could be decreased significantly by increasing temperature (Dlugolecki et al. 2010b), and it is mainly determined by the compartment with the lowest external salt concentration (Galama et al. 2014a). Contrary, the membrane permselectivity decreases when the salt concentration in the solution increases, due to a weakened Donnan exclusion of co-ions (Dlugolecki et al. 2010a). The nature of co-ions also influences the membrane permselectivity; co-ions that are charge dense and have low polarizability tend to provide a higher membrane permselectivity (Geise et al. 2014b).

It has been found that a high ion-exchange capacity and a low swelling degree lead to an increased membrane selectivity, but do not have a clear effect on the membrane area resistance (Guler et al. 2012, 2013). In order to decrease the membrane resistance, a more appropriate step passes through decreasing membrane thickness as the counter-ion transport becomes faster for thinner membranes (Guler et al. 2012, 2013). The gross power density of a RED stack with membranes prepared following the mentioned indications was found to be the highest one in comparison with stacks equipped with commercial membranes (Guler et al. 2012, 2013).

4.3.3.2 Ion Transport and Concentration Polarization

In a RED stack, the potential difference established across the membranes due to the difference in salinity of the two aqueous streams leads to transport of ions. The amount of current transported by an ion is characterized by its effective transport number (Dlugolecki et al. 2010a; Kozmai et al. 2010; Nikonenko et al. 2010, 2014; Strathmann et al. 2013). In the membrane, the current is preferentially due to the respective counter-ion, and if the external solutions are sufficiently diluted, the transport numbers of a counter-ion and co-ion in the membrane are close to 1 and 0, respectively. In the solution, both ions transport the current and each ion transport number in the solution depends from their mobility. For example for 1 g/L NaCl solution, the transport numbers in the solution are 0.392 and 0.608 for Na^+ and Cl^-, respectively (Pawlowski et al. 2014b). Due to this ion transport numbers difference in the solution and in the membrane, the salt concentration in the concentrated stream is lower close to the membranes than in the bulk, while in the dilute stream compartments an opposite situation occurs (Brauns 2009; Kim et al. 2012; Ramon et al. 2011; Tedesco et al. 2015). Development of a solute concentration gradient across the liquid DBL close to a membrane is a phenomenon known as concentration polarization, which in RED leads to a decrease in the potential difference over the membranes (4.9) and, as a consequence, to a decrease in the obtainable gross power density (Ramon et al. 2011). Figure 4.4 shows a schematic representation of concentration profiles in a RED cell pair.

In practice, the concentration gradient at the solution bulk–DBL interface is continuous, but by assuming the Nernst film model, the DBL thickness (δ) can be estimated from the tangents intersection of the two concentration profiles (across the DBL and in the bulk). Such an approximation allows for a reasonable description of the diffusion phenomena in electromembrane processes (Brauns 2009; Choi et al. 2002; Galama et al. 2013, 2014b; Geraldes and Afonso 2010; Kozmai et al. 2010; Nikonenko et al. 2010, 2014; Sistat and Pourcelly 1997). Furthermore, it is commonly assumed that the transport of a target counter-ion (J) in the membrane takes place predominantly by migration, i.e., due to electric field:

$$J = \frac{iT}{zF} \tag{4.11}$$

while in the DBL the same counter-ion moves due to both migration and diffusion caused by its concentration difference across the boundary layer:

$$J = \frac{it}{zF} + D\left(\frac{\partial C}{\partial x}\right)_x \tag{4.12}$$

where i is the current density, T is the counter-ion transport number in the membrane, z is its valence, F is the Faraday constant, t is the counter-ion transport number in solution, D is salt diffusivity, C is salt concentration, and x is the

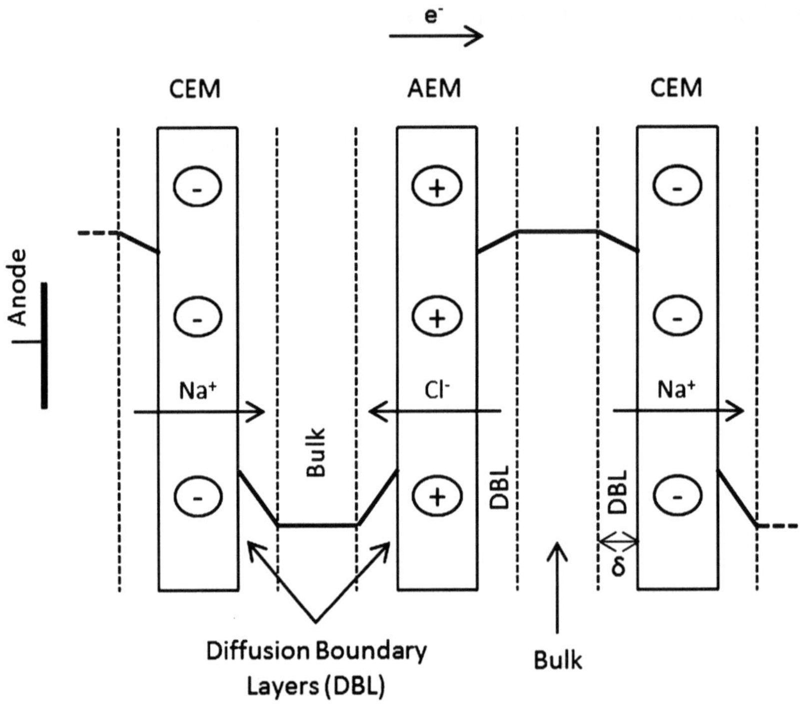

Fig. 4.4 Schematic representation of electrical potential and concentration profiles in a *RED* cell pair (*CEM* cation-exchange membrane, *AEM* anion-exchange membrane)

direction perpendicular to the membrane surface (Galama et al. 2013, 2014b; Geraldes and Afonso 2010; Kozmai et al. 2010; Nikonenko et al. 2010, 2014). Integration of equations (4.11) and (4.12) across the diffusion liquid boundary layer in the concentrated stream compartment gives:

$$i = \frac{zF(C_b - C_m)D}{\delta(T - t)} \tag{4.13}$$

where C_b is salt concentration in the bulk, C_m is salt concentration at the membrane surface, and δ is the DBL thickness.

In electrodialysis, where a potential difference is externally imposed between the electrodes, it is possible to reach a current density value (limiting current density), for which the salt concentration in the aqueous phase contacting the membrane surface becomes zero (Geraldes and Afonso 2010; Nikonenko et al. 2010, 2014). Determination of limiting current density by the Cowan-Brown method from current–voltage curves, allows for estimation of the DBL thickness (Choi et al. 2002; Dlugolecki et al. 2010a; Geraldes and Afonso 2010). In RED however, the ionic current is due to internally established potential difference (due to the salinity gradient between the two aqueous saline streams), so the current density values reached are much lower than the limiting current density value.

4.3.3.3 Chronopotentiometry

Chronopotentiometry consists in applying a constant current during a defined period of time, until a steady-state potential difference is achieved, and then the current supply is interrupted (Fig. 4.5a). Chronopotentiometry is a simple, fast and accurate electrochemical tool since it allows for a direct access of the voltage contributions in polarized and non-polarized solution-membrane systems and can be applied for different saline concentrations in the underlimiting current density range (Krol et al. 1999; Pawlowski et al. 2014b; Pismenskaia et al. 2004; Sistat and Pourcelly 1997). Under RED-operating conditions, we have found that estimation of the DBL thickness using chronopotentiogram data relative to the final (diffusion relaxation) part of the chronopotentiogram (see Fig. 4.5b) was more accurate, in comparison with its initial region (Pawlowski et al. 2014b; Vermaas et al. 2014b). This was because the concentration polarization was already fully developed at the end of the measurement, thus avoiding a possible initial non-uniform current distribution due to the relatively large ion-exchange membrane exposed area and its surface inhomogeneity (Pawlowski et al. 2014b).

Assuming a unidimensional counter-ion transport (described through the Nernst-Planck formalism within the two DBLs adjacent to the cation-exchange membrane), no co-ion transport and the same electrolyte concentration on both sides of the membrane, the transmembrane potential ($\Delta\varphi_{\text{tot}}$) is (Belova et al. 2006; Kozmai et al. 2010; Pismenskaya et al. 2007):

$$\Delta\varphi_{\text{tot}} = 2\frac{RT}{F}\ln\left(\frac{C_{m\ (\text{enriched})}}{C_{m\ (\text{depleted})}}\right) + iR_{m+\text{sol}} \tag{4.14}$$

where the considered salt concentrations are those on the membrane surfaces (C_m) contacting the enriched and depleted in salt compartments, $R_{m+\text{sol}}$ is the electric resistance of the membrane and of the bulk solutions (with salt concentration C_b),

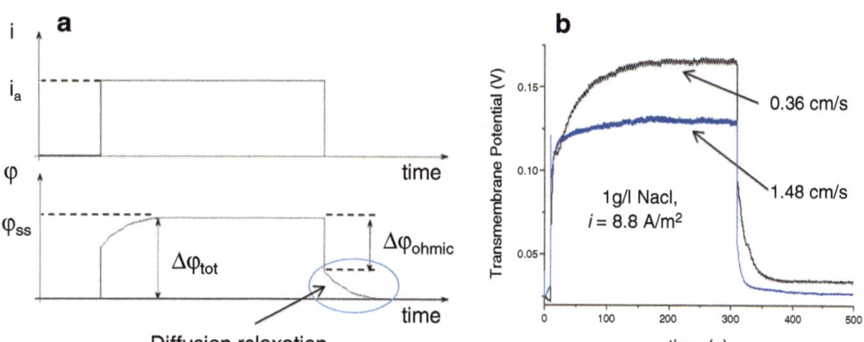

Fig. 4.5 Chronopotentiograms: (**a**) schematic current density-time curve and the resulting voltage-time curve; (**b**) chronopotentiograms obtained in the underlimiting current density regime over the central Ralex cation-exchange membrane with surface area of 64 cm^2, in a two-cell pair, spacer-free RED stack, fed with 1 g/L NaCl solution at two different linear flow velocities

and i is the applied current density. At steady state, for an underlimiting current density, $i < i_{\text{lim}}$, a linear salt concentration profile is expected across the DBLs (Belova et al. 2006; Choi et al. 2002; Kozmai et al. 2010; Pismenskaya et al. 2007):

$$C_{\text{m}} = C_{\text{b}}\left(1 \pm \frac{i}{i_{\text{lim}}}\right) \tag{4.15}$$

and, in such a way, the transmembrane potential measured at the steady state ($\Delta\varphi_{\text{SS}}$) becomes:

$$\Delta\varphi_{\text{tot}} = 2\frac{RT}{F}\ln\left(\frac{1+\frac{i}{i_{\text{lim}}}}{1-\frac{i}{i_{\text{lim}}}}\right) + iR_{m+\text{sol}} \approx 4\frac{RT}{F}\frac{i}{i_{\text{lim}}} + iR_{m+\text{sol}} = \Delta\varphi_{\text{SS}} \tag{4.16}$$

When the current application is interrupted, the potential difference ($\Delta\varphi_{\text{SS}}$–$\Delta\varphi_{\text{ohmic}}$) measured at that instance is only due to concentration polarization ($4(RT/F)(i/i_{\text{lim}})$) developed during the time of current application (Pawlowski et al. 2014b). Knowing that i_{lim} can be obtained by (4.13) for the case of $C_m = 0$, the DBL thickness (δ) becomes:

$$\delta = \frac{C_b F^2 \text{D}(\Delta\varphi_{\text{SS}} - \Delta\varphi_{\text{ohmic}})}{4RTi(T-t)} \tag{4.17}$$

Figure 4.5b shows that the potential difference $\Delta\varphi_{\text{SS}}$–$\Delta\varphi_{\text{ohmic}}$ increases with the decrease in linear flow velocity in the compartment channel, what means that the DBL thickness increases. In result, the obtainable gross power density decreases (Dlugolecki et al. 2009; Vermaas et al. 2011b). The DBL thickness can be decreased by intensifying the hydrodynamic conditions, such as an increase in the linear flow velocity, as can be seen in Fig. 4.5, or by introducing turbulence promoters inside the channels (Dlugolecki et al. 2009, 2010a, b; Vermaas et al. 2011a, 2014a). However, both measures lead also to an increase of the pressure drop inside the stack (Pawlowski et al. 2014a; Tedesco et al. 2012; Veerman et al. 2009; Vermaas et al. 2011a, 2014a) and, consequently, may result in even negative values of the net power density (4.6) and (4.7), because of spending more power for pumping of the solutions that the one generated by the RED stack itself. A possible solution to this problem could pass through utilizing enhancing flow entrance effects on mass transfer, as in the entrance of the channels, the concentration polarization is minimal while it increases along the channel length until the salt concentration profile becomes fully developed (Pawlowski et al. 2014b).

4.3.3.4 Monovalent Versus Multivalent Ions

During the development of the RED technology, in order to investigate and model the process performance most frequently NaCl electrolyte solutions of different concentrations have been considered. However, in natural seawater and river water a number of other ions, including multivalent such as Mg^{2+} and SO_4^{2-} can be present in both feed solutions (Guler et al. 2014a; Post et al. 2009; Vermaas et al. 2014c).

Contrary to monovalent ions such as Na^+ and Cl^-, which are the major contributors for water salinity (Vermaas et al. 2014c) multivalent ions could be transported from the dilute to the concentrated in NaCl stream compartments (Guler et al. 2014a; Post et al. 2009; Vermaas et al. 2014c). This is caused by the Donnan equilibrium condition, which requires a valence-interdependent-redistribution of counter-ions, which, for a monovalent "i" and divalent counter-ions "j" in the concentrated (*c*) and dilute (*d*) saline solutions is written as follows:

$$\Delta\varphi_D = \frac{RT}{F}\ln\left(\frac{a_{i,\mathrm{c}}}{a_{i,\mathrm{d}}}\right)^{1/1} = \frac{RT}{F}\ln\left(\frac{a_{j,\mathrm{c}}}{a_{j,\mathrm{d}}}\right)^{1/2} \tag{4.18}$$

Until the Donnan equilibrium is not achieved, multivalent ions might continue to be transported, even again their own concentration gradient (uphill transport), from the dilute to the concentrated saline compartment (Guler et al. 2014a; Post et al. 2009; Vermaas et al. 2014c), as can be seen in Fig. 4.6.

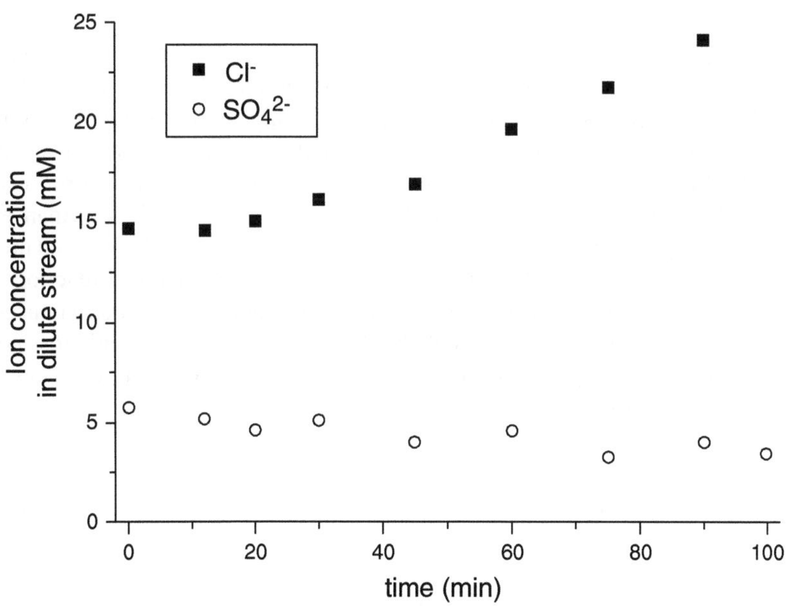

Fig. 4.6 *Cl*$^-$ and SO_4^{2-} concentrations evolution in dilute stream in 10-cell pair *RED* stack, with Ralex ion-exchange membranes. The initial solutions compositions were: dilute stream (4 mM $MgSO_4$, 13 mM NaCl) and concentrated stream (50 mM $MgSO_4$, 460 mM NaCl). The solutions volume was 1 dm^3 and they were fed at 72 l/h flow rate. The anion concentrations were followed by HPLC measurements using AS9 column (Dionex) connected to a Dionex conductivity detector

Depending on the initial ionic compositions, membrane properties, and other process parameters, such as temperature and hydrodynamic conditions, etc., the uphill ion transport may occur in the time scale of tens of minutes up to several hours (Vermaas et al. 2014c), thus causing a decrease in the salinity gradient across the membranes, lowering the stack voltage and, consequently, reducing the obtainable power density in RED (Guler et al. 2014a; Post et al. 2009).

The use of ion-exchange membranes with selectivity for monovalent counter-ions may overcome this limitation as they offer steric hindering and/or charged repulsion of multivalent counter-ions (Guler et al. 2014a; Post et al. 2009; Vermaas et al. 2014c). This would be beneficial especially if the dilute water composition is rich in multivalent ions (Post et al. 2009), since, for example, the power density was reduced by 63 % when artificial multi-ionic solutions mimicking real brackish water and an exhausted brine were used instead of NaCl solutions (Tufa et al. 2014). For river–sea systems, in which the presence of multivalent ions in the feed solutions has led to a 9–20 % lower power density than when sodium chloride has been the single ions source (Hong et al. 2013), utilization of monovalent membranes did not significantly improve the obtainable power (Guler et al. 2014a; Post et al. 2009).

4.3.3.5 Redox Couples

In RED, the ionic current is converted into electron current by an appropriate redox couple contained in the solution recirculated in the electrode compartments (Scialdone et al. 2012). Since the conversion is mainly controlled by mass transfer in the electrolyte rather than by the electrocatalytic properties of the electrode material (Burheim et al. 2012), an inexpensive graphite carbon material can be used as electrodes (Burheim et al. 2012; Scialdone et al. 2012; Veerman et al. 2010b). As redox couples, homogeneous redox systems, such as those based on iron compounds, are preferable, because they do not cause net chemical reactions and the potential difference needed for reduction on the cathode is counterbalanced by oxidation on the anode (Veerman et al. 2010b). The OCV, current, and gross power density obtained in experiments performed under identical conditions, but with different redox couples, can be consulted in Table 4.3.

Table 4.3 Redox couple influence on performance of a 10-cell pair RED stack, with Ralex ion-exchange membranes, fed with 0.5 and 0.017 M NaCl solutions at 72 l/h flow rate

Redox couple electrolytes	OCV (V)	I (mA)	P_{gross} (mW)
Na_2SO_4	1.31	12.5	4.1
$K_3Fe(CN)_6$ and $K_4Fe(CN)_6$	1.33	38.4	12.8
$KMnO_4$ and K_2MnO_4	1.38	39.0	13.4

The redox solution is 0.015 M electrolyte in 0.26 M bulk of NaCl, recirculated at 280 mL/min

The hexacyanoferrate(III)/hexacyanoferrate(II) system has a low toxicity (except if in contact with strong acids, when a toxic gas is released) and proved to be stable in the absence of light and oxygen (Scialdone et al. 2012) as well as for long-term experiments (more than 2 months) performed by the authors in our ongoing research.

Power generation by RED does not have to be necessarily the only objective of the process. Simultaneous power generation and treatment of recalcitrant water contaminants (Cr(VI)) or organic pollutant resistant to conventional biological processes (Acid Orange 7) has been suggested to be done in the electrode compartments by redox reactions (Scialdone et al. 2014, 2015). Also, as mentioned in Sect. 4.2.3, biological wastes can be converted into electricity, hydrogen gas, methane, and other chemicals by coupling microbial electrochemical technologies with RED (Hatzell et al. 2014; Logan and Rabaey 2012; Luo et al. 2014; Nam et al. 2012). In such systems, appropriate microorganisms may oxidize and reduce the organic compounds at the anode and the cathode, respectively (Logan and Rabaey 2012).

An alternative to the use of redox couples could be utilization of capacitive electrodes, what eliminates redox reactions at the electrodes and avoids possible small leakages of redox couple solution to the feed water compartments, which may pollute the environment via the brackish effluent (Vermaas et al. 2013b). Furthermore, in case of internal leakages, hexacyanoferrate anions may poison the AEM in the RED stack through irreversible binding, thus leading to its decreased performance (Lacey 1980).

However, since the feed waters (dilute and concentrated saline streams) must be periodically switched when using capacitive electrodes, the obtainable gross power density is lower in a RED stack with capacitive electrodes in comparison with a stack equipped with conventional Ti/Pt electrodes and using a hexacyanoferrate system as the redox couple (Vermaas et al. 2013b).

4.4 Challenges and Perspectives

4.4.1 Profiled Membranes

The RED performance depends significantly on the stack design and hydrodynamic conditions inside its flow channels. As mentioned in Sect. 4.3.3, introducing turbulence promoters inside the channels can reduce the DBL thickness, but on the cost of a higher pressure drop inside the channel. Moreover, the turbulence promoters are usually non-conductive spacers, which partially cover the ion-exchange membranes area (the so-called spacer "shadow" effect) and, consequently, increase the ohmic electric resistance of the stack (Balster et al. 2010; Dlugolecki et al. 2009, 2010c; Pawlowski et al. 2014b; Veerman et al. 2009; Vermaas et al. 2014a). An explored possibility for overcoming this limitation is the use of conductive spacers, which still promotes solutions mixing, but without taping a part of the membrane surface for counter-ion transport (Dlugolecki et al. 2010c). However, since the pressure drop remains relatively high in such a design, a more viable alternative seems to be utilization of profiled membranes (Guler et al. 2014b; Vermaas et al. 2014a).

Profiled membranes, also known as corrugated membranes, provide channels for the feed streams, while the relief formed on their surfaces keeps the membranes separated (Brauns 2009; Racz et al. 1986; Strathmann 2010; Vermaas et al. 2011b). In such a way, the intermembrane distance is still maintained constant even without spacers. The use of profiled membranes is very promising since, if used, the obtainable net power density can be almost two times higher than that obtained in a RED stack with spacers due to both the increase of available membrane area and a pressure drop reduction in the channels (Vermaas et al. 2014a). However, there is still a need to design new corrugation forms, since the actual ones do not yet grant the same reduction of the DBL resistance as spacers do (Guler et al. 2014b; Vermaas et al. 2011b, 2014a).

4.4.2 Pressure Drop

Minimization of the pressure drop in a RED stack is one of the most important and challenging issues for optimizing the RED performance as it is proportional to the power spent for solutions pumping, which may consume up to 25 % of the generated power (Veerman et al. 2009). To predict the pressure drop in a RED stack, the classical approach of considering a single cell pair as a repeating unit in the stack does not take into account the possible non-uniform fluid distribution due to pressure variation inside the stack (Kodym et al. 2012). Utilization of Hagen–Poiseuille or Darcy–Weisbach equations to calculate pressure drop in flow compartments assuming a fully developed, uniform flow (Veerman et al. 2011; Vermaas et al. 2011a, b, 2012) and, therefore, equal pressure drop in all compartments gives values which are up to 20 times lower than the total pressure drop in the stack (Vermaas et al. 2011a, b). Since, for example, in a 50 cell-pair RED stack, pressure drop in manifolds constituted 80 % of the total pressure drop (Veerman et al. 2009), only when pressure drops in distribution ducts, branches, and beams are all considered, the total pressure drop can be accurately predicted (Pawlowski et al. 2014a). As the number of cell pairs increases, the partial pressure drop in the distribution duct and, especially the partial pressure drop in the branches become dominant and are the principal causes of a non-uniform fluid flow distribution inside a RED stack (Pawlowski et al. 2014a). Therefore, the development of more efficient feed water distribution designs is a challenge, not only for RED, but also for other membrane technologies with plate-and-frame stacks.

4.4.3 Fouling

Fouling may cause a significant decrease in the obtained power density (Vermaas et al. 2013c, 2014d). This decrease can be manifested by an increase of the non-ohmic overpotential in chronopotentiometric measurements, which can be used to decide on and selectively apply cleaning in RED (Vermaas et al. 2014b). Periodically switching the feed waters (i.e., changing seawater for river water and vice versa) allows maintaining a high power density in the first hours of operation, probably due to a removal of multivalent ions and organic foulants (Vermaas et al. 2014d). However in long term, colloidal fouling is also observed, which can be only partly reversible. Periodic air sparking grants a minimum of colloidal fouling, resulting in a higher power density in long-term operation (Vermaas et al. 2014d).

The mentioned anti-fouling strategies has, however, associated energy costs, and are therefore applied after preferential channeling has been detected by chronopotentiometry (Vermaas et al. 2014b) or when a pronounced pressure drop increase has been registered (Vermaas et al. 2013c, 2014d). Since a decrease in obtained power density is, most likely, to be caused by organic fouling (Vermaas et al. 2013c), process monitoring should focus on organic compounds, with an objective of an early detection of membrane fouling. 2D fluorescence spectroscopy, which is a sensitive and non-invasive technique, can provide rapid information about the composition of complex biological media (Galinha et al. 2011, 2012) appears to be a feasible analytical tool for the RED monitoring.

4.4.4 Net Power Density

Currently, the maximal obtained power density is close to 2.2 W/m^2 at 35 % of energy efficiency for mixing seawater (~0.5 M NaCl) and river water (~0.01 M NaCl) (Vermaas et al. 2011a) and a theoretical goal of 3.5 W/m^2 is expected to be eventually achievable (Yip et al. 2014). Even a higher power density (up to 6.7 W/m^2) can be obtained if NaCl 0.01 and 5 M solutions, at 60 °C, are used (Daniilidis et al. 2014b). The power density rise is not directly proportional to the increase of concentration difference since the open circuit voltage (4.9) has a logarithmic dependence on the solution concentrations ratio, as well as high solution saline concentrations may weaken the Donnan exclusion of co-ions by ion-exchange membranes (Yip and Elimelech 2014).

Since the membranes are the major component of the initial capital outlay (~80 %) (Kempener and Neumann 2014) as they are the "heart" of a RED stack, as well as due to their relatively high current cost (Daniilidis et al. 2014a), the final challenge for RED implementation in practice is to improve the cost-effectiveness of this technology by developing more affordable ion-exchange membranes (below 4.3 €/m^2 of membrane (Daniilidis et al. 2014a)), for the current performance efficiency (Yip et al. 2014). It is expected that at such conditions, electricity generation by RED can become competitive with other energy sources, resulting in a LCOE of around 0.16 €/kWh (Daniilidis et al. 2014a). However, several factors such as costs for fouling prevention and investments for feed water flow distribution are still highly uncertain regarding the financial feasibility of RED. Recently (on November 26, 2014), a RED pilot plant (50 kW of capacity) had been inaugurated at the Afsluitdijk (The Netherlands) by Fujifilm, RED stack B.V., and Wetsus consortium (http://www.fujifilm.eu/dk/nyheder/article/news/fujifilm-membranes-in-blue-energy-test-power-station). Its performance is expected to provide more relevant data for further development of the RED technology and for its comparison with other possible renewable energy sources.

Acknowledgment Sylwin Pawlowski would like to acknowledge the *Fundação para a Ciência e a Tecnologia* (FCT), Lisbon, Portugal, for his Ph.D. research grant SFRH/BD/68649/2010.

References

Alvarez-Silva O, Osorio AF (2015) Salinity gradient energy potential in Colombia considering site specific constraints. Renew Energy 74:737–748

Balster J, Stamatialis DF, Wessling M (2010) Membrane with integrated spacer. J Membr Sci 360:185–189

Belova E, Lopatkova G, Pismenskaya N, Nikonenko V, Larchet C, Pourcelly G (2006) Effect of anion-exchange membrane surface properties on mechanisms of overlimiting mass transfer. J Phys Chem B 110:13458–13469

Brauns E (2009) Salinity gradient power by reverse electrodialysis: effect of model parameters on electrical power output. Desalination 237:378–391

Burheim OS, Seland F, Pharoah JG, Kjelstrup S (2012) Improved electrode systems for reverse electro-dialysis and electro-dialysis. Desalination 285:147–152

Choi J, Park J, Moon S (2002) Direct measurement of concentration distribution within the boundary layer of an ion-exchange membrane. J Colloid Interface Sci 251:311–317

Chu S, Majumdar A (2012) Opportunities and challenges for a sustainable energy future. Nature 488:294–303

Daniilidis A, Herber R, Vermaas DA (2014a) Upscale potential and financial feasibility of a reverse electrodialysis power plant. Appl Energy 119:257–265

Daniilidis A, Vermaas DA, Herber R, Nijmeijer K (2014b) Experimentally obtainable energy from mixing river water, seawater or brines with reverse electrodialysis. Renew Energy 64:123–131

Dlugolecki P, Nymeijer K, Metz SJ, Wessling M (2008) Current status of ion exchange membranes for power generation from salinity gradients. J Membr Sci 319:214–222

Dlugolecki P, Gambier A, Nijmeijer K, Wessling M (2009) Practical potential of reverse electrodialysis as process for sustainable energy generation. Environ Sci Technol 43:6888–6894

Dlugolecki P, Anet B, Metz SJ, Nijmeijer K, Wessling M (2010a) Transport limitations in ion exchange membranes at low salt concentrations. J Membr Sci 346:163–171

Dlugolecki P, Ogonowski P, Metz SJ, Saakes M, Nijmeijer K, Wessling M (2010b) On the resistances of membrane, diffusion boundary layer and double layer in ion exchange membrane transport. J Membr Sci 349:369–379

Dlugolecki P, Dabrowska J, Nijmeijer K, Wessling M (2010c) Ion conductive spacers for increased power generation in reverse electrodialysis. J Membr Sci 347:101–107

Emami Y, Mehrangiz S, Etemadi A, Mostafazadeh A, Darvishi S (2013) A brief review about salinity gradient energy. Int J Smart Grid Clean Energy 2:295–300

European Commission (2011) Background on energy in Europe, Information prepared for the European Council, 4 Feb 2011

Eurostat (2014) The EU in the world 2014. A statistical portrait, Publications Office of the European Union, 2014

Forster J (2014) Environment and energy—water use in industry. Eurostat—statistics in focus 14/2014

Galama AH, Post JW, Cohen Stuart MA, Biesheuvel PM (2013) Validity of the Boltzmann equation to describe Donnan equilibrium at the membrane-solution interface. J Membr Sci 442:131–139

Galama AH, Vermaas DA, Veerman J, Saakes M, Rijnaarts HHM, Post JW, Nijmeijer K (2014a) Membrane resistance: the effect of salinity gradients over a cation exchange membrane. J Membr Sci 467:279–291

Galama AH, Daubaras G, Burheim OS, Rijnaarts HHM, Post JW (2014b) Seawater electrodialysis with preferential removal of divalent ions. J Membr Sci 452:219–228

Galinha CF, Carvalho G, Portugal CAM, Guglielmi G, Reis MAM, Crespo JG (2011) Two-dimensional fluorescence as a fingerprinting tool for monitoring wastewater treatment systems. J Chem Technol Biotechnol 86:985–992

Galinha CF, Carvalho G, Portugal CAM, Guglielmi G, Reis MAM, Crespo JG (2012) Multivariate statistically-based modelling of a membrane bioreactor for wastewater treatment using 2D fluorescence monitoring data. Water Res 46:3623–3636

Geise GM, Curtis AJ, Hatzell MC, Hickner MA, Logan BE (2014a) Salt concentration differences alter membrane resistance in reverse electrodialysis stacks. Environ Sci Technol Lett 1:36–39

Geise GM, Cassady HJ, Paul DR, Logan BE, Hickner MA (2014b) Specific ion effects on membrane potential and the permselectivity of ion exchange membranes. Phys Chem Chem Phys 16:21673–21681

Geraldes V, Afonso MD (2010) Limiting current density in the electrodialysis of multi-ionic solutions. J Membr Sci 360:499–508

Guler E, Zhang Y, Saakes M, Nijmeijer K (2012) Tailor-made anion-exchange membranes for salinity gradient power generation using reverse electrodialysis. Chem Sus Chem 5:2262–2270

Guler E, Elizen R, Vermaas DA, Saakes M, Nijmeijer K (2013) Performance-determining membrane properties in reverse electrodialysis. J Membr Sci 446:266–276

Guler E, Baak W, Saakes M, Nijmeijer K (2014a) Monovalent-ion-selective membranes for reverse electrodialysis. J Membr Sci 455:254–270

Guler E, Elizen R, Saakes M, Nijmeijer K (2014b) Micro-structured membranes for electricity generation by reverse electrodialysis. J Membr Sci 458:136–148

Hatzell M, Zhu X, Logan BE (2014) Simultaneous hydrogen generation and waste acid neutralization in a reverse electrodialysis system. Sustain Chem Eng 2(9):2211–2216

Helfer F, Sahin O, Lemckert CJ, Anissimov YG (2013) Salinity gradient energy: a new source of renewable energy in Australia. Water Utility J 5:3–13

Hong J, Zhang W, Luo J, Chen Y (2013) Modeling of power generation from the mixing of simulated saline and freshwater with a reverse electrodialysis system: the effect of monovalent and multivalent ions. Appl Energy 110:244–251

http://ec.europa.eu/clima/policies/roadmap/perspective/index_en.htm. Accessed 10 Nov 2014

http://www.fujifilm.eu/dk/nyheder/article/news/fujifilm-membranes-in-blue-energy-test-power-station/. Consulted in 5 Jan 2015

Huckerby J, Jeffrey H, Sedgwick J, Jay B, Finlay L (2012) An international vision for Ocean energy–version II. Ocean Energy Systems Implementing Agreement (OESIEA)

Jones AT, Finley W (2003) Recent developments in salinity gradient power. OCEANS 2003 Proc 4:2284–2287

Kempener R, Neumann F (2014) Salinity gradient energy—technology brief. IRENA Ocean Energy Technology

Kim K, Ryoo W, Chun M, Chung G, Lee S (2012) Transport analysis in reverse electrodialysis with pulsatile flows for enhanced power generation. Korean J Chem Eng 29(2):162–168

Kodym R, Panek P, Snita D, Tvrznık D, Bouzek K (2012) Macro homogeneous approach to a two-dimensional mathematical model of an industrial-scale electrodialysis unit. J Appl Electrochem 42:645–666

Kozmai A, Nikonenko V, Pismenskaya N, Pryakhina O, Sistat P, Pourcelly G (2010) Diffusion layer thickness in a membrane system as determined from voltammetric and chronopotentiometric data. Russ J Electrochem 46:1383–1389

Krol JJ, Wessling M, Strathmann H (1999) Chronopotentiometry and overlimiting ion transport through monopolar ion exchange membranes. J Membr Sci 162:155–164

Kuleszo J, Kroeze C, Post JW, Fekete BM (2010) The potential of blue energy for reducing emissions of CO_2 and non-CO_2 greenhouse gases. J Integr Environ Sci 7(S1):89–96

Lacey RE (1980) Energy by reverse electrodialysis. Ocean Eng 7:1–47

Logan BE, Elimelech M (2012) Membrane-based processes for sustainable power generation using water. Nature 488:313–319

Logan BE, Rabaey K (2012) Conversion of wastes into bioelectricity and chemicals by using microbial electrochemical technologies. Science 337:686–690

Luo X, Cao X, Mo Y, Xiao K, Zhang X, Liang P, Huang X (2012) Power generation by coupling reverse electrodialysis and ammonium bicarbonate: implication for recovery of waste heat. Electrochem Commun 19:25–28

Luo X, Zhang F, Liu J, Zhang X, Huang X, Logan BE (2014) Methane production in microbial reverse-electrodialysis methanogenesis cells (MRMCs) using thermolytic solutions. Environ Sci Technol 48(15):8911–8918

Merz CR, Moreno WA, Barger M, Lipka SM (2012) Salinity gradient power (SGP): a developmental roadmap covering existing generation technologies and recent investigative results into the feasibility of bipolar membrane-based salinity gradient power generation. Technol Innov 14:249–275

Murray C, Blackledge J, Kearney D (2013) The feasibility of salinity gradient technology for Ireland: an initial case study by the river Suir. J Dev Appl Ocean Eng 1–11

Nam J, Cusick RD, Kim Y, Logan BE (2012) Hydrogen generation in microbial reverse-electrodialysis electrolysis cells using a heat-regenerated salt solution. Environ Sci Technol 46:5240–5246

Neuman F, Hamelers B, Siebers R, Helsen J, Schroeder R, Stenzel P (2012) Report of the meeting on salinity gradient power generation. INES, Brussels 20 of June 2012

Nikonenko VV, Pismenskaya ND, Belova EI, Sistat P, Huguet P, Pourcelly G, Larchet C (2010) Intensive current transfer in membrane systems: modelling, mechanisms and application in electrodialysis. Adv Colloid Interface Sci 160:101–123

Nikonenko VV, Kovalenko AV, Urtenov MK, Pismenskaya ND, Han J, Sistat P, Pourcelly G (2014) Desalination at overlimiting currents: state-of-the-art and perspectives. Desalination 342:85–106

Pattle RE (1954) Production of electric power by mixing fresh and salt water in the hydroelectric pile. Nature 174:660

Pawlowski S, Crespo JG, Velizarov S (2014a) Pressure drop in reverse electrodialysis: experimental and modeling studies for stacks with variable number of cell pairs. J Membr Sci 462:96–111

Pawlowski S, Sistat P, Crespo JG, Velizarov S (2014b) Mass transfer in reverse electrodialysis: flow entrance effects and diffusion boundary layer thickness. J Membr Sci 471:72–83

Pismenskaia N, Sistat P, Huguet P, Nikonenko V, Pourcelly G (2004) Chronopotentiometry applied to the study of ion transfer through anion exchange membranes. J Membr Sci 228:65–76

Pismenskaya N, Nikonenko V, Belova E, Lopatkova G, Sistat P, Pourcelly G, Larshe K (2007) Coupled convection of solution near the surface of ion exchange membranes in intensive current regimes. Russ J Electrochem 43:307–327

Post JW, Veerman J, Hamelers HVM, Euverink GJW, Metz SJ, Nymeijer K, Buisman CJN (2007) Salinity-gradient power: evaluation of pressure-retarded osmosis and reverse electrodialysis. J Membr Sci 288:218–230
Post JW, Hamelers HVM, Buisman CJN (2008) Energy recovery from controlled mixing salt and fresh water with a reverse electrodialysis system. Environ Sci Technol 42:5785–5790
Post JW, Hamelers HVM, Buisman CJN (2009) Influence of multivalent ions on power production from mixing salt and fresh water with a reverse electrodialysis system. J Membr Sci 330:65–72
Racz IG, Groot Wassink J, Klaassen R (1986) Mass transfer, fluid flow and membrane properties in flat and corrugated plate hyper filtration modules. Desalination 60:213–222
Ramon GZ, Feinberg BJ, Hoek EMV (2011) Membrane-based production of salinity-gradient power. Energy Environ Sci 4:4423–4434
Scialdone O, Guarisco C, Grispo S, Angelo A, Galia A (2012) Investigation of electrode material—redox couple systems for reverse electrodialysis processes. Part I: iron redox couples. J Electroanal Chem 681:66–75
Scialdone O, Angelo A, Lume E, Galia A (2014) Cathodic reduction of hexavalent chromium coupled with electricity generation achieved by reverse-electrodialysis processes using salinity gradients. Electrochim Acta 137:258–265
Scialdone O, Angelo A, Galia A (2015) Energy generation and abatement of Acid Orange 7 in reverse electrodialysis cells using salinity gradients. J Electroanal Chem 738:61–68
Sistat P, Pourcelly G (1997) Chronopotentiometric response of an ion-exchange membrane in the under limiting current-range. Transport phenomena within the diffusion layers. J Membr Sci 123:121–131
Slagt M (2012) Blue energy: current status and progress of RED. 3rd Osmosis membrane Summit Barcelona, April 26–27, 2012
Smalley RE (2005) Future global energy prosperity: the terawatt challenge. Mater Res Soc Bull 30:412–417
Strathmann H (2010) Electrodialysis, a mature technology with a multitude of new applications. Desalination 264:268–288
Strathmann H, Grabowski A, Eigenberger G (2013) Ion-exchange membranes in the chemical process industry. Ind Eng Chem Res 52:10364–10379
Tedesco M, Cipollina A, Tamburini A, Baak W, Micale G (2012) Modelling the reverse electrodialysis process with seawater and concentrated brines. Desalin Water Treat 49:404–424
Tedesco M, Cipollina A, Tamburini A, Bogle IDL, Micale G (2015) A simulation tool for analysis and design of reverse electrodialysis using concentrated brines. Chem Eng Res Des 93:441–456
Tufa R, Curcio E, Baak W, Veerman J, Grasman S, Fontananova E, Profio G (2014) Potential of brackish water and brine for energy generation by salinity gradient power-reverse electrodialysis (SGP-RE). RSC Adv 4:42617–42623
Veerman J, Post JW, Saakes M, Metz SJ, Harmsen GJ (2008) Reducing power losses caused by ionic shortcut currents in reverse electrodialysis stacks by a validated model. J Membr Sci 310:418–430
Veerman J, Saakes M, Metz SJ, Harmsen GJ (2009) Reverse electrodialysis: performance of a stack with 50 cells on the mixing of sea and river water. J Membr Sci 327:136–144
Veerman J, Saakes M, Metz SJ, Harmsen GJ (2010a) Electrical power from sea and river water by reverse electrodialysis: a first step from the laboratory to a real power plant. Environ Sci Technol 44:9207–9212
Veerman J, Saakes M, Metz SJ, Harmsen GJ (2010b) Reverse electrodialysis: evaluation of suitable electrode systems. J Appl Electrochem 40:1461–1474
Veerman J, Saakes M, Metz SJ, Harmsen GJ (2011) Reverse electrodialysis: a validated process model for design and optimization. Chem Eng J 166:256–268
Vermaas DA, Saakes M, Nijmeijer K (2011a) Doubled power density from salinity gradients at reduced inter membrane distance. Environ Sci Technol 45:7089–7095

Vermaas DA, Saakes M, Nijmeijer K (2011b) Power generation using profiled membranes in reverse electrodialysis. J Membr Sci 385–386:234–242
Vermaas DA, Guler E, Saakes M, Nijmeijer K (2012) Theoretical power density from salinity gradients using reverse electrodialysis. Energy Proc 20:170–184
Vermaas DA, Veerman J, Yip NY, Elimelech M, Saakes M, Nijmeijer K (2013a) High efficiency in energy generation from salinity gradients with reverse electrodialysis. Sustain Chem Eng 1 (10):1295–1302
Vermaas DA, Bajracharya S, Sales B, Saakes M, Hamelers B, Nijmeijer K (2013b) Clean energy generation using capacitive electrodes in reverse electrodialysis. Energy Environ Sci 6:643–651
Vermaas DA, Kunteng D, Saakes M, Nijmeijer K (2013c) Fouling in reverse electrodialysis under natural conditions. Water Res 47:1289–1298
Vermaas DA, Saakes M, Nijmeijer K (2014a) Enhanced mixing in the diffusive boundary layer for energy generation in reverse electrodialysis. J Membr Sci 453:312–319
Vermaas DA, Saakes M, Nijmeijer K (2014b) Early detection of preferential channeling in reverse electrodialysis. Electrochim Acta 117:9–17
Vermaas DA, Veerman J, Saakes M, Nijmeijer K (2014c) Influence of multivalent ions on renewable energy generation in reverse electrodialysis. Energy Environ Sci 7:1434–1445
Vermaas DA, Kunteng D, Veerman J, Saakes M, Nijmeijer K (2014d) Periodic feed water reversal and air sparging as antifouling strategies in reverse electrodialysis. Environ Sci Technol 48:3065–3073
Yip N, Elimelech M (2014) Comparison of energy efficiency and power density in pressure retarded osmosis and reverse electrodialysis. Environ Sci Technol 48:11002–11012
Yip N, Vermaas DA, Nijmeijer K, Elimelech M (2014) Thermodynamic, energy efficiency, and power density analysis of reverse electrodialysis power generation with natural salinity gradients. Environ Sci Technol 48:4925–4936

Chapter 5
The Kinetic Parameters Evaluation for the Adsorption Processes at "Liquid–Solid" Interface

Svetlana Lyubchik, Elena Lygina, Andriy Lyubchyk, Sergiy Lyubchik, José M. Loureiro, Isabel M. Fonseca, Alexandra B. Ribeiro, Margarida M. Pinto, and Agnes M. Sá Figueiredo

5.1 Introduction

Adsorption is one of the most important processes for the practical applications in industry and environmental protection. Adsorption at various interfaces ("liquid to solid," "gas to solid," or "gas to liquid") has concerned scientists since the beginning of twenty-first century (Dabrowski 2001). When the adsorption process is concerned, the adsorption kinetics and thermodynamics are of great significance to gain insight into the adsorption mechanisms and to evaluate the given adsorbent performance.

S. Lyubchik (✉) • E. Lygina • I.M. Fonseca
REQUIMTE, Departamento de Química, Faculdade de Ciências e Tecnologia, Universidade Nova de Lisboa, Caparica 2829-516, Portugal
e-mail: sve_lubchik@yahoo.com

A. Lyubchyk
CENIMAT, Departamento de Engenharia de Materiais e Cerâmica, Faculdade de Ciências e Tecnologia, Universidade Nova de Lisboa, Caparica 2829-516, Portugal

S. Lyubchik
CQE, Departamento de Engenharia Química, Instituto Superior Técnico de Lisboa, Lisbon, Portugal

J.M. Loureiro
Faculdade de Engenharia, Universidade do Porto, Porto, Portugal

A.B. Ribeiro
CENSE, Departamento de Ciências e Engenharia do Ambiente, Faculdade de Ciências e Tecnologia, Universidade Nova de Lisboa, Caparica 829-516, Portugal

M.M. Pinto
ISQ-R&D Department, Instituto de Soldadura e Qualidade, Porto Salvo, Portugal

A.M.S. Figueiredo
Instituto de Microbiologia Paulo de Góes, Universidade Federal do Rio de Janeiro, Rio de Janeiro, Brazil

© Springer International Publishing Switzerland 2016

A.B. Ribeiro et al. (eds.), *Electrokinetics Across Disciplines and Continents*,
DOI 10.1007/978-3-319-20179-5_5

Analysis of the kinetic data allows evaluating (a) the rate of the adsorbate uptake by given adsorbent and (b) the factors affecting the reaction rate, i.e., to estimate the parameters which determine the time required to complete the adsorption process. The kinetic parameters estimation is necessary for design and scale/performance determination of the industrial adsorption apparatus (such as the fixed-beds, flow-through columns, and adsorbers) in a batch- and dynamic-modes. There are numbers of widely used adsorption kinetic models to describe adsorption process at "gas–solid" interface. However, these models are often failed or insufficient to fit the kinetic data at "liquid–solid" adsorption interface. This is mainly due to the complicity of the chemisorptions reactions and diffusion processes of the adsorbate from liquid phase into the porous matrix of the adsorbents.

There are abundant publications on heavy metal adsorption on activated carbons with different oxygen functionalities covering wide-range conditions (solution pH, ionic strength, initial sorbate concentrations, carbon loading, etc.). Although much has been accomplished in this area, the principal problem remains in interpretation of obtained results for the adsorption process at "liquid–solid" interface. A basic understanding of the scientific principles is far behind; in part because the study of interfaces requires extremely careful experimentation if meaningful and reproducible results are obtained (Dabrowski 2001). The thermodynamics and kinetics data are still under continuing debates. The main problem concerns relatively low comparability of the data obtained by different research groups. That is happening not only due to the differences in chosen adsorbents nature or operating conditions for the adsorption runs, but in part because of the choice of the methodological approaches for the adsorption data analysis. More far as, chemisorptions from the solution into porous media is much more complex process than gas phase physisorptions.

Therefore, the adsorption kinetics and dynamics at solid–liquid interface remain the open questions in the heterogeneous surface science. There is vast literature on the kinetics and thermodynamics parameters evaluation for gas-phase adsorption process into a porous media, while analytical base for the adsorption kinetics and thermodynamics at solid–liquid interface is less developed area. The common methodology for the adsorption data analysis at liquid–solid interface either does not exist or its development is only in infancy. That is mainly due to the complicity of chemisorptions reactions of the adsorbate on the adsorbent active centers and mass transfer diffusion restriction at the solid–liquid adsorption interface.

In general, the molecules attachment from liquid phase to the solid surface by adsorption is a broad subject. Therefore the only complex investigation of adsorbate/adsorbent surfaces interactions at the aqueous–solid interface can help to understand the metals adsorption mechanism, which is an important point in optimization of the conditions of their environmental applications.

While the adsorption thermodynamics highlights the equilibrium conditions in the solution between the adsorption phases, another area of the debates remains on an optimum contact time to reach the adsorption equilibrium in the adsorption systems at "liquid–solid" interface. The majority of studies on the adsorption kinetics have revealed two-step behavior of the adsorption systems (Carrott et al. 1997; Kumar et al. 2000; Raji and Anirudhan 1998) with fast initial uptake and much slower gradual uptake afterwards, which might take days or even months. Some of the authors reported the optimum contact time of minutes (Ajmal et al. 2001; Lakatos et al. 2002), while other extreme suggested 100 h (Csobán et al. 1998; Brigatti et al. 2000) to get equilibrium in the adsorption systems. Furthermore, regardless of the adsorption run conditions (initial and equilibrium pH, temperature regime, etc.) the solutes adsorption by given adsorbent could be completed in a quite different contact time.

Thus, regardless of the solution pHs, differences in metal ions speciation, temperature dependence of the adsorption process, variability of the active centers on the carbon surface, surface charge, and potential, a comparison of the adsorption runs results for the metal ions capture by activated carbons is complicated. Where investigation of the adsorption kinetics (time evolution of the adsorption processes) has the remarkable ability to connect seemingly unrelated parameters.

It was suggested that known mathematical models for the adsorption kinetics data description could be classified as *adsorption reaction models* and *adsorption diffusion models*, which are quite different in nature (Lazaridis and Asouhidou 2003). Namely, the adsorption diffusion models are always constructed of three consecutive steps (Qiu et al. 2009): (1) external (or film) diffusion across the liquid film surrounding the adsorbent particles; (2) internal (or intra-particle) diffusion in liquid contained in the pores and/or into the pore texture; and (3) adsorption/desorption between the adsorbate and active sites, i.e., mass action. The adsorption reaction models, originating from chemical reaction kinetics, are based on the adsorption process as a whole, i.e., without considering the abovementioned steps (Qiu et al. 2009; Dabrowski 2001).

In a current practice, for the kinetic parameters evaluation, the adsorption reaction models is widely developed and used. However, it is not fully correct to use the adsorption reaction models only (such as the pseudo-first-order, pseudo-second-order models, or/and chemisorptions kinetic models) to describe complicity of the adsorption kinetics at "solid–liquid" interface. Therefore, the mass transfer diffusion models (both external and intra-particle) should to be also taken into consideration.

The present chapter is focused on the kinetics parameters evaluation for the adsorption process at "liquid–solid" interface. Kinetics parameters have been evaluated through the sets of time-based experiments for the chromium(III) ions adsorption under varying temperatures and adsorbent loading for two sets of the commercially available activated carbons and their post-oxidized forms of different surface oxygen functionality and porous texture. Several kinetic models, such as the pseudo-first-order, pseudo-second-order, chemisorptions kinetics model, external and intra-particle mass transfer diffusion models and the *systems dynamics* modeling (derivatives in order to time) were applied to follow the adsorption process. This approach served the dual purpose: (a) gained deep insight into effect of the porous matrixes' structural characteristics on the adsorption kinetics of the metal ions from liquid phase into porous solid surface and (b) gained insight, which is often very difficult or impossible to obtain by other mean, on overall dynamic behavior of the adsorption systems at "liquid–solid" interface aiming at overall adsorption process control.

5.2 Experimental

Two commercially available activated carbons GR MERCK 2518 and GAC Norit 1240 Plus (A-10128) have been used as supplied (parent or initial carbons) and after their post-chemical treatment with 1 M nitric acid to enhance the surface oxygen functionality and to vary the carbon's porous texture for the kinetics testing experiments. The carbon's textural parameters were analyzed using gas adsorption technique and surface functionalities were evaluated by temperature-programmed desorption (TPD). The pH at the point of zero charge (pH_{PZC} values) were measured using the pH drift method (Sontheimer et al. 1988). The details of the oxidation posttreatment and carbons characterization procedures/techniques are described elsewhere (Lyubchik et al. 2005) and summary results of the carbon surface characteristics and texture are presented in Tables 5.1 and 5.2.

Chemical posttreatment resulted in an introduction of the surface oxygen functional groups to the carbon surface; it also changed the carbon's porous texture. In general, posttreatment by nitric acid destroys the micropores structure, thus led to the reduction of the total pore volume and apparent surface area of the resulted carbons. This effect is more pronounced for the samples post-treated by 13 M nitric acid [data published elsewhere (Lyubchik et al. 2005)]. For the posttreatment by 1 M nitric acid this effect is evident for the parent GR MERCK 2518, which is more microporous, then parent GAC Norit 1240 Plus (Table 5.1). Furthermore, due to the

Table 5.1 Textural and surface characteristics of the studied activated carbons

Carbons	S_{BET} (m^2/g)	V_{total} (cm^3/g)	V_{micro} (cm^3/g)	S_{meso} (m^2/g)	S_{micro} (m^2/g)	pH_{PZC}
Merck_initial	1017	0.59	0.55	40	977	7.02
Merck_1 M HNO_3	755	0.33	0.31	41	714	3.41
Norit_initial	770	0.40	0.32	41	729	6.92
Norit_1 M HNO_3	945	0.43	0.41	72	873	4.41

Adapted from Lyubchik et al. (2005)

Table 5.2 Surface groups of the studied activated carbons

Sample	Carboxylic (μmol/g)		Lactones (μmol/g)	Phenols (μmol/g)	Quinones (μmol/g)
	Hydrous	Anhydrous			
Parent Merck	–	–	0.19	0.48	0.24
Merck_1 M HNO_3	1.12	0.12	1.77	4.53	3.14
Parent Norit	–	–	0.21	0.50	0.19
Norit_1 M HNO_3	1.15	0.14	0.76	2.52	1.82

oxidative posttreatment the carbon surface acquired strong acidic character (carbons pH_{PZC} drastically decreased, Table 5.1), and the content of the active centers of different nature significantly increases (Table 5.2). This effect was more pronounced for post-treated GR MERCK 2518 parent carbon; however the same tendency was also observed for GAC Norit 1240. Both carboxylic hydrous and anhydrous groups (which are responsible for efficient metal ions chemisorptions) appeared on the carbon surface (total of ca. 1.30 μmol/g). Weakly acidic functional groups, such as phenolic, lactones, and quinones, are also well presented on the surfaces (Table 5.2).

It was established that the nitric acid bonding to active edges is thermodynamic favorable process ($\Delta H = 431.82$ kcal/mol) resulted in cluster formation followed by transfer of the hydroxyl groups to the neighbor carbon atoms and formation of nitrophenol compounds, that also is energetically favorable ($\Delta H = 349.43$ kcal/mol). Water presence during oxidation results in its linkage by weak intermolecular interactions ($\Delta H = 294.25$ kcal/mol) followed by water molecules regrouping to more stable compounds that is clearly visible on enthalpy reduction ($\Delta H = 227.95$, 171.01, 170.92, and 121.71 kcal/mol) of the cluster formation at their identical chemical composition.

According to the XPS data the differences between parents and modified by acids samples were observed for carbon, oxygen, and nitrogen contents, which reveal their oxidation ability.

The carbon post-treated by HNO_3 is rich in aromatic carbons (aliphatic/aromatic carbons ratio is 0.27) and the oxidation reactions slightly change initial aromaticity of the carbons. Fitted at (287.9 ± 0.1) eV and at ca. 289.3 eV peaks could be attributed to the carbonyl and carboxylic bonds. The presence of those bonds is considerable confirming a strong oxidation of the materials during chemical posttreatment.

The conclusion, extracted from the carbon peak analysis, is confirmed by the analysis of O 1s region. The following oxygen functionalities: carbonyl/quinine (C=O), alcohol/ether (C–O), and carboxylic acid/esther (O=C–O*), contribute with peaks centered at 531.1 eV, 532.3 eV, and 533.3 eV, respectively, chemisorbed water contributes with peak at ca. 535.9 eV. The global quantitative results confirm the efficiency of HNO_3 treatment. The degree of surface oxidation expressed as the ratio of total oxygen to total carbon, O_{total}/C_{total}, is of ca. 0.10. In fact, the O 1s region in post-treated by HNO_3 samples is shifted towards lower binding energy. This is compatible both with oxygen included in groups bound to aromatic groups or included in ether groups instead of alcohol groups confirming that activated surface is richer in aromatic carbons and C–X–R groups.

From the XPS data, a consequent amount of nitrogen is present in samples post-treated by HNO_3. The N 1s core level signal also confirms the presence of carbon/nitrogen terminated groups of O–C(=O)N, CN(H)C. After fitting, peaks could be attributed to protonated amine functions (398.4 eV), O–C(=O)N or C(=O)NC(=O) (400.6 eV), $-N(CH_3)^{3+}$ (402.7 eV) and nitrous groups (NO_2) (405.7 eV). The contribution of protonated amine functions is weak (up to 0.19 at.%), whereas the presence of nitrogen/oxygen functionalities is remarkable (up to 1.20 at.%).

The ratio of oxidized carbon/electronegative elements was ~2 for post-treated samples. It means that the number of groups C–X–H [X is O, N] is high in post-treated samples.

The adsorption kinetics experiment was carried out to evaluate the rate and the activation energy of the Cr(III) adsorption on the given activated carbons, and to evaluate the rate limiting steps for the overall adsorption process at "solid–liquid" interface.

As a sources of trivalent chromium, the tanning industry reagent—salt $Cr_2(SO_4)_2OH_2$, was used. The chromium solution was always freshly prepared and used within a day in order to avoid its aging.

The influence of temperature and carbon dosage on the kinetics of the Cr(III) adsorption in the system "Cr(III)-activated carbon" was evaluated by time-based analyses. The experimental runs were performed for four samples: parent and oxidized by 1 M HNO_3 forms of GR MERCK 2518 and GAC Norit 1240 Plus (A-10128), at four different temperatures: 22, 30, 40, and 50 °C. The adsorption runs were performed at a fixed adsorbate concentration of 200 ppm, for three different carbon loadings fixed at 2.4, 4.8, and 16 g/L. Tests were performed at initial pH of 3.2 of the Cr(III) solution; i.e., without adding any buffer to control the pH to prevent introduction of any new electrolyte into the systems.

Batch laboratory technique was used for the adsorption kinetics testing experiments. The batch tests were conducted by fixed adsorbent loading to the 250 mL Erlenmeyer flasks containing 25 mL of the 200 ppm chromium(III) solution, which were shaking on a gyratory shaker at 200 rev/min for certain discrete time and, depending on the temperature of the experiment, until adsorbate concentration in the solution became fully unchanging. Each experiment was performed for two parent and two post-oxidized forms of Norit and Merck activated carbons for three different carbon loadings, thus generated ca. of $20 \times 4 \times 3 = 240$ samples for each experimental temperature [thus total of $240 \times 4 = 960$ for all kinetics tests]. Experiments were duplicated for quality control and statistical purposes. The percent standard deviation of the sorption parameters was less than 5.0 %. At the end of the time-based experiments, the adsorbent was removed by filtration through the membrane filters with a pore size of 0.45 μm, and chromium concentration was estimated spectrophotometrically using standard procedure (measuring absorbance at 371 nm).

5.2.1 Supporting Theory

With the development of the theory of adsorption equilibria from liquid phase on heterogeneous solid surfaces, the theory of adsorption kinetics on heterogeneous surfaces was also developed. There are three mass transfer processes (i.e., diffusion processes), where the *adsorption diffusion models* are appropriate to describe the kinetics, and one more, so-called mass action process, where the *adsorption reaction models* are appreciable to describe the kinetics. Namely, adsorption kinetics is determined by the following stages (Dabrowski 2001): (a) diffusion of molecules from the bulk phase towards the interface space, so-called external diffusion; (b) diffusion of molecules inside the pores, so-called internal diffusion; (c) diffusion of molecules in the surface phase, so-called surface diffusion; and (d) adsorption/desorption elementary processes inside the porous structure, which can be combined process of physisorptions and chemisorptions. Furthermore, for the adsorption kinetics of metal ions on microporous solids and/or inside the porous matrixes reached by surface functional groups (which are serving as the active centers of the adsorption), other, then physisorptions and chemisorptions, mechanisms may additionally take place, for instance capillary condensation, ion-exchange, complexation of the ions with the surface active centers, etc. It is also assumed that the total rate of the kinetic process is determined by the rate of the slowest process.

5.3 Adsorption Diffusion Models

Various models of diffusion were studied, including single steps of external or intra-particle diffusion or combined phenomena.

5.3.1 External Mass Transfer Diffusion Model

External mass transfer model, or external diffusion model, is an application of the Fick's laws to the adsorption process at "liquid–solid" interface. The mass transfer is generally expressed as (Srivastava et al. 1989):

$$C_0 - C_t = D \cdot e^{k_0 t} \tag{5.1}$$

where C_0 is the initial metal ion concentration (mmol/L), C_t is the metal ion concentration at time t (shaking time (min)), D is a fitting parameter, k_0 is the adsorption constant which is related to the mass transfer adsorption coefficient. A linearized form of (5.1) gives (5.2):

$$\ln(C_0 - C_t) = \ln D + k_0 t \tag{5.2}$$

If the adsorption process follows the mass transfer model, the plot of $\ln(C_0–C_t)$ vs. time (t) should give a linear relationship; where constants $\ln D$ and k_0 can be determined from the slope and intercept, respectively.

Model expresses thc cvolution the liquid phase adsorbate concentration in the solution at a time $t(C_t$, mmol/L) as a function of the difference of the adsorbate concentrations in the solution (C_t) and it's liquid phase concentration at the adsorbent surface (C_s, mmol/L) (5.3) according to Weber and Morris (1963), Ho et al. (2000):

$$\frac{dC_t}{dt} = -\beta \cdot S \cdot (C_t - C_S) \tag{5.3}$$

where β is external mass transfer coefficient (m/min); S is surface area per unit solution volume (m^{-1}).

Equation (5.3) can be simplified to (5.4) by substituting the following boundary conditions: $C_t \to C_0$ and $C_s \to 0$ when $t \to 0$; where $C_0 =$ initial metal ion concentration (McKay and Poots 1984, McKay et al. 1986; Weber and Morris 1962). Thus, making assumptions that (a) adsorbate surface concentration (C_s) is negligible at $t = 0$; i.e., (b) its concentration in solution (C_t) tending to the initial one (C_0); and (c) intra-particle diffusion is also negligible; the external mass transfer coefficients could be determined using (5.4):

$$\left[\frac{dC_t/C_0}{dt}\right]_{t \to 0} = -\beta \times S \tag{5.4}$$

where the external mass transfer rate $[-\beta S]$ ($\min^{-1}$) could be approximated by the initial slope of the C_t/C_0 vs. time (t) plot, and then can be calculated either by assuming a polynomial linearization of $[C_t/C_0]$ and subsequent derivation at $t = 0$ or by assuming that the relationship $[C_t/C_0$ vs. $t]$ is linear for first initial slop.

For this model, (a) the surface area is approximated as external surface of adsorbent particles and (b) the particles are supposed to be the spherical ones. Their surface area S could be calculated according to (5.5):

$$S = \frac{6 \cdot m}{d_p \cdot \rho_{app} \cdot V_{sol}} \tag{5.5}$$

where m is adsorbent mass in solution (g); d_p is particle size diameter (m); ρ_{app} is apparent density of adsorbent (g/m^3); V_{sol} is solution volume (m^3); $S_p = 6/(d_p \cdot \rho_{app})$ is specific surface area of particles (m^2/g).

5.3.2 The Intra-particle Mass Transfer Diffusion

A common empirically functional relationship for the adsorption processes is that the adsorbate uptake varies almost proportionally with $t^{1/2}$, rather than with the contact time, t (Ho et al. 2000):

$$q_t = K_i t^{1/2} \tag{5.6}$$

where q_t is the amount (in mmol/g) of solute adsorbed at certain time t (min) and K_i is the intra-particle diffusion rate constant.

When the adsorption mechanism follows the intra-particle diffusion process, the Weber Morris plot of q_t vs. $t^{1/2}$ (according to (5.6)), should be a straight line with a slope K_i(mmol/g × min$^{-0.5}$) and intercept C.

The slop value gives the intra-particle diffusion rate constant, and intercept gives an idea about the boundary layer thickness: the larger the intercept, the greater the boundary layer effect (Unnithan et al. 2002).

The intra-particle diffusion coefficient (D_i, m^2/min) could be calculated from Weber and Morris equation (5.7):

$$K_i = \left(\frac{12q_{eql}}{d_p}\right) \times \left(\frac{D_i}{\pi}\right)^{0.5} \tag{5.7}$$

and Urano and Tachikawa equation (5.8), i.e., from the slope of the plot of log[1−$(q_t/q_{eql})^2$] vs. time (t):

$$-\log\left[1 - \left(q_t/q_{eql}\right)^2\right] = \frac{4(\pi)^2 \cdot D_i \cdot t}{2.3d_p^2} \tag{5.8}$$

5.4 Adsorption Reaction Models

The most commonly used chemical kinetic expressions to explain the solid/liquid adsorption processes are the pseudo-first-order kinetics and pseudo-second-order kinetic models.

5.4.1 Pseudo-First-Order Rate Model

The earliest first-order rate equation was proposed by Lagergren (1898) to evaluate the adsorption rate based on the adsorption capacity for the kinetics of oxalic and malonic acid adsorption on charcoal (5.9):

$$\frac{dq_t}{dt} = k_1\left(q_{eql} - q_t\right) \tag{5.9}$$

where q_{eql} and q_t are the adsorption capacity at equilibrium and at certain time t, respectively (mmol/g), k_1 is the rate constant of pseudo-first-order adsorption (min^{-1}).

After integration and applying boundary conditions $t=0$ to $t=t$ and $q_t=0$ to $q_t=q_t$, the integrated form of (5.9) becomes (5.10):

$$\log\left(q_{eql} - q_t\right) = \log\left(q_{eql}\right) - \frac{k_1}{2.303}t \tag{5.10}$$

When the values of $\log(q_{\text{eql}} - q_t)$ are linearly correlated with t; the plot of log $(q_{\text{eql}} - q_t)$ vs. t should give a linear shape from which the rate constants (k_1) and the adsorption capacity at equilibrium (q_{eql}) can be determined from the slope and intercept of the plot, respectively.

Lagergren's first-order rate equation has been called the pseudo-first-order equation, and currently this model is widely used to describe the adsorption kinetics of different pollutants from liquid phase into the porous media. However, it has to be noticed, no adsorption mechanisms could be reasonably assigned to the kinetics data obtained from the pseudo-first-order rate model application.

5.4.2 *Pseudo-Second-Order Rate Model*

The second-order rate equation was proposed by Ho (1995) to describe the adsorption kinetics of divalent metal ions onto peat, where cation-exchange between metal ions and adsorbent's polar functional groups (aldehydes, ketones, acids, and phenolic) is the adsorption mechanisms (Cheung et al. 2001). The main assumption was that the adsorption may be second-order, and rate limiting step may be chemical adsorption involving valent forces through sharing or electrons exchange between the peat and divalent metal ions.

Ho's second-order rate equation has been called pseudo-second-order rate. The pseudo-second-order equation was also based on the sorption capacity of the solid phase (Ho and McKay 1998). Currently, this model is successfully applied to the adsorption processes of metal ions, dyes, herbicides, oils, and organics from solutions. However, it has to be noticed that the pseudo-second-order rate equation based on chemical adsorption cannot be suitably applied to the processes where the driving forces is physical adsorption (case of the organics adsorption by nonpolar adsorbents) and/or combined physisorptions/chemisorptions (case of metal ions adsorption by carbon materials).

The pseudo-second-order adsorption kinetic rate equation is expressed as (Ho et al. 2000):

$$\frac{\mathrm{d}q_t}{\mathrm{d}t} = k_2\left(q_{\text{eql}} - q_t\right)^2 \tag{5.11}$$

where k_2 is the rate constant of pseudo-second-order adsorption ($\text{g} \cdot \text{mmol}^{-1} \cdot \text{min}^{-1}$).

For the boundary conditions $t = 0$ to $t = t$ and $q_t = 0$ to $q_t = q_t$, the integrated form of (5.11) becomes (5.12), which is the integrated rate law for a pseudo-second-order reaction:

$$\frac{1}{\left(q_{\text{eql}} - q_t\right)} = \frac{1}{q_{\text{eql}}} + k_2 t \tag{5.12}$$

Equation (5.12) can be rearranged to obtain (5.13), which has a linear form:

$$\left(\frac{t}{q_t}\right) = \frac{1}{k_2 q_{\text{eql}}^2} + \frac{1}{q_{\text{eql}}}(t) \tag{5.13}$$

The initial sorption rate constant, h (mmol/g min^{-1}), as $q_t/t \rightarrow 0$ at $t = 0$ can be defined as (Ho and McKay 1999):

$$h = k_2 q_{\text{eql}}^2 \tag{5.14}$$

5.4.3 Chemisorptions Kinetics Model

A kinetic equation of chemisorptions was established by Zeldowitsch (1934) to describe the rate of gas adsorption on solids that decreases exponentially with an increase in the amount of gas adsorbed. That is so-called Elovich equation (5.15) (Sparks 1986):

$$q_{\text{t}} = \alpha_{\text{e}} + \beta_{\text{e}} \ln(t) \tag{5.15}$$

where q_{t} (in mmol/g) is the amount adsorbed at time t, α_{e} and β_{e} are Elovich constants (in mmol/g min^{-1}).

The Elovich equation assumes that the solid surface active sites are heterogeneous in nature and therefore, exhibit different activation energies for chemisorptions (Cheung et al. 2001). In Elovich model, (α_{e}) is related to rate of chemisorptions at zero coverage (initial adsorption rate) and (β_{e}) is related to the extent of surface coverage (desorption constant) and the activation energy of chemisorptions. The application of the Elovich equation is rapidly gaining popularity: it was successfully used, for instance to describe metal ions adsorption from liquid phase into solid matrixes (Rudzinski and Panezyk 2002).

The differential form of (5.15) gives (5.16):

$$\frac{\text{d}q_t}{\text{d}t} = \alpha_{\text{e}} \exp(-\beta_{\text{e}} q_t) \tag{5.16}$$

The constant α_{e} (in mmol/g min^{-1}) can be regarded as initial rate since $\text{d}q_t/\text{d}_t \rightarrow \alpha$ as $q_t \rightarrow 0$.

Integration of (5.16) assuming the initial boundary condition $q_t = 0$ at $t = 0$ gives (5.17):

$$q_{\text{t}} = (1/\beta_{\text{e}}) \ln(1 + \alpha_{\text{e}} \beta_{\text{e}} \cdot t) \tag{5.17}$$

To simplify the Elovich equation, Chien and Clayton (1980) assumed that $[\alpha_e\beta_e t] >> 1$ and applying the boundary conditions: $q_t = 0$ at $t = 0$ and $q_t = q_t$ at $t = t$, (5.17) gives (5.18) (Ho and McKay 2004):

$$q_t = (1/\beta_e)\ln(\alpha_e\beta_e) + (1/\beta_e)\ln(t) \tag{5.18}$$

The Elovich constants can be obtained from the slope and intercept of a linearized plot of q_t against $\ln(t)$.

5.5 Results and Discussion

Kinetics experimental data obtained for the Cr(III) adsorption on studied activated carbons are presented in Figs. 5.1 and 5.2 as a function of studied carbons loading (data presented for the adsorption runs at 40 °C). The carbon dosage was varied for 2.4, 4.8, and 16 g/L.

The results of chromium adsorbed on activated carbons were quantified by mass balance. The following parameters were used:

Adsorption capacity of the carbon (q_{eql}) expressed in terms of metal amount adsorbed on the unitary adsorbent mass (mmol/g), i.e., [Cr uptake, mmol/g]:

$$q_{eql} = \frac{(C_{init} - C_{eql})}{m} \tag{5.19}$$

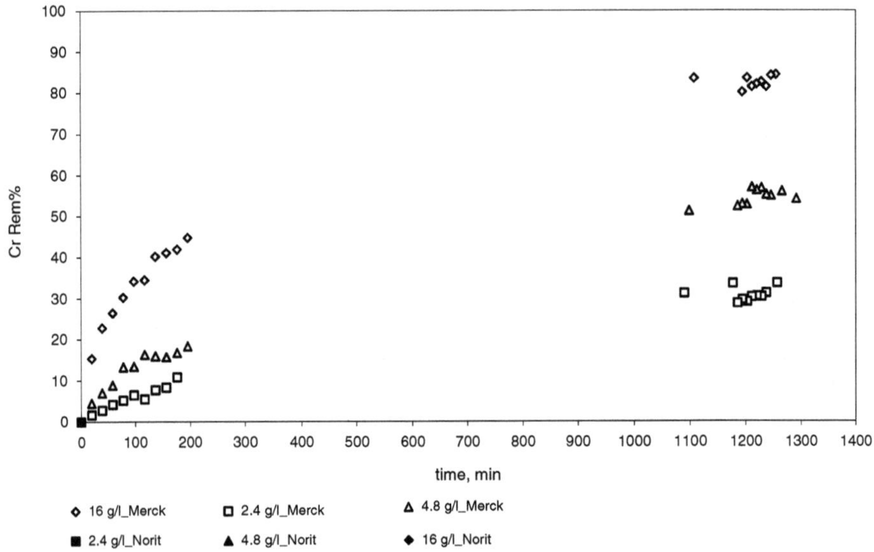

Fig. 5.1 Kinetics (adsorption efficiency vs. time) of Cr III adsorption on modified by 1 M HNO_3 Merck and 1 M HNO_3 Norit activated carbons at 40 °C (data are presented for different carbon loading)

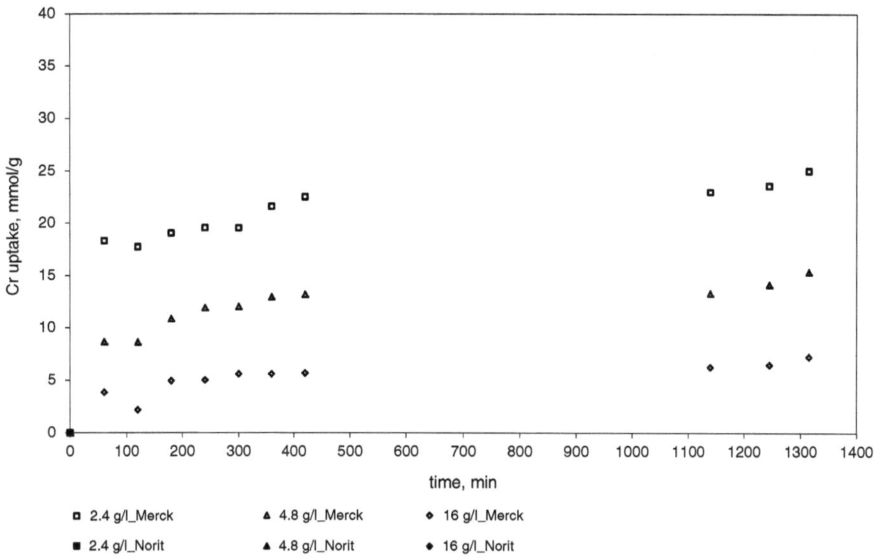

Fig. 5.2 Kinetics (adsorption capacity vs. time) of Cr III adsorption on parent Merck and Norit activated carbons at 40 °C (data are presented for different carbon loading)

Adsorption efficiency of the system (R %) indicated from the percentage of removed metal ions relative to their initial amount, i.e., [Cr *Rem%*]:

$$R\ \% = \frac{\left(C_{\text{init}} - C_{\text{eql}}\right)}{C_{\text{eql}}} 100 \tag{5.20}$$

where C_{init} and C_{eql} are, respectively, the initial and equilibrium concentrations of metal ions in solution (mmol/L) and m is the carbon dosage (g/L).

It may be concluded that the increasing of the carbon dosage increased the sorption efficiency [Cr *Rem %*], but decreased sorption capacity [Cr uptake, mmol/g] (c.f. Figs. 5.1 and 5.2). The decrease in adsorption capacity may be due to the fact that some adsorption sites may remain unsaturated during the adsorption process whereas the number of sites available for adsorption site increases by increasing the adsorbent doses and that results in the increase of removal efficiency.

5.6 Uptake of Metal Ions on the Carbon Active Sites

5.6.1 Pseudo-First-Order Model Application to the Experimental Kinetics Data

The rate constants (k_1) were calculated from the slope of the plot of $\log(q_{\text{eql}} - q_t)$ vs. time (t) of the Lagergren plot (5.10). The (q_{eql}) equilibrium adsorption capacity values are taken from the experimental testing on thermodynamics (data published by authors elsewhere, Lyubchik et al. 2005).

The straight-line plots of $\log(q_{\mathrm{eql}}-q_t)$ vs. t (Figs. 5.3 and 5.4) at different temperatures indicate the validity of Lagergren equation for the present systems and it indicates that the process follows first-order kinetics. The rate constants k_1 were calculated from the slopes of the linear plots and are presented in Table 5.3 for different carbon loading at different temperatures.

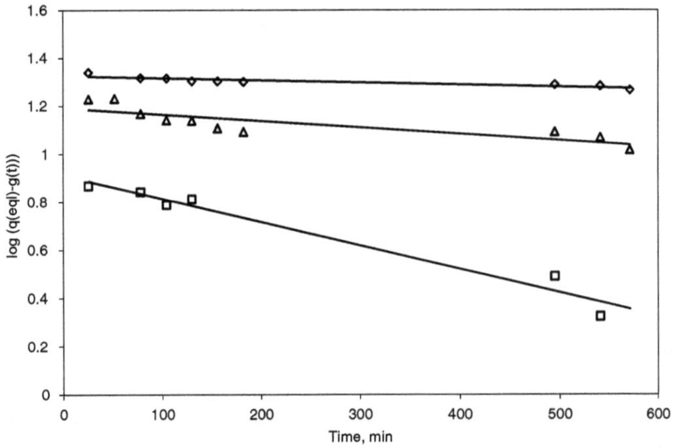

Fig. 5.3 Lagergren plots for the adsorption of Cr III on initial Merck activated carbon at 40 °C (data are presented for different carbon loading: *open square*—2.4 g/L; *triangle*—4.8 g/L; *diamond*—16 g/L)

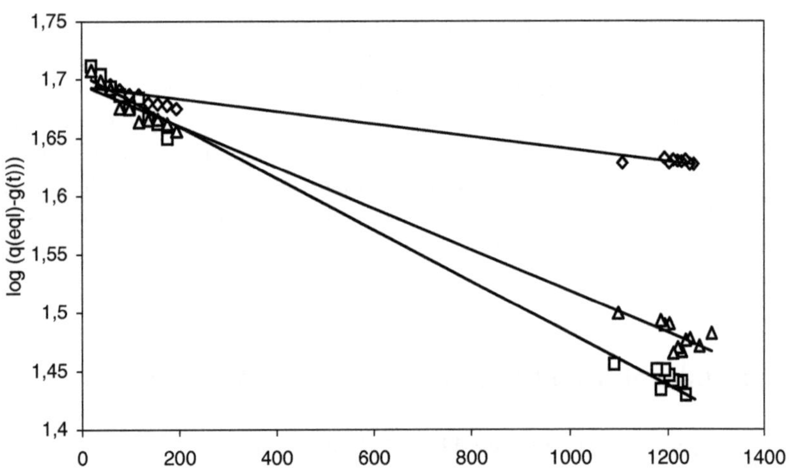

Fig. 5.4 Lagergren plots for the adsorption of Cr III on modified by 1 M HNO_3 Merck activated carbon at 40 °C (data are presented for different carbon loading: *open square*—2.4 g/L; *triangle*—4.8 g/L; *diamond*—16 g/L)

Table 5.3 Pseudo-first-order rate constants for Cr III adsorption with various carbon loadings

Influence of temperature			
T °C	$k_1 \times 10^{-2}$, min^{-1} [Carbon] = 2.4 g/L	$k_1 \times 10^{-2}$, min^{-1} [Carbon] = 4.8 g/L	$k_1 \times 10^{-2}$, min^{-1} [Carbon] = 16 g/L
	Norit treated by 1M HNO_3		
50	0.1678	0.1243	0.0692
40	0.7162	0.1673	0.0465
30	0.3050	0.0692	0.0236
22	0.0046	0.0023	0.0014
	Norit_initial		
50	0.2844	0.2326	0.1451
40	0.1846	0.1857	0.1476
30	0.0697	0.1264	0.0046
	Merck treated by 1 M HNO_3		
50	0.4412	0.1437	0.0453
40	0.1842	0.0697	0.0465
30	0.1456	0.0681	0.0230
22	0.0046	0.0023	0.0016
	Merck_initial		
50	0.2212	0.0827	0.0381
40	0.1178	0.0297	0.0120
30	0.0803	0.0175	0.0062

From the results it was observed that there are deviations in the k_1 values with the carbon loading. The adsorption rate increases with decrease of the carbon loading at given temperature and rise in temperature for given carbon loading (Table 5.3). It is evident that the adsorption of chromium on parent Merck and Norit carbons is higher at higher temperatures, which appears to indicate that the diffusion reactions play an important role in controlling of the adsorption process.

However, close inspection of the time-based data reveals that the situation is not entirely straightforward. For instance, for oxidized sample Norit, the rate of the adsorption was maximal at 40 °C (Table 5.3). This fact indicates the importance of the chemisorptions reactions for the carbons rich by surface functionality.

5.6.2 *Pseudo-Second-Order Model Application to the Experimental Kinetics Data*

A plot of t/q_t vs. t was used to calculate the second-order rate constant k_2 [in g/mmol min^{-1}], the initial adsorption rate, h [mmol/g min^{-1}], and q_{eql}. [mmol/g]. Data are presented in Table 5.4 for the studied adsorption systems for different temperatures and carbon loadings.

Table 5.4 Pseudo-second-order rate constants for Cr III adsorption with various carbon loadings

Influence of temperature									
	[Carbon] = 2.4 g/L			[Carbon] = 4.8 g/L			[Carbon] = 16 g/L		
T °C	k_2 (g/mmol min^{-1})	q_{eql} (mmol/g)	$h \times 10^{-2}$ (mmol/g min^{-1})	k_2 (g/mmol min^{-1})	q_{eql} (mmol/g)	$h \times 10^{-2}$ (mmol/g min^{-1})	k_2 (g/mmol min^{-1})	q_{eql} (mmol/g)	$h \times 10^{-2}$ (mmol/g min^{-1})
	Norit treated by 1M HNO_3								
50	0.1531	0.4824	3.56	0.1725	0.2846	1.40	0.3425	0.1241	0.53
40	0.0641	0.5266	1.78	0.1107	0.3110	1.07	0.1900	0.1409	0.38
30	0.0354	0.5163	0.94	0.0490	0.3164	0.49	0.1149	0.1807	0.38
	Norit initial								
50	0.0381	0.8333	2.64	0.0179	0.6223	0.66	0.0181	0.3190	0.18
40	0.0292	0.5173	0.78	0.0121	0.4864	0.29	0.0730	0.2345	0.40
30	0.0277	0.3252	0.29	0.0059	0.3738	0.08	–	–	–
	Merck treated by 1M HNO_3								
50	0.1203	0.4817	2.79	0.0959	0.3201	0.98	0.1407	0.1669	0.39
40	0.0714	0.4572	1.49	0.1910	0.2223	0.94	0.1656	0.1382	0.32
30	0.0293	0.5583	0.91	0.0567	0.3542	0.71	0.0778	0.1739	0.24
	Merck initial								
50	0.0224	0.3194	0.23	0.1274	0.1146	0.39	0.0336	0.1628	0.09
40	0.0384	0.4798	0.88	0.0497	0.1747	0.42	0.0543	0.1425	0.11
30	0.0558	0.3766	0.79	0.0592	0.2210	0.29	–	–	–

From the results it was observed that there are deviations in the k_2 values with the carbon loading. The adsorption rate [h, mmol/g min^{-1}] increases with decrease of the carbon loading at given temperature for all studied carbons (Table 5.4). The tendency for the adsorption rate with rise in temperature for given carbon loading is different for Norit and Merck activated carbons. For parent and post-treated Norit samples chromium adsorption is higher at higher temperatures, which indicates the diffusion role in the adsorption process control. However, this observation is not so evident in a case of Merck carbons (what so ever of the posttreatment) (Table 5.4). For the parent Merck carbon, it looks like, that the adsorption rates do not change (even slightly decrease) with rise in temperature for given carbon loading, and for post-oxidized Merck carbon the rate of the adsorption is maximal at 40 °C. It means that for highly microporous carbon (the case of parent Merck) or for carbon with developed surface oxygen functionalities (the case of post-oxidized Merck), the pseudo-second-order model is not fully appropriated. It is happening due to the fact that based on chemical adsorption this model cannot be suitably applied to the processes where the driving forces is physical adsorption, including capillary condensation (case of the highly microporous adsorbents) and/or more complex the chemical reactions with surface oxygen groups (such as ion-exchange, chelation, and lone pair electron sharing); when rise in temperature sets off a wide range of different adsorption mechanisms.

The adsorption capacity [q_{eql} mmol/g] of the studied carbons towards chromium (III) ions adsorption at equilibrium conditions were also estimated based on the results of an application of the pseudo-second-order models to the adsorption kinetics in the studies "liquid-to-solid" systems. It was observed that adsorption capacity decreases with increase of the carbon loading at given temperature for all studied carbons; and it changes depending on carbon nature with rise in temperature at given carbon loading. For the parent carbons the maximal adsorption capacities were observed for given carbon loading at 40°C for Merck and at 50 °C for Norit; and for their oxidized forms at 30 °C (Table 5.4).

5.6.3 Chemisorptions Elovich Model Application to the Experimental Data

Further, the kinetic data was fitted to the chemisorptions model based on the Elovich equation (5.15), which assumes that the solid surface active sites are heterogeneous in nature and therefore, exhibit different activation energies for chemisorptions. Simplified the Elovich equation (5.18) was used to test the applicability of the Elovich equation to the chemisorptions kinetics description in studied "liquid-solid" adsorption systems.

The Elovich constants, such as (α_e) in mmol/g min^{-1}, which is related to rate of the chemisorptions and (β_e) in mmol/g min^{-1}, which is desorption constant related to the extent of surface coverage, were obtained from the slope and intercept of the linearized plots of q_t against ln(t). Summary results are presented in Table 5.5.

Table 5.5 Summary of the adsorption kinetics analyses for the studied systems at "liquid–solid" adsorption interface

Sample	Pseudo first-order		Pseudo second-order		Elovich equation			Intra-particle diffusion model			External mass transfer		
	$k_1 \times 10^{-2}$, min^{-1}	R^2	k_2 g/mmol (min^{-1})	R^2	a_e (mmol/g min^{-1})	β_e (mmol/g min^{-1})	R^2	K_i (mmol/g $min^{-0.5}$)	D_i, $\times 10^{-6}$ (m^2 min^{-1})	R^2	$\beta \times 10^{-2}$ (m min^{-1})	$[\beta \times S]$, min^{-1}	R^2
	Merck post-treated by 1 M HNO_3: [Carbon loading] = 2, 4.8 and 16 g/L; *Adsorption runs are at different temperatures*												
Merck_1M_room_2.4	0.0046	0.890	0.03×10^{-4}	0.630	–	–	n/a	0.0039	0.0502	0.867	0.0550	0.996	0.688
Merck_1M_room_4.8	0.0023	0.947	0.0001	0.673	47.62	0.0266	0.630	0.0022	0.0350	0.898	0.0275	0.997	0.845
Merck_1M_room_16	0.0016	0.919	0.0001	0.695	–	–	n/a	0.0011	0.0243	0.947	0.0083	0.997	0.868
Merck_1M_30°_2.4	0.1456	0.898	0.0293	0.964	0.0813	15.63	0.920	0.0101	0.1480	0.952	0.0442	0.801	0.992
Merck_1M_30°_4.8	0.0681	0.959	0.0567	0.980	0.0488	27.03	0.942	0.0058	0.1212	0.963	0.0199	0.721	0.993
Merck_1M_30°_16	0.0230	0.989	0.0778	0.987	0.0298	37.04	0.950	0.0048	0.3445	0.994	0.0054	0.658	0.994
Merck_1M_40°_2.4	0.1842	0.949	0.0714	0.782	0.0503	20.00	0.916	0.0097	0.2035	0.924	0.0515	0.934	0.978
Merck_1M_40°_4.8	0.0697	0.965	0.1410	0.945	0.0492	20.41	0.953	0.0075	0.5147	0.939	0.0263	0.954	0.984
Merck_1M_40°_16	0.0465	0.959	0.1656	0.984	0.0301	33.33	0.980	0.0047	0.5230	0.987	0.0074	0.888	0.986
Merck_1M_50°_2.4	0.4412	0.976	0.1202	0.990	0.0844	18.52	0.939	0.0124	0.2996	0.942	0.0431	0.781	0.977
Merck_1M_50°_4.8	0.1437	0.968	0.0959	0.990	0.0502	20.83	0.977	0.0114	0.5735	0.985	0.0233	0.843	0.968
Merck_1M_50°_16	0.0453	0.834	0.1407	0.909	0.0231	43.48	0.962	0.0067	0.7287	0.897	0.0067	0.814	0.942
	Parent GR MERCK 2518: [Carbon loading] = 2, 4.8 and 16 g/L; *Adsorption runs are at different temperatures*												
Merck_init_30°_2.4	0.0803	0.977	0.0558	0.996	0.0488	37.04	0.881	0.0023	0.0169	0.861	0.0332	0.810	0.914
Merck_init_30°_4.8	0.0175	0.978	0.0592	0.990	0.0255	47.62	0.860	0.0017	0.0268	0.854	0.0174	0.850	0.922
Merck_init_30°_16	0.0062	0.989	–	n/a	–	–	n/a	0.0008	0.0645	0.649	0.0050	0.813	0.809
Merck_init_40°_2.4	0.1178	0.979	0.0384	0.999	1.475	23.26	0.901	0.0043	0.0363	0.967	0.0330	0.806	0.995
Merck_init_40°_4.8	0.0297	0.989	0.0497	0.998	0.048	26.32	0.909	0.0036	0.1920	0.939	0.0174	0.847	0.995
Merck_init_40°_16	0.0120	0.918	0.0543	0.958	0.018	58.82	0.937	0.0022	0.1078	0.915	0.0051	0.828	0.993
Merck_init_50°_2.4	0.2212	0.957	0.0224	0.774	0.0441	23.26	0.766	0.0056	0.1390	0.798	0.0306	0.748	0.999
Merck_init_50°_4.8	0.0827	0.968	0.1274	0.910	0.0345	29.41	0.969	0.0042	0.6073	0.892	0.0153	0.749	0.999
Merck_init_50°_16	0.0381	0.969	0.0336	0.897	0.0291	34.48	0.862	0.0028	0.1338	0.956	0.0045	0.736	0.983

	Norit post-treated by 1 M HNO_3: [Carbon loading] = 2, 4.8 and 16 g/L; *Adsorption runs are at different temperatures*												
Norit_1M_room_2.4	0.0046	0.99	0.0002	0.65	–	–	n/a	0.0016	0.0072	0852	0.0350	0.793	0.877
Norit_1M_room_4.8	0.0023	0.97	0.0019	0.71	–	–	n/a	0.0010	0.0058	0.836	0.0175	0.795	0.738
Norit_1M_room_16	0.0014	0.95	0.0022	0.74	–	–	n/a	0.0007	0.0061	0.876	0.0047	0.704	0.759
Norit_1M_30°_2.4	0.3050	0.97	0.0356	0.997	0.0758	22.73	0.962	0.0051	0.0441	0.979	0.0344	0.780	0.988
Norit_1M_30°_4.8	0.0692	0.96	0.0490	0.995	0.0395	34.48	0.965	0.0034	0.0522	0.985	0.0163	0.740	0.977
Norit_1M_30°_16	0.0236	0.95	0.1149	0.995	0.0183	71.43	0.975	0.0017	0.0400	0.974	0.0034	0.508	0.997
Norit_1M_40°_2.4	0.7162	0.98	0.0641	0.997	0.0863	25.64	0.975	0.0085	0.1178	0.978	0.0343	0.778	0.984
Norit_1M_40°_4.8	0.1673	0.93	0.1107	0.999	0.0405	40.00	0.954	0.0058	0.1519	0.943	0.0160	0.724	0.966
Norit_1M_40°_16	0.0465	0.98	0.1900	0.999	0.0164	73.43	0.981	0.0031	0.2189	0.989	0.0043	0.652	0.993
Norit_1M_50°_2.4	0.1678	0.93	0.1531	0.990	0.2206	15.38	0.937	0.0132	0.3386	0.937	0.0386	0.875	0.949
Norit_1M_50°_4.8	0.1243	0.96	0.1725	0.992	0.0365	29.41	0.941	0.0075	0.8547	0.959	0.0175	0.795	0.956
Norit_1M_50°_16	0.0692	0.95	0.3425	0.992	0.0202	52.63	0.971	0.0040	0.4698	0.974	0.0049	0.739	0.990
	Parent GAC Norit 1240 Plus: [Carbon loading] = 2, 4.8 and 16 g/L; *Adsorption runs are at different temperatures*												
Norit_init_30°_2.4	0.0697	0.969	0.0277	0.973	0.0379	33.33	0.868	0.0076	0.2470	0.956	0.0478	0.884	0.971
Norit_init_30°_4.8	0.1264	0.976	0.0059	0.958	0.1770	12.35	0.680	0.0086	0.2393	0.971	0.0220	0.814	0.987
Norit_init _30°_16	0.0046	0.978	–	n/a	–	–	n/a	0.0039	0.1905	0.854	0.0080	0.983	0.834
Norit_init_40°_2.4	0.1846	0.975	0.0292	0.997	0.0734	23.81	0.840	0.0108	0.1971	0.941	0.0419	0.775	0.971
Norit_init_40°_4.8	0.1857	0.974	0.0121	0.996	0.0477	22.22	0.908	0.0081	0.1254	0.982	0.0210	0.775	0.982
Norit_init_40°_16	0.1476	0.939	0.0730	0.932	0.0290	43.48	0.680	0.0023	0.0435	0.929	0.0051	0.632	0.869
Norit_init_50°_2.4	0.2844	0.936	0.0381	0.989	0.1950	19.23	0.859	0.0152	0.1504	0.969	0.0206	0.380	0.998
Norit_init_50°_4.8	0.2326	0.985	0.0179	0.996	0.0831	12.33	0.921	0.0125	0.1824	0.967	0.0071	0.263	0.981
Norit_init_50°_16	0.1451	0.959	0.0181	0.936	0.0533	19.61	0.966	0.0089	0.3520	0.959	0.0031	0.381	0.994

Parameters evaluated: (a) External mass transfer coefficient [β, m min^{-1}]; (b) External mass transfer rate [($\beta \times S$), min^{-1}]; (c) Intra-particle diffusion rate [K_i, mmol/g $min^{-0.5}$]; (d) Intra-particle diffusion coefficient [D_i, m^2 min^{-1}]; (e) Chemisorptions rate [a_e, mmol/g min^{-1}]; (f) Pseudo first-order rate constant [k_1 min^{-1}]; (g) Pseudo second-order rate constant [k_2 g/mmol min^{-1}]

With increase in the temperature for given carbon loading, the constants α_e and β_e increase showing that both the rate of chemisorptions and the available adsorption surface would increase (Table 5.5). However, in reality the Elovich kinetic model provided excellent model fits for the Norit parent and oxidized carbons, while for the Merck parent and oxidized carbons the kinetic process was much more complex.

Namely, the above mentioned tendency is evident in the case of the Norit parent and oxidized carbons. For the parent Merck carbons, where the main mechanisms of the adsorption are physisorptions and capillary condensation, the Elovich model shows a low degree of correlation for the adsorption kinetics results; and for the oxidized Merck carbons, the constants α_e seems to be unchangeable with temperature rise (Table 5.5). Taking into account a very complex chemical nature of the oxidized carbons, which are rich by surface oxygen functionalities (Table 5.2) it is not surprising and connected to the huge numbers of others then chemisorptions reactions on carbon surface (such as ion-exchange, complexes formations via metal ions chelation with the surface active centers, and lone pair electron sharing) where the Elovich model is not working. With decrease of the carbon loading at given temperature, for all studied carbons, the constants α_e increase and constants β_e decrease, showing that the rate of chemisorptions would increase, while the available adsorption surface decrease (Table 5.5).

5.7 Adsorption Diffusion

The external mass transfer was analyzed applying the external diffusion model to the experimental data on adsorption kinetics. The plots (C_t/C_0) against time (t) allowed to evaluate the external mass transfer rate ($\beta \times S$), min^{-1} by tracing the slopes at $t=0$, and then recalculate the external mass transfer coefficients β, m min^{-1}. Summary data are presented in Table 5.5.

For oxidized Norit and Merck carbons, the external mass transfer coefficients β, m min^{-1} is unchangeable with rise in temperature at given carbon loadings; while it is increased with temperature rise in the cases of parent Norit and Merck carbons (Table 5.5). For all studied carbons (what so ever of posttreatment), the external mass transfer coefficients β, m min^{-1} increase with carbon loading increase for given temperature.

For the oxidized Norit and Merck carbons, the external mass transfer rates [$\beta \times S$], min^{-1} are in the range of [0.7–0.8] and [0.9–1.0] min^{-1} for all studies temperatures [20, 30, 40, and 50 °C]. For the parent carbons, the external mass transfer rates [$\beta \times S$], min^{-1} slightly decrease with rise in temperature for Merck parent carbon [from 0.8 to 0.7 min^{-1} with temperature rise from 30 to 50 °C] and drastically decrease for parent Norit carbon [from 0.9 to 0.3 min^{-1} with temperature rise from 30 to 50 °C] (Table 5.5).

The intra-particle mass transfer diffusion was analyzed using the Weber and Morris and Urano and Tachikawa equations (5.6)–(5.8). The rate constant of intra-particle diffusion (K_i, mmol/g × min$^{-0.5}$) was calculated from the slope of the plot of q_t vs. $t^{0.5}$ according to (5.6). The intra-particle diffusion coefficient (D_i, m^2/min) was calculated from Urano and Tachikawa equation (5.8), i.e., from the slope of the plot of $\log[1-(q_t/q_{\mathrm{eql}})^2]$ vs. t. Summary data are presented in Table 5.5.

From analysis of the results, it could be concluded that for all studied samples the intra-particle diffusion rate constants [K_i, mmol/g min$^{-0.5}$] increase with decrease of the carbon loading at given temperature and rise in temperature at given carbon loadings (Table 5.5). The intra-particle diffusion coefficients (D_i, m^2/min) increase with rise in temperature for given carbon loadings, and it seems that D_i increase with increase of the carbon loadings at given temperature. However, there are some deviations from this tendency for the data presented in Table 5.5. That is most probably due to the fact that for the D_i, recalculation the values of q_{eql} for different carbon loadings were taken from an application of the pseudo-second-order model (Table 5.4).

5.8 Energy of Activation

The energy of activation E_{a} for the overall chemisorptions reactions of the metal ions into carbon's porous structure was determined using the Arrhenius equation (5.21).

$$\ln k_{\mathrm{ads}} = \ln z - \frac{E_{\mathrm{a}}}{RT} \tag{5.21}$$

where k_{ad} is the adsorption rate constant (could be k_1 (pseudo-first-order rate constant) k_2 (pseudo-first-order rate constant) or (α_{e})) and the other terms have their usual meaning.

The E_{a} value was calculated from the slope of the plot ln k_{ads} vs. temperature. The plots of ln k_{ads} vs. $1/T$ were found to be linear for the parent Merck and Norit carbons only, while there were the deviation for their oxidized forms (c.f. Figs. 5.5 and 5.6, example is given for the calculation based on the rate constant of the pseudo-first-order rate constant).

The E_{a} values calculated from the slope of the plot are summarized in Table 5.4. The relatively low activation energy for the adsorption on parent Merck and Norit activated carbons indicated that Cr(III) adsorption on the carbon surface of low functional groups are coupled diffusion-controlled and physisorptions process.

Whereas, relatively high E_{a} values for the oxidized samples are related to the adsorption process controlled by chemical reactions. However, it has been noted that due complex physisorptions/chemisorptions operation mechanisms for the metal ions adsorption inside the porous matrixes rich by functional active sites (i.e., nonlinear behavior presented in Fig. 5.6), the given E_{a} values for the post-treated are approximated, i.e., have just a tentative, indicative character (Table 5.6).

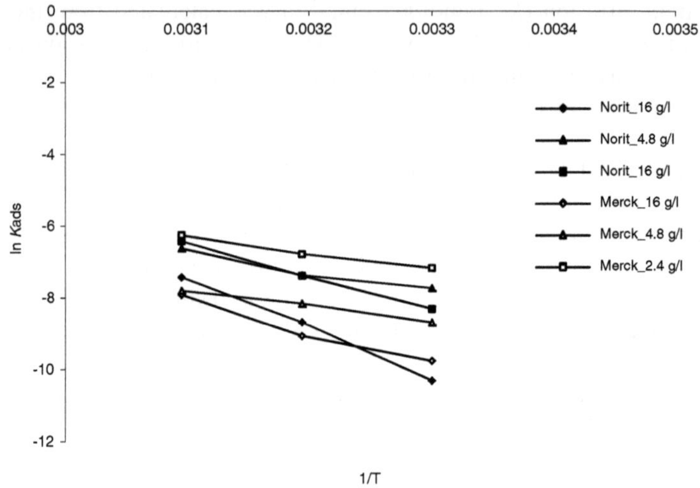

Fig. 5.5 Arrhenius plots for the Cr III adsorption on parent Norit and Merck activated carbons

Fig. 5.6 Arrhenius plots for the Cr III adsorption on modified by 1 M HNO_3 Norit and 1 M HNO_3 Merck activated carbons

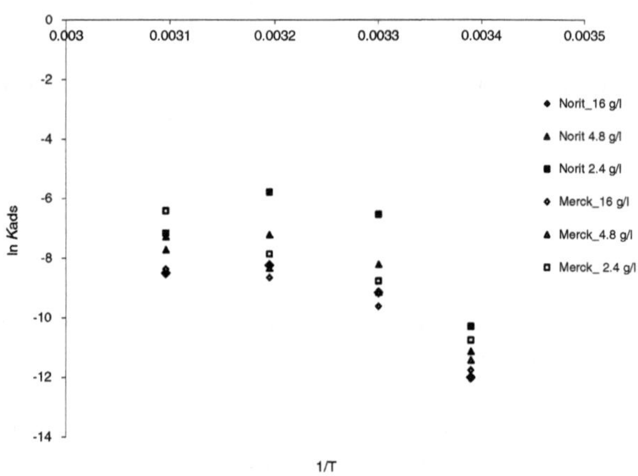

Table 5.6 The activation energy of Cr III adsorption by studied activated carbons

	E_{akt} (kJ mol^{-1})		
	Carbon loading		
	2.4 g/L	4.8 g/L	16 g/L
Merck_1M HNO_3	116.5345	110.5814	93.2209
Norit_1M_HNO_3	83.2353	92.9049	94.8173
Merck_init	37.0448	36.3422	75.0108
Norit_init	84.3079	51.3081	94.6427

5.9 Adsorption Dynamic Models

In general sense, the *adsorption dynamics* deals with the time evolution of adsorption processes in industrially used adsorption units (such as the adsorbers, fixed-beds, and flow-through columns).

The systematic study of the dynamic behavior of systems is the object of *Systems Dynamics* (SD) aimed at the process control. When the dynamic behavior is concern, the accumulation terms (derivatives in order to time) are never dropped from the model equations (balances). Then, the differential equations (ordinary or partial) could be obtained.

In present work, the effects of perturbations on input variables in the systems responses were studied to devise control actions which are able to react to the undesired perturbations in the system. According to the systems complexity (which is related with their dynamic behavior), they were classified into two categories:

- *Lumped parameter systems*. The intensive properties are independent of the position within extensive property of the system (commonly, that is volume), i.e., the intensive properties within the system change with time only. In this case, ordinary differential equations (ODE's) could be obtained. According to the order of the ODE's model, the system itself is called of first order, second order, etc.
- *Distributed parameter systems*—the intensive properties are changed both with the system position and time. In this case, partial differential equations (PDE's) could be obtained for the system model.

Some examples of both systems are given below with an explanation how the respective systems can be classified and what is their expected dynamic behavior.

For any of the conservative property (mass, energy, momentum, etc.) in any system, the general framework for balances is postulated as:

$$\text{(amount in)} = \text{(amount out)} + \text{(accumulation)} \tag{5.22}$$

The foregoing equation is valid for total mass. It is know (since Lavoisier) that, although total mass is conserved, it is not always true for the mass of a given component, since if a chemical reaction is going inside the system, the component can be consumed (if it is a reactant) or produced (if it is a product).

Then, more useful way of writing the general conservation balance, including component mass balances when chemical reactions are present, is the following:

$$\text{(amount in)} + \text{(production)} = \text{(amount out)} + \text{(accumulation)} \tag{5.23}$$

The general (5.23) was applied to model the adsorption dynamics in the studied systems at "liquid–solid" interface taking into account the following approximation: conservation equations; equilibrium relations; kinetic (physical, chemical) rate expressions; initial and boundary conditions.

The studied systems were considered as *Distributed parameter systems* and the following two models were proposed to describe adsorption kinetics and dynamics at "solid–liquid" interface:

1. *The heterogeneous model for batch experiment including intra-particle mass transfer diffusion via* Partial differential equation (5.22):

$$\varepsilon_{\mathrm{p}} D_{\mathrm{p}} \frac{\partial^2 c_{\mathrm{p}}}{\partial r^2} + \frac{2\varepsilon_{\mathrm{p}} D_{\mathrm{p}}}{r} \frac{\partial c_{\mathrm{p}}}{\partial r} = \varepsilon_{\mathrm{p}} \frac{\partial c_{\mathrm{p}}}{\partial t} + \rho_{\mathrm{ap}} \frac{\partial q}{\partial t} \tag{5.24}$$

with initial conditions:	$t = 0$	$c_{\mathrm{p}} = q = 0$
Boundary conditions:	$r = 0$	$\frac{\partial c_{\mathrm{p}}}{\partial r} = 0$
	$r = R$	$c_{\mathrm{p}} = c$ $V_{\mathrm{liq}} \left. \frac{\partial c_{\mathrm{p}}}{\partial t} \right\|_{r=R} = -\frac{3 m_c}{R \rho_{\mathrm{ap}}} \varepsilon_{\mathrm{p}} D_{\mathrm{p}} \left. \frac{\partial c_{\mathrm{p}}}{\partial r} \right\|_{r=R}$

2. *The heterogeneous model for dynamic mode including external and intra-particle mass transfer diffusions* via Partial differential equations (5.24) and (5.25):

$$U_0 \frac{\partial c}{\partial \zeta} + \varepsilon \frac{\partial c}{\partial t} + (1 - \varepsilon) k_f a_{\mathrm{p}} (c - c_s) = \varepsilon D_{\mathrm{ax}} \frac{\partial^2 c}{\partial \zeta^2} \tag{5.25}$$

$$\varepsilon_{\mathrm{p}} D_{\mathrm{p}} \left(\frac{2}{r} \frac{\partial c_{\mathrm{p}}}{\partial r} + \frac{\partial^2 c_{\mathrm{p}}}{\partial r^2} \right) = \varepsilon_{\mathrm{p}} \frac{\partial c_{\mathrm{p}}}{\partial t} + \rho_{\mathrm{ap}} \frac{\partial q}{\partial t} \tag{5.26}$$

Conditions		
Initial conditions:	$t = 0$	$c(0,\zeta) = \begin{cases} c_0, & \zeta = 0 \\ 0, & 0 < \zeta \le L \end{cases}, \; c_{\mathrm{p}}(0,\zeta,r) = \begin{cases} c_0, & \zeta = 0, r = R \\ 0, & \text{Else} \end{cases},$ $q = 0$
Boundary conditions:	$\zeta = 0$	$U_0 c_0 = U_0 c - \varepsilon D_{\mathrm{ax}} \frac{\partial c}{\partial \zeta}$
	$\zeta = L$	$\frac{\partial c}{\partial \zeta} = 0$
	$r = 0$	$\frac{\partial c_{\mathrm{p}}}{\partial r} = 0$
	$r = R$	$k_f (c - c_s) = \varepsilon_{\mathrm{p}} D_{\mathrm{p}} \left. \frac{\partial c_{\mathrm{p}}}{\partial r} \right\|_{r=R}$

where ε_{p} is the particle porosity, ε is the bed porosity, D_{p} is the diffusivity, cm^2 s^{-1}, D_{ax} is the dispersion coefficient in flow direction, cm^2 s^{-1}, c_{p} is the solute concentration in particle, g cm^{-3}, c is the solute concentration in bulk, g cm^{-3}, c_0 – initial (inlet) solute concentration, g cm^{-3}; c_s is the solute concentration in film, g cm^{-3}, ρ_{ap} is the apparent density of carbon, g cm^{-3}, q is the solute concentration

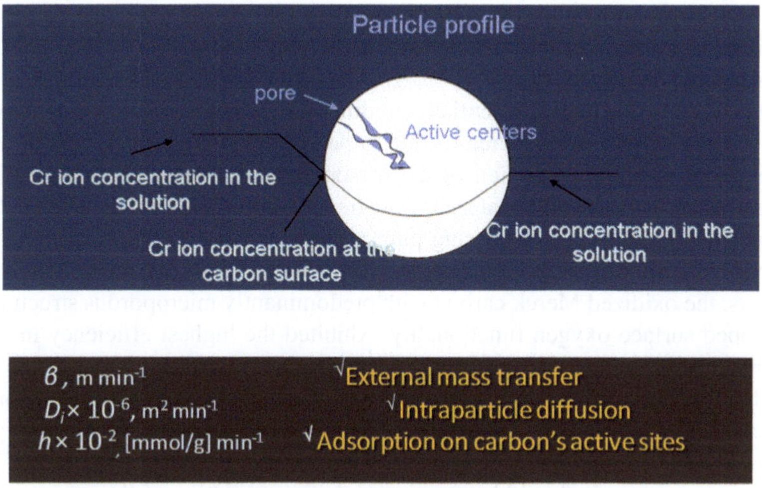

Fig. 5.7 KINETICS of the ADSORPTION at "LIQUID-SOLID" INTERFACE

in adsorbed phase, g/g, t is the time, s, V_{liq} is the volume of solute, cm^3, R is the particle radius, cm, u_0 is the superficial velocity, cm s^{-1}, k_f is the film mass transfer coefficient, cm s^{-1}, and a_{ap} is the particle specific area, m^2/g.

Based on the results obtained from the analysis of the experimental data on the adsorption kinetics studies by applying the pseudo-first-order, pseudo-second-order, chemisorptions kinetics, external and intra-particle mass transfer diffusion models and by using additional systems dynamics modeling approach, the main limiting step for an overall adsorption process was evaluated for given adsorption systems at "liquid–solid" interface.

It was concluded that the metal ions transfer from the adsorbent surface to the intra-particle actives sites is the rate limiting step for the adsorption process at "liquid–solid interface" (Fig. 5.7).

5.10 General Remarks

All the studied carbons adsorbed appreciable amount metal ions under chosen experimental conditions, but the adsorption efficiency of the carbons is strongly influenced by their texture and surface functionalities.

The effect of the texture is related to the carbon porous structure, particularly to the accessibility of the adsorbent internal surface for large-size high-hydrated metal ions (e.g., 0.922 nm for $[Cr(H_2O)_6]^{3+}$). The diameters of the micropores less than 1 nm are inaccessible for high-hydrated metals ions adsorption. The absolute value of the apparent surface area seems to be less important factor, which has the highest surface area of ca. 1020 m^2/g (case of the parent Merck carbon) was less efficient for Cr(III) ions adsorption. This fact emphasizes the dominating role of pore size distribution in the adsorption kinetics for the activated carbons of poor surface oxygen functionality.

Oxidation by nitric acid essentially modifies the structure of the initial activated. The specific surface area decreases due to partial destruction of the initial micropores along with the formation of the mesoporous structure. This observation is more evident in case of the materials post-treated with 13 M HNO_3 when a degree of oxidation is deeper. Considering purely physisorptions process, an increase of the pore size should increase an uptake of the metal ions. Among all the studied carbons, the oxidized Merck carbon with predominantly microporous structure and developed surface oxygen functionality exhibited the highest efficiency in metals adsorption. In this case, along with the development of the mesoporosity, one could see the dominant influence of the surface oxygen functionality on the adsorption process. The adsorption increases when the carbon surface is oxidized, although the nature and the amount of the surface oxygen groups play an important role in the adsorption process. It should be mentioned that an oxidation of the parent carbons with 1 M HNO_3 increased significantly the concentration of functional oxygen groups, especially carboxylic ones and oxidized carbons are more efficient for the metal ions adsorption, confirming the importance of chemisorptions processes.

It has been observed that the adsorption efficiency of the studied carbons increases and their adsorption capacity decreases with carbon loading. The reduction of the sorption capacity is related to the fact that some of the active centers remain unsaturated during adsorption process. However, with increase in carbon loading, the absolute amount of those active centers increases resulting in increase of the overall sorption efficiency of the systems. The established fact specifies that in case of the static mode of the adsorption (batch mode) there is an optimum carbon loading which limits the metals uptake/removal. For the studied systems, the adsorption efficiency of the activated carbon grows continuously up to 6–8 g/L of carbon loading. Further an increase of the carbon loading did not affect the adsorption process and the chromium removal [Cr *Rem%* parameter] remains practically constant. Adsorption capacity [Cr uptake, mmol/g parameter] is sharply falling with increase in carbon loading up to 8 g/L, and then it's smoothly changes for the higher carbon content in the adsorption systems.

Therefore, in a present work a number of physicochemical characteristics of the activated carbons, such as pore size and shape; pore volume, surface area; particle size and bulk density; functionality of surface; chemical inertness are compared with the results of the adsorption kinetics testing. The following parameters for the optimal removal of metal ions by activated carbons are established:

1. The presence of the oxygen functional groups, mainly carboxylic ones.
2. Optimum balance of the micro/mesopores for easy access to the adsorbent's internal surface for high hydrated metal ions. (It is especially significant for the carbons with a lack of oxygen functional groups.)

3. Sufficient time of contact of the components for establishment of real adsorption equilibrium in the systems. This requirement is especially significant for carrying out adsorption process in a static mode.
4. For given adsorbents, there is an optimum content in the system (loading) when further increase of its concentration does not result in increase of its adsorption capacity towards heavy metals adsorption.

Optimization of conditions of the adsorption process in accordance to the suggested parameters allows making favorable the activated carbons usage in real industrial scale at "solid–liquid" interface, when the balance of cost/efficiency for heavy metals removal should be provided.

5.11 Conclusion

Due to the unquestionable fact of enormous complexity, which is inherent to the adsorption phenomena at various interfaces, the kinetics studies and appropriate models choice to the data analysis would help to gain insight, which is often very difficult or impossible to obtain by other means, on overall dynamic behavior of the adsorption systems at different interface aiming at the overall process control.

It was concluded that in a case of "liquid–solid" interface for the heavy metal ions adsorption on activated carbons the adsorption diffusion and chemisorptions reaction models should be considered for the adsorption kinetic parameters evaluation; namely: (a) external mass transfer of the adsorbate from the liquid phase to the adsorbent surface; (b) internal (or interparticle) diffusion of the adsorbate from the surface of the adsorbent into the porous structures; (c) physisorptions of the adsorbate on the adsorbent surface; and (iv) chemisorptions of the adsorbate on active sites of the adsorbent, i.e., chemical reactions between adsorbate and adsorbent inside porous structure of the adsorbent. The chemical kinetics data allowed evaluating the rate of metal ions chemisorptions inside the porous structure and also determining the structural factors of the carbon surface, which are affected the chemical reactions rate. It was observed that the adsorbent's physical (porous texture) and chemical (surface functionality) characteristics drastically affected the adsorption kinetics along with the adsorption runs conditions (in present work temperature and the adsorbents loading).

Based on the obtained results on the adsorption kinetics data analysis and on data obtained using the systems dynamics modeling approach (derivatives in order to time) it was also concluded that intra-particle mass transfer diffusion of the metals ions form liquid phase into microporous carbon structure is the rate limiting step for the overall adsorption process at "liquid–solid" interface.

Acknowledgements Authors are thankful for the financial support of the work by the European Commission, Marie Curie 7FP projects FP7-PEOPLE-2010-IRSES/MC-IRSES-269138 NANOGUARD and FP7-PEOPLE-2010-IRSES/MC-IRSES-269289 ELECTROACROSS.

References

Ajmal M, Rao RAK, Ahmad R, Ahmad J, Rao LAK (2001) Removal and recovery of heavy metals from electroplating wastewater by using Kyanite as an adsorbent. J Hazard Mater 87:127–137

Brigatti MF, Franchini G, Lugli C, Medici L, Poppi L, Turci E (2000) Interaction between aqueous chromium solutions and layer silicates. Appl Geochem 15:1307–1316

Carrott PJM, Carrott MML, Nabais JMV, Ramalho JP (1997) Influence of surface ionization on the adsorption of aqueous zinc species by activated carbons. Carbon 35:403–410

Cheung CW, Porter JF, Mckay G (2001) Sorption kinetic analysis for the removal of cadmium ions from effluents using bone char. Water Res 35(3):605–612

Chien SH, Clayton WR (1980) Application of Elovich equation to the kinetics of phosphate release and sorption on soils. Soil Sci Amer J44:265

Csobán K, Párkányi-Berka M, Joó P, Behra P (1998) Sorption experiments of Cr(III) onto silica. Colloid Surface A 141:347–364

Dabrowski A (2001) Adsorption from theory to practice. Adv Colloid Interfac Sci 93:135–224

Ho YS (1995) Adsorption of heavy metals from waste streams by peat. Ph.D. thesis. University of Birmingham

Ho YS, McKay G (1998) A two stage batch sorption optimized design for dye removal to minimize contact time. Trans I Chem E 76:313

Ho YS, McKay G (1999) Pseudo-second order model for sorption process. Process Biochem 34:451–465

Ho YS, McKay G (2004) Sorption of copper(II) from aqueous solution by peat. Water Air Soil Pollut 158:77–97

Ho YS, Ng JCY, McKay G (2000) Kinetics of pollutant sorption by biosorbents: review. Sep Purif Method 29(2):189–232

Kumar A, Rao NN, Kaul SN (2000) Alkali-treated straw and insoluble straw xanthate as low cost adsorbents for heavy metal removal—preparation, characterization and application. Biores Technol 71:133–142

Lagergren S (1898) About the theory of so-called adsorption of soluble substances. Kungliga Svenska Vetenskapsakademiens Handlingar 24(4):1–39

Lakatos J, Brown SD, Snape CE (2002) Coals as sorbents for the removal and reduction of hexavalent chromium from aqueous waste streams. Fuel 81:691–698

Lazaridis NK, Asouhidou DD (2003) Kinetics of sorptive removal of chromium(VI) from aqueous solutions by calcined Mg-Al-CO_3 hydrotalcite. Water Res 37(12):2875–2882

Lyubchik SB, Perepichka II, Galushko OL, Lyubchik AI, Lygina ES, Fonseca IM (2005) Optimization of the conditions for the Cr(III) adsorption on activated carbon. Adsorption 11:581–593

McKay G, Poots VJP (1984) Kinetics and diffusion processes in colour removal from effluent using wood as an adsorbent. J Chem Technol Biotechnol 30:279–282

McKay G, Blair HS, Findon A (1986) Sorption of metal ions by chitosan. In: Heccles H, Hunt S (eds) Immobilization of ions by biosorption. Ellis Horwood, Chichester, pp 59–69

Qiu H, Lv L, Pan B, Zhang Q, Zhang W, Zhang Q (2009) Critical review in adsorption kinetic models. J Zhejiang Univ Sci A 10(5):716–724

Raji C, Anirudhan TS (1998) Batch Cr(VI) removal by polyacrylamide-grafted sawdust: kinetics and thermodynamics. Water Res 32:3772–3780

Rudzinski W, Panezyk T (2002) The Langmuirian adsorption kinetics revised: a farewell to the XXth century theories. Adsorption 8:23

Sontheimer H, Crittenden JC, Summers RS (1988) Activated carbon for water treatment. DVGW ForschungsstelleEngler-BunteInstitut, Karlsruhe

Sparks DL (1986) Kinetics of reaction in pure and mixed systems. In: Sparks DL (ed) Soil physical chemistry. CRC, Boca Raton, pp 83–145

Srivastava SK, Tyagi R, Pant N (1989) Adsorption of heavy metal ions on carbonaceous materials developed from water-slurry generated in local fertilizer plant. Water Res 13:1161–1165

Unnithan MR, Vinod VP, Anirudhan TS (2002) Ability of iron(III) loaded carboxylatedpolysacrylamide-grafted sawdust to remove phosphate ions from aqueous solution and fertilizer industry wastewater: adsorption kinetics and isotherm studies. J Appl Polymer Sci 84(13):2541–2553

Weber WJ, Morris GC (1962) Removal of biologically-resistant pollutants from waste waters by adsorption. In: Advances in water pollution research, Pergamon Press, New York, pp 231–266

Weber WJ, Morris JC (1963) Kinetics of adsorption on carbon from solution. J Sanit Eng Div Amer Soc Eng 89:31–59

Zeldowitsch J (1934) Über den mechanismus der katalytischen oxydation von CO an MnO2. Acta Physicochim URSS 1(3–4):449–464

Part II
Remediation of Contaminants and Recovery of Secondary Resources with Socio-Economical Value

Chapter 6
Electrochemical Process for Phosphorus Recovery from Water Treatment Plants

Nazaré Couto, Margarida Ribau Teixeira, Paula R. Guedes, Eduardo P. Mateus, and Alexandra B. Ribeiro

6.1 Introduction

6.1.1 Background

Waste strategy from EU Member States goes to the enhanced utilization of secondary resources avoiding, at the same time, its landfill disposal. However, available technologies to fulfill the requirements of this strategy are scarce.

Assuring the levels of staple food that accomplish the population growth is secured by sufficient levels of nutrients on agricultural soil. Phosphorus is an essential, limited resource that cannot be replaced by any other element.

Eutrophization in aquatic ecosystems has several environmental and socio-economical impacts such as loss of biodiversity and suitability of aquatic systems for drinking water and recreation use. But, and although phosphorus can be considered a contaminant, it is also an element of vital importance for the environment. Thus, the development of technologies for phosphorus recycling is an inevitable consequence of the non-renewability of rock phosphate and the non-substitutability of phosphates in agricultural production. Phosphate stocks

N. Couto (✉) • P.R. Guedes
CENSE, Departamento de Ciências e Engenharia do Ambiente, Faculdade de Ciências e Tecnologia, Universidade Nova de Lisboa, Caparica 2829-516, Portugal

Key Laboratory of Soil Environment and Pollution Remediation, Institute of Soil Science, Chinese Academy of Sciences, Nanjing 210008, China
e-mail: md.couto@fct.unl.pt

M.R. Teixeira
CENSE, Faculdade de Ciências e Tecnologia, Universidade do Algarve, Campus de Gambelas, Edifício 7, Faro 8005-139, Portugal

E.P. Mateus • A.B. Ribeiro
CENSE, Departamento de Ciências e Engenharia do Ambiente, Faculdade de Ciências e Tecnologia, Universidade Nova de Lisboa, Caparica 2829-516, Portugal

© Springer International Publishing Switzerland 2016
A.B. Ribeiro et al. (eds.), *Electrokinetics Across Disciplines and Continents*,
DOI 10.1007/978-3-319-20179-5_6

(economically minable deposits divided by actual annual consumption) are estimated for one century (Steen 1998; Cordell et al. 2009).

To deal with the future economic and environmental problem, the ability to understand where phosphorus losses can be avoided or where is viable to recover it will be a huge step in reducing not only its impacts in the environment, but also to allow its recovery for further reuse guaranteeing future food security.

6.1.2 Algal Blooms and Production of Cyanotoxins

The increased fluxes of nutrients arriving to water bodies is one of the main factors to promote the existence of cyanobacterial blooms (Corbel et al. 2014). One fundamental cause of phosphorus loading is the inefficient fertilizer use in crop production. When phosphorus fertilization exceeds its removal by the crop, most of the surplus will remain in the soil and added to the phosphorus reserve (Hooda et al. 2001). Soils with excessive phosphorus reserves in turn pose the highest risk to the environment (Yli-Halla et al. 1995), mainly linked to the hydrological cycle, where rain events can erode the soil enriched with this element into water bodies.

In 2001, it was determined that for Portugal, although the trophic status is only known to 55 % of the larger reservoirs, 4 % proved to be oligotrophic (P total > 10 mgP/m^3), 28 % mesotrophic (10 mgP/m^3 < P total < 35 mgP/m^3), and 23 % eutrophic (P total > 35 mgP/m^3). In the case of Portuguese Azores islands, from their 24 lakes, almost 50 % are eutrophic, 9 mesotrophic, and 4 oligotrophic (SRAM, Secretaria Regional do Ambiente e do 2011). Considering that most lakes indicate a consistent increase in the level of the trophic state, the ability to succeed in combating lake eutrophication by the involvement and cooperation of local inhabitants, small factories, and farmers in reducing phosphorus discharge is very important.

Several freshwater bodies in the world (including reservoirs, lakes, and rivers), used for recreational or drinking purposes, have been found to have hepatotoxic blooms with production of microcystins (MC) such as MC-LR, MC-YR, and MC-RR, mostly related to the dominance of *Microcystis aeruginosa* (Meriluoto 1997). The presence of high concentration of MC-LR in freshwater sources is a worldwide problem (Sivonen and Jones 1999). Australia, Brazil, China, Ireland, and Italy are some of the countries with MC-LR concentrations above legislation threshold from drinking water supply (Corbel et al. 2014). For example, the highest value reported in China was of 1556 μg/L (Ueno et al. 1996). World Health Organization (WHO) recommends a guideline value of 1 μg/L for MC-LR in drinking water (WHO 2004). Cyanotoxins may enter in the food chain through many routes. By the use of water from sources with cyanobacterial blooms and toxins for spray irrigation leading to uptake into the food chain and toxin accumulation on the external surfaces of edible plant material (Corbel et al. 2014). Cyanobacterial blooms have been used as organic fertilizers in some countries (Chen et al. 2006a, b) which may represent an additional source of microcystins in the food chain.

Cyanobacteria are prokaryotic cells, phototropic and oxygenic that exists symbiotically with other organisms in natural environments (Carmichael 1994). If nutrient and climatic conditions are optimal, toxic cyanobacteria blooms develop

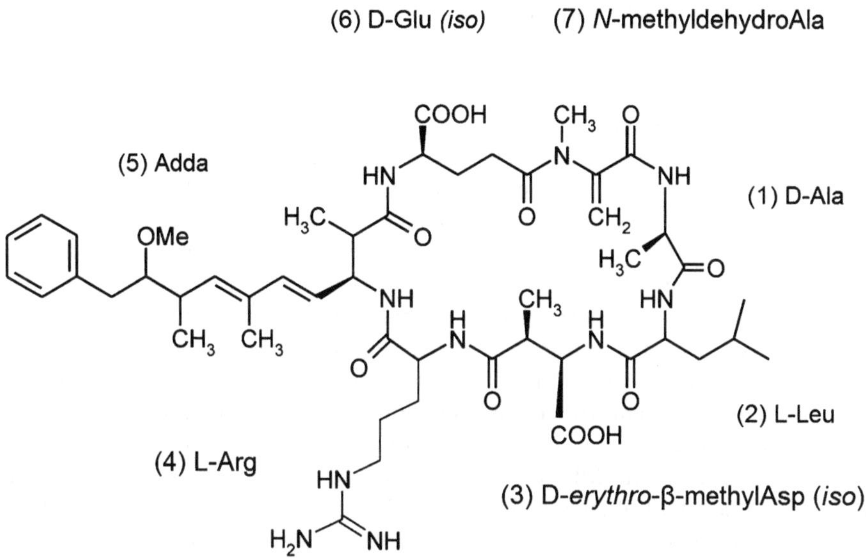

Fig. 6.1 Structure of MC-LR (Meriluoto 1997)

in a fast way. They may compromise the water quality in lakes, rivers, and estuaries, thus affecting human consumption, support of aquatic life, recreational activities, and agriculture (Carmichael 1994). Some freshwater cyanobacteria release toxins when cells died or lyse that irreversibly inhibit serine/threonine protein phosphatases 1 and 2A (Yoshizawa et al. 1990).

MCs are peptides with seven amino acids connected via peptide bonds in a cyclic configuration. This high molecular weight toxin has a hydrophilic character. MCs contain one unique amino acid in the C_{20} position, 3-amino-9-methoxy-2,6, 8-trimethyl-10-phenyldeca-4,6-dienoicacid (Adda) (Meriluoto 1997). The Adda moiety is not toxic if alone, but the hydrophobicity and stereochemistry of the Adda chain is critical in the biological activity of MC (Harada et al. 1990; Antoniou et al. 2008).

Different types of toxins exist depending on the radical groups, due to the variations of the a-amino acids found only at positions 2 and 4. For the most common and toxic variant of MCs (Sharma et al. 2012), MC-LR (Fig. 6.1), LR stands for leucine (L) and arginine (R) (Antoniou et al. 2008). MC-LR has an octanol/water distribution coefficient logarithm (log D_{ow}) between 2.18 for pH 1 and −1.76 at pH 10 (Maagd et al. 1999; Ribau Teixeira and Rosa 2006).

The charge of MC depends on the structure of each derivate (Antoniou et al. 2008). MC-LR has a D-methylaspartic acid and a D-glutamic acid both being ionized to the anionic carboxylate forms (p*K*a of 2.09 and 2.19) and also contains an L-arginine unit with a basic amino group (p*K*a of 12.48) (Ribau Teixeira and Rosa 2006; Liao et al. 2014). As so, MC-LR is singly positively charged in a pH lower than 2.09. With increasing pH (between 2.19 and 12.48), MC-LR loses the two protons from the carboxylic groups, presenting an overall charge of −1 whereas at extremely basic pH (>12.48), MC-LR loses the proton from the protonated basic group presenting an overall charge of −2 (Maagd et al. 1999; Antoniou et al. 2008; Liu et al. 2008).

6.1.3 Removal of Cyanotoxins

After being released from the cells, microcystins may persist in aquatic environment for weeks. But cyanotoxins may also be retained on sediments or suspended particles (Corbel et al. 2014 and references therein). Thirumavalavan et al. (2012) reported that the adsorption of MC-LR onto natural organic matter (NOM) and suspended solids was influenced by turbidity, humic acid, organic matter content, and other pollutants. The MCs were reported to adsorb to suspended particulate matter at acidic and neutral pH (3 and 7) with adsorption decreasing at elevated pH (pH 13), fact explained by the pH-dependent hydrophobicity (Liu et al. 2008).

In the absence of other substrates, exposure to solar irradiation presents low effect on MC degradation as MC-LR are stable due to their cyclic structure. When the MC is exposed to a UV light with a maximum absorption of 238 nm, degradation may occur. However, the presence of humic substances, dissolved organic matter or pigments may transform MCs through photosensitized reaction (Tsuji et al. 1994; Robertson et al. 1999; Welker and Steinberg 2000; Hayakawa and Sugiyama 2008; Wörmer et al. 2010; Thirumavalavan et al. 2012; Yan et al. 2014). This reaction involves the photolysis of photosensitizers forming singlet stats and then crossing to the triplet excited state which will react with dissolved oxygen forming reactive oxygen species ($^1O_2$, $^{\cdot}OH$, and O_2^{-}) (Burns et al. 2012; Yan et al. 2014) that may oxidize organic contaminants. The excited states may also directly decompose organic contaminants (Wenk et al. 2011; Yan et al. 2014). Several literature report the photodegradation of MC-LR. Thirumavalavan et al. (2012) reported that at 254 nm the turbidity only affected the pattern of photodegradation curves but not the rate of photodegradation whereas at 365 nm high turbidity decreases the rate of photodegradation. Yan et al. (2014) studied the effect of adsorption on indirect photodegradation and reported the photosensitized transformation of MC-LR in solutions enriched with dissolved organic matter under solar simulated irradiation (MC-LR adsorbed on the dissolved organic matter). It was also observed an increased phototransformation with decreasing pH, fact explained by the adsorptive interaction of MC-LR with the dissolved organic matter (Yan et al. 2014). As at alkaline condition, the surface of dissolved organic matter is dominated by negative charges and the glutamic and methyl aspartic acids of MC-LR are deprotonated, the repulsive force between both will increase leading to a reduced effect of sorption whereas at acidic pH the charges of organic matter are neutralized (Yan et al. 2014). Jacobs et al. (2013) also reported photocatalytic degradation of MC-LR in aqueous solution. Despite MC-LR being degraded by photolysis under UV-C radiation, photolysis usually only causes small conformational changes in the functional groups that are located on the Adda residue whereas photocatalysis induces higher degrees of degradation (Jacobs et al. 2013).

For drinking water purposes, the removal of MC-LR may be carried out using chemical oxidants such as chlorine dioxide, chlorine, monochloroamine, ozone, and permanganate (Sharma et al. 2012; Chang et al. 2014) that can reduce toxicity namely through the oxidation of the Adda side chain (Chang et al. 2014 and references therein). But some oxidation intermediates and by-products were already identified in some oxidation processes (Antoniou et al. 2008; Chang et al. 2014) which may compromise the total effectiveness of this methodology.

Other way to decrease MCs levels is through the electrogeneration of highly reactive hydroxyl radicals ($^{\cdot}OH$) (Liao et al. 2013, 2014; Wang et al. 2013). Advanced oxidation processes such as ultraviolet radiation alone or combined with hydrogen peroxide, Fenton reaction, ultrasonic irradiation, and photocatalytic oxidation in the presence of titanium dioxide and ferrate have also been used (Sharma et al. 2012; Jacobs et al. 2013; Fang et al. 2014; Liao et al. 2014; Fotiou et al. 2015). For example, TiO_2 photocatalysis (Loo et al. 2012) is based on electron/hole pairs and highly oxidizing $^{\cdot}OH$ radicals which leads to oxidation and mineralization of organic pollutants. TiO_2 photocatalysis is a "green" technology as it is cost-effective, inert photocatalyst do not require chemical additives and do not generate hazardous waste (Antoniou et al. 2008).

Other combined technologies have been developed. Fraga et al. (2009) promoted in situ generation of chlorine by direct oxidation of chloride on an electrode based on Ti/TiO_2 by applying a controlled current density from 5 to 30 mA/cm^2 and UV irradiation and reported that photoelectrochemical oxidation of chloride indicates to be effective to degrade microcystins from surface waters. Nevertheless, the author suggests further studies to test the formation of chlorinated by-products (Fraga et al. 2009). The electrochemical degradation behavior of MC-LR in aqueous electrolyte at boron doped diamond (BDD) anode under neutral pH condition was also reported (Liao et al. 2014). The removal of MC-LR through an electro-oxidation process was also reported by Tran and Drogui (2013) that tested parameters such as current density, reaction time, anode material, and type of supporting electrolyte. It was reported that current density and type of anode material were important for MC-LR degradation efficiency whereas the type of sodium salts influenced the removal efficiency. Direct anodic oxidation was the main responsible for MC-LR degradation and 98 % removal was achieved for 60 min, 38 mA/cm^2 and recycling rate of 0.1 L/min, Ti/BDD anode and sodium sulfate.

Membrane filtration has also been used to remove micropollutants like cyanotoxins and pharmaceuticals (Ribau Teixeira and Rosa 2005; Nghiem and Hawkes 2007; Verliefde et al. 2009), as well as multivalent ions and NOM (Ribau Teixeira and Rosa 2005, 2006). The membrane acts as a physical barrier, allowing the separation from the water the suspended solids and dissolved materials, depending on its type and operational conditions (Mulder 1997). For microcystins removal, membrane pressure-driven processes like ultrafiltration (UF) and nanofiltration (NF) have been used (Ribau Teixeira and Rosa 2005, 2006; Gijsbertsen-Abrahamse et al. 2006; Lee and Walker 2008; Dixon et al. 2010; Teixeira and Rosa 2012). NF with lower pore diameter than UF ($<$2 nm) retains organic compounds with low molecular weight. Results indicate high removals of microcystins, especially using NF membranes (Ribau Teixeira and Rosa 2005, 2006; Gijsbertsen-Abrahamse et al. 2006). For UF, removals varied between 35 and 66 % (Lee and Walker 2008) without any pre-treatment, and 82 % using the hydride membrane process PAC/UF for 10 mg/L PAC (Campinas and Rosa

2010). The main microcystins removal mechanism for NF was size exclusion (Ribau Teixeira and Rosa 2005, 2006), while for UF adsorption played an important role due to hydrophobic interactions or hydrogen bonding (Lee and Walker 2008).

In membrane processes, the feed stream is divided in two streams, the permeate and the concentrate streams. In a Water Treatment Plant (WTP), the permeate stream is the treated water. The concentrate stream has a large organic fraction and/or high toxic compounds. As in NF toxins are not destroyed, this toxin-enriched stream may represent an environmental problem if not adequately treated or disposed (Sharma et al. 2012). In fact, most of the studies in drinking water focus on quality and effectiveness of the water treatment plants (Van Hege et al. 2004), and few studies focus on the concentrate streams. Additionally, cyanobacterial blooms are associated with nutrient enrichment of surface waters which may be removed by NF membranes when this treatment is used, as already stated (Niewersch et al. 2008; Leo et al. 2011; Cathie Lee et al. 2014; Santos et al. 2014).

6.1.4 Electrokinetic and Electrodialytic Processes to Promote Phosphorus Recovery

Electrokinetic (EK) process is induced in porous media through the application of a low-level electric field and will promote the movement of species towards one of the electrode compartments, from where they can be removed. The main mechanisms responsible for the movement are electromigration, electroosmosis, electrophoresis and, in the specific case of electrodialytic (ED) process, also electrodialysis. In ED process, ion exchange membranes are placed as separators between the processing solutions surrounding the electrodes and the matrix to be remediated/upgraded. By having a main transport of ions out of the porous media, ED may enhance separation significantly. The electrode reactions will conduct to the production of H^+ ions at the anode and OH^- ions at the cathode changing pH in these compartments, but the pH of the central compartment should not change significantly due to the unidirectional flux of ions towards the electrode compartments.

The recovery of phosphorus using EK/ED process is based on the principle that phosphate exists as anionic species unless the pH is strongly acid. Phosphorus has four speciation species corresponding to acidity constants of 2.12, 7.2, and 12 (p*K*a, 298 K), according to (6.1)–(6.3).

$$H_3PO_{4(aq)} + H_2O_{(l)} \rightleftharpoons H_3O^+{}_{(aq)} + H_2PO_4^-{}_{(aq)} \quad pK_{a1} = 2.12 \tag{6.1}$$

$$H_2PO_4^-{}_{(aq)} + H_2O_{(l)} \rightleftharpoons H_3O^+{}_{(aq)} + HPO_4^{2-}{}_{(aq)} \quad pK_{a2} = 7.21 \tag{6.2}$$

$$HPO_4^{2-}{}_{(aq)} + H_2O_{(l)} \rightleftharpoons H_3O^+{}_{(aq)} + PO_4^{3-}{}_{(aq)} \quad pK_{a3} = 12.67 \tag{6.3}$$

When phosphoric acid is present, there is no electric charge and phosphorus is unaffected by the electric field. At higher pH, ionic species predominates. As so, in the presence of an electric current, phosphate, and other anions selectively accumulate in the anolyte.

EK and ED processes had been used and proved to be successful for the recovery of phosphorus from other matrices such as sewage sludge (Guedes et al. 2015) and sewage sludge ashes (Guedes et al. 2014; Ottosen et al. 2014a).

6.2 Case Study

6.2.1 *Aims and Scope*

Water is considered the foundation of sustainable development as it is the common denominator of all global challenges. For this reason, it becomes necessary to implement integrated and safe sequences in WTP that promote water treatment while searching for phosphorus recovery. The study focused on the development of alternatives to recover from slurries compounds with value as fertilizer together with the separation or removal of unwanted compounds, so they disposal costs could be diminished, opening new possibilities for waste minimization.

A process to recover phosphorus from NF concentrate was tested using ED process.

6.2.2 *Experimental*

6.2.2.1 Production of Membrane Concentrate

Two Portuguese Dam reservoirs Amoreiras (101.2 km^2 and 11 hm^3, Alentejo) and Funcho (200 km^2 and 42.8 hm^3, Algarve) were used as NF feed water. Microcystins, extracted from cultures of *M. aeruginosa* grown in laboratory (supplied by Pasteur Culture Collection), were added to the surface water (10 μg/L) since no occurrence of cyanobacterial blooms was reported during the experimental period. Potassium dihydrogen phosphate (2.2 mg/L) of KH_2PO_4 was also added. The production of membrane concentrate was carried out using a plate-and-frame unit (Lab-unit M20) (Ribau Teixeira and Rosa 2005) and a thin film composite NF/RO membrane, NF90 (DowFilmtec), at 10 bar.

6.2.2.2 ED Experiments

The experiments were carried out in laboratory cells, running experiments with different conditions. The design in Fig. 6.2a (Ribeiro 1998) is divided into three

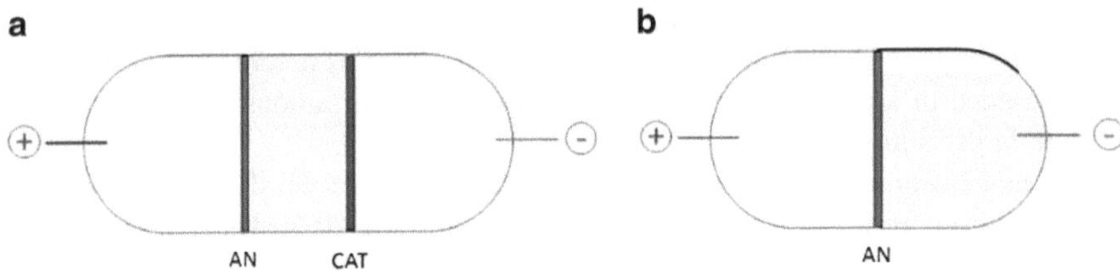

Fig. 6.2 Electrodialytic cell designs used: (**a**) three compartments, 3C and (**b**) two compartments, 2C (adapted from Couto et al. 2015)

compartments (3C cell), consisting of two electrode compartments (each one with an electrode) and a central one, in which the membrane concentrate was placed. The three compartments were separated by ion exchange membranes (cation exchange membrane, CAT, IC1-61CZL386, and anion exchange membrane, AN, IA1-204SXZL386, both from Ionics Inc., Massachusetts, USA). The design in Fig. 6.2b (Ottosen et al. 2014b) is divided in two (electrode) compartments (2C cell), separated by an anion exchange membrane (AN similar to the previous one). Membrane concentrate was placed in the cathode compartment. Platinized titanium bars (3 mm diameter and 5 cm length; Bergsøe Anti Corrosion A/S, Denmark) were used as electrodes. Electrolyte solution (10^{-2} M $NaNO_3$; pH 7) was circulated in a closed circulation system. A power supply (Hewlett Packard E3612A) was used to maintain a constant DC current.

Control experiments with no current applied and to assess MC-LR photodegradation, and electrodegradation were carried out.

6.2.2.3 Analytical Methodologies

Phosphorus was analyzed by ICP-OES at 178 nm. MC-LR extraction was carried out according to the method described elsewhere (Ribau Teixeira and Rosa 2005) and followed the operating procedure developed by Meriluoto and Spoof (2005). MC-LR was determined by HPLC-PDA and chromatograms were analyzed between 180 and 900 nm, with a main detection at 238 nm to the typical absorption spectra of microcystins.

6.2.3 Discussion

6.2.3.1 Membrane Concentrates

The concentrate from NF presented cyanotoxins (MC-LR), phosphorus, and the water background inorganic and organic matrix (NOM). It was found that phosphorus removal using NF is high, assuring a high-quality water regarding this parameter (Santos et al. 2014). Table 6.1 summarizes the characteristics of treated some of the produced membrane concentrates (Santos et al. 2014).

Table 6.1 Characteristics of some of the produced membrane concentrates (adapted from Couto et al. 2013, 2015)

Memb conc.	Dam		Number	Cond (μS/cm)	pH	Turbidity (NTU)	DOC (mg C/L)	UV 254 nm (1 cm^{-1})	P (μg/L)	MC-LR (μg/L)
1	Funcho	w/	1	1476	8.3	188	32.0	0.36	1494.6	–
2		w/	2	1276	8.3	146	26.2	0.31	1244.5	42.9
3		w/o	3	2630	8.4	30.0	21.0	0.39	1429.1	–

w/o without occurrence of precipitation, *w/* with occurrence of precipitation

Other membrane concentrates were produced and the details as well as the characteristics can be seen elsewhere (Couto et al. 2013, 2015; Santos et al. 2014). High removals of cyanotoxins were achieved by NF and the quality of the permeate was guaranteed. Good removal efficiency of phosphorus had also been reported using NF elsewhere (Niewersch et al. 2010; Leo et al. 2011; Blöcher et al. 2012). In fact, this process could be combined with other technologies to recover phosphorus like, for example, low pressure wet oxidation for sewage sludge decomposition and phosphorus dissolution plus NF to separate phosphorus from heavy metals (Blöcher et al. 2012).

6.2.3.2 Suitability of ED Process to Recover Phosphorus and Decrease MC-LR Levels

The duration of ED experiments were related to the applied current intensity and the characteristics of membrane concentrate (Table 6.2). Figure 6.3 shows the percentage of phosphorus recovery at the anode compartment.

Table 6.2 Characteristics and duration of ED experiments

Exp.	Conc. number	Current (mA)	Duration (h)
A	1	10	7
B	1	20	5
C	2	10	8
D	3	20	6.5

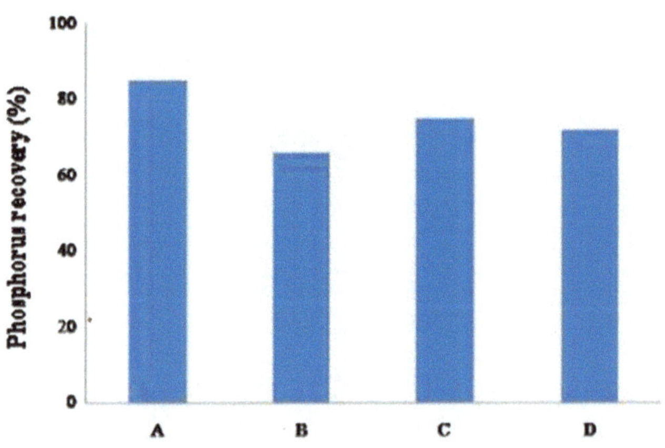

Fig. 6.3 Percentage of phosphorus recovered in the anode compartment in different experiments (adapted from Couto et al. 2013, 2015)

In the presence of a 3C cell design, the pH after ED process remained similar (around 8) or slightly decreased as anion exchange membranes are not perfect rectifiers, allowing the passage of some H^+ ions. Also, water splitting at the anion exchange membrane may also have contributed to the supply of H^+ to the central compartment (Simons 1979, 1984; Ottosen et al. 2000). In the 2C cell design pH of membrane concentrate was up to 12. Under pH between 7 and 9, the forms $H_2PO_4^-$ (97 g/mol and a Stokes radius of 0.256 nm) and HPO_4^{2-} (96 g/mol; 0.323 nm) are present, with an increase of HPO_4^{2-} fraction with higher pH of membrane concentrates. When in its anionic forms, phosphorus electromigrate from the membrane concentrate (either in central or in cathode compartment) towards the anode compartment (Couto et al. 2013, 2015).

The phosphorus migration was observed after a lag period in both cases (Couto et al. 2013, 2015). In general, in the tested conditions almost no movement of phosphorus was observed during the first hours (Couto et al. 2013, 2015). The observed lag time period was related to the time needed to promote the movement of charged particles and thus the time required for the appearance of phosphorus in the anode compartment. In 6.5 h ca. 72 % of phosphorus (Couto et al. 2013) was found in the anolyte. The behavior of phosphorus migration may help a cost-benefit analysis of the best time to perform the nutrient recovery, according to, e.g., the characteristics of membrane concentrates (e.g., conductivity) (Couto et al. 2015).

In all experiments with the presence of electric field, the conductivity in the central compartment decreased (in comparison with the initial value) due to the movement of ions towards the electrode compartments (Couto et al. 2013, 2015). In electrode compartments the conductivity increased, namely in the anode compartment due to the amount of phosphate ions and/or other species like carbonates, sulfates, chlorides, etc. Phosphorus migration towards the anode compartment had a maximum of 84 %. In most cases, migration was around 70–80 % but was also low such as 32 % (Couto et al. 2013, 2015). The different removals were mainly dependent on the applied current, membrane concentrate characteristics, and time to perform ED process. NF concentrates with similar characteristics presented comparable amount of phosphorus recovery. But membrane concentrates with lower conductivity accelerated phosphorus mobilization as they most likely have less "competing species" to phosphorus ions thus benefiting its migration (Couto et al. 2015). When no current was applied, phosphorus mobilization towards the electrode compartments was very low.

In all cases, as the ED process was applied without conditioning, electrolysis reactions at the electrodes generated an acidic medium (H^+) at the anode compartment and an alkaline medium (OH^-) at the cathode compartment. The use of ion exchange membranes allowed the migration of anions to the anode compartment and cations to the cathode compartment. The severe pH decrease at the anode compartment puts phosphorus in the forms $H_3PO_4/H_2PO_4^-$.

ED process also showed potential to decrease MC-LR concentrations in the NF concentrate (Couto et al. 2015). The pH of membrane concentrates at the beginning of the experiments (approx. 8) puts MC-LR in the form of a negatively charged ion that can migrate towards the anode compartment, where it may degrade.

In the absence of an electric field MC-LR migrated, probably by diffusion, to the lateral cell compartments, either in 2C or 3C cell designs. In the electrodegradation control carried out with a different cell design (beakers and saline bridge), MC-LR concentration was negligible in all compartments and hydrogen peroxide was detected (10 mg/L). Indirect photodegradation also appears as a mechanism to decrease MC-LR levels, as observed in the photodegradation control.

Summing-up, in the tested conditions, the ED process showed potential to produce a clean phosphorus product while promoting a significant decrease in MC-LR concentrations in the NF concentrate. In some cases, this value was below the WHO guideline for MC-LR.

6.3 Conclusions and Final Remarks

NF treatment has already been applied to water treatment. This process produces a concentrate stream that can be dangerous if it contains contaminants, such as microcystins. However, this stream could also be a valuable phosphorus source that cannot be directly applied in agricultural soils due to the presence on unwanted compounds. Consequently, phosphorus has to be removed for further reuse. For this, NF coupled to ED can be a feasible option to recover phosphorus as here described. The presence of electric field also promotes the movement/degradation of MC-LR, decreasing the dangerousness of the concentrate slurry.

This work demonstrated that, in tested condition, ED recovered up to 84 % of phosphorus and reduced MC-LR content, mainly through its electromigration towards the anode compartment where it may degrade. Still, more tests are needed to achieve optimum conditions for phosphorus recovery and MC-LR degradation, namely to scale-up the process that may start to be part of water treatment plants. Other strategies should also be taken into account balancing parameters such as life cycle costs, energy consumption, availability, resource use and pollution, in order to make the process economically viable.

The results of this study may be used worldwide in water treatment. The removal of contaminants from water and slurries is a transversal and broad approach to increasing environment quality and consequently human health.

Acknowledgments Financial support for the work was provided by projects FP7-PEOPLE-2010-IRSES-269289-*ELECTROACROSS—Electrokinetics across disciplines and continents: an integrated approach to finding new strategies for sustainable development* and PTDC/ECM/111860/2009—*Electrokinetic treatment of sewage sludge and membrane concentrate: Phosphorus recovery and dewatering*. Authors also thank RIARTAS-Red Iberoamericana de Aprovechamiento de Residuos Industriales para el Tratamiento de Suelos y Aguas Contaminadas, Programa Iberoamericano de Ciencia y Tecnología para el Desarrollo (Cyted) and N. Couto acknowledges *Fundação para a Ciência e a Tecnologia* for her Post-Doc fellowship (SFRH/BPD/81122/2011).

References

Antoniou MG, Shoemaker JA, de la Cruz AA, Dionysiou DD (2008) LC/MS/MS structure elucidation of reaction intermediates formed during the TiO_2 photocatalysis of microcystin-LR. Toxicon 51:1103–1118

Blöcher C, Niewersch C, Melin T (2012) Phosphorus recovery from sewage sludge with a hybrid process of low pressure wet oxidation and nanofiltration. Water Res 46:2009–2019

Burns J, Cooper W, Ferry J, King D, DiMento B, McNeill K, Miller C, Miller W, Peake B, Rusak S, Rose A, Waite T (2012) Methods for reactive oxygen species (ROS) detection in aqueous environments. Aquat Sci 74:683–734

Campinas M, Rosa M (2010) Removal of microcystins by PAC/UF. Sep Purif Technol 71:114–120

Carmichael WW (1994) The toxins of cyanobacteria. Sci Am 270:78–86

Cathie Lee W, Mah S-K, Leo C, Wu T, Chai S-P (2014) Performance studies of phosphorus removal using cross-flow nanofiltration. Desalin Water Treat 52:5974–5982

Chang J, Chen Z-l, Wang Z, Shen J, Chen Q, Kang J, Yang L, Liu X-W, Nie C-X (2014) Ozonation degradation of microcystin-LR in aqueous solution: Intermediates, byproducts and pathways. Water Res 63:52–61

Chen W, Li L, Gan N, Song L (2006a) Optimization of an effective extraction procedure for the analysis of microcystins in soils and lake sediments. Environ Pollut 143:241–246

Chen W, Song L, Gan N, Li L (2006b) Sorption, degradation and mobility of microcystins in Chinese agriculture soils: risk assessment for groundwater protection. Environ Pollut 144:752–758

Corbel S, Mougin C, Bouaïcha N (2014) Cyanobacterial toxins: modes of actions, fate in aquatic and soil ecosystems, phytotoxicity and bioaccumulation in agricultural crops. Chemosphere 96:1–15

Cordell D, Drangert J-O, White S (2009) The story of phosphorus: global food security and food for thought. Glob Environ Chang 19:292–305

Couto N, Guedes P, Mateus EP, Santos C, Ribau Teixeira M, Nunes LM, Hansen HK, Gutierrez C, Ottosen LM, Ribeiro AB (2013) Phosphorus recovery from a water reservoir-potential of nanofiltration coupled to electrodialytic process. Waste Biomass Valoriz 4:675–681

Couto N, Guedes P, Ferreira AR, Ribau Teixeira M, Mateus EP, Ribeiro A (2015) Electrodialytic process of nanofiltration concentrates—phosphorus recovery and microcystins removal. Electrochim Acta, in press, doi:10.1016/j.electacta.2015.04.081

Dixon M, Falconet C, Ho L, Chow C, O'Neill B, Newcombe G (2010) Nanofiltration for the removal of algal metabolites and the effects of fouling. Water Sci Technol 61:1189–1199

Fang Y, Zhang Y, Ma W, Johnson DM, Huang Y-P (2014) Degradation of microcystin-LR in water: hydrolysis of peptide bonds catalyzed by maghemite under visible light. Appl Catal B Environ 160–161:597–605

Fotiou T, Triantis TM, Kaloudis T, Hiskia A (2015) Evaluation of the photocatalytic activity of TiO2 based catalysts for the degradation and mineralization of cyanobacterial toxins and water off-odor compounds under UV-A, solar and visible light. Chem Eng J 261:17–26

Fraga LE, Anderson MA, Beatriz MLPMA, Paschoal FMM, Romão LP, Zanoni MVB (2009) Evaluation of the photoelectrocatalytic method for oxidizing chloride and simultaneous removal of microcystin toxins in surface waters. Electrochim Acta 54:2069–2076

Gijsbertsen-Abrahamse AJ, Schmidt W, Chorus I, Heijman SGJ (2006) Removal of cyanotoxins by ultrafiltration and nanofiltration. J Membr Sci 276:252–259

Guedes P, Couto N, Ottosen LM, Ribeiro AB (2014) Phosphorus recovery from sewage sludge ash through an electrodialytic process. Waste Manag 34:886–892

Guedes P, Magro C, Couto N, Mosca A, Mateus EP, Ribeiro AB (2015) Potential of the electrodialytic process for emerging organic contaminants remediation and phosphorus separation from sewage sludge. Electrochim Acta, in press, doi:10.1016/j.electacta.2015.03.167

Harada K, Ogawa K, Matsuura K, Murata H, Suzuki M, Watanabe M, Itezono Y, Nakayama N (1990) Structural determination of geometrical isomers of microcystins LR and RR from cyanobacteria by two-dimensional NMR spectroscopic techniques. Chem Res Toxicol 3:473–481

Hayakawa K, Sugiyama Y (2008) Spatial and seasonal variations in attenuation of solar ultraviolet radiation in lake Biwa, Japan. J Photochem Photobiol B 97:121–133

Hooda P, Truesdale V, Edwards A, Withers P, Aitken M, Miller A, Rendell A (2001) Manuring and fertilization effects on phosphorus accumulation in soils and potential environmental implications. Adv Environ Res 5:13–21

Jacobs LCV, Peralta-Zamora P, Campos FR, Pontarolo R (2013) Photocatalytic degradation of microcystin-LR in aqueous solutions. Chemosphere 90:1552–1557

Lee J, Walker H (2008) Mechanisms and factors influencing the removal of microcystin-LR by ultrafiltration membranes. J Membr Sci 320:240–247

Leo CP, Chai WK, Mohammad AW, Qi Y, Hoedley AF, Chai SP (2011) Phosphorus removal using nanofiltration membranes. Water Sci Technol 64:199–205

Liao W, Zhang Y, Zhang M, Murugananthan M, Yoshihara S (2013) Photoelectrocatalytic degradation of microcystin-LR using Ag/AgCl/TiO_2 nanotube arrays electrode under visible light irradiation. Chem Eng J 231:455–463

Liao W, Murugananthan M, Zhang Y (2014) Electrochemical degradation and mechanistic analysis of microcystin-LR at boron-doped diamond electrode. Chem Eng J 243:117–126

Liu G, Qian Y, Dai S, Feng N (2008) Adsorption of microcystin LR and LW on suspended particulate matter (SPM) at different pH. Water Air Soil Pollut 192:67–76

Loo S-L, Fane AG, Krantz WB, Lim T-T (2012) Emergency water supply: a review of potential technologies and selection criteria. Water Res 46:3125–3151

Maagd P, Hendriks A, Seinen W, Sijm D (1999) pH dependent hydrophobicity of the cyanobacteria toxin microcystin-LR. Water Res 33:677–680

Meriluoto J (1997) Chromatography of microcystins. Anal Chim Acta 352:277–298

Meriluoto J, Spoof L (2005) SOP: solid phase extraction of microcystins in water samples. Abo Akademi University, Finland

Mulder M (1997) Basic principles of membrane technology, 2nd edn. Kluwer Academic, Dordrecht

Nghiem L, Hawkes S (2007) Effects of membrane fouling on the nanofiltration of pharmaceutically active compounds (PhACs): mechanisms and role of membrane pore size. Sep Purif Technol 57:176–184

Niewersch C, Koh C, Wintgens T, Melin T, Schaum C, Cornel P (2008) Potentials of using nanofiltration to recover phosphorus from sewage sludge. Water Sci Technol 57:707–714

Niewersch C, Meier K, Wintgens T, Melin T (2010) Selectivity of polyamide nanofiltration membranes for cations and phosphoric acid. Desalination 250:1021–1024

Ottosen LM, Hansen HK, Hansen CB (2000) Water splitting at ion-exchange membranes and potential differences in soil during electrodialytic soil remediation. J Appl Electrochem 30:1199–1207

Ottosen LM, Jensen PE, Kirkelund GM (2014a) Electrodialytic separation of phosphorus and heavy metals from two types of sewage sludge ash. Sep Sci Technol 49:1910–1920

Ottosen LM, Jensen PE, Kirkelund GM, Ebbers B (2014b) Electrodialytic recovery and purification of phosphorous from sewage sludge ash, sewage sludge and wastewater. Patent Version Number PCT/EP2014/068956

Ribau Teixeira M, Rosa MJ (2005) Microcystins removal by nanofiltration membranes. Sep Purif Technol 46:192–201

Ribau Teixeira M, Rosa MJ (2006) Neurotoxic and hepatotoxic cyanotoxins removal by nanofiltration. Water Res 40:2837–2846

Ribeiro AB (1998) Use of electrodialytic remediation technique for removal of selected heavy metals and metalloids from soils. PhD thesis, Technical University of Denmark, Denmark

Robertson P, Lawton L, Cornish B (1999) The involvement of phycocyanin pigment in the photodecomposition of the cyanobacterial toxin, microcystin-LR. J Porphyrins Phthalocyanines 3:544–551
Santos C, Ribeiro A, Ribau Teixeira M (2014) Phosphorus recovery from waters using nanofiltration. Desal Water Treat 1–8. http://dx.doi.org/10.1080/19443994.2014.925831
Sharma VK, Triantis TM, Antoniou MG, He X, Pelaez M, Han C, Song W, O'Shea KE, de la Cruz AA, Kaloudis T, Hiskia A, Dionysiou DD (2012) Destruction of microcystins by conventional and advanced oxidation processes: a review. Sep Purif Technol 91:3–17
Simons R (1979) The origin and elimination of water splitting in ion exchange membranes during water demineralisation by electrodialysis. Desalination 28:41–42
Simons R (1984) Electric field effects on proton transfer between ionizable groups and water in ion exchange membranes. Electrochim Acta 29:151–158
Sivonen K, Jones G (1999) Cyanobacterial toxins. In: Chorus I, Bartram J (eds) Toxic cyanobacteria in water: a guide to their public health consequences, monitoring, and management. E&FN Spon, London, pp 41–111
SRAM, Secretaria Regional do Ambiente e do (2011) Plano de Gestão da Região Hidrográfica dos Açores. Technical report (in Portuguese)
Steen P (1998) Phosphorus availability in the 21st century. Management of a nonrenewable resource. Phosphorus Potassium 217:25–31
Teixeira M, Rosa M (2012) How does the adsorption of microcystins and anatoxin-a on nanofiltration membranes depend on their co-existence and on the water background matrix. Water Sci Technol 66:976–982
Thirumavalavan M, Hu Y-L, Lee J-F (2012) Effects of humic acid and suspended soils on adsorption and photo-degradation of microcystin-LR onto samples from Taiwan reservoirs and rivers. J Hazard Mater 217–218:323–329
Tran N, Drogui P (2013) Electrochemical removal of microcystin-LR from aqueous solution in the presence of natural organic pollutants. J Environ Manage 114:253–260
Tsuji K, Naito S, Kondo F, Ishikawa N, Watanabe M, Suzuki M, Harada K-I (1994) Stability of microcystins from cyanobacteria: effect of light on decomposition and isomerization. Environ Sci Technol 28:173–177
Ueno Y, Nagata S, Tsutsumi T, Hasegawa A, Watanabe M, Park HD, Chen G-C, Chen G, Yu SZ (1996) Detection of microcystins, a blue-green algal hepatotoxin, in drinking water sampled in haimen and fusui, endemic areas of primary liver cancer in china, by highly sensitive immunoassay. Carcinogenesis 17:1317–1321
Van Hege K, Verhaege M, Verstraete W (2004) Electro-oxidative abatement of low salinity reverse osmosis membrane concentrates. Water Res 38:1550–1558
Verliefde A, Cornelissen E, Heijmana S, Petrinic I, Luxbacher T, Amy G, Van der Bruggen B, van Dijk J (2009) Influence of membrane fouling by (pretreated) surface water on rejection of pharmaceutically active compounds (PhACs) by nanofiltration membranes. J Membr Sci 330:90–103
Wang X, Utsumi M, Yang Y, Shimizu K, Li D, Zhang Z, Sugiura N (2013) Removal of microcystins (-LR, -YR, -RR) by highly efficient photocatalyst Ag/Ag_3PO_4 under simulated solar light condition. Chem Eng J 230:172–179
Welker M, Steinberg C (2000) Rates of humic substance photosensitized degradation of microcystin-LR in natural waters. Environ Sci Technol 34:3415–3419
Wenk J, von Gunten U, Canonica S (2011) Effect of dissolved organic matter on the transformation of contaminants induced by excited triplet states and the hydroxyl radical. Environ Sci Technol 45:1334–1340
WHO (2004) Guidelines for drinking water quality, vol 1, 3rd edn. World Health Organization, Geneva
Wörmer L, Huerta-Fontela M, Cirés S, Carrasco D, Quesada A (2010) Natural photodegradation of the cyanobacterial toxins microcystin and cylindrospermopsin. Environ Sci Technol 44:3002–3007

Yan S, Zhang D, Song W (2014) Mechanistic considerations of photosensitized transformation of microcystin-LR (cyanobacterial toxin) in aqueous environments. Environ Pollut 193:111–118

Yli-Halla M, Hartikainen H, Ekholm P, Turtola E, Puustinen M, Kallio K (1995) Assessment of soluble phosphorus load in surface runoff by soil analyses. Agr Ecosyst Environ 56:53–62

Yoshizawa S, Matsushima R, Watanabe M, Harada K-I, Ichihara A, Carmichael W, Fujiki H (1990) Inhibition of protein phosphatases by microcystis and nodularin associated with hepatotoxicity. J Cancer Res Clin Oncol 116:609–614

Chapter 7
Electrochemical Process for Phosphorus Recovery from Wastewater Treatment Plants

Paula R. Guedes, Nazaré Couto, Eduardo P. Mateus, and Alexandra B. Ribeiro

7.1 Introduction

7.1.1 Problem Statement

Current population growth rates require an increased supply of staple foods and to guarantee it, a sufficient nutrient level of agricultural soils needs to be maintained by application of soil fertilizers. One indispensable nutrient for plant growth is phosphorus (P) and phosphate rock, its primary source, is becoming progressively limited (Cordell et al. 2011). Currently, 178.5 Mt of phosphate rock, equivalent to 23 Mt of P, is being mined every year, of which 90 % is used for food production as fertilizer (approx. 82 %) and feed additives (approx. 7 %) (IFA 2011). Phosphorus peak has been estimated to occur by 2035 (Cordell et al. 2009) and published data on the lifetime of the exploitable high quality reserves of phosphate rock vary to a great extent, between one hundred and several hundreds of years (EFM 2000; IFDC 2010; Survey USG 2012).

The European Union (EU) is almost entirely dependent upon phosphate imports, with China, Jordan, Morocco, South Africa, and the USA controlling 85 % of global phosphate reserves (Smit et al. 2009). As a consequence, EU is vulnerable to geopolitical tensions in the countries that export phosphate and to its volatile prices (as demonstrated during the 800 % spike in the price of phosphate rock in 2008)

P.R. Guedes (✉) • N. Couto
CENSE, Departamento de Ciências e Engenharia do Ambiente, Faculdade de Ciências e Tecnologia, Universidade Nova de Lisboa, Caparica 2829-516, Portugal

Key Laboratory of Soil Environment and Pollution Remediation, Institute of Soil Science, Chinese Academy of Sciences, Nanjing 210008, China
e-mail: p.guedes@campus.fct.unl.pt

E.P. Mateus • A.B. Ribeiro
CENSE, Departamento de Ciências e Engenharia do Ambiente, Faculdade de Ciências e Tecnologia, Universidade Nova de Lisboa, Caparica 2829-516, Portugal

© Springer International Publishing Switzerland 2016
A.B. Ribeiro et al. (eds.), *Electrokinetics Across Disciplines and Continents*,
DOI 10.1007/978-3-319-20179-5_7

(Schröder et al. 2010). This makes the development new strategies for P recovery for further reuse one of the world challenges.

One possible secondary P source are the wastes generated at Wastewater Treatment Plants (WWTP). A recent study has reported that, if collected, the P available from human urine and feces could account for 22 % of the global P fertilizer demand (Mihelcic et al. 2011).

There are already a variety of very different approaches (see reviews Cordell et al. 2011; Rittmann et al. 2011; Sartorius et al. 2011) which differ by the origin of the used matter (wastewater, sewage sludge, and its ashes) and the type of process (e.g., precipitation, wet chemical extraction, and thermal treatment). The electrokinetic (EK) process can also be an effective technique for P recovery from these waste streams.

This chapter provides an overview on the WWTPs treatments and the potential of EK process for P recovery from sewage sludge and its ashes. Simultaneous contaminants removal will also be discussed.

7.1.2 Wastewater Treatment Plants

The water and sewage industry is at a transformation point. Nowadays, there is a challenge to develop and extend conventional centralized water and sewage systems on the existing planning parameters (Mitchell et al. 2011). Factors including climate change impacts, changing of hydrological conditions, population growth, resource scarcity, aging infrastructure, economic constraints in financing large scale systems, and changing expectations for water quality (e.g., European Union water directives) contribute for need of the plants to adapt to these uncertainties. This approach is based on the idea that an environmentally, economically, and socially sustainable sanitation system requires sewage to be viewed as a set of resources to be recovered, recycled, and reused (water, energy, nutrients) rather than a waste product to be treated to successively higher standards before release to the environment (Mitchell et al. 2011).

In terms of P, globally, it has been estimated that there are 0.3–1.5 million metric tons reused annually. From this, 3–3.3 million metric tons of P is generated in human excreta (i.e., feces and urine) and greywater (Liu et al. 2008; Cordell et al. 2009). For example, in Australia it has been estimated that annually 8000 metric tons of P are discharged with human excreta, and 40–50 % of the P that reach a treatment plant is applied to agricultural soils as biosolids (Cordell and White 2009). In Europe, the potential production of P from urine has been estimated in 0.3 kg P per person per year (Lienert et al. 2003). The available P from human excreta is reported to also be split near equally between rural and urban areas (i.e., 1.6 million metric tons excreted in urban environments, 1.7 million metric tons excreted in rural environments) (Liu et al. 2008). This split is likely to become more urban in the future because an estimated 70 % of the world's population is expected to reside in cities by the year 2050 (UNFPA 2007). If collected, the P available from human excreta could account for 22 % of the global P fertilizer demand (Mihelcic et al. 2011).

At the global scale, it is estimated that only 10–50 % of P in human excreta is recycled to agricultural soils (Liu et al. 2008; Cordell et al. 2009) leading to large P losses through wastewater or municipal solid waste. Many studies have reported low efficiencies in the current waste management systems in recovering and recycling P from waste (Antikainen et al. 2005; Kalmykova et al. 2012). It can be concluded that progress in this domain would help to achieve a better closed-loop P cycle and to meet future P demand (Cordell et al. 2011).

7.1.3 Phosphorus Recovery Potential

Nowadays, concepts to recover P within the wastewater treatment scheme are particularly attractive, if they promise to obtain a product which is free from contaminants and with a high quality as fertilizer (Blöcher et al. 2012).

The effluent of WWTPs has a large quantity of phosphate (Carpenter and Bennett 2011) and, to limit eutrophication potential of wastewaters, P removal has been in the toolbox of wastewater-treatment engineers for several decades (Rittmann et al. 2011). Several methods were developed in order to remove P from wastewater (e.g., chemical or biological P precipitation). But these approaches do not recycle it as a truly sustainable product as it is removed from liquid phase together with various other waste products like organic chemicals (Bright and Healey 2003; Clarke and Smith 2011), metals (Basta et al. 2005) or even pathogens (Gerba and Smith 2005; Elliott and O'Connor 2007; Sidhu and Toze 2009).

In a common WWTP, there are several potential locations for P recovery from the liquid phase: the effluent of the WWTP, the supernatant liquor from side-stream treatment, and the sludge liquor (Cornel and Schaum 2009). The theoretical recovery potential from liquid phase in common activated sludge plants is limited to <55 % (Fig. 7.1).

In WWTPs without P removal, 90–95 % of the incoming P load is contained in the sewage sludge (Cornel and Schaum 2009; Blöcher et al. 2012). The use of biosolids (treated sewage sludge) is approved by Member States in several situations, e.g., short-rotation plantations, plantations for growing energy crops (European Commission 2010). As there is an agronomic interest for nutrients or for the improvement of the content of organic matter in soil, biosolids can be used when the legislation values for organics and heavy metals are met. Still, concerns have been raised about municipal biosolids applications, composed of waste streams from residential, commercial, and industrial sources, which increase the probability of including potential chemicals, solvents, and pharmaceuticals in the crops. The presence of pathogens can also be problematic in terms of land contamination, and it was already shown that their concentrations increase over time in biosolids, even in samples that indicated no measureable pathogens after the application of destruction methods (Gibbs et al. 1997). High concentrations of heavy metals in biosolids are also of concern, as they can contaminate and reduce the productivity of land used for disposal by reducing the bacterial diversity within

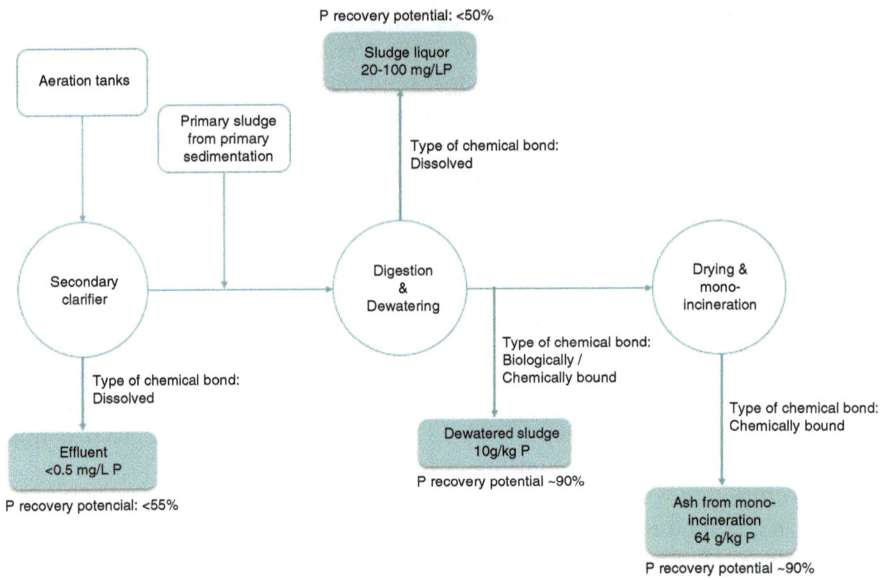

Fig. 7.1 Different possibilities for phosphorus recovery at a wastewater treatment plant, typical concentrations of phosphorus and the recovery potential (recovery potential related to the concentration present in the influent)

the soil (Moffett et al. 2003) among contamination. Besides the above-mentioned factors, direct use of biosolids in agriculture may be difficult due to public perception, odors or difficulties of transport and storage. To avoid direct use of treated sewage sludge as fertilizer, a side recovery technology may be the answer and there are several potential applications aiming P recovery from sewage sludge, i.e., primary, excess and raw sludge, stabilized sludge before and after dewatering (Cornel and Schaum 2009). In these cases, the theoretical recovery potential is significantly higher than with separation processes from the aqueous phase, approx. more 40 % (Fig. 7.1).

Another option is P recovery from ashes. These ashes are generally considered a waste material to be disposed into landfill (Donatello et al. 2010) but it can also be reused as adsorbents (Pan et al. 2003), in geotechnical applications or in construction materials (Al Sayed et al. 1995; Anderson et al. 2002; Lin et al. 2005). However, none of the above applications make use of the valuable P resource in the ashes. The application of ash in soils is still a controversial subject as it raises questions of toxicity that must be carefully addressed (Ferreira et al. 2003), prior to its application, due to the high levels of heavy metals and salts present in the ash. Major concerns are generally about the heavy metals, which can accumulate in the soil over time, and enter the food chain or groundwater systems. However, the risk for this occurring depends primarily on metal concentrations in ash, their antecedent concentrations in the soil, their mobility from the ash and subsequently from the soil, and their uptake by plants.

Phosphorus recovery from sewage sludge ash (SSA) first involves the re-dissolution of the bound P followed by its separation, e.g., electrokinetic process followed by precipitation processes. The advantage of treating SSA is their exclusively inorganic formation, a fact which, in contrast to sewage sludge, facilitates P recovery. Furthermore, the incineration of organic matter causes the enrichment of P in the ashes (Cornel and Schaum 2009). In this case, P recovery potential from the ashes is the same as from the sewage sludge, approx. 90 % (Fig. 7.1).

Taking into consideration that the increased utilization of secondary resources is an important issue in the European Union countries waste strategies, it makes sense to search for the upgrade of these waste matrices while recycling P. This offers the immediate advantage of avoiding the environmental impacts associated with primary production from phosphate rock, complemented with the reuse of an essential nutrient that is being wasted. For these reasons, the technologies that promote P reuse within the wastewater treatment scheme are particularly attractive. In contrast to fertilizers, quality standards are much higher to industries, requiring further purification of the product to a level where all or most of the trace impurities are removed.

7.2 Electrokinetic P recovery

ED is an extraction technology which selectively separates anions and cations across an ion-exchange membrane driven by an applied electrical field between electrodes. Cationic species move towards the cathode passing through cation-exchange membranes (CEM) which allow only positively charged species to pass through while rejecting negatively charged species. Anions (e.g., PO_4^{3-}) move towards the anode passing through anion-exchange membranes (AEM) which allow only negatively charged species to pass through while rejecting positively charged species. Through this process, cations and anions are obtained separately in concentrated solutions (Mehta et al. 2015).

In the case of P recovery through electromigration, four speciation states, namely H_3PO_4, H_2PO_4, HPO_4^{2-}, and PO_4^{3-} need to be considered. The corresponding acidity constants (pKa, 298 K) are 2.12, 7.2, and 12. At pH below 2, phosphate dominantly exists as phosphoric acid and has no electric charge. Thus, it is unaffected by the electric field. Under alkaline conditions (pH 7), the bi- and trivalent species prevail, which implies a double or triple amount of energy to move one phosphate ion (Sturm et al. 2010).

7.2.1 Sewage Sludge Ashes

As previously referred, and contrary to most heavy metals, phosphate in the pore water exists as anionic species unless the pH is strongly acidic. Thus, when a direct current is applied to water saturated SSA, phosphate, among other anions may selectively accumulate in the anolyte. The pore water pH may influence the efficacy of an electrokinetic phosphate recovery from SSA in multiple ways. The dissolution of phosphate minerals is increased under acidic conditions due to the pH-dependent solubility of P compounds. The pH also determines the affinity of dissolved phosphate towards variable charge surfaces and the transport velocity in an electric field is influenced by ionic charge and thereby dependent on the chemical speciation.

Acid extraction using HNO_3 and H_2SO_4 for P recovery has been compared for two different SSA from mono-incineration, where either Al or Fe originally was used for precipitation P from the wastewater (Ottosen et al. 2013). Sulfuric acid is the cheapest mineral ash and is thus often suggested for P extraction from SSA. The use of this acid though caused the formation of a high quantity of gypsum crystals in the remaining ash. The gypsum must be taken into account when handling the remaining ash in, e.g., construction materials or as an increased volume to be deposited (Ottosen et al. 2013). The Al rich ash contained activated carbon and required significantly more acid for full extraction of P meaning that in order to obtain the lowest acid consumption for P recovery from SSA, it is preferable to use Fe instead of Al for precipitation of P at the WWTP (Ottosen et al. 2013).

Although the heavy metal concentrations in the two ashes were low, simultaneously to the release of P by acidification, a fraction of the heavy metals was also released. In order to produce a high grade P-product from the extracted solution, a separation step is thus needed (Ottosen et al. 2013).

Phosphorus extraction from SSA using an electrokinetic process in a packed bed has been tested and the investigation showed that it was possible to concentrate on a small part of the P in processing solutions at the anodes, but also that the setup used was not feasible from the point of energy demand (Sturm et al. 2010). One option to improve the efficiency is to combine ED process with acid extraction by suspending the ash in acid just prior to the treatment (Ottosen et al. 2014). Ideally during the process, P is concentrated in the processing solution in the anode compartment (anolyte) and the heavy metals in the processing solution in the cathode compartment (catolyte).

An experimental screening of EDS P recovery and simultaneous heavy metals removal has been conducted with a 3 compartment (3C) cell, Fig. 7.2a (Ebbers et al. 2015; Guedes et al. 2014; Ottosen et al. 2014) and 2 compartment (2C) cell, Fig. 7.2b (Ebbers et al.).

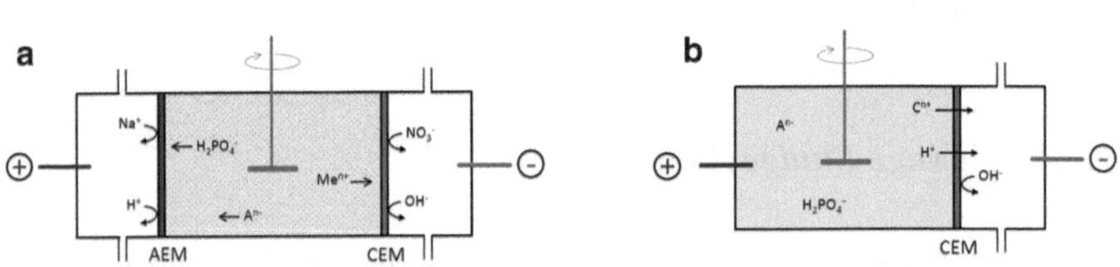

Fig. 7.2 Principle in electrodialytic separation (EDS) using (**a**) three compartment cell design and (**b**) two compartment cell (*AEM* anion-exchange membrane, *CEM* cation-exchange membrane)

Ottosen et al. (2014) conducted experiments with two different ashes: rich in Fe or Al. In EDS experiments with the most Al rich ash, only a minor part of P was transported into the anolyte and the major part stayed in solution of the ash suspension likely due to the formation of uncharged species between P and Al. At a high acid addition (11.4 mol H^+/kg ash), P was transported equally into the anolyte and catholyte and thus no separation was obtained in the Al rich ash. At lower acid additions the separation was better, but here the major part of P was in the ash suspension (in the ash or dissolved) and separation was not obtained during the week the experiments lasted. In the Fe rich ash, the separation was better as Fe was mainly present in insoluble particles, speciation between Fe and P in the ash suspension was not a complicating factor as in the case of Al. The results for the Fe-rich ash showed that it was possible to separate P into one processing solution, heavy metals (Cu, Zn, Ni, Pb) into another, keeping the ash suspended in a third solution (that still contained P after 1 week of EDS).

Guedes et al. (2014) also applied the EDS to SSA aiming at P recovery. Two SSA rich in Fe were sampled, one immediately after incineration and the other from an open deposit. After 14 days, P had been mainly mobilized to the anolyte (between 60 and 70 %), whereas heavy metals mainly electromigrated towards the cathode end. Still, during the EDS process, other elements present in the ash may also move towards the anode end (e.g., metals of negative standard reduction potential). At the end of the experiments, the anolyte presented a composition of 98 % of P, mainly as orthophosphate, and 2 % of heavy metals.

Ebbers et al. (2015) made adjustments to the 3C ED cell setup by reducing the number of compartments and introducing the anode directly into the ash suspension (Fig. 7.2b). This new approach uses the acid produced by electrolysis at the anode combined with initial acidification of the suspension by H_2SO_4 to promote a faster mobilization of the P and heavy metals. Mobilized heavy metals will then electromigrate from the suspension liquid and concentrate into the cathode compartment.

The combination of ED in the 2C setup and initial acidification of the stirred suspension with H_2SO_4 was more effective in dissolving P and separating the heavy metals. In this setup, up to 96 % of the P in the ash was dissolved after 7 days (Ebbers et al. 2015). Using the 3C setup and initially suspending the ash in distilled water, resulted in 53 % dissolution of the total recovered P after 7 days (Ebbers et al. 2015). This shows that the 2C setup is a good approach to avoid the use of mineral acids as it will avoid the precipitation of secondary minerals such as gypsum (Ebbers et al. 2015).

In terms of heavy metals, the release in the 2C and 3C cell using H_2SO_4 was significantly larger than the release observed in 2C and 3C with distilled water (Ebbers et al. 2015). This indicated that acidification of the ash suspension through a mineral acid at the start of the experiment (in this case, 2 h) was more important for dissolution of these heavy metals than acidification through half reactions at the anode during early stages of the experiments. The amount of heavy metals dissolved remained constant after 2 h for up to 14 days (Ebbers et al. 2015).

7.2.1.1 Case Study

For this study, SSA were collected at Lynettefællesskabet, Copenhagen, Denmark, in June 2012. This plant mono-incinerates sewage sludge from about 500,000 PE. In this WWTP, the process of P precipitation is initially done in a Bio-P tank followed by the addition of iron salt. The sludge is incinerated in a fluid bed oven at about 800 °C. After incineration, the SSA is stored (until it finds use) in an open air deposit at the site of Lynettefællesskabet close to the sea. Two SSAs were sampled for this investigation: one immediately after the incineration process (SA) and the other from the deposit (SB). The ashes were then subjected to EDS separation using a 2C cell design and H_2SO_4 (0.08 M) for 7 days to compare the results with the previous conducted study were a 3C cell was used (Guedes et al. 2014). All analytical methodologies and SSA characterization can be seen in Guedes et al. (2014).

The obtained results (Table 7.1) showed that similar total removals were achieved after 7 days for the SB ashes between the 2C and 3C setup, approx. 70 %. After filtrating the ash suspension, the liquid phase of the 2C setup contained 56 % of the P and for the 3C setup the liquid phase contained 50 % of P. The remaining P in the 3C was collected in the anolyte (19 %) whereas in the 2C it migrated towards the cathode (14 %).

The type of ash influenced the total P removal by the different setups. In this case, the use of the 2C setup allowed to remove 88 % of the P from the SSA whereas in the 3C setup, only 62 % were extracted. A major difference was also found in the distribution of P within the cell as the amount collected in the anolyte in the 2C setup was higher when SA ash was used (77 % vs. 56 % for the SB ash). In the 3C setup, the anolyte contained 4 % vs. 19 % for the SA and SB ash, respectively. In this study, the storage of the ash influenced P removal from these ashes.

In terms of heavy metals, in general, higher removals were achieved using the 3C cell setup comparing to the 2C setup (Table 7.2). For example, 71 % of Cd was removed from the new ashes comparing to the 18 % removal achieved with the 2C setup. However, 32 % of the mobilized Cd is in the liquid phase of the central compartment where 54 % of the recovered P is also present. In the 2C setup, only 8 % of Cd was detected in the anolyte where 77 % of the recovered P is present. The same was observed for the SB ashes although with different percentages.

Table 7.1 Percentage of phosphorus determined in the different cell compartment subjecting the sewage sludge ashes to electrodialytic separation for 7 days

	Recently collected ashes (SA)				Deposited ashes (SB)			
Exp.	Anolyte[a]	Central[b]	Catholyte	Ash	Anolyte[a]	Central[b]	Catholyte	Ash
2C	77	–	11	12	56	–	14	30
3C[c]	4	54	4	38	19	50	1	30

[a]Liquid in the anode compartment: electrolyte solution in the 3C cell; liquid phase (acid) in the 2C cell

[b]Liquid phase (acid) in the 3C cell

[c]Results reported in Guedes et al. (2014)

Table 7.2 Percentage of heavy metals determined in the different cell compartment subjecting the sewage sludge ashes to electrodialytic separation for 7 days

		Recently collected ashes (SA)							
Exp.	Compartment	Al	Cd	Cr	Cu	Fe	Ni	Pb	Zn
3C[a]	Anolyte[b]	1	3	3	1	0	1	0	1
	Central[c]	34	32	13	23	5	6	1	13
	Catholyte	12	36	2	45	0	19	1	49
2C	Anolyte[b]	35	8	9	0	9	0	4	0
	Catholyte	18	10	1	55	1	16	3	64
		Deposited ashes (SB)							
3C[a]	Anolyte[b]	0	1	1	1	0	2	0	1
	Central[c]	32	24	14	14	5	5	1	10
	Catholyte	12	38	1	6	0	19	1	58
2C	Anolyte[b]	25	10	10	0	10	0	4	0
	Catholyte	17	13	2	44	1	17	5	51

[a]Results reported in Guedes et al. (2014)
[b]Liquid in the anode compartment: electrolyte solution in the 3C cell; liquid phase (acid) in the 2C cell
[c]Liquid phase (acid) in the 3C cell

The use of a 2C EDS setup increased overall efficiency of P recovery after 7 days of experiment, under the conditions tested here. Dissolution of P was faster in the 2C experiments than in the 3C experiments. However, relative to the total amount of heavy metals released from the ash, 3C setup shows better results.

7.2.2 Sewage Sludge Case Study

The advantage of treating SSA is their exclusively inorganic forms, a fact which in contrast to sewage sludge facilitates P recovery (Cornel and Schaum 2009). Recently, the group conducted a series of experiments aiming P recovery from sewage sludge (Guedes et al. 2015). For that, sewage sludge collected at a WWTP from Simarsul located in Quinta do Conde, Sesimbra, Portugal. The samples were collected from the secondary settling tank in January 2014.

The samples were then subjected to EDS in a 2C setup. In total, three designs were tested: (a) sludge placed in the anode compartment and separated by a CEM; (b) sludge placed in the cathode compartment and separated by a CEM; (c) sludge placed in the cathode compartment and separated by an AEM.

Overall, as observed in Fig. 7.3, the most promising results were achieved for the cell where the sludge was directly placed in the cathode compartment with an AEM and 50 mA applied. In this experiment, P was mobilized towards the anode end due to the high pH achieved in the cathode compartment (pH 13), and after 5 days, 78 % of the P was recovered (Guedes et al. 2015). However, when investigating P recovery from sewage sludge, a special attention must be made to all the

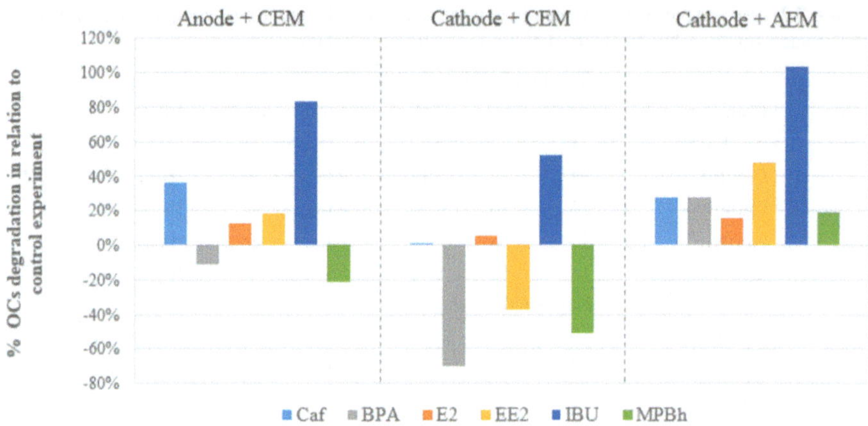

Fig. 7.3 Percentage of organic contaminants degraded in the electrodialytic separation experiments comparing to the control experiment

contaminants present in the matrix. In this case, the study also assessed the removal of six emerging organic contaminants (OCs), caffeine (CAF), ibuprofen (Ibu), 17β-oestradiol (E2), 17α-ethinyloestradiol (EE2), bisphenol A (BPA), and oxybenzone (MBPh). In general, ED showed potential to enhance the degradation of the contaminants being the process mainly dependent on the intensity of the applied DC field and the cell design which highly influences the pH changes. Comparing to the control experiment (0 mA), the design that presented the best results for P recovery was also the most promising design for increasing OCs removal (Guedes et al. 2015). In this design, all compound degradations were increased comparing to the control experiments. In terms of percentage, Ibu presented the highest degradation, e.g., 100 % whereas in the control experiment no degradation was detected for this compound.

Still, in this study, the removal of heavy metals was not assessed, and it should also be considered a topic for further research. Taking into consideration the results obtained with the ashes, it is expected that the heavy metals will be mainly detected in the cathode compartment. When the sludge is placed in the cathode compartment the pH increased to 13, and in this case most heavy metals will be precipitated in the sludge. Some exceptions can occur, namely with metals like As, that are also mobilized at high pHs.

7.3 Conclusions

The development of a technology for P removal offers the opportunity for recycling P, making its use more sustainable. There are a number of technologies, both established and under development, which can be used to remove P from wastewater and can potentially be used within a sustainability strategy. The major aim in using the electrodialytic process to recover P from the waste matrices is the possibility to separate it from contaminants. This overview shows that, in the case of the SSA, the separation is highly dependent on whether P was precipitated with Fe- or Al-based chemicals at the WWTP. In terms of cell design, the new 2C cell setup seems promising. This setup may help to increase the overall efficiency of the EDS. Introducing the anode into the suspension can substitute the acidification that is usually carried out by mineral acids and effectively help in the removal and concentration of most of heavy metals during this process. In terms of P recovery from the sewage sludge, the results are still scarce. The preliminary study conducted by the team showed that EDS may also be an attractive method for P recovery from the sludge especially as it may also improve the degradation of several emerging organic contaminants.

Acknowledgements Financial support was provided by PTDC/ECM/111860/2009—*Electrokinetic treatment of sewage sludge* and *membrane concentrate: Phosphorus recovery and dewatering and* FP7-PEOPLE-2010-IRSES-269289-*ELECTROACROSS—Electrokinetics across disciplines and continents: an integrated approach to finding new strategies for sustainable development.* RIARTAS-Red Iberoamericana de Aprovechamiento de Residuos Industriales para el Tratamiento de Suelos y Aguas Contaminadas, Programa Iberoamericano de Ciencia y Tecnología para el Desarrollo (Cyted). N. Couto acknowledges Fundação para a Ciência e a Tecnologia for her Post-Doc fellowship (SFRH/BPD/81122/2011).

References

Al Sayed MH, Madany IM, Buali ARM (1995) Use of sewage sludge ash in asphaltic paving mixes in hot regions. Construct Build Mater 9:19–23

Anderson M, Elliott M, Hickson C (2002) Factory-scale proving trials using combined mixtures of three by-product wastes (including incinerated sewage sludge ash) in clay building bricks. J Chem Technol Biotechnol 77:345–351

Antikainen R, Lemola R, Nousiainen JI, Sokka L, Esala M, Huhtanen P, Rekolainen S (2005) Stocks and flows of nitrogen and phosphorus in the Finnish food production and consumption system. Agr Ecosyst Environ 107:287–305

Basta NT, Ryan JA, Chaney RL (2005) Trace element chemistry in residual-treated soil: key concepts and metal bioavailability. J Environ Qual 34(1):49–63

Blöcher C, Niewersch C, Melin T (2012) Phosphorus recovery from sewage sludge with a hybrid process of low pressure wet oxidation and nanofiltration. Water Res 46:2009–2019

Bright DA, Healey N (2003) Contaminant risks from biosolids land application: contemporary organic contaminant levels in digested sewage sludge from five treatment plants in Greater Vancouver, British Columbia. Environ Pollut 126:39–49

Carpenter SR, Bennett EM (2011) Reconsideration of the planetary boundary for phosphorus. Environ Res Lett 6(6):014009. doi:10.1088/1748-9326/6/1/014009

Clarke BO, Smith SR (2011) Review of 'emerging' organic contaminants in biosolids and assessment of international research priorities for the agricultural use of biosolids. Environ Int 37:226 247

Cordell D, White S (2009) The Australian story of phosphorus: sustainability implications of global phosphate scarcity for a net food-producing nation (in review). Food Policy 23

Cordell D, Drangert J-O, White S (2009) The story of phosphorus: global food security and food for thought. Glob Environ Chang 19:292–305
Cordell D, Rosemarin A, Schröder JJ, Smit AL (2011) Towards global phosphorus security: a systems framework for phosphorus recovery and reuse options. Chemosphere 84:747–758
Cornel P, Schaum C (2009) Phosphorus recovery from wastewater: needs, technologies and costs. Water Sci Technol 59:1069–1076
Donatello S, Tyrer M, Cheeseman CR (2010) EU landfill waste acceptance criteria and EU hazardous waste directive compliance testing of incinerated sewage sludge ash. Waste Manag 30:63–71
Ebbers B, Ottosen LM, Jensen PE (2015) Comparison of two different electrodialytic cells for separation of phosphorus and heavy metals from sewage sludge ash. Chemosphere 125:122–129
EFM (2000) Phosphorus essential element for food production. European Fertilizer Manufacturers Association, Brussels
Elliott HA, O'Connor GA (2007) Phosphorus management for sustainable biosolids recycling in the United States. Soil Biol Biochem 39:1318–1327
European Commission (2010). Sewage sludge. Environment, European Commission, Brussels. http://ec.europa.eu/environment/waste/sludge/. Accessed 22 Mar 2015
Ferreira C, Ribeiro A, Ottosen L (2003) Possible applications for municipal solid waste fly ash. J Hazard Mater 96:201–216
Gerba CP, Smith JE (2005) Sources of pathogenic microorganisms and their fate during land application of wastes. J Environ Qual 34:42–48
Gibbs RA, Hu CJ, Ho GE, Unkovich I (1997) Regrowth of faecal coliforms and salmonellae in stored biosolids and soil amended with biosolids. Water Sci Technol 35:269–275
Guedes P, Couto N, Ottosen LM, Ribeiro AB (2014) Phosphorus recovery from sewage sludge ash through an electrodialytic process. Waste Manag 34:886–892
Guedes P, Magro C, Couto N, Mosca A, Mateus EP, Ribeiro A (2015) Potential of the electrodialytic process for emerging organic contaminants remediation and phosphorus separation from sewage sludge. Electrochim Acta (in press). http://dx.doi.org/10.1016/j.electacta.2015.03.167
IFA (2011) Feeding the earth—global phosphate rock production trends from 1961 to 2010: reasons for the temporary set-back in 1988–1994. International Fertilizer Industry Association, Paris
IFDC (2010) Sufficient phosphate rock resources available for years. IFDC report 35
Kalmykova Y, Harder R, Borgestedt H, Svanäng I (2012) Pathways and management of phosphorus in urban areas. J Ind Ecol 16:928–939
Lienert J, Haller M, Berner A, Stauffacher M, Larsen TA (2003) How farmers in Switzerland perceive fertilizers from recycled anthropogenic nutrients (urine). Water Sci Technol 48:47–56
Lin D-F, Luo H-L, Sheen Y-N (2005) Glazed tiles manufactured from incinerated sewage sludge ash and clay. J Air Waste Manag Assoc 55:163–172
Liu Y, Villalba G, Ayres RU, Schroder H (2008) Global phosphorus flows and environmental impacts from a consumption perspective. J Ind Ecol 12:229–247
Mehta CM, Khunjar WO, Nguyen V, Tait S, Batstone DJ (2015) Technologies to recover nutrients from waste streams: a critical review. Crit Rev Env Sci Technol 45:385–427
Mihelcic JR, Fry LM, Shaw R (2011) Global potential of phosphorus recovery from human urine and feces. Chemosphere 84:832–839
Mitchell C, Fam D, Cordell D (2011) Effectively managing the transition towards restorative futures in the sewage industry: a phosphorus case study. IWA, London
Moffett BF, Nicholson FA, Uwakwe NC, Chambers BJ, Harris JA, Hill TCJ (2003) Zinc contamination decreases the bacterial diversity of agricultural soil. FEMS Microbiol Ecol 43:13–19
Ottosen LM, Kirkelund GM, Jensen PE (2013) Extracting phosphorous from incinerated sewage sludge ash rich in iron or aluminum. Chemosphere 91:963–969
Ottosen LM, Jensen PE, Kirkelund GM (2014) Electrodialytic separation of phosphorus and heavy metals from two types of sewage sludge ash. Sep Sci Technol 49:1910–1920

Pan S-C, Lin C-C, Tseng D-H (2003) Reusing sewage sludge ash as adsorbent for copper removal from wastewater. Resour Conserv Recycl 39:79–90

Rittmann BE, Mayer B, Westerhoff P, Edwards M (2011) Capturing the lost phosphorus. Chemosphere 84:846–853

Sartorius C, Horn J, Tettenborn F (2011) Phosphorus recovery from wastewater—state-of-the-art and future potential. In: Water Environment Federation IWA (ed) International Recovery and Management, Miami, FL, USA

Schröder JJ, Cordell D, Smit AL, Rosemarin A (2010) Sustainable use of phosphorus, EU Tender ENV.B.1/ETU/2009/0025

Sidhu JPS, Toze SG (2009) Human pathogens and their indicators in biosolids: a literature review. Environ Int 35:187–201

Smit AL, Bindraban PS, Schröder JJ, Conjin JG, Van der Meer HG (2009) Phosphorus in agriculture: global resources, trends and developments. Plant Research International B.V., Wageningen Report 282, The Netherlands

Sturm G, Weigand H, Marb C, Weiß W, Huwe B (2010) Electrokinetic phosphorus recovery from packed beds of sewage sludge ash: yield and energy demand. J Appl Electrochem 40:1069–1078

Survey USG (2012) Mineral commodity summaries—phosphate rock

UNFPA (2007) State of world population, 2007: unleashing the potential of urban growth. United Nations Population Fund

Pan S-C, Lin C-C, Tseng D-H (2003) Reusing sewage sludge ash as adsorbent for copper removal from wastewater. Resour Conserv Recycl 39:79–90

[illegible] (2011) [illegible] phosphorus. [illegible]

[illegible]

[illegible]

[illegible]

[illegible]

[illegible]

Chapter 8
Electrokinetic Remediation of Copper Mine Tailings: Evaluating Different Alternatives for the Electric Field

Henrik K. Hansen, Adrián Rojo, Claudia Gutiérrez, Pernille E. Jensen, and Lisbeth M. Ottosen

8.1 Introduction

Chile has the largest mineral reserves of copper and about 32 % of the world production of this metal (COCHILCO 2008), from which 80 % is obtained by grinding and flotation processes. Annually, this mineral processing generates large quantities of waste which are transported and disposed as pulps forms in conditioned sites called mine tailing ponds. In these places, mine tailings not only have a damaging effect on water resources by the natural leaching of heavy metals and chemicals, but also generate effects on flora, fauna, and air quality by the generation of fugitive emissions of fine particles (Dold and Fontbote 2002; Hansen et al. 2005a).

Electrokinetic remediation (EKR) is a technology used to remove contaminants from soils (Lageman 1993; Acar and Alshawabkeh 1993; Probstein and Hicks 1993). In recent years, the technology has been of research interest for stabilizing mine tailings from the copper industry (Hansen et al. 2005b; Rojo et al. 2006). Due to the characteristics of the mine tailing minerals, EKR treatment with conventional DC electric fields had that inconveniency that a concentration polarization was buildup with time. This increased remediation times and power consumption due to the higher electrical resistance (Hansen et al. 2012; Li et al. 2013).

One way to avoid or reduce the effect of concentration polarization is to change the electric field in fashion that would allow mass transfer processes in the solid matrix during EKR to occur more efficient. The electric field enhancements include use of: (a) pulsed electric DC fields, (b) low frequency sinusoidal electric fields, and

H.K. Hansen (✉) • A. Rojo • C. Gutiérrez
Departamento de Ingeniería Química y Ambiental, Universidad Técnica Federico Santa Maria, Avenida España 1680, Valparaíso, Chile
e-mail: henrik.hansen@usm.cl

P.E. Jensen • L.M. Ottosen
Department of Civil Engineering, Technical University of Denmark, B118, Kgs. Lyngby 2800, Denmark

© Springer International Publishing Switzerland 2016
A.B. Ribeiro et al. (eds.), *Electrokinetics Across Disciplines and Continents*,
DOI 10.1007/978-3-319-20179-5_8

(c) high frequency electric fields. This work gives experiences and an evaluation of advantages and disadvantages of the different options.

8.2 Difference in Mine Tailings Characteristics vs. Remediation Results

This work shows results and experiences obtained during years of study of how the electric field influences the copper removal from copper mine tailings. The characteristics of the tailings were different taking the overall period of study into account but the same tailings were used when analyzing one type of applied electric field. The difference are due to the fact that (a) the mine tailings were sampled at different times, and (b) the mine tailings were sampled at two different tailings dumps. In addition, in some experimental series ion-exchange membranes are used (EDR) (Hansen et al. 1997)—in others only nylon mesh/filter paper (EKR)—as separators between the tailings and the electrolytes. Therefore for all experimental EKR series the results should be compared to a conventional DC remediation experiment, where a constant current or voltage drop is applied to the mine tailing sample without any modification in the electric field. This baseline experiment is done for all experimental series and any modification in the electric field should be compared to this experiment when analyzing if improvement is obtained in the copper removal. So it is difficult to compare directly results between different modifications of the electric field. In conclusion, in this book chapter the effect of the modification of the electric field should be taken as a guideline of tendencies when using the modifications suggested in this work.

8.2.1 Mine Tailings

Several mine tailings have been used in this work. Here the tailings are described and numbered, and in the following the EKR results are referred to that number.

1. The mine tailings from the Cauquenes impoundment at Codelco-El Teniente copper mine in VI Region of Chile.
2. Mine tailings sampled directly from the canal, which transports the tailings to the Caren impoundment from the El Teniente copper mine in the VI region of Chile.
3. The mine tailings from the Caren impoundment at Codelco-El Teniente copper mine in VI Region of Chile. As an example of composition, Table 8.1 gives characteristics of these mine tailings, determined by X-Ray Diffraction Analysis.

Table 8.1 Mineral composition in the mine tailings sample from Caren impoundment at Codelco-El Teniente copper mine in VI Region of Chile

Quartz	SiO_2
Muscovite	$KAl_2(Si_3Al)O_{10}(OH,F)_2$
Ferric clinochlore	$(Mg,Fe)_6(Si,Al)_4O_{10}(OH)_8$
Calcic albite	$(Na,Ca)Al(Si,Al)_3O_8$
Anorthite	$CaAl_2Si_2O_8$
Hydrated calcium sulfate	$CaSO_4{\cdot}0.6H_2O$
Chalcocite	Cu_2S
Brochantite	$Cu_4SO_4(OH)_6$
Chalcopyrite	$CuFeS_2$
Ramsbeckite	$Cu_{15}(SO_4)_4(OH)_{22}{\cdot}6H_2O$
Wroewolfeite	$Cu_4(SO_4)(OH)_6{\cdot}2H_2O$
Guildite	$CuFe(SO_4)_2(OH){\cdot}4H_2O$

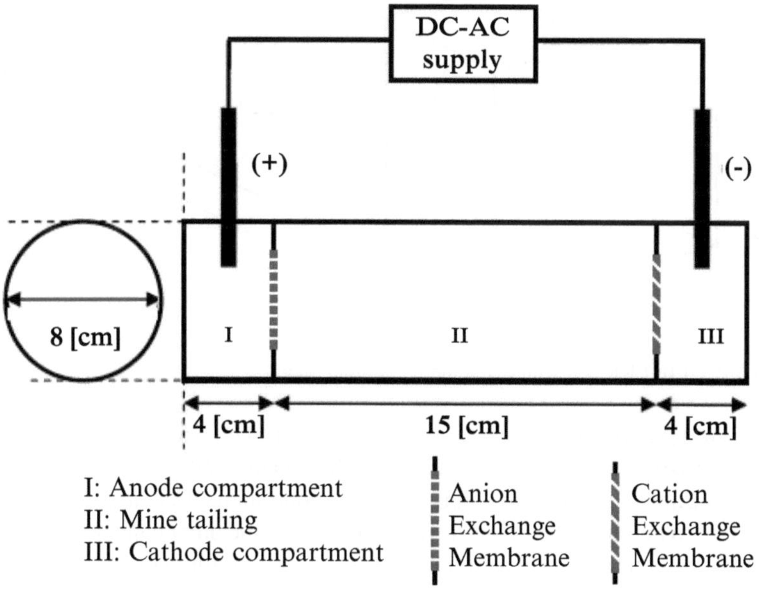

Fig. 8.1 Cell A: EKR cell with ion-exchange membranes

8.2.2 *EKR Cells*

Two experimental cells have been used in this work. The main difference between the cells is the use of ion-exchange membranes (Cell A) or nylon mesh/filter paper (Cell B) as separation between the mine tailings and electrolytes. Figures 8.1 and 8.2 show the principles in Cell A and Cell B, respectively.

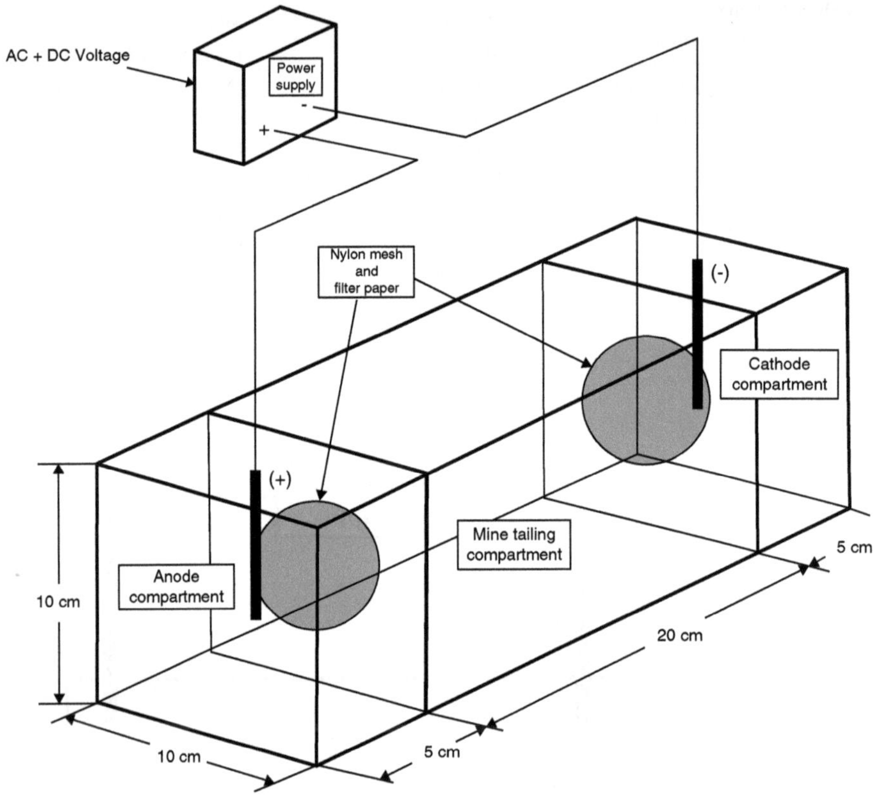

Fig. 8.2 Cell B: EKR cell with nylon mesh and filter paper

The pre-treated mine tailings were placed in central compartment. Initially, in the anode compartment, the electrolyte was distilled water, and in the cathode compartment was dilute sulfuric acid solution, later a continuous drop addition of concentrated acid to maintain pH below 4 was supplied. On the other hand, to control pH in the cathode compartment, a sample was taken each day for pH monitoring.

After the experiments were carried out, mine tailings sample were segmented into three slices of equal size, where copper concentration was measured. The anode zone is defined as the slice closest to the anode, center zone the slice in the middle, and cathode zone the slice closest to the cathode.

8.2.3 Analytical and Tailings Preparation

In this work, the main parameters when evaluating the effect of modifications in the electric field are copper removal and pH in the tailings during EKR. For all experimental series, the same analytical procedure for copper and pH measurements was used.

The total and soluble copper was determined according to the following methods. In both cases, the analysis was done in triplicate, and an average was used.

The total copper content of the tailings was determined by adding 20 mL 1:1 HNO_3 to 1.0 g of dry material and treating the sample in autoclave, according to the Danish Standard DS 259:2003 (30 min at 200 kPa (120 °C)). The liquid was separated from the solid particles by vacuum through a 0.45 μm filter and diluted to 100.0 mL. The metal content was determined by AAS in flame.

The soluble copper content of the tailings was determined by adding 50 mL H_2SO_4 5 % (v/v) to 5.0 g of dry material, and stirring the sample in an 250 mL Erlenmeyer flask for 30 min. The liquid was separated from the solid particles by vacuum through a 0.45 μm filter and diluted to 100.0 mL by adding 10 mL concentrated HCl and distilled water. The metal content was determined by AAS in flame.

pH was measured by mixing 5.0 g dry matter and 25.0 mL distilled water. After 1 h of contact time, pH was measured using a pH electrode.

Before remediation experiments, the tailings were stove dried for 2 days at 70 °C. Once dried, the material was pulverized in a mortar and sieved with meshes #4 and #20, until a homogeneous sample was obtained. For all experiments, 1 M sulfuric acid was added to the tailings until an average humidity of 20 % was reached.

8.2.4 Modifications of the Electric Field

In conventional EKR of copper mine tailings, where a constant voltage or constant current applied to a humid solid waste product, some inconvenient side effects are generated. First of all, a concentration polarization is built up with time due to the general characteristics of the tailings. The result is an increase in the electric resistance as can been graphically in Fig. 8.3. This would mean either longer

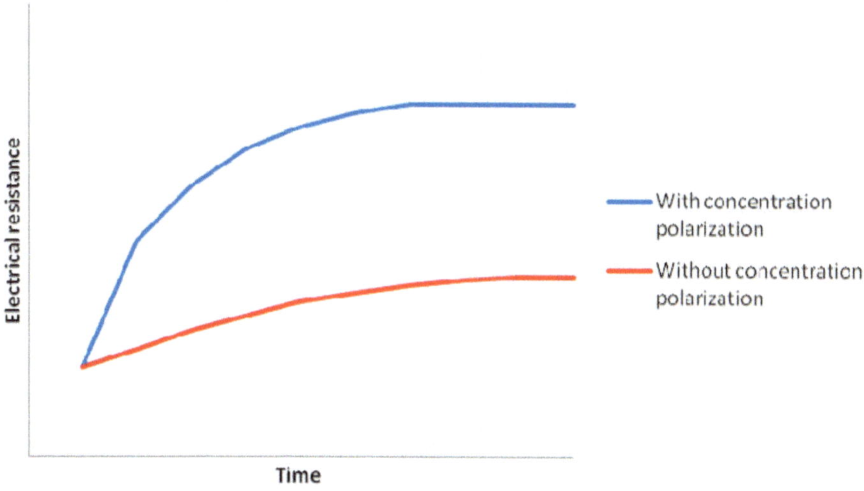

Fig. 8.3 Typical development of electrical resistance during EKR with and without concentration polarization

remediation times (when working at constant voltage) or higher power consumption (when working at constant current).

The main goal is therefore to decrease the electrical resistance over the solid waste product during EKR by avoiding the concentration polarization buildup. This can be done by giving the waste product "some time to relax," which in practice means that the applied constant voltage or current should "be broken" by manipulating the electric field. The manner of breaking the constant electric field investigated in this work is either by:

1. Applying a pulsed electric DC field
2. Applying a sinusoidal electric field at low frequencies
3. Applying a sinusoidal electric field at high frequencies

8.2.5 *Pulsed Electric DC Fields*

The idea of pulsed electric DC fields maintains a constant voltage or current for a determined time period (T_{on}) and then turns off the power for a short time (T_{off}); this allows the process to have a short period where the concentration polarization can be reduced. Figure 8.4 shows the voltage/time behavior when applying pulsed electric field. The duration of one cycle (T_{cycle}) is defined as $T_{on}+T_{off}$, and the frequency is given as number of cycles in a given period of time (f. ex. cycles per day).

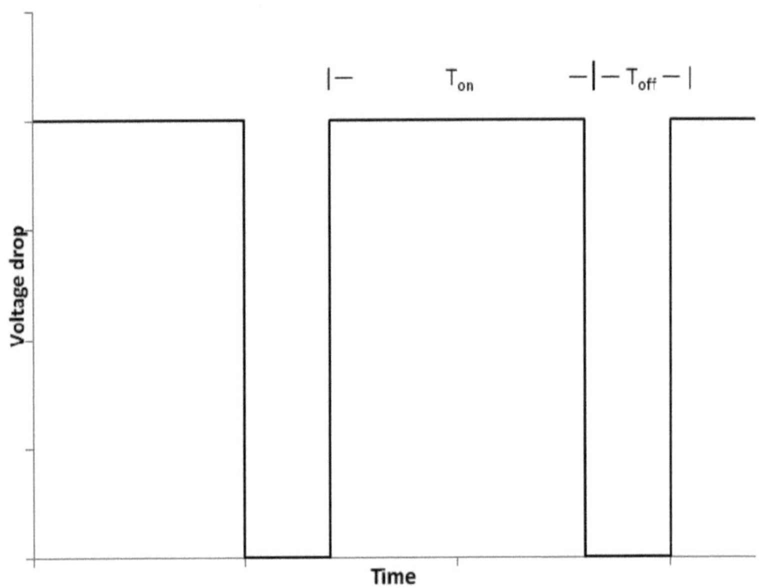

Fig. 8.4 Voltage vs. time when applying pulsed electric fields

During electroremediation processes with DC electric fields, the transport phenomena of ionic migration and electroosmosis do not favor the dissolution of ionic species. In this case, the chemical mechanisms of dissolution are slow, due to the fact that the electric field generates a negative effect on diffusion gradients of the leaching reagents (e.g., H^+ or Fe^{3+}). Using pulsed electric fields, during the time with current "OFF," the diffusion gradients (produced during the time in "ON") can be diminished. So, the chemical mechanisms of dissolution can progress and the species concentration in solution increases. In electroremediation, it is expected that the reduction of the polarization improves the efficiency of heavy metal removal and reduces the time of the process.

In the publication of Hansen and Rojo (2007), the effect of applying pulsed electric fields to copper mine tailings were first analyzed. From this work, the main results are discussed in the following. Eleven EDR experiments were conducted in both cell A (exp. 1–7, mine tailings type 2) and cell B (exp. 8–11, mine tailings type 2) with the conditions and results given in Table 8.2. Here, the initial concentration (C_0) and final concentrations (C_f) for the anode side, middle, and cathode side are given. In all experiments, copper removal from the anode side, middle, and cathode side represents the difference between the copper leaving and entering the section. Only at the anode side, the amount of copper entering is always zero. pH in the tailings after remediation was around 4 (as the initial value) in all sections and experiments.

Figure 8.5 shows the normalized concentration of copper for EDR experiments with time in the anode side. It can be seen from the figure that for 72 h of remediation time the use of pulsed electric fields improves the copper removal. With 14 cycles/day, the copper removal would be equivalent to a continuous EDR time in the order of 120 h. Increasing the frequency to 28 or 55 cycles/day, the equivalent EDR time, in terms of copper removal, would correspond to around 270 h. In the case of EKR results (see Table 8.2), for 120 h remediation the copper removal is independent of frequency and even when doubling time, the copper removal is practically the same for both frequencies.

In addition, Fig. 8.6 shows the normalized concentration of copper in the anode side as a function of frequencies in cycles/day for the EDR experiments as a result of the pulsed electric field applied to the cell. The continuous electric field experiment (experiment 1) is fixed to 0 cycles/day. This is explained by: DC remediation is equivalent to pulsed electric field remediation where $T_{on} \rightarrow \infty$, so $T_{cycle} \rightarrow \infty$, and consequently since frequency $f \propto 1/T_{cycle}$, $f \rightarrow 0$.

With EKR and pulsed electric fields, the same improvement could be expected with respect to DC remediation, and the copper removal seems to be independent of frequency (in the range of 14–55 cycles/day) and remediation time (120–240 h).

With both EDR and EKR of 120 h, for 14 cycles/day, the copper removal is practically uniform in all three sections of the cell. Increasing the frequency ≥ 28 cycles/day, the copper removal in the anode side is improved, and in the cathode side an accumulation of copper is observed. With an EKR of 240 h, the obtained results are nearly independent of the frequency, observing a major accumulation in the cathode side.

Table 8.2 Electrodialytic remediation conditions and initial (C_0) and final (C_f) copper concentrations (mg kg^{-1}) in mine tailings

					Anode side		Middle section		Cathode side	
Exp.	Remediation process	Duration (h)	Frequency (cycles/day)	C_0	C_f	C_f/C_0	C_f	C_f/C_0	C_f	C_f/C_0
1	EDR	72	–	1130	938	0.83	1074	0.95	1086	0.96
2	EDR	96	–	1130	893	0.79	1097	0.97	1130	1.00
3	EDR	196	–	1130	870	0.77	1029	0.91	1052	0.93
4	EDR	504	–	1130	531	0.47	713	0.63	836	0.74
5	EDR	72	14	1130	904	0.80	926	0.82	904	0.80
6	EDR	72	28	1130	758	0.67	916	0.81	1086	0.96
7	EDR	72	55	1130	734	0.65	904	0.80	1096	0.97
8	EKR	120	14	1760	1300	0.74	1510	0.86	1300	0.74
9	EKR	120	55	1720	1230	0.72	1470	0.85	1900	1.10
10	EKR	240	14	1670	1160	0.69	1600	0.96	1870	1.12
11	EKR	240	55	1620	1210	0.75	1430	0.88	1850	1.14

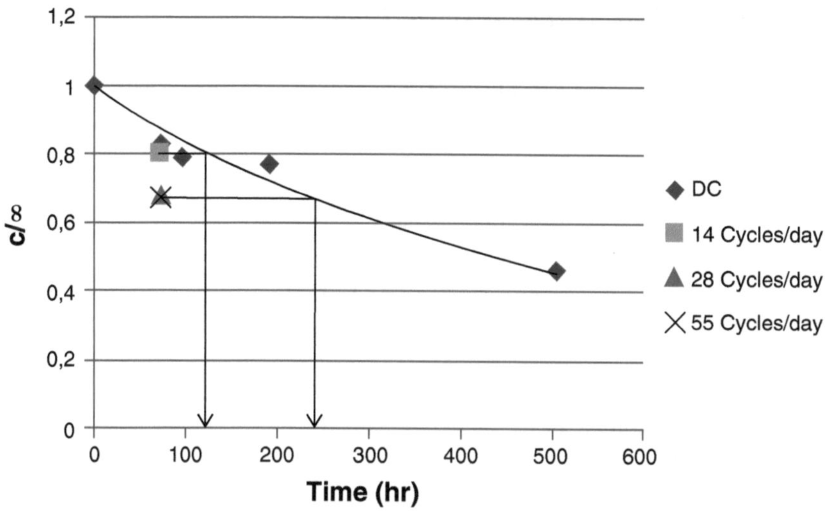

Fig. 8.5 Normalized concentration of copper with time in the anode side with continuous and pulsed electric fields applied to acidic mine tailings in EDR

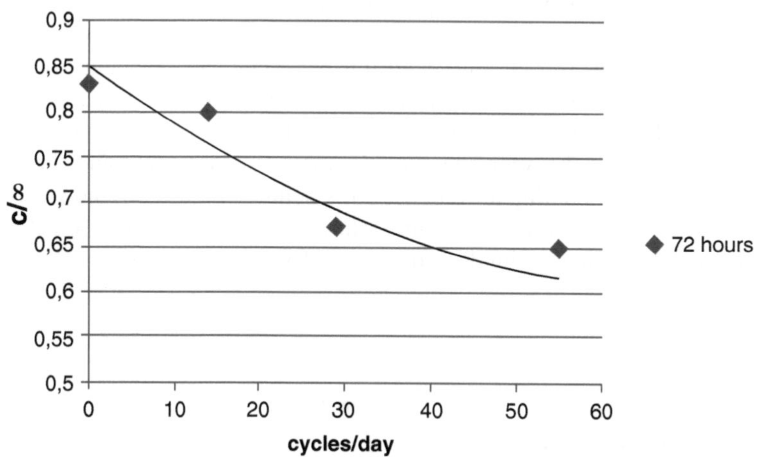

Fig. 8.6 Normalized concentration of copper in the anode side as a function of frequencies in cycles/day for EDR experiments

In EDR with pulsed electric field the transport across the cation-exchange membrane becomes the rate determining step. This fact can be deduced from: (1) the accumulation of copper observed in the cathode side when the frequency was increased and (2) the total copper removal was nearly the same for the three frequencies studied. In EKR, the diaphragm allows free alkali diffusion into the cell, with the consequent copper precipitation process, which reduces the effect of the pulsed electric field on total copper removal.

8.2.6 Sinusoidal AC/DC Electric Field

A sinusoidal electric field is obtained by applying simultaneously continuous-alternating (DC–AC) voltages, where the effective voltage applied to the cell is determined by:

$$V = V_{\mathrm{DC}} + V_{\mathrm{AC}} \cdot \sin(2\pi f t)$$

$$V_{\mathrm{effective}} = \sqrt{\frac{1}{T} \cdot \int_0^T V^2 \, dt}$$

where:

V_{DC} = continuous voltage (V)
V_{AC} = alternating voltage (V)
f = frequency (min^{-1})
t = time (min)
T = period (min)
$V_{\mathrm{effective}}$ = effective voltage (V)

One main characteristic when applying sinusoidal electric fields is that short periods of time where the direction of the electric field is opposite of overall electric field can be obtained. This current reversal is beneficial when reducing the concentration polarization and to have zone where dissolution of copper happens due to acidification. To compare the results of experiments with a reversal polarity of the cell, the time ratio ρ between the time of normal polarity (t+) and the time of polarity reversal ($t-$) during a cycle (t_{cycle}) is defined. Figure 8.7 shows schematically how the different times are defined during a cycle with a sinusoidal electric field obtained by applying simultaneously AC–DC voltages.

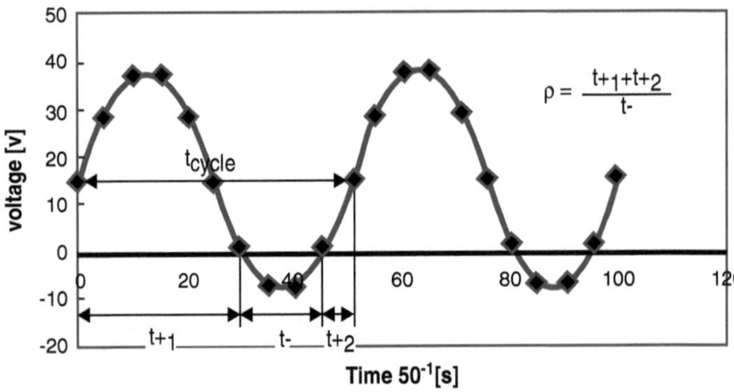

Fig. 8.7 Schematic representation of a sinusoidal electric field with a frequency of 50 Hz and a DC field of 15 V and definition of the time ratio ρ

Table 8.3 Experimental conditions during EKR of copper mine tailings applying low frequency sinusoidal electric fields

Exp	Pre-treatment [acid]	*t* time [days]	V_{DC} (V)	V_{AC} (V)	$V_{effective}$ (V)	*T* period AC (min)
1	Sulfuric	7	20	0	20	0
2	Sulfuric	7	15	20	20.6	120
3	Sulfuric	7	15	20	20.6	60
4	Sulfuric	7	20	20	24.5	120
5	Sulfuric	7	20	20	24.5	60
6	Sulfuric	7	10	20	17.3	30
7	Sulfuric	7	10	20	17.3	120
8	Sulfuric	14	10	20	17.3	120

8.2.7 *Low Frequency Sinusoidal Electric Fields*

When applying sinusoidal fields with a cycle in magnitude of minutes or even hours this is generally referred to as low frequency sinusoidal fields. For example, a cycle of 60 min gives a frequency of 1/(60 min × 60 s/min) = 0.28 mHz.

In the work of Rojo et al. (2010), the application of sinusoidal electric fields with cycles of 30–120 min, V_{DC} of 10–20 V, and V_{AC} of 20 V was first reported. Here, the main findings are given. The cell was type A, and the tailings type 3. Table 8.3 shows the remediation conditions for eight EDR experiments applying constant voltage (exp. 1) or sinusoidal fields (exp. 2–8).

Table 8.4 includes general EDR results in terms of: (a) total and soluble copper concentration ratios, C_F/C_0 (final/initial) in the anode, center, and cathode zone, (b) the total net electric charge passed through the mine tailing, and (c) the effective current for the experiment. The net electric charge and effective current were calculated from current-time recording.

It can be seen from Table 8.4, all experiments with sinusoidal electric field (experiments 2–8) show better results than the reference experiment (experiment 1) with continuous electric field applied to the cell, since in the case of anode zone the C_F/C_0 concentration ratios of total and soluble copper using sinusoidal electric field are always less than with continuous electric field. These show that the reduction of polarization by applying sinusoidal electric fields improves the remediation performance.

According to Table 8.4, with a remediation time of 7 days, a period for the alternating voltage of 120 min and sulfuric acid addition in the mine tailing pre-treatment, when increasing the effective voltage, from 17.3 to 24.5 V, the total and soluble copper removal in the anode zone and the copper accumulation in the cathode zone decreases.

From the EDR results, the improvement could be explained by the positive effect of the polarity reversal, which allows the system to depolarize, for example, in the experiment 7 for 40 min each period. In experiment 4, with the highest effective voltage, the polarity reversal phenomenon is not produced. The positive effect of

Table 8.4 General remediation results of low frequency sinusoidal EKR in terms of concentration ratios [–]

	Total copper				Soluble copper					
	C_0	Anode	Center	Cathode	C_0	Anode	Center	Cathode		
Exp	(ppm)	C_F/C_0	C_F/C_0	C_F/C_0	(ppm)	C_F/C_0	C_F/C_0	C_F/C_0	Total net charge passed (C)	$I_{effective}$ (mA)
1	995	0.94	1.01	1.05	423	0.86	1.02	1.12	342	0.6
2	1070	0.90	0.97	1.13	427	0.75	0.92	1.33	3921	6.5
3	1070	0.90	1.00	1.16	443	0.76	1.00	1.39	5162	8.5
4	1080	0.92	0.95	1.13	470	0.82	0.89	1.30	6919	11.4
5	1075	0.92	1.02	1.14	451	0.81	1.05	1.33	7147	11.8
6	1030	0.83	1.14	1.23	457	0.62	1.32	1.52	3107	5.1
7	1060	0.83	1.15	1.18	515	0.65	1.31	1.37	2257	3.7
8	1100	0.82	1.18	1.21	530	0.63	1.37	1.44	3983	6.6

depolarization during the polarity reversal of the system is obtained although in each period electrokinetic phenomena occur in the opposite direction with an energy consumption associated.

Certainly in this kind of process, a lowest net charge through the mine tailings is expected, and the best removal observed is produced by the better use of current through the tailings. In this context, the lowest effective voltage will be favorable in terms of current efficiency and energy consumption.

In all cases, the total and soluble copper removal, for 7 days experiments with sulfuric pre-treated mine tailings, does not seem to be affected by changing the AC period from 30 (or 60) to 120 min. However, it must be stated that all experiments were carried out at relative high periods—or low frequencies. It was chosen to work with low frequencies according to previous experiences applying pulsed electric fields with similar frequencies. The total net electric charge and effective current obtained in each pairs are relative similar as expected due to effective voltage applied to the cell.

Higher remediation times would increase the total and soluble copper removal in the center and the accumulation in the cathode zone of the cell, and only a greater amount of remediated material is obtained. On the other hand, in the anode zone, the removal is only significant during the first 7 days. As expected, the electrical parameters are almost duplicated with the remediation time.

So, polarity reversal of the system reduces polarization during the remediation, and by this way enhances the process mainly in terms of increasing the copper removal. The remediation seems to be favored by low effective voltage in spite of the relatively long period of polarity reversal, which affects the power consumption. The disadvantage of higher power consumption, which normally is produced in processes with polarity reversal, seems to be compensated by increased net removal efficiency in EDR/EKR systems.

8.2.8 High Frequency Sinusoidal Electric Fields

When applying sinusoidal fields with a frequency of 50 Hz, or higher, one can consider the sinusoidal field as a high frequency field. This means that the duration of the cycle is 0.020 s or lower. In this chapter, only experiences with frequencies of 50 Hz will be discussed despite that higher frequencies have been used lately by the authors (Rojo et al. 2014).

Rojo et al. (2011) demonstrated the improvements when using 50 Hz sinusoidal fields compared to conventional DC EKR. Seven experiments were reported using tailings of type 3 in an EKR cell B. The general experimental conditions are given in Table 8.5, where exp. 7 is performed at a constant DC voltage as a reference experiment.

Table 8.6 includes a summary with general EKR results in terms of removal (positive values) and/or accumulation (negative values) of total copper from the already mentioned zones and the whole cell.

Table 8.5 Remediation conditions applying high frequency sinusoidal electric fields

Exp	Applied voltage DC (V)	AC (V)	$V_{Effective}$ (V)	V_{max} (V)	$V_{min}{}^{a}$ (V)
1	7.2	23.0	17.8	30.2	−15.9
2	15.3	23.0	22.3	38.3	−7.8
3	23.3	23.0	28.4	46.3	0.3
4	29.0	23.0	33.2	52.0	6.0
5	23.3	31.0	31.8	54.3	−7.7
6	29.0	31.0	36.4	60.0	−2.0
7	23.0	–	–	–	–

[a]Negative values, experiment with periodic polarity reversal

Table 8.6 General remediation results applying high frequency sinusoidal electric fields in terms of removal and/or accumulation [%]

Exp	Anode $\left(1-\frac{C_E}{C_0}\right)\times 100$	Center	Cathode	Cell
1	19.6	4.3	33.7	19.2
2	54.3	31.5	38.0	42.2
3	8.5	6.5	−16.8	1.6
4	47.8	4.3	−25.0	8.4
5	50.0	−3.3	8.7	18.7
6	46.7	0.0	33.7	28.2
7	51.1	32.6	−8.7	24.6

EKR with sinusoidal electric field obtained by applying DC and AC voltages simultaneously can achieve a copper removal from the cell better than EKR with continuous electric field. In this context, comparing experiment 2 and 7, which were made at a similar voltage, the use of a sinusoidal electric field represents a 70 % improvement in the copper removal from the cell. In all experiments, no significant electroosmotic flow was observed, meaning that the main mechanism for copper removal is electromigration.

Moreover, experiments with sinusoidal electric field in which a polarity reversal occurs during the cycle, indicated by the negative minimum voltage shown in Table 8.6 (Experiments 1, 2, 5 and 6), copper removal was also observed in the cathode zone. In addition, experiment 2 with sinusoidal electric field reached a copper removal from the cell which involved virtually all soluble copper in the mine tailings sample.

In the analysis of copper removal and/or accumulation, the effect of the effective voltage ($V_{Effective}$) was evaluated by considering the following combinations: (a) $V_{AC} = 23.0$ V fixed and V_{DC} varying between 7.2 and 29.0 V, and (b) $V_{DC} = 23.3$ and 29.0 V fixed and V_{AC} varying between 23.0 and 31.0 V in both cases.

Comparing the experiments with fixed V_{AC} and variable V_{DC}; it can be seen that copper removal from the cell increases proportionally with the effective voltage in the range 17.8–22.3 V and 28.4–33.2 V. In the first range, the removals are greater because the polarity of the cell during the cycle is reversed, this phenomenon does not occur in the second range of 28.4–33.2 V.

Table 8.7 Removal and/or accumulation [%] vs. time ratio ρ

Exp	ρ [–]	Anode	Center	Cathode	Cell
		$\left(1-\frac{C_F}{C_0}\right)\times 100$			
1	1.50	19.6	4.3	33.7	19.2
5	1.50	50.0	−3.3	8.7	18.7
6	1.86	46.7	0.0	33.7	28.2
2	2.33	54.3	31.5	38.0	42.2

On the other hand, comparing the experiments in which V_{DC} remained fixed at 23.3 V (experiments 3 and 5) and 29.0 V (experiments 4 and 6) and V_{AC} was variable, it can be seen that the copper removal from the cell follows the same trend. In this case, for both fixed DC voltage the removal increase is due to the reversal polarity of the cell during the cycle. Consequently, the reversal of the polarity of the cell is the decisive phenomenon in improving the copper removal from the whole cell.

Table 8.7 shows the copper removal and/or accumulation of the four experiments with periodical reversal polarity of the cell, characterized by the time ratio ρ. The results show that the copper removal from the cell increases proportionally to the ratio of time, ρ. This is consistent with the experiments, since the time ratio is increased by reducing the time of reversal polarity, and accordingly to the frequency of applied AC voltage of 50 Hz two positive effects are obtained: (a) times greater of normal polarity for the copper removal and (b) reduction of polarization by the periodical polarity reversal of the cell.

According to the highest copper removal efficiency of 42.2 % obtained in experiment 2, it should be noted that in fresh mine tailings, copper could be expected to be found as residual insoluble copper sulfide, which was not liberated in the grinding process prior to flotation. The apparent low copper removals obtained here were due to the soluble copper content in the tailings of 44.9 %, so the highest removal efficiency of 42.2 % of the total copper contents seems to be promising.

On the other hand, the content of soluble copper is variable due to the heterogeneous origin of the mine tailings in the ponds, among other reasons: copper grades depend on the original characteristics of the tails disposed, aging of the tailing in the ponds as consequence of physical–chemical changes due to weathering and bacterial actions in time. The use of this remediation technology will imply the periodic application of the method in order to remove the additional soluble copper that will be generated with time. Therefore, the remediation action for this heterogeneous solid waste is to remove the soluble copper in the tailings and in this way making the final residue more stable.

8.3 Conclusions

One disadvantage when remediating soil or another solid porous waste material by a constant electric DC field is the buildup of a concentration polarization, which will increase the electrical resistance of the whole process. Therefore, it is important to find ways to diminish this effect. The main key to do this is to manipulate the electric field without affecting the overall remediation efficiency.

In this work, two possible options to reduce the polarization—and thereby the electrical resistance—were shown. One way was to apply the DC field in pulses, with a long period with the electric field on, followed by a sort period with the electric field off. The second possibility was to apply a combined AC–DC field in a manner that the net electric field was in the direction of the cathode but during the process, short periods with current reversal occurred.

In both cases, the remediation time decreased remarkably and thereby showing their potential for further development. Of course, these findings are based on laboratory experiments done on one type of waste (copper mine tailings), but still it is expected that EKR on other waste materials will follow the same trends as found here.

References

Acar YB, Alshawabkeh AN (1993) Principles of electrokinetic remediation. Environ Sci Technol 27:2638–2647

Chilean Copper Commission (COCHILCO) (2008) Yearbook: copper and other mineral statistics 1988–2007

Dold B, Fontbote L (2002) A mineralogical and geochemical study of element mobility in sulfide mine tailings of Fe oxide Cu–Au deposits from the Punta del Cobre belt, northern Chile. Chem Geol 189(3–4):135–163

Hansen HK, Rojo A (2007) Testing pulsed electric fields in electroremediation of copper mine tailings. Electrochim Acta 52:3399–3405

Hansen HK, Ottosen LM, Hansen L, Kliem BK, Villumsen A (1997) Electrodialytic soil remediation—a derivative from electrokinetic soil remediation. In: Proceeding of the workshop on electromigration applied to soils remediation, Albi, France, May 15–16, 1997

Hansen HK, Yianatos J, Ottosen LM (2005a) Speciation and leachability of copper in mine tailings from porphyry copper mining: influence of particle size. Chemosphere 60(10):1497–1503

Hansen HK, Rojo A, Ottosen LM (2005b) Electrodialytic remediation of copper mine tailings. J Hazard Mater 117(2–3):179–183

Hansen HK, Rojo A, Ottosen LM (2012) Electrodialytic remediation of copper mine tailings. In: Procedia engineering (euromembrane conference 2012), vol 44, pp 2053–2055

Lageman R (1993) Electro reclamation. Environ Sci Technol 27:2648–2650

Li D, Niu Y-Y, Fan M, Xu D-L, Xu P (2013) Focusing phenomenon caused by soil conductance heterogeneity in the electrokinetic remediation of chromium (VI)-contaminated soil. Sep Purif Technol 120:52–58

Probstein RF, Hicks RE (1993) Removal of contaminants from soil by electric fields. Science 260:498–503

Rojo A, Hansen HK, Ottosen LM (2006) Electrodialytic remediation of copper mine tailings: comparing different operational conditions. Miner Eng 19:500–504

Rojo A, Hansen HK, del Campo J (2010) Electrodialytic remediation of copper mine tailings with sinusoidal electric field. J Appl Electrochem 40:1095–1100

Rojo A, Hansen HK, Agramonte M (2011) Electrokinetic remediation with high frequency sinusoidal electric fields. Sep Purif Technol 79:139–143

Rojo A, Hansen HK, Monárdez O, Jorquera C (2014) Electrical behavior of copper mine tailings during EKR with modified electric fields. In: Book of abstracts of 13th symposium on electrokinetic remediation (EREM 2014), September 7–10, 2014, Malaga, Spain, pp 147–149

Chapter 9
Electrokinetics as an Alternative for Soil and Compost Characterization

Alejandro Serna González, Lucas Blandón Naranjo, Jorge Andrés Hoyos, and Mario Víctor Vázquez

9.1 Introduction

Electrokinetic remediation is an environmental technique, especially designed for the in situ restoration of contaminated soil (Reddy and Cameselle 2009). The technique is based on the application of a direct electric current to the contaminated soil through two main electrodes (anode and cathode). The application of the electric current induces a variety of reactions in contaminants and soil, which results in the mobilization of the contaminants and their transportation towards the main electrodes and out of the contaminated soil. The main two transportation mechanisms are called electromigration and electroosmosis. Electromigration is the movement of ionic species in the electric field towards the electrode of opposite charge. Thus, cations (positive ions) move towards the cathode (the negatively charged electrode), while anions (negative ions) move towards the anode (the positively charged electrode). Electroosmosis is the net flux of water induced by the electric field through the porous structure of the soil. The electroosmotic flux is the result of the combining effects of the electric field and the electric charge in the surface of the soil particles. Commonly, soil particles are negatively charged, and that surface charge is neutralized by positive ions. When an electric field is applied to a moisten soil specimen, the positive ions, that neutralize the surface charge, start to move towards the cathode, dragging the solvated water molecules around. Thus, the water in the soil pores moves towards the cathode due to electroosmosis (Cameselle and Reddy 2012).

The electric field also induces some reactions upon the main electrodes and into the soil specimen. Those reactions include the electrolysis of water, adsorption/desorption of contaminants on the solid particle surfaces, redox reactions, and acid/base reactions. The global effects of the chemical reactions during the

A. Serna González • L. Blandón Naranjo • J.A. Hoyos • M.V. Vázquez (✉)
Interdisciplinary Group of Molecular Studies (GIEM), Chemistry Institute, Faculty of Exact and Natural Sciences, University of Antioquia, A.A.1226, Medellín, Colombia
e-mail: mario.vazquez@udea.edu.co

© Springer International Publishing Switzerland 2016

A.B. Ribeiro et al. (eds.), *Electrokinetics Across Disciplines and Continents*,
DOI 10.1007/978-3-319-20179-5_9

electrokinetic treatment are a dramatic change in soil pH, ion concentration in soil solution, contaminant speciation, and contaminant dissolution or precipitation. In electrokinetic remediation, the operating conditions and the addition of chemicals are specially designed to favor the dissolution of contaminants and its transportation out of the soil. In some cases, the oxidation of the contaminants is favored into the soil to eliminate them by in situ chemical oxidation.

Electrokinetic technology has been used for the last 20 years in the removal of contaminants from contaminated soils. The first reports on electrokinetics were focused on the removal of the heavy metals from soils. Then, a lot of attention was paid to the removal of recalcitrant and hydrophobic organic compounds. Finally, some experimental works focused in the restoration of contaminated soil with both heavy metals and hydrophobic organics. Despite the efforts in the last 20 years in the study and application of this technique, there is not still a standardized electrokinetic technology that can be applied to any soil or any type and concentration of contaminants. This is due to the large number of variables that affect the process, including electrical parameters, geochemistry of soils, type and concentration of the contaminants, and interaction of the soil-contaminant system. Furthermore, the electric field induces temporary variations in the soil, including pH, ionic concentration, moisture content, conductivity, and others. In fact, during the electrokinetic treatment, the evolution of the voltage distribution along the soil sample and the electric current intensity is the result of the changes in chemical speciation, variation in concentrations, pH changes, and moisture content in the soil. The relation between the electric parameters and the geochemistry of the system has not been explored significantly in the literature, and there is very limited information about the relation between the variation of electric parameters (voltage and current intensity) and the physical–chemical characteristics of the soil-contaminant system.

A conventional electrokinetic experiment at laboratory scale consists in applying a continuous electric field (120 V m^{-1}) to a wet paste of soil using a pair of electrodes, typically graphite (Vázquez et al. 2007, 2008, 2009). A schema of an electroremediation laboratory cell is shown in Fig. 9.1.

This system can be analyzed, in a simplified way, as an electric circuit with elements that show different electrical resistance and capacitance, as shown in Fig. 9.2.

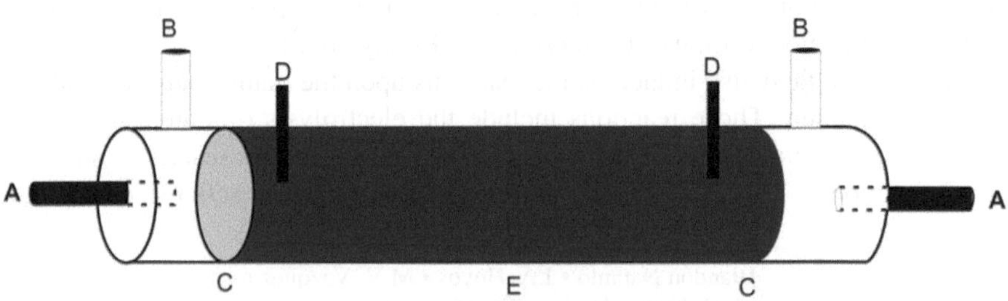

Fig. 9.1 Scheme of an electroremediation cell: *A* electrodes, *B* gas exit, *C* separator, *D* auxiliary electrodes, *E* material under study (adapted from Vásquez et al. 2009)

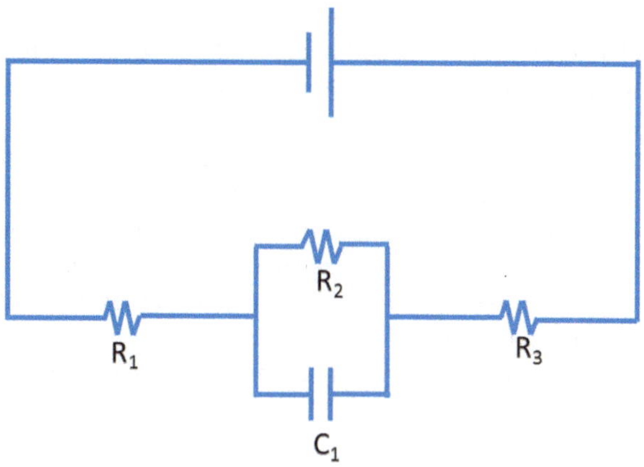

Fig. 9.2 Equivalent circuit of an electrokinetic system. R_1 and R_3 can be assigned as the resistance in the electrodes chamber, R_2 the resistance of the soil paste, and C_1 the capacitance of this material

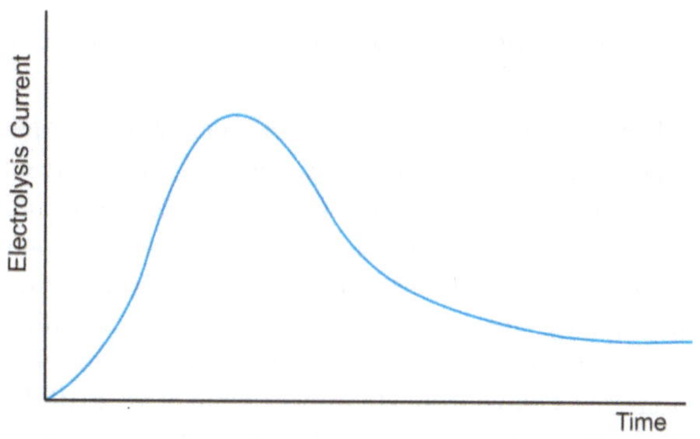

Fig. 9.3 Electrolysis current versus treatment time during an electrokinetic treatment

Within all experimental parameters that can be analyzed during and after the electrokinetic treatment, the electrolysis current is highlighted. The magnitude thereof depends on the availability of charged species that ensure electrical conductivity within the material.

The *I–t* variation obtained depends on the initial conditions of the experiment. For example, if deionized water is used in the chambers of the electrodes, it is expected to have a gradual increase in current as the free ionic species can be directed to the corresponding electrode. Once the conductivity values in the chambers are enough, oxygen and hydrogen evolution begins with the pH gradient corresponding to each chamber. At this point, the electrolysis current increases to the maximum value that the conductivity of the paste, object of study, allows. The paste conductivity is the parameter that begins to control the process.

After reaching the maximum value, the current usually decreases to an approximately constant value; this behavior can be attributed to the motion of charged species roughly retained inside the paste, i.e., there is a competition between the electric field effect and the established equilibriums within the material. This behavior can be observed in Fig. 9.3.

The behavior of the current throughout time may be, therefore, related to the characteristics of the soil under study, especially with the presence of ionic species weakly retained. This implies that soils with different clay and/or organic material content should display different behavior in the change of current with time.

To use the current profile as a characterization tool is complicated as far as is a transient, and the objective is to make a comparison with specific characterization parameters obtained in conventional soil analysis. A more appropriate option is the use of electric charge associated with that current. This value, obtained by integrating the response i vs t, may, in some cases, be related to some characteristics of the materials studied.

In the following examples, these types of electrokinetic characterizations are presented.

9.2 Characterization of Soils Subjected to Thermal Shock

When a soil is subjected to thermal shock, either intentionally or naturally, depending on the intensity, i.e., the temperature reached and the impact duration, different types of structural changes occur, from moisture loss, decomposition of organic matter to the most extreme case, destruction of the clay material.

Given the importance of knowing the behavior of different types of soil when affected by fire, a significant number of scientific studies have been made by analyzing, in field or laboratory, these structural changes (Certini 2005; González-Pelayo et al. 2006; Arcenegui et al. 2010; Salgado et al. 2004; Carrington 2010; Esque et al. 2010; Santin et al. 2008; Ferreira et al. 2008; Notario del Pino et al. 2007).

To analyze, through an electrokinetic technique, the changes that a thermal treatment produces, a volcanic soil with the following characteristics was selected (Table 9.1).

Table 9.1 Characteristics of the soil sample

Classification, USDA, 1999	RhodicPaleustalfs	
pH (H_2O)	6.50	
Carbon content (C)	21.73	g kg^{-1}
Na	3.12	cmol kg^{-1}
K	2.38	cmol kg^{-1}
Mg	5.81	cmol kg^{-1}
Ca	4.19	cmol kg^{-1}
Cation exchange capacity (CEC)	51.2	cmol kg^{-1}
Granulometry		
Sand	44	g kg^{-1}
Silt	95	g kg^{-1}
Clay	701	g kg^{-1}

Table 9.2 Correlations between characterization parameters and total electric charge

Characterization variables	Correlation parameter (r^2)
CEC	0.725
C	0.893
Clay	0.959
C + CEC	0.963
C + CEC + Clay	0.984

This soil was subjected at 300 and 500 °C for 12 and 24 h, respectively, and after the heat treatment, it was introduced as a wet paste, in the electrokinetic characterization cell. The test was performed for 28 h by applying an electric field of 120 V m^{-1}

Given the characteristics of the applied heat treatment, particle size analysis indicates that the percentage of clay decreased from 62.02 %, for the untreated soil, to 17.70 % in the case of the higher temperature. Similarly, the percentage of silt varies from an initial value of 31.74–12.44 %. Another parameter that is expected to change with heat treatment is the organic carbon content. Thereby the content decreases from an initial value of 21.73 g kg^{-1}, for untreated soil, to a value of 1.48 g kg^{-1} after the higher temperature treatment.

These changes are reflected in variations in cation exchange capacity (from 51.2 to 32.8 cmol kg^{-1}), so it is expected that the lower ionic availability is also evidenced in lower values of electrolysis current. While this current can be monitored during the electrokinetic treatment, it is convenient to use the corresponding electric charge in order to obtain a unique numeric value to compare with other parameters.

Table 9.2 shows different correlations between electrolysis charge and some soil characterization parameters.

As seen, the electric charge obtained in the experiment can be related to the variation of some soils' characterization parameters when they are heat treated.

9.3 Electrokinetic Characterization of Compost

Compost is defined as organic matter that has been stabilized to become a product with similar characteristics to soil humic substances (Haug 1993). As in soil characterization, to determine the quality of this material, various parameters are analyzed, such as: color, smell, texture, heavy metal content, C/N ratio, ash content, organic carbon, pH, and cation exchange capacity among others.

Considering that the compost contains ionic species weakly retained, these may confer conductive properties to the compost and therefore facilitate and promote its electrokinetic characterization.

The same way as soils with different taxonomies have different behavior when subjected to an electrokinetic treatment, compost from different sources has

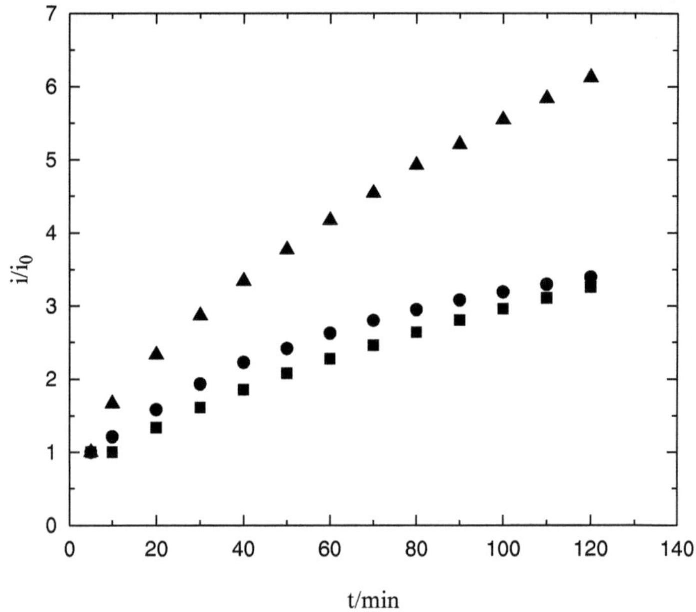

Fig. 9.4 Normalized current during the electrokinetic experiment with compost obtained from: *filled circles*—chicken manure, *filled squares*—mushroom crop waste, and *filled triangles*—municipal solid waste

different availability of charged species, affecting their ability to be mobilized by an electric field.

This is evidenced by the results shown in Fig. 9.4.

For the compost obtained from the treatment of municipal solid waste, a distinct variation is observed in regard to materials obtained from chicken manure or mushroom crop waste. These results were obtained in laboratory experiments using the same experimental conditions as those reported in the previous point with the ground subjected to thermal shock.

Another aspect that could be relevant for considering the electrokinetic technique as a characterization tool is what happens when the application of the electric field on the compost sample is interrupted. Similarly to what happens when a capacitor is discharged, the current decays exponentially as presented in Fig. 9.5.

This current variation or better yet, the electric charge involved could be used to characterize compost samples without the need for multiple analyses. If this charge can be correlated with usual characterization parameters, it would be a relatively easy way to monitor the compost maturation and/or characterization.

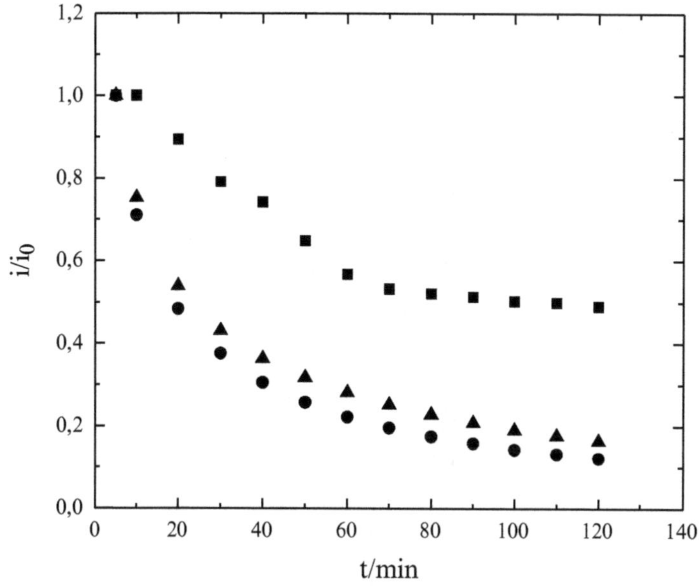

Fig. 9.5 Normalized "discharge" current after the switch off the power sources for compost obtained from: *filled circles*—chicken manure, *filled squares*—mushroom crop waste, and *filled triangles*—municipal solid waste

9.4 Application of Artificial Neural Networks in Electrokinetic Treatment

The variation of the electrolysis current can be used, as presented in the previous paragraphs, to, fundamentally, evaluate corresponding electrical charge, and then, this parameter be correlated with one or several commonly used for characterization of soil and compost. Another possibility offered by the electrokinetic technique is to analyze a set of responses obtained from a number of samples and try to predict characterization parameters from these results. For this purpose, it is possible to use artificial neural networks (ANNs).

ANNs represent sophisticated computational modelling tools, which can be used to solve a wide variety of complex problems, therefore their actual potential in science is high. The attractiveness of ANNs in soil science comes from their capability to learn and/or model very complex systems, which allows for the possibility of them being used as a tool for classification (Anagu et al. 2009; Baker and Ellisom 2008). ANN's theory has been widely discussed in the previous literature. Good overview of ANN principles can be found in the monographs (Bishop 1995; Fausett 1994); the theory of different networks has been reviewed by (Zupan and Gasteiger 1991). ANN is a computational model formed from a certain number of single units, artificial neurons, or nodes connected with coefficients (weights), w_{ij}, which constitute the neural structure. Many different neural

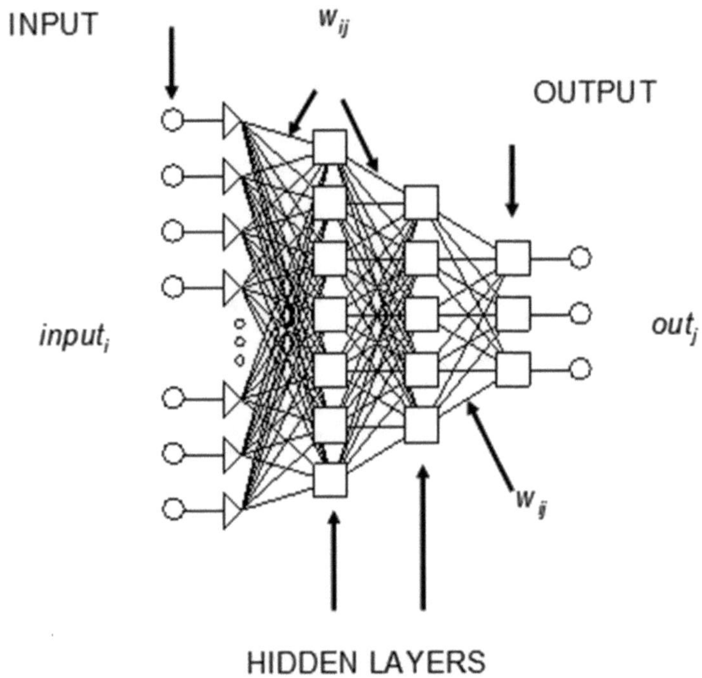

Fig. 9.6 General scheme of artificial neural network architecture

network architectures can be used. One of the most common is the feed forward supervised neural network of multilayer perceptrons (MLP). The MLP is conventionally constructed with three or more layers, i.e., input, output, and hidden layers. A general scheme of an ANN is shown in Fig. 9.6.

Each layer has a different number of nodes. The input layer receives the information about the system (the nodes of this layer are simple distributive nodes, which do not alter the input value at all). The hidden layer processes the information initiated at the input, while the output layer is the observable response or behavior. The inputs, $input_i$, multiplied by connection weights, w_{ij}, are first summed and then passed through a transfer function to produce the output, out_i. The determination of the appropriate number of hidden layers and number of hidden nodes in each layer is one of the most critical tasks in ANN design. Unlike the input and output layers, one starts with no prior knowledge of the number and size of hidden layers.

The use of ANN consists of two steps: "Training" and "Prediction." The "Training" consists first of defining input and output data to the network. It is usually necessary to scale the data or normalize it to the networks paradigm. This data is referred to as the training set. In this training phase, where actual data must be used, the optimum structure, weight coefficients, and biases of the network are searched for. The training is considered complete when the neural networks

Table 9.3 Comparison of experimental and ANN predicted CEC values concerning volcanic soils from Tenerife (Spain)

Soil	CEC_{exp}	CEC ANN_{calc} (47 h)	Difference	CEC ANN_{calc} (2 h)	Difference
Ravelo	53.90	53.80	−0.10	53.72	−0.18
Junquito	32.00	32.00	0.00	30.08	−1.92
Pachona	44.40	44.40	0.00	49.27	4.87
Los Lirios	53.40	53.39	−0.01	49.08	−4.31
La Laja	23.70	23.81	0.11	24.03	0.32
Erques	26.70	26.70	0.00	29.33	2.63

achieved the desired statistical accuracy as it produces the required outputs for a given sequence of inputs. A good criterion to find the correct network structure and therefore to stop the learning process is to minimize the root mean square error (RMS), where y_{ij} is the element of the matrix ($N \times M$) for the training set or test set, and out_{ij} is the element of the output matrix ($N \times M$) of the neural network, where N is the number of variables in the matrix and M is the number of samples. RMS gives a single number which summarizes the overall error.

For this analysis, six soil samples with different characteristics from the island of Tenerife (Spain) were studied. The samples were subjected to an electrokinetic treatment, applying an electric field of 120 V m^{-1}.

ANN computation was performed using Trajan Neural Network Simulator, Release 3.0 D (*Trajan Software Ltd., 1996–1998, UK*). All computation was performed on a standard PC computer with operating system Microsoft Windows Professional XP 2000.

For the study, currents obtained during 47 h or in the first 2 h of treatment were selected, and corresponding values of cation exchange capacity were calculated, obtaining the results shown in Table 9.3.

It is noted that, except on the soils *Pachona* and *Erques*, the values obtained using the ANN to calculate CEC do not differ substantially from the value obtained experimentally. It is also noted that, using the data obtained from the first 2 h of treatment, a very good approximation to the real value of CEC is obtained, which would make it unnecessary to use the data obtained after 47 h of treatment.

9.5 Conclusions

The possibility of mobilizing charged species by the action of an electric field, foundation of electrokinetic technique, in compost or soil samples may be considered as an option for the characterization thereof. Through simple experiments, it is possible to simultaneously obtain an important set of experimental parameters, which can then be correlated with physicochemical characteristics of the materials studied. Electrokinetic characterization could be seen as closest to the naturally occurring processes disturbance when ionic species in soil or compost move. This would be a point to be considered in comparison to conventional methods of analysis, for example, for determining the content of organic material or cation exchange capacity, where extreme changes are applied.

Acknowledgements The authors wish to acknowledge to Dr. Josef Havel, Dr. Claudio Cameselle, and Dr. Felipe Hernández-Luis, for the valuable contribution to the realization of this document.

References

Anagu I, Ingwersen J, Utermann J, Streck T (2009) Estimation of heavy metal sorption in German soils using artificial neural networks. Geoderma 152:104–112

Arcenegui V, Mataix-Solera J, Zornoza R, Pérez-Bejarano A, Mataix-Beneyto J, Gómez I (2010) Estimation of the maximum temperature reached in burned soils using near-infrared spectroscopy: effects of soil sample pre-treatments. Geoderma 158:85–92

Baker L, Ellison D (2008) Optimisation of pedotransfer functions using an artificial neural network ensemble method. Geoderma 144:212–224

Bishop CM (1995) Neural networks for pattern recognition. Oxford University Press, Oxford

Cameselle C, Reddy KR (2012) Development and enhancement of electro-osmotic flow for the removal of contaminants from soils. Electrochim Acta 86:10–22. doi:10.1016/j.electacta.2012.06.121

Carrington ME (2010) Effects of soil temperature during fire on seed survival in Florida Sand Pine Scrub. Int J Forest Res 2010:1–10

Certini G (2005) Effects of fire on properties of forest soils: a review. Oecologia 143:1–10

Esque TC, Young JA, Tracy CR (2010) Short-term effects of experimental fires on a Mojave Desert seed bank. J Arid Environ 74:1302–1308

Fausett LV (1994) Fundamentals of neural networks: architectures, algorithms and applications. Prentice Hall, New York

Ferreira AJA, Coelho COA, Ritsema CJ, Boulet AK, Keiser JJ (2008) Soil and water degradation processes in burned areas: lessons learned from a nested approach. Catena 74:273–285

González-Pelayo O, Andreu V, Campo J, Gimeno-García E, Rubio JL (2006) Hydrological properties of a Mediterranean soil burned with different fire intensities. Catena 68:186–193

Haug RT (1993) The practical handbook of compost engineering. Lewis, Boca Raton

Notario del Pino J, Dorta I, Navarro F, Rodríguez-Rodríguez A, Arbelo C, Armas C, Guerra JA, Mora JL (2007) Temporal evolution of organic carbon and nitrogen forms in volcanic soils under broom scrub affected by a wildfire. Sci Total Environ 378:245–252

Reddy KR, Cameselle C (2009) Electrochemical remediation technologies for polluted soils, sediments and groundwater. Wiley, Hoboken

Salgado J, Mato MM, Vázquez-Galiñanes A, Paz-Andrade MI, Carballas T (2004) Comparison of two calorimetric methods to determine the loss of organic matter in Galician soils (NW Spain) due to forest wildfires. Thermochim Acta 410:141–148

Santín C, Knicker H, Fernández S, Menéndez-Duarte R, Alvarez MA (2008) Wildfires influence on soil organic matter in an Atlantic mountainous region (NW of Spain). Catena 74:286–295

Vázquez MV, Hernández-Luis F, Benjumea D, Grandoso D, Lemus M, Arbelo CD (2007) Electrokinetic determination of the buffer capacity of Andisols. Sci Total Environ 378:214–217

Vázquez MV, Vasco DA, Hernández-Luis F (2008) Effect of time and electrical potential gradient on the buffer capacity of soils, as measured using an electrokinetic method. Land Contam Reclamat 16(3):249–260

Vázquez MV, Vasco DA, Hernández-Luis F, Grandoso D, Lemus M, Benjumea DM, Arbelo CD (2009) Electrokinetic study of the buffer capacity of some soils from Tenerife. Comparison with a volumetric technique. Geoderma 148:261–266

Zupan J, Gasteiger J (1991) Neural networks in chemistry and drug design. Wiley-VCH, Weinheim

Chapter 10
Life Cycle Assessment of Soil and Groundwater Remediation: Groundwater Impacts of Electrokinetic Remediation

Luís M. Nunes, Helena I. Gomes, Margarida Ribau Teixeira, Celia Dias-Ferreira, and Alexandra B. Ribeiro

10.1 Introduction

Life cycle assessment (LCA) methodologies have well-developed methods for assessing environmental impacts for emissions to air, surface water, and surface soil, but deep soil emissions and emissions to groundwater have received little attention (Cappuyns 2013; Lemming et al. 2010). Fluoride emissions to groundwater were included in the LCA of a landfill containing spent pot lining by Godin et al. (2004); concentrations of fluoride as a function of time at a point of discharge to the (surface) aquatic environment were modeled using the 3D saturated/unsaturated

L.M. Nunes (✉)
CERIS—Civil Engineering Research and Innovation for Sustainability, Faculdade de Ciências e Tecnologia, Universidade do Algarve, Campus de Gambelas, Edifício 7, Faro 8005-139, Portugal
e-mail: lnunes@ualg.pt

H.I. Gomes
CENSE, Departamento de Ciências e Engenharia do Ambiente, Faculdade de Ciências e Tecnologia, Universidade Nova de Lisboa, Caparica 2829-516, Portugal

CERNAS—Research Center for Natural Resources, Environment and Society, Escola Superior Agraria de Coimbra, Instituto Politecnico de Coimbra, Coimbra, Portugal

Department of Civil Engineering, Technical University of Denmark, Lyngby, Denmark

M.R. Teixeira
CENSE, Faculdade de Ciências e Tecnologia, Universidade do Algarve, Campus de Gambelas, Edifício 7, Faro 8005-139, Portugal

C. Dias-Ferreira
CERNAS—Research Center for Natural Resources, Environment and Society, Escola Superior Agraria de Coimbra, Instituto Politecnico de Coimbra, Bencanta, Coimbra 3045-601, Portugal

A.B. Ribeiro
CENSE, Departamento de Ciências e Engenharia do Ambiente, Faculdade de Ciências e Tecnologia, Universidade Nova de Lisboa, Caparica 2829-516, Portugal

© Springer International Publishing Switzerland 2016

A.B. Ribeiro et al. (eds.), *Electrokinetics Across Disciplines and Continents*,
DOI 10.1007/978-3-319-20179-5_10

FRAC3DVS software (Therrien and Sudicky 1996). Given that LCA classification, characterization, normalization, and weighting were made using EDIP97 method (Wenzel et al. 1997), impacts to groundwater were in fact not considered. These were introduced in the EASEWASTE model (DTU 2012; Kirkeby et al. 2006) as an additional impact category, designated "Spoiled Groundwater Resource". The impact quantification is made, simply calculating the amount of groundwater necessary to dilute the leachate, so that it meets drinking water standards (in m^3/person/year). Leachate may reach groundwater without attenuation; or a constant attenuation factor (AFR) may be introduced by the modeler, in fact reducing by a constant factor, the amount of a substance that reaches groundwater. The model does not consider the transport of the substance nor its substance transformation. In addition, the exploitation of groundwater resources is also not considered. In fact, groundwater as a natural resource is seldom considered in LCA remediation studies, with the exception of process water (i.e., excluding other indirect impacts), and the volume lost due to local contamination. Diamond et al. (1999) is an exception, though no indication is given as to the specific characterization factors. Table 10.1 presents an updated summary of LCA-based evaluation of soil remediation interventions.

Regional impact of groundwater extractions has not largely been considered in LCA as well, nor their relation to surface water bodies and groundwater-dependent ecosystems. These aspects are now compulsory in the management of water resources in Europe, under EU Water Framework Directive (2000/60/EC of the European Parliament and of the Council of 23 October 2000). Future developments in LCA will necessarily have to include these relations in the characterization of the impacts. Recent developments along this line are discussed in Sect. 10.2.

Given the strong retardation that many pollutants undertake in the soil, the temporal factor is very relevant in the groundwater compartment, as contamination may extend for decades. Moreover, groundwater contamination due to industrial sources tends to be spatially concentrated, dispersing from the point of origin depending on hydrogeological conditions, soil retention capacity, pollutant's degradation rates, and time. However, due to temporal and spatial aggregation during the LCI phase, a simultaneous exposure is assumed in LCA (Flemström et al. 2004), which may be limiting for the correct assessment of impacts, in particular those related to health and ecological impacts. As stressed by Owens (1996), in the case of local and transient categories, the impact indicators used in LCA are not a plausible means for detailed understanding the impacts. This problem is particularly relevant in human toxicity assessment, since the probability of surpassing thresholds is the main point of attention, while LCA only provides concentration increases rather than environmental concentrations (Potting et al. 1999).

Methodologies combining environmental risk assessment (ERA) with LCA methods have been proposed as a way to introduce a more in-depth toxicological and ecotoxicological analysis in the latter. It is relevant here to make the distinction between ERA and LCA. While the objective of LCA is to reduce the overall impact on the environment of a product or service, in a "cradle-to-grave" perspective by evaluating the total release of toxic substances and use of resources, ERA focuses exclusively on toxicological and ecotoxicological impacts. Impact categories considered in LCA include (Olsen et al. 2001): *Input categories* (abiotic resources,

Table 10.1 LCA based evaluation of soil remediation interventions

Case study	Contaminated medium	Remediation technology (options studied)	Dominant causes of environmental impact	Time boundary	Reference
Landfill of an industrial area with 460,000 m^3 of waste mix and 100,000 m^3 of spent pot lining (SPL), a waste product of aluminum refining containing cyanide, fluoride, iron, and aluminum	Groundwater	(1) Leaving landfill in place	(1) Significant ecotoxicological impacts	50 years	Godin et al. (2014)
	Surface water	(2) Excavation and on-site disposal	(4) Reduces the overall environmental impacts and the local primary impacts associated with the contaminants left in place		
		(3) Excavation and treatment of SPL fraction			
		(4) Excavation and incineration of SPL fraction			
Multimetal-contaminated site (60 m^2; 90 m^3), 1 km of a refinery plant (As, Cu, Pb), in South Korea	Soil	Electrokinetic (EK) treatment installed as a pilot-scale system	Major impacts: higher GHG emissions and total energy consumption in site preparation, closure and dismantling system installed phase than in system operation and maintenance phase	1 year	Kim et al. (2014)
Site is contaminated with trichloroethene (TCE) located in Tommerup, Denmark, which poses a risk to the drinking water abstraction wells, where water is extracted from the regional groundwater aquifer	Groundwater	(1) Long-term monitoring	(2) Preferable management options as it resulted in the lowest secondary environmental impacts, if it can be ensured that it does not cause any significant leaching of VC to the groundwater	ISCO: 80 years	Lemming et al. (2012)
	Soil	(2) Remediation by enhanced reductive dechlorination (ERD)	(1) Causes the groundwater quality criterion for TCE to be exceeded for 800 years	ERD: 90–200 years	
		(3) Remediation by in-situ substance oxidation (ISCO)	(4) Larger secondary environmental impacts than ERD, but will still be preferable to the ISCO scenario		
		(4) Long-term monitoring combined with activated carbon treatment at the waterworks			

(continued)

Table 10.1 (continued)

Case study	Contaminated medium	Remediation technology (options studied)	Dominant causes of environmental impact	Time boundary	Reference
1.6 ha of contaminated area with oil and fat from a processing plant and with BTEX (benzene, toluene, ethylbenzene, xylene) from the leakage of fuel tanks	Groundwater	(1) Soil excavation	(1) Impact categories most affected freshwater ecotoxicology, climate change affecting human health, nonrenewable energy use and human toxicology	Not specified	Cappuyns and Kessen (2012)
	Soil	(2) Soil vapor extraction	(2) Lower environmental impacts than option 1; impact categories more affected fossil fuel depletion and climate change affecting human health		
Restore soil PCB concentration (200 mg kg^{-1}) to levels to waste acceptance criteria (50 mg kg^{-1} of soil) in hazardous waste landfill sites in France	Soil	(1) Excavation and incineration	(1) Incineration: responsible for the majority of the impacts	265 days	Busset et al. (2012)
		(2) Excavation and bioremediation	(2) Bioremediation processes: more environmental friendly especially for global warning and depletion of abiotic resources impacts		
Remediation of ~3400 m^3 of contaminated soil with VOCs in northeastern US	Soil Groundwater	In situ electrothermal dynamic stripping process	Major impacts: the electricity use in operation stage associated with the fossil fuels, and the mining activities associated with extracting the fossil fuels from the ground	180 days	Fisher (2012)
Treatment of 10,000 t of PCB contaminated soil from 800 to 1000 ppm to less than 5 ppm	Soil	(1) Infrared incineration (2) Base Catalyzed Decomposition (BCD) technology	BCD with less environmental impacts than incineration. Energy consumption from nonrenewable resources in operation phase was	Not specified	Hu et al. (2011)

			the highest impact in both technologies		
5000 m^2 of an area with 6500 m^3 of contaminated soil by organic aliphatic compounds, some aromatic compounds, and BTEX	Soil	(1) Biofuel remediation by *Salix viminalis* cultivation	Biofuel remediation environmental impacts negligible compared with excavation and refilling, except for land use. Transports were the higher cause of impacts	40 days–20 years	Suer and Andersson-Skold (2011)
		(2) Excavation and refilling			
140 m^2 of a contaminated site with TCE (700 m^3) near a catchment for water supply, in central Copenhagen	Soil	(1) Enhanced reductive dechlorination	Options (1) with higher primary toxic impacts due to leaching TCE	Not specified	Lemming et al. (2010a)
	Groundwater	(2) In situ thermal desorption	Options (2) and (3) with human toxic secondary impacts even higher than primary impacts in option (1)		
		(3) Excavation with off-site soil treatment and disposal			
Remediation of a site in Denmark contaminated with 10 t of tetrachloroethene (PCE), 7.5 m below the ground surface, and located within the groundwater catchment that is the source drinking water in this area	Soil	(1) Excavation with off-site soil treatment	Option (1) highly diesel consuming, which results in higher toxic impacts and consumption of more crude oil	30 years	Lemming et al. (2010b)
	Groundwater	(2) Soil vapor extraction	Options (2) and (3) with similar environmental impacts		
		(3) Thermally soil vapor extraction by electrical heating			
Remediation a contaminated site with VOCs in Dover	Soil	(1) Permeable reactive barrier	Option (2) presented higher primary environmental impact in all impact categories	30 years	Higgins and Olson (2009)
		(2) Pump-and-treat system			

(continued)

Table 10.1 (continued)

Case study	Contaminated medium	Remediation technology (options studied)	Dominant causes of environmental impact	Time boundary	Reference
Remediation of a 375 m^3 diesel-contaminated site to Quebec B criterion in soil (700 mg kg^{-1}) and to the detectable limit of C_{10}–C_{50} for potable ground and surface waters (0.1 mg L^{-1})	LNAPL	(1) Pump and treat	Option (2) with the least secondary and primary impacts	2–300 years depending on technology used	Cadotte et al. (2007)
	Soil	(2) Bioslurping, bioventing, and biosparging	Option (3) with the most secondary impacts, while scenario (1) the highest primary impacts		
	Groundwater	(3) Bioslurping, bioventing, and substance oxidation	Option (2) took 38 years to remediate, while (3) required 11 years		
		(4) Ex situ treatment using biopiles	Most affected impact category was water ecotoxicology		
Two different intervention options: rehabilitation of 50.5 ha in urban core at Montreal (Canada) contaminated with heavy industrial activity from the railroad sector (petroleum hydrocarbons, metals, PAH) for residential redevelopment or covering the site with clean soil	Soil	(1a) Excavation and disposal of contaminated soils and wastes	Dominance of tertiary impacts superior to secondary impact by factors of 3–6		Lesage et al. (2007)
		(1b) Infrastructure demolition and recycling of recoverable materials			
		(1c) Landscaping			
		(2) Minimizing exposure			
Remediation of a site of a former manufactured gas plant in the city of Karlsruhe, Germany	Groundwater	(1) Pump-and-treat (PTS)	FGS was more disadvantages since it required a higher level of stability, and more resistant materials for the in-situ construction		Bayer and Finkel (2006)
		(2) Funnel-and-gate systems (FGS)	PTS is much more accessible, flexible, and less demanding on technological element		

Diesel-contaminated site (16,900 m^2) in the province of Quebec (Canada), 8000 m^3 of soil with petroleum hydrocarbon (6145 mg C_{10}–C_{50}/kg soil). The target was 700 mg C_{10}–C_{50}/kg soil (Quebec's B criterion)	Soil	Excavation and treating using an aboveground biopile treatment	Soil itself was responsible for an important fraction of the system's total impact because the toxicity associated with the petroleum hydrocarbons is very important	2 years	Toffoletto et al. (2005)
			Treatment efficiency did not influence the level of secondary impact		
			Off-site transport and the biotreatment process did not contribute notably to the level of environmental impact		
Remediation of a contaminated site in Baden-Wurttemberg, Germany with PAH, mineral oil and chromium	Soil	(1) Excavation and on-site redumping contaminated soil	(2) Unfavorable for primary impacts compared with (1)	Not specified	Volkwein et al. (1999)
		(2) Sealing of the surface by asphalt	(3) More favorable if the remediation type is neglected		
		(3) Excavation and decontamination			
Site contaminated with lead and, secondarily, with arsenic, cadmium, and polyaromatic hydrocarbons (PAHs), located Toronto, Canada	Soil	Excavation and disposal	Major impacts: toxicity, energy consumption, solid waste production, and land use. Energy consumption, with resultant resource depletion and air pollution impacts, was the primary impact due to off-site transportation	25 years	Page and Diamond (1999)

(continued)

Table 10.1 (continued)

Case study	Contaminated medium	Remediation technology (options studied)	Dominant causes of environmental impact	Time boundary	Reference
Generic	Soil	(1) Encapsulation	Options no action and (1) with potential impacts related with land use and, ecosystem and human health	25 years	Diamond et al. (1999)
	Groundwater	(2) Excavation and off-site disposal	Option (2) leads to land consumption		
			Options (3) and (5) impacts in ecosystem and human health		
		(3) In situ bioremediation	Options (3) and (4) on-site aquifer water quality and off-site water quality may be affected		
		(4) Soil washing	Option (4) impacts in the off-site land use possibilities		
		(5) Vapor extraction			

LNAPL Light Nonaqueous Phase Liquid

biotic resources, land); *output related categories* (global warming, depletion of stratospheric ozone, human toxicological impacts, ecotoxicological impacts, photo-oxidant formation, acidification, eutrophication, odor, noise, radiation, casualties); *pro memoria*, which include flows not followed to the system boundary (input-related, such as energy, materials; output related, such as solid waste). The scales at which these categories are evaluated are very different, though. Some can only be evaluated at the global scale, such as global warming, depletion of stratospheric ozone, and some biotic resources. Others are local, such as land, odor, and casualties. Human toxicological and ecotoxicological impacts may, however, be evaluated at different scales, reason why LCA complements well with ERA. Category end points are also similar or convertible. Most methods include (eco)toxicological analysis, though with variable depth and assumptions (EC 2010a; Lehtinen et al. 2011).

LCA is prone to many uncertainty sources, namely arising from: (1) the use of different spatial and temporal scales for different impact categories; (2) assuming linear relationships between pressures and impacts, even for nonlinear processes, such as acidification, photosubstance smog, pollutant transport in the subsoil, and ecotoxicity; (3) not considering natural thresholds, since the evaluation takes into account only the change in the evaluated parameter, disregarding the background value; (4) the lack of standardized characterization factors and category end points; (5) data quality and availability. Scaling factors have been proposed to scale large-scale processes down to regions, which may help reduce the scale effect. These are numerical scores used to indicate ranges of the degree of sensitivity that a particular region has for the selected impact category (Potting and Hauschild 1997; Tolle 1997). Unfortunately, some uncertainty is also introduced through the sensitivity scaling factors. The inaccuracies and uncertainties raised by the use of differing data and methods in LCA, sometimes with contradictory results when different methods were used (Ayres 1995; Owens 1996; Potting et al. 1999; Villanueva and Wenzel 2007), led international agencies, in particular in the European Union, to set minimum reference data and recommended methods for more reliable LCA studies (EC 2010a, b, 2011, 2012), though the benefits of such normalization can only be evaluated within a few years. Unfortunately, for natural resources, no recommendation was made, which leaves out a very broad range of impacts, including those on surface and groundwater resources.

The present chapter discusses the inclusion of a "Groundwater" category in LCA, along with an adaptation to existing methods for impact assessment to include the mobility and fate of substances.

10.2 Remediation with Electrokinetic Methods

Electrokinetic remediation (EKR) also called electrokinetics, electroremediation, or electroreclamation is a solid technology that has been successfully used since the late 1980s to treat contaminated soils, especially low-permeability soils (Lageman et al. 1989; Pamukcu and Wittle 1992; Acar and Alshawabkeh 1993; Probstein and Hicks 1993; Ottosen et al. 1997). Electrokinetics can be defined as the application

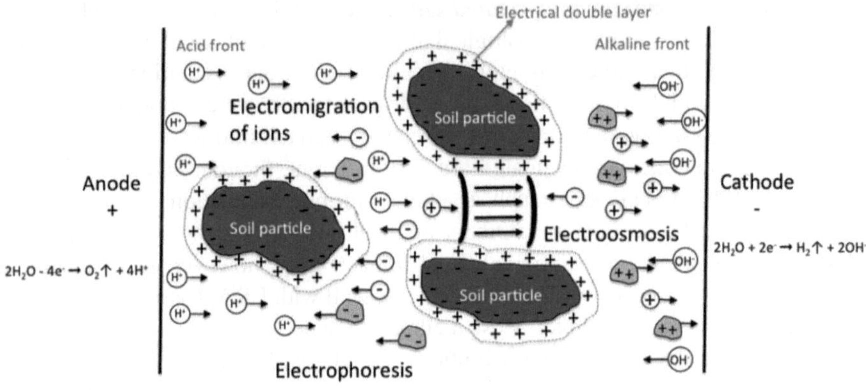

Fig. 10.1 Representation of the main phenomena occurring in electrokinetics

or induction of a low-level direct current in the order of mA cm^{-1} of cross-sectional area between the electrodes or an electric potential difference about few V cm^{-1} across a soil mass containing fluid or a high fluid content slurry/suspension, causing or caused by the motion of electricity, charged soil and/or fluid particles (Sun 2013). In this method, the current acts as the "cleaning agent", generating transport processes (as electroosmosis, electromigration, and electrophoresis) and electrochemical reactions (electrolysis and electrodeposition) (Acar and Alshawabkeh 1993) as shown in Fig. 10.1.

Early EKR models assumed that *electroosmosis* (field-induced convection of water through a porous medium towards the positive or negative electrode depending on the overall surface charge of the porous material) was the only significant transport process (Page and Page 2002). This approach was inadequate for explaining the movement of ionic species (ions and ionic complexes) in the aqueous pore solution, which mainly depends on *electromigration* (field-induced transport of ions in an electrolyte towards the electrode of opposite charge). Electromigration occurs in two steps: (1) desorption of ions from the soil mineral surface to the pore fluid, and its (2) subsequent migration through the pore fluid (Yuan et al. 2008). The electromigration flux is determined by the ionic mobility, tortuosity, porosity of the material, and charge of the ions (Acar and Alshawabkeh 1993). In the case of colloidal suspended charged particles (colloids, clay particles, and organic particles), electrophoresis is the prime transport process. *Diffusion* is the movement of the ionic species in soil solution caused by concentration gradients formed by the electrically induced mass transport. Diffusion is often ignored when studying EK as the ionic mobility of a species is much higher than its diffusion coefficient (Acar and Alshawabkeh 1993).

Electrochemical reactions are also important in electroremediation. Electrolysis reactions prevail at the electrodes, due to the oxidation occurring at the anode and the reduction at the cathode, resulting in water electrolysis, generating an acid front from the anode, whereas reduction at the cathode produces an alkaline front:

$$2H_2O - 4e^- \rightarrow O_2 \uparrow + 4H^+ \; E_0 = -1.229 \text{ V(anode)} \tag{10.1}$$

$$2H_2O + 2e^- \rightarrow H_2 \uparrow + 2OH^- \ E_0 = -0.828 \ \mathrm{V(cathode)} \quad (10.2)$$

The migrating fronts influence the physicochemical properties of the soil–water–electrolyte mixture, thus influencing the remediation effectiveness. Secondary reactions may occur depending on the concentration of available species like metals M_e, for example (Acar and Alshawabkeh 1993):

$$H^+ + e^- \rightarrow \tfrac{1}{2} H_2 \uparrow \quad (10.3)$$

$$Me^{n+} + ne^- \rightarrow Me \quad (10.4)$$

$$Me(OH)_{n(s)} + ne^- \rightarrow Me + nOH^- \quad (10.5)$$

The removal of heavy metals from soils is one of the most-studied processes in electroremediation (Virkutyte et al. 2002). Despite the success at bench scale, there are limited full-scale applications for soil remediation (Fig. 10.2) due to the treatment cost, technical challenges, complex geochemical conditions, and in some cases the need for acidification to induce desorption (Alshawabkeh 2009).

The use of EKR in situ has some advantages, such as (1) the minimization of the site disturbance and posttreatment volume of waste material; (2) it is well suited for fine-grained, heterogeneous media, where other techniques are ineffective; (3) it can remove organic and inorganic pollutants from soil simultaneously; and (4) there is a lower risk of exposure to adsorbed contaminants through groundwater to the workers and surrounding population (Alshawabkeh 2009; Kim et al. 2011). In field applications, EKR involves the installation of electrodes into multiple wells within a contaminated site, followed by the application of a low electric potential across the electrodes. The contaminants are extracted and treated as they migrate and reach the electrode wells.

The use of EKR for remediating soils with organic contaminants was first tested with phenol (Acar et al. 1995), chlorinated solvents (Rabbi et al. 2000; Rohrs et al. 2002), petroleum hydrocarbons (Park et al. 2005; Murillo-Rivera et al. 2009), herbicides (Ribeiro et al. 2005, 2011), creosote (Mateus et al. 2010), calmagite (Agarwal et al. 2008), and polycyclic aromatic hydrocarbons (PAH) (Maini et al. 2000; Niqui-Arroyo and Ortega-Calvo 2007; Alcántara et al. 2010; Pazos et al. 2010; Lima et al. 2011, 2012a, b). The use of different enhancement methods, such as the use of surfactants to enhance EKR of organic compounds, was reviewed by Gomes et al. (2012) and Cameselle et al. (2013).

Challenges for upgrading EKR to full scale include: (1) conditions tested in the lab may not be suitable for real sites, (2) real contamination may be complex and difficult to reproduce at lab scale, (3) the remediation time is too long and the whole operation process is too complicated to be controlled and/or simulated, (4) the electrode conditioning solutions are difficult to inject, (5) the technology may be not sustainable, (6) the treatment cost is too high (although electricity costs often account for only 15 % of the total cost, with major cost drivers being electrodes,

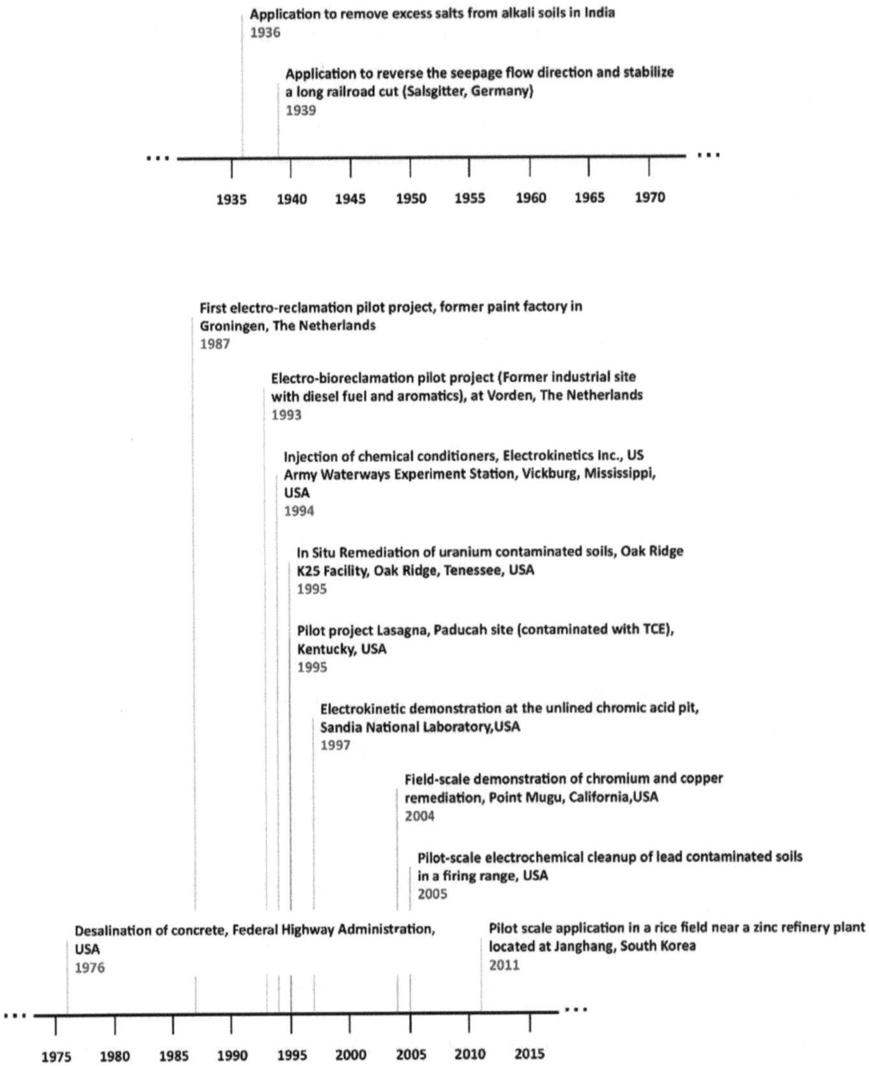

Fig. 10.2 Timeline of the main pilot and full-scale applications of electrokinetics (Page and Page 2002; Lageman et al. 2005; Virkutyte et al. 2002; Ottosen et al. 2008; Yeung 2011; Alshawabkeh 2009; Kim et al. 2012)

electrolyte, anode/cathode reservoir agents, monitoring, and waste management) (Ren et al. 2014). Although the transport processes are well defined, the contaminant removal is complex, highly dependent on variables such as soil mineralogy and organic constituents, the pollutant type, treatment time, electrolyte solution, applied voltage, and resulting current (Herrada et al. 2014; Alshawabkeh 2009).

In EKR field applications, the type of electrode is very important. The electrodes should be electrically conductive, chemically inert, porous, and hollow (Ribeiro and Rodríguez-Maroto 2006). The hollow can work as an outlet of solutions from the subsurface or as an inlet for enhancing agents. The anode operates under highly corrosive conditions and needs to be very resistant. Materials such as graphite, coated titanium, or platinum have been used as anode at lab scale, to prevent dissolution of electrodes and generation of undesirable corrosion products in an acidic-oxidant environment. When upscaling steel and stainless steel pipes are frequently used due to the lower cost. Sacrificial electrodes (in iron, for example) can also be used as the anode, and any conductive material that does not corrode in a basic environment can be used as the cathode (Kim et al. 2011). A hollow stainless steel electrode system was slightly more effective than one with PVC casing, even though the total energy consumption was 5.3 times higher (Kim et al. 2013a). However, this system failed to circulate the anolyte to enhance the metals desorption, due to leakages. An effective anolyte circulation system could enhance the overall EKR performance.

The electrode configuration is also important. One-dimensional and two-dimensional systems of electrodes have been used. In the simplest configuration, a group of sheet electrodes, electrode trenches, or lines of rod electrodes placed in boreholes produce a one-dimensional approach (Ribeiro and Rodríguez-Maroto 2006). When two-dimensional systems (triangular, square, hexagonal, or circular) are employed, the main objective is the generation of a radial or axial-symmetrical flow from the peripheral electrodes to the central one. If the selected contaminants to be removed are positive ions of heavy metals, the peripheral electrodes will be anodes and the central, a cathode, favoring high concentrations of cations around the central zone and their easier and faster removal. If anions are to be removed, the polarity of electrodes reverses this position (Ribeiro and Rodríguez-Maroto 2006). Square and hexagonal configurations with 2 m of electrode spacing showed the higher removal efficiency of As, Cu, and Pb (Kim et al. 2013b). A different approach is proposed in an in situ electroremediation technology called Lasagna™ (Ho et al. 1995, 1997, 1999a, b) that employs the horizontal layering of electrodes and treatment zones for sorption, immobilization, and/or degradation of contaminants.

Dense electrode configurations can result in high current under a constant voltage gradient, which raises soil temperature and causes unnecessary electrical energy consumption (Kim et al. 2012). Additionally, the soil temperature increase transported pore water from the bottom to the top layer, which caused metal accumulation in the topsoil layer (Kim et al. 2012).

EKR is highly dependent on site-specific geochemical conditions such as soil composition, native electrolytes, contaminant aging, and contaminant mixtures (Reddy 2010). An important limitation of the process is the "focusing" effect (Probstein and Hicks 1993) that happens when the acid and alkaline front meet, favoring the metal precipitation, as the pH of minimum solubility generally occurs at the isoelectric point (Li et al. 2012). Solutions like approaching anodes have been proposed to minimize this effect and optimize EKR effectiveness (Li et al. 2012; Zhang et al. 2014). On the other hand, the groundwater flow, gravitational force, and electroosmotic flow, combined together, affect complexly the transport of mobilized contaminants (Kim et al. 2012).

The sustainability assessment of three different remediation technologies in a contaminated site in Illinois, USA, showed that EKR was more advantageous than phytoremediation and excavation (Reddy 2013). The evaluation was made with GREM (Green Remediation Evaluation Matrix) and SiteWise™. However, in this case study, the groundwater was not contaminated, nor the impacts of the remediation process in the groundwater quality were considered.

Another case study of an in situ pilot scale EKR treatment in South Korea considered the environmental impacts of the major phases: remedial investigations, remedial action construction, remedial action operation, and long-term monitoring using SiteWise™ ver. 2 (Kim et al. 2014). The major focuses were greenhouse gas (GHG) emissions, total energy used, air emissions of criteria pollutants, such as NO_x, SO_x, and PM10, and water consumption, and the potential impacts on groundwater were also overlooked.

The efforts to turn EKR more sustainable are concentrated in the use of alternative materials, the reuse/recycling of electrodes, and alternative energy sources such as solar energy (Yuan et al. 2009; Zhou et al. 2013) for the system operation.

10.3 Assessing Impacts in Groundwater Resources

Lemming et al. (2010) justify the inexistence of a "groundwater" category in LCA studies due to the fact that the medium is not included as an emission in LCA models, along with the fact that, in many countries, groundwater is not an important source of drinking water. We may also add that groundwater is seen by LCA modelers as a local category, acting more as a sink (boundary) for contamination than as a relevant environmental medium, naturally connected with surface water and on which many ecosystems depend upon. It is worth noting that the end points currently used in LCA for ecological assessment aren't without critique, namely because scale considerations are largely absent; there is a disproportionate focus on indicators reflecting changes in compositional aspects of biodiversity, mainly changes in species richness, and functional and structural attributes of biodiversity are largely neglected (Curran et al. 2011).

The most frequent approach for assessing impacts in groundwater is essentially volumetric, since what is accounted is the amount of water pumped out and not returned to the medium. For instance, U.S. EPA (USEPA 2008) recommends that water withdrawn from groundwater and subsequently discharged to a surface water body should be included, because the groundwater is not replaced to maintain its beneficial purposes; however, water quantity to be estimated is net consumptive usage, i.e., which either is incorporated into the product, coproducts, or wastes, or is evaporated. The end point for the water use category is exclusively the loss of water (volume). When contaminants' load is considered, LCIA focuses traditionally on the emitted mass, not on the concentration, which disregards of the temporal course of the emissions, and the slow movement and substance transformations in the subsoil.

To the authors' knowledge, the first article outlining methodologies to calculate characterization factors that address the effects of groundwater extraction on the species richness of terrestrial vegetation was first proposed by Zelm et al. (2011) for the Netherlands. The method is data intensive and difficult to adapt to other case studies. Simultaneously, the team involved in EU project LC-IMPACT—"Improved Life Cycle Impact Assessment Methods for Better Sustainability Assessment of Technologies" was also developing characterization factors, in particular for wetlands with varying degrees of complexity, to better adapt to a broader range of practical applications (Amores et al. 2013; Verones et al. 2012, 2013a, b). Both methodologies focused in the hydraulic balance of the system and in the impacts that water deficits may cause on biodiversity (usually as potentially disappeared fraction of species or potentially disappeared number of species). The synergic effects of water quantity and water quality are, however, absent. To the moment, no characterization factors were developed for substance impacts on groundwater-dependent ecosystems due to the transport, or removal/accumulation, of substances in groundwater. Human toxic impacts due to contaminant leaching into groundwater that is used for drinking water have been quantitatively determined using the methods of environmental risk assessment (RA) (Lemming et al. 2012). This approach has some issues, though, namely regarding time, space, boundaries, and data requirements. Conceptually, time could be the same for LCA and RA, but the choice of time frame for LCA in soil remediation processes has proven difficult to establish. Several periods have been considered, namely 20 years (Suer and Andersson-skold 2011), 25 years (Diamond et al. 1999), 30 years (Suèr et al. 2004), 50 years (Bender et al. 1998), and over 800 years (Lemming et al. 2012). Many of these periods are lower than the reference exposure period of 70 years considered in chronic health risk assessment studies, limiting the use of health risk assessment in the context of LCA. In what concerns space, there are important differences, since the space for LCA is defined by following consequences up and downstream, while space for RA is a downstream analysis. The latter has not the problems with well-defined system boundaries and limitations since all consequences should be found (Flemström et al. 2004). The majority of existing models for assessing groundwater concentrations require detailed information about soil and substances properties, which are usually not available, nor are they possible to obtain in the scope of LCA studies. At present, no simple generic model has been proposed for the evaluation of pollutant transport and fate in the subsoil, and how these concentrations convert into impacts. The most common approaches for LCIA are: (1) the amount of a substance leached into groundwater is divided by the accepted concentration limit in groundwater, estimating how much groundwater could be contaminated to the limit values by the leaching (Christensen et al. 2007), and this is designated as the critical-volume approach (Heijungs et al. 2004); (2) to assume that all mass entering the aquifer is captured by supply wells and that no further degradation or purification takes place after the contaminants enter the aquifer (Lemming et al. 2010, 2012). In this latter approach, the exposure concentration in the drinking water, C (g/m^3), at any given time, t, is estimated by dividing the mass discharge, J (g/year), by the annual pumping rate at the waterworks, Q (m^3). The accumulated intake of the substances by the exposed population supplied from the waterworks is estimated by integrating over time the drinking water concentrations multiplied by the daily-ingested groundwater volume (op. cit.). LCIA methods vary significantly on the assessment of effect and severity. Some just evaluate exposure, while others go into detailing specific health end points. Readers are referred to EC (2011) for a detailed comparison of models.

LCA has been shown to complement well with RA, as it helps to identify significant environmental issues (i.e., significant ecotoxicological potential impact) and therefore identify the necessity to proceed to a complete ERA (Godin et al. 2004). While LCA's objective is to reduce the overall pressure on the environment of an entire product system from cradle to grave, being environmental loading focused, the objective of an ERA is to guarantee the environmental safety of a product, being receptor focused (Flemström et al. 2004).

In the characterization phase of LCIA, the value of potential impact indicators is calculated, which result from applying a characterization model to the inventory data. The usual procedure is to multiply each environmental intervention with the corresponding characterization factor, and finally aggregating by impact category (10.6). Indicators may subsequently be grouped, normalized and weighted (ISO 2006).

$$I_{\mathrm{gw},i} = \sum_k \mathrm{CF}_{\mathrm{gw},k,i} \cdot m_{k,i} \tag{10.6}$$

where $I_{\mathrm{gw},i}$ is the indicator result for impact category groundwater for substance i; $\mathrm{CF}_{\mathrm{gw},k,i}$ are the characterization factors relating the emission of substance i in intervention k with the impact category; and $m_{k,i}$ is the magnitude of intervention k, for instance, the amount of the substance i that is emitted and may reach groundwater.

Characterization factors may be obtained from characterization models, which may be based on (1) the critical-volume approach, (2) in more evolved fate-eco (toxicological) models; or, as proposed here, (3) in a combination of the former with soil and groundwater fate parameters.

In the critical-volume approach, the characterization factors are calculated by

$$\mathrm{CF}_{\mathrm{gw},i} = \frac{1}{\mathrm{TV}_{\mathrm{gw},i}} \tag{10.7}$$

where $\mathrm{TV}_{\mathrm{gw},i}$ are the threshold values for groundwater quality for substance i. For instance, if a substance has a threshold of 5 µg L^{-1}, then a kg can pollute a volume of $1/5 \times 10^6 = 2 \times 10^5$ m^3. Instantaneous mixing of a conservative substance is assumed, which is a very strong assumption for groundwater pollution given the large times a substance may take to travel and the many physical–chemical processes with the solid matrix that retard the movement. The values of $\mathrm{TV}_{\mathrm{gw},i}$ depend exclusively on the effect of the substance on the receiving human or ecological end point, having no relation to environmental conditions, such as medium permeability, (ad)sorption, physical and biochemical conditions; nor to the properties of the substance.

Fate-eco (toxicological) models already contemplate the fact that substances will react in the environment, while the ecological and toxicological effects are considered (frequently in the form of (10.7)). These are commonly associated to multimedia transport models to estimate environmental concentrations in the different environmental compartments, though less frequent in groundwater studies, where analytical solutions and methods based on partial differential equations are usually used. Characterization factors are in the form

$$\mathrm{CF}_{\mathrm{gw},i} = \mathrm{EF}_{\mathrm{gw},i} \cdot \mathrm{FF}_{\mathrm{gw},i} \tag{10.8}$$

where $\mathrm{EF}_{\mathrm{gw},i}$ is the effect factor in groundwater for substance i; and $\mathrm{FF}_{\mathrm{gw},i}$ is the fate factor for substance i, which includes all transformations affecting the substance beginning in the location where it entered the environment. When very evolved and detailed environmental fate models are used, LCA becomes very similar to RA, though applied in a larger domain due to a broader definition of system boundaries. For instance, the concentration of substance i in environmental compartment j, in equilibrium with compartment g is generically given for multimedia equilibrium models by (Heijungs et al. 2004)

$$C_{i,j} = \sum_{j \neq g} T_{gj,i} \frac{m_{g,i}}{t} + T_{j,i} \frac{m_{j,i}}{t} \tag{10.9}$$

where $T_{gj,i}$ is the transport coefficient from environmental compartment g to j (for instance, from surface water to groundwater or from soil to groundwater); $T_{j,i}$ is the removal of the substance from compartment j by transport and degradation; t is an arbitrary modeling time necessary to accommodate the steady-state flow and emission rates. Any other modeling method may be used to obtain the predicted environmental concentrations, $C_{\mathrm{gw},i}$. Once calculated, the indicator becomes

$$I_{\mathrm{gw},i} = \sum_i \frac{C_{\mathrm{gw},i}(m_i)}{\mathrm{TV}_{\mathrm{gw},i}} \tag{10.10}$$

Unfortunately, in many instances, the data necessary for calibration and validation of the models isn't available, so more expedite solutions are required. The simplest approach is the critical-volume, as seen above. However, for groundwater, more than for other environmental media, the complete instantaneous mixing of conservative substances is far too large a simplification, which may lead to large overvaluation of impacts, in particular if many substances are included in the assessment.

10.3.1 Proposed Characterization Factor

Here we propose the introduction of a fate factor in the characterization factor for groundwater, which is simpler than the modeling approaches and is introduced to weight the contribution of different substances to the indicator value. The proposed scheme differs from the weighting elements of LCA (ISO 2006) in that it is not converting nor aggregating indicator results across impact categories. The objective of the weighting is here to distinguish substances by their mobility and degradability, attributing a larger weight to those that are expected to contaminate a larger aquifer volume, in a shorter period of time. The new characterization factor is altered from the critical-volume approach by introducing a rank order weight in (10.7):

$$\mathrm{CF}_{\mathrm{gw},k,i} = \sqrt[\mathrm{wi}]{\frac{1}{\mathrm{TV}_{\mathrm{gw},i}}} \tag{10.11}$$

Being W_i the rank order of the substance according to a chosen scoring and ranking methodology, as described below. W_i is the position in an ordered series of risk a substance poses to groundwater. The indicator is then calculated using (10.6). This altered CF will make that less mobile and/or less persistent substances contribute less to the value of the indicator, as they should.

The altered CF may help correct the unrealistic instantaneous, complete dilution assumption discussed above. Substances ranking higher will dissolve more, degrade less, and retard less, contaminating large volumes; in opposition to low ranking substances, which are more reactive, retarding and degrading more, and contaminating smaller aquifer volumes. The method, at the moment, can only be applied to organic contaminants, as no ranking system has yet been proposed for metals.

10.3.2 Ranking Order of Substances

A substance is highly mobile in subsoil if its soil organic carbon–water partitioning coefficient, as Log K_{OC} [Log(mL g^{-1})], is lower than 1; and immobile if it is higher than 5. In between these two extremes, several other classes of mobility can be established, as presented in Table 10.2.

Biodegradation reactions depend on environmental conditions, in particular, soil pH, temperature, presence of electron donors and acceptors, and presence of specific microorganisms. Biodegradation is limited by (Schwarzenbach et al. 2003): (1) bioavailability of the contaminant, namely the delivery to the organisms' metabolic apparatus capable of transforming the chemical, (2) the enzyme's ability to mediate the initial transformation of the chemical, or (3) the growth of a population of microorganisms in response to the presence of a new

Table 10.2 Classification of mobility according to FAO (2000)

Log K_{OC}	Classification
<1	Highly mobile
1–2	Mobile
2–3	Moderately mobile
3–4	Slightly mobile
4–5	Hardly mobile
>5	Immobile

substrate. In general, biodegradation is conditioned by the rate of microbial population increase.

Depending on the factors limiting population growth, different mathematical formulations have been proposed, namely: (1) the first-order biodegradation (removal) model; (2) the *Monod* model (also known as Michaelis–Menten kinetics); (3) the *instantaneous reaction model* (electron-acceptor-limited) (Borden and Bedient 1986; Borden et al. 1986; Koussis et al. 2003); and (4) *biogeochemical models* (McNab and Narasimhan 1994; Postma and Jakobsen 1996). The first-order removal model is described by a power function given by

$$\frac{dC}{dt} = -bC \tag{10.12}$$

where b is the biodegradation rate constant (T^{-1} $(\mathrm{M/L^3})^{1-d}$), which after integration results in

$$\frac{C(t)}{C_0} = \mathrm{e}^{[-\mathrm{b}\,dt]} \tag{10.13}$$

The value of the biodegradation rate when the concentration is half of the initial is

$$b = \frac{\ln[2]}{t_{1/2}} \tag{10.14}$$

The parameter $t_{1/2}$ is the half-life (T) of the contaminant in a given environment. Table 10.3 shows FAO's classification for degradability according to half-life.

The ranking W_i is an ordered series of scores calculated with one of the models presented below, where the substances are ranked from the highest mobility to the lowest. Hence, the substance with the highest mobility will have a $W_i = 1$. Substances with equal scores receive the same rank.

The revised AFR proposed by Li et al. (1998) is based on conceptual models of water flow and chemical fate, based on the value of net recharge rate, q, groundwater depth, d, water content at field capacity, θ_{FC}, chemical half-life in soil, $t_{1/2}$, soil bulk density, ρ_b, fraction organic carbon content, f_{OC}, and K_{OC}.

Table 10.3 Classification of degradability (FAO 2000)

$t_{1/2}$ (days)	Classification
<20	Readily degradable
20–60	Fairly degradable
60–180	Slightly degradable
>180	Very slightly degradable

$$\mathrm{AFR} = \ln\left[\frac{d\theta_{\mathrm{FC}}(1+\rho_b f_{\mathrm{OC}} K_{\mathrm{OC}})/\theta_{\mathrm{FC}}}{q t_{1/2}}\right] + k \qquad (10.15)$$

With k, a constant big enough to insure that AFR is greater than unity. More mobile substances receive lower AFR values.

The groundwater ubiquity score (GUS) (Gustafson 1989) is one of the best well-known models to rank mobility of chemicals in soil. It is based on only two properties of the chemical, half-life in soil and K_{OC}

$$\mathrm{GUS} = \log\left(t_{1/2}\right)\left(4 - \log(K_{\mathrm{OC}})\right) \qquad (10.16)$$

The equation separates substances considered as leachers, i.e., that may migrate down to groundwater, from those that are retained in the soil, or that degrade fast in the soil. GUS > 2.8 identifies very mobile substances (leachers); GUS < 1.8 identifies substances that do not leach; substances with GUS values in between indicate substances for which evidence, at the date, of leaching was either incomplete or contradictory. Being dependent only on properties of the chemical, it is less susceptible to uncertainties introduced with the soil matrix and recharge, but, is on the other hand, not suitable for case-study application.

Following the works of Rao (1985) and Jury (1987), Meeks and Dean (1990) introduced the Leaching Potential Index (MPI), which is obtained by inverting the exponential in the AF model.

$$\mathrm{LPI} = \frac{1000 q t_{1/2}}{R d \ln(2)} \qquad (10.17)$$

With $R = \rho_b f_{\mathrm{OC}} K_{\mathrm{OC}}$.

Nunes et al. (2011) proposed the Naphthalene Exposure Risk Index (NERI) for ranking the potential risk of exposure to organic contaminants in industrial areas in the saturated zone. It may be seen as a complement to GUS, which evaluates the mobility of a substance in the unsaturated zone toward the groundwater. NERI is a weighted sum of the logarithms of solubility, half-life, density, and K_{OC}. The weights are logarithms of the values of the respective properties for naphthalene, chosen because it has a GUS value of 1.8, being at exactly the limit between a leacher and a non-leacher chemical. The first two parameters enter as positive values, while the latter two as negative, reflecting the fact that higher solubility and persistence contribute to increased mobility; whereas density and K_{OC} have the opposite effect. The ranking of chemicals is made from the more mobile (highest NERI) in class

(i) to the least mobile (lowest NERI) in class (iii). The method correlates well with scores obtained with GUS, with the exception of chemicals with more extreme properties, as, e.g., 1,2-dibromoethane due to its high density, and benzo(k) fluoranthene due to its low solubility. NERI is an empirical index which considers that higher solubility and half-life times contribute to increasing the probability of a contaminant to migrate between the source and the receptor; contaminants denser than water will migrate more vertically, therefore increasing their vertical mobility, but decreasing the horizontal; higher adsorption will always cause retardation, thereby decreasing the probability of the contaminant reaching the receptor by providing more time for degradation. The constant on the last term of the second equation serves for zeroing the index for naphthalene. The NERI is obtained by the following equations for the horizontal and vertical directions. NERI values above zero indicate mobile substances; while below zero indicate the opposite. Note that the units of solubility and half-life must be chosen such that the logarithm of the values for naphthalene is not negative to avoid inversing the ranking.

$$\mathrm{NERI}_H = \frac{\mathrm{Log}(S_i)}{\mathrm{Log}(S_{\mathrm{Naph}})} + \frac{\mathrm{Log}(t_{1/2_i})}{\mathrm{Log}(t_{1/2\mathrm{Naph}})} - \frac{\rho_i}{\rho_{\mathrm{Naph}}} - \frac{\mathrm{Log}(K_{\mathrm{OC}i})}{\mathrm{Log}(K_{\mathrm{OCNaph}})} \tag{10.18}$$

10.4 Example of Application of the Adapted Characterization Factor

To illustrate the method proposed above, the following substances were chosen: polycyclic aromatic hydrocarbons naphthalene (Naph) and benzo(α)pyrene (BaP), phenol (Phe), and trichloroethylene (TCE). Their water solubility, mobility, and persistence in soil and groundwater are indicated in Table 10.4. The soil and local properties were: $q = 0.000822$ m/d; $d = 5$ m; sandy soil with $\theta_{\mathrm{FC}} = 0.3$; $\theta = 0.43$; $\rho_{\mathrm{b}} = 1.5$ g/cm^3; $f_{\mathrm{OC}} = 0.006$.

Given the large range of values published for many of the chemical and fate properties of substances, the probabilistic quantitative assessment was made with Monte Carlo simulation (MCS) to evaluate the impact of the uncertainties on the

Table 10.4 Chemical properties of selected substances found in contaminated soils. Medians of published values (Mackay et al. 2006). TV_i are the Portuguese groundwater quality guidelines (INAG 2009, 2011)

	Substance			
Property	B(a)P	Naphthalene	Phenol	TCE
Log K_{OC}	5.79	3.02	1.56	2.13
$t_{1/2}$ in soil (d)	1 103	115	8.5	161
$t_{1/2}$ in gw (d)	1 060	258	7.0	720
S (g m^{-3})	0.00266	31.17	102,442.6	1338.2
ρ (g cm^{-3})	1.24	1.03	1.05	1.62
TV_i (mg m^{-3})	0.01	2.4	0.182	0.2

estimated scores. During simulation values are picked from probability density functions (pdf) obtained from published data, and used to determine the scores until the limit number of trials has been reached. At the end of the simulation, there will be as many score results as trials, allowing the determination of statistics for the score based on the trials. Hence, unlike a deterministic approach, for which only one score value is calculated, in probabilistic risk assessment, a posterior pdf for the score is obtained. During simulation, numerical stability was achieved after 50,000 trials, as indicated by a difference lower than 1 % on the calculated values of the 50th and 95th percentiles. To better reproduce the pdf and increase the accuracy of the estimates, Latin hypercube sampling was used. Simulations were performed with Crystal Ball®. Chemical and fate properties of substances were obtained from the "Handbook of Physical-Chemical Properties and Environmental Fate for Organic Chemicals" (Mackay et al. 2006). Pdf were obtained by fitting continuous statistical distributions (normal, lognormal, and Weibull) to the chemical property data using the chi-square test when the number of data values was superior to fifteen; and a triangular function with extremes on the minimum and maximum and centered on the median, when lower.

The median scoring calculated by the MCS and the rankings thus obtained are shown in Table 10.5. Interestingly, in spite of the different formulations, the scoring methods provide similar rankings when the median score is used. B(a)P ranks as the substance with the lowest mobility, therefore with the potential to contaminate a lower volume of groundwater; it is followed in increasing potential to contaminate groundwater by naphthalene, phenol, and TCE. Only in the case of NERIh, which is specific for the saturated zone, is the ranking slightly different, being phenol in this case more mobile than TCE. This difference is justified by the different densities: TCE is a dense nonaqueous phase hydrocarbon with a larger tendency to migrate vertically, dispersing less with the flowing groundwater; whereas phenol has a density similar to that of water, dispersing and diluting more in the flowing groundwater.

The scorings calculated with the different methods are, in general, much in agreement with each other, making very clear distinction between substances with very disparate K_{OC} and $t_{1/2}$, such as TCE and B(a)P (Fig. 10.3), if the mode or median score values are used. Only NERI scores of substances are different when their K_{OC} and $t_{1/2}$ are similar, due to the inclusion of water solubility and density, where pdf curves superpose the methods cannot make the distinction between substances, meaning that for a large set of chemical parameter values, the

Table 10.5 Scores and ranks calculated with different methods

	Substance			
Scoring and ranking method	B(a)P	Naphthalene	Phenol	TCE
GUS	−5.32 [4]	1.93 [3]	2.10 [2]	4.05 [1]
AFR	10.42 [4]	6.36 [3]	6.29 [2]	4.17 [1]
LPI	0.04 [4]	2.59 [3]	5.29 [2]	30.07 [1]
NERIv	−3.18 [4]	0.00 [3]	2.22 [2]	2.17 [1]
NERIh	−3.60 [4]	0.00 [3]	2.17 [1]	1.01 [2]

Ranks are shown in parenthesis

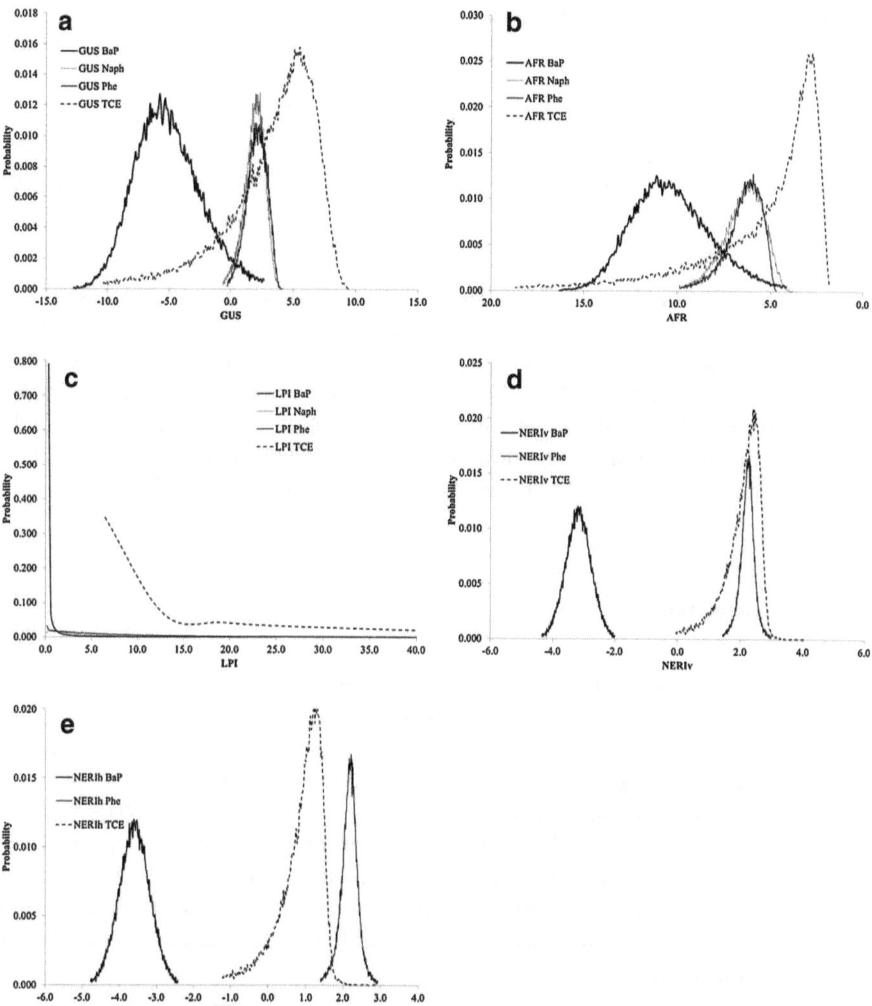

Fig. 10.3 Posterior probability density functions of scores. The x-axis in AFR is shown in inverse order such that more mobile substances are always shown to the *right* of the graphs

substances have an equal potential for groundwater contamination. NERI is more resilient to data uncertainty, providing a better classification. Sensitivity analysis showed that K_{OC} contributes to over 95 % of the variability of these scores, meaning that small differences in the value of the parameter will have a large impact on the estimated score (and rank). NERI is less much less sensitive to the parameter, with a weight of around 55 %.

The characterization factors obtained by substituting the ranks indicated in Table 10.5 and TV_i presented in Table 10.4 are shown in Table 10.6. The altered CF is lower than the original CF by a large amount. This difference is justified by the

Table 10.6 Calculated characterization factors for groundwater

	Substance			
	B(a)P	Naphthalene	Phenol	TCE
R_c	38,088	12	1.6	3.6
Original CF (m^3)	100×10^6	400,000	5,500,000	5,000,000
Altered CF (m^3)	100	75	2 344	5,000,000

cumulative effect of chemical degradation and retardation. Given the high sensitivity of the models and of real environmental mobility to K_{OC}, soil retardation plays a fundamental role in the estimation of the volumes of contaminated groundwater. The retardation coefficient, $R_c = \rho_b f_{OC} K_{OC} + 1$, for the selected substances is shown in Table 10.6. R_c for B(a)P is about three orders of magnitude higher than for Naph, indicating that the volume of groundwater contaminated by Naph will be about 3000 times larger than by B(a)P. The original CF disregards completely this effect.

10.5 Conclusions

Exposure to contaminated groundwater is frequently assessed by modeling, but the necessary parameter data is many times impossible to obtain in the context of LCA. Simpler methods are therefore required. The present chapter discussed some of these alternatives, in particular applied to soil remediation with electrokinetic methods. A characterization factor for groundwater was proposed based on an existing one. This altered CF receives as exponent, the inverse of the ranking order given by a chosen scoring and ranking method. The altered CF reflects a substance's fate in the environment through ratios of volumes of contaminated groundwater by different substances, which is in line with practice in groundwater contamination studies.

References

Acar YB, Alshawabkeh AN (1993) Principles of electrokinetic remediation. Environ Sci Technol 27(13):2638–2647

Acar YB, Gale RJ, Alshawabkeh AN, Marks RE, Puppala S, Bricka M, Parker R (1995) Electrokinetic remediation: basics and technology status. J Hazard Mater 40(2):117–137. doi:10.1016/0304-3894(94)00066-P

Agarwal S, Cluxton P, Kemper M, Dionysiou DD, Al-Abed SR (2008) Assessment of the functionality of a pilot-scale reactor and its potential for electrochemical degradation of calmagite, a sulfonated azo-dye. Chemosphere 73:837–843

Alcántara MT, Gómez J, Pazos M, Sanromán MA (2010) Electrokinetic remediation of PAH mixtures from kaolin. J Hazard Mater 179(1-3):1156–1160. doi:10.1016/j.jhazmat.2010.03.010

Alshawabkeh AN (2009) Electrokinetic soil remediation: challenges and opportunities. Sep Sci Technol 44:2171–2187

Amores MJ, Verones F, Raptis C, Juraske R, Pfister S, Stoessel F, Antón A, Castells F, Hellweg S (2013) Biodiversity impacts from salinity increase in a coastal wetland. Environ Sci Technol 47:6384–6392

Ayres RU (1995) Life cycle analysis: a critique. Resour Conserv Recycl 14:199–223

Bayer P, Finkel M (2006) Life cycle assessment of active and passive groundwater remediation technologies. J Contam Hydrol 83:171–199. doi:10.1016/j.jconhyd.2005.11.005

Bender A, Volkwein S, Battermann G, Hurtig H-W, Kopffer W, Kohler W (1998) Life cycle assessment method for remedial action techniques: methodology and application. In: Telford, T. (Ed.), Contaminated Soil 98—Sixth International FZK/TNO Conference. Edinburgh, pp 367–376

Borden RC, Bedient PB (1986) Transport of dissolved hydrocarbons influenced by oxygen-limited biodegradation: 1: theoretical development. Water Resour Res 22:1973–1982

Borden RC, Bedient PB, Lee MD, Ward CH, Wislon JT (1986) Transport of dissolved hydrocarbons influenced by oxygen-limited biodegradation: 2: field application. Water Resour Res 22:1983–1990

Busset G, Sangely M, Montrejaud-Vignoles M et al (2012) Life cycle assessment of polychlorinated biphenyl contaminated soil remediation processes. Int J Life Cycle Assess 17:325–336

Cadotte M, Deschênes L, Samson R (2007) Selection of a remediation scenario for a diesel-contaminated site using LCA. Int J Life Cycle Assess 12:239–251

Cameselle C, Gouveia S, Akretche DE, Belhadj B (2013) Advances in electrokinetic remediation for the removal of organic contaminants in soils. In: Rashed MN (ed) Soils, organic pollutants—monitoring, risk and treatment. http://www.intechopen.com/books/organic-pollutants-monitoring-risk-and-treatment/advances-in-electrokinetic-remediation-for-the-removal-of-organic-contaminants-in-soils. doi:10.5772/54334

Cappuyns V, Kessen B (2012) Evaluation of the environmental impact of Brownfield remediation options: comparison of two life cycle assessment-based evaluation tools. Environ Technol 33:2447–2459

Cappuyns V (2013) LCA based evaluation of site remediation. Opportunities and limitations. Chem Today 31:18–21

Christensen TH, Bhander G, Lindvall H, Larsen AW, Fruergaard T, Damgaard A, Manfredi S, Boldrin A, Riber C, Hauschild M (2007) Experience with the use of LCA-modelling (EASEWASTE) in waste management. Waste Manag Res 25:257–262

Curran M, De Baan L, De Schryver AM, Van Zelm R, Hellweg S, Koellner T, Sonnemann G, Huijbregts MAJ (2011) Toward meaningful end points of biodiversity in life cycle assessment. Environ Sci Technol 45:70–79

Diamond ML, Page CA, Campbell M, Mckenna S, Lal R (1999) Life-cycle framework for assessment of site remediation options: case study. Environ Toxicol Chem 18:801–810

DTU (2012) EASEWASTE user manual. Technical University of Denmark, Kongens Lyngby

EC (2010a) ILCD handbook: analysing of existing environmental impact assessment methodologies for use in life cycle assessment. European Commission, Brussels

EC (2010b) ILCD handbook: framework and requirements for LCIA models and indicators first edition. European Commission, Brussels

EC (2011) ILCD handbook: recommendations for life cycle impact assessment in the European context. European Commission, Brussels

EC (2012) The International Reference Life Cycle Data System (ILCD) handbook. European Commission, Brussels

FAO (2000) Assessing soil contamination. A reference manual. FAO Pesticide Disposal Series 8. Food and Agriculture Organization of the United Nations, Rome

Fisher A (2012) Life-cycle assessment of in situ thermal remediation. Autumn 22(4):75–92

Flemström K, Carlson R, Erixon M (2004) Relationships between life cycle assessment and risk assessment—potentials and obstacles, technology. Industrial Environmental Informatics (IMI), Chalmers University of Technology, Stockholm

Godin J, Ménard J, Hains S (2004) Combined use of life cycle assessment and groundwater transport modeling to support contaminated site management. Hum Ecol Risk Assess An Int J 10:37–41

Gomes HI, Dias-Ferreira C, Ribeiro AB (2012) Electrokinetic remediation of organochlorines in soil: enhancement techniques and integration with other remediation technologies. Chemosphere 87(10):1077–1090. doi:10.1016/j.chemosphere.2012.02.037

Gustafson DI (1989) Groundwater ubiquity score: a simple method for assessing pesticide leachability. Environ Toxicol Chem 8:339–357

Herrada RA, Perez-Corona M, Shrestha RA, Pamukcu S, Bustos E (2014) Electrokinetic remediation of polluted soil using nano-materials: nano-iron case In: Peralta-Hernández JM, Rodrigo-Rodrigo MA, Martínez-Huitle CA (eds) Evaluation of electrochemical reactors as a new way to environmental protection. Research Signpost, Trivandrum, pp 41–57

Heijungs R, de Koning A, Ligthart T, Korenromp R (2004) Improvement of LCA characterization factors and LCA practice for metals, Cycle. TNO Environment, Energy and Process Innovation, Apeldoorn

Higgins M, Olson T (2009) Life cycle case study comparison of permeable reactive barrier versus pump-and-treat remediation. Environ Sci Technol 43:9432–9438

Ho SV, Athmer C, Sheridan PW, Hughes BM, Orth R, McKenzie D, Brodsky PH, Shapiro A, Sivavec TM, Salvo J, Schultz D, Landis R, Griffith R, Shoemaker S (1999a) The Lasagna technology for in situ soil remediation. 2. Large field test. Environ Sci Technol 33:1092–1099

Ho SV, Athmer C, Sheridan PW, Hughes BM, Orth R, McKenzie D, Brodsky PH, Shapiro A, Thornton R, Salvo J, Schultz D, Landis R, Griffith R, Shoemaker S (1999b) The Lasagna technology for in situ soil remediation. 1. Small field test. Environ Sci Technol 33:1086–1091

Ho SV, Athmer CJ, Sheridan PW, Shapiro AP (1997) Scale-up aspects of the Lasagna process for in situ soil decontamination. J Hazard Mater 55:39–60

Ho SV, Sheridan PW, Athmer CJ, Heitkamp MA, Brackin JM, Weber D, Brodsky PH (1995) Integrated in situ soil remediation technology: the Lasagna process. Environ Sci Technol 29:2528–2534

Hu X, Zhu J, Ding Q (2011) Environmental life-cycle comparisons of two polychlorinated biphenyl remediation technologies: incineration and base catalyzed decomposition. J Hazard Mater 191:258–268

INAG (2009) Estabelecimento de limiares nas águas subterrâneas. Instituto Nacional da Água, Ministério do Ambiente, Lisboa

INAG (2011) Estabelecimento de limiares para hidrocarbonetos nas águas subterrâneas. Massa de água de Sines. Instituto Nacional da Água, Ministério do Ambiente, Lisboa

ISO (2006) ISO 14044—Environmental management—life cycle assessment—requirements and guidelines. International Organization for Standardization, Geneva

Jury WA, Focht DD, Farmer WJ (1987) Evaluation of pesticide groundwater pollution potential from standard indices of soil-chemical adsorption and biodegradation. J Environ Qual 16:422–428

Kim B-K, Baek K, Ko S-H, Yang J-W (2011) Research and field experiences on electrokinetic remediation in South Korea. Sep Purif Technol 79(2):116–123. doi:10.1016/j.seppur.2011.03.002

Kim B-K, Park G-Y, Jeon E-K, Jung J-M, Jung H-B, Ko S-H, Baek K (2013a) Field application of in situ electrokinetic remediation for As-, Cu-, and Pb-contaminated paddy soil. Water Air Soil Pollut 224(9):1–10

Kim D-H, Yoo J-C, Hwang B-R, Yang J-S, Baek K (2014) Environmental assessment on electrokinetic remediation of multimetal-contaminated site: a case study. Environ Sci Pollut Res 21(10):6751–6758. doi:10.1007/s11356-014-2597-1

Kim W-S, Jeon E-K, Jung J-M et al (2013b) Field application of electrokinetic remediation for multi-metal contaminated paddy soil using two-dimensional electrode configuration. Environ Sci Pollut Res 21:1–10. doi:10.1007/s11356-013-2424-0

Kim W-S, Park G-Y, Kim D-H, Jung H-B, Ko S-H, Baek K (2012) In situ field scale electrokinetic remediation of multi-metals contaminated paddy soil: influence of electrode configuration. Electrochim Acta 86:89–95. doi:10.1016/j.electacta.2012.02.078

Kirkeby JT, Birgisdottir H, Hansen TL, Christensen TH, Bhander GS (2006) Environmental assessment of solid waste systems and technologies: EASEWASTE. Waste Manag Res 24:3–15

Koussis AD, Pesmajoglou S, Syriopoulou D (2003) Modelling biodegradation of hydrocarbons in aquifers: when is the use of the instantaneous reaction approximation justified? J Contam Hydrol 60:287–305

Lageman R, Clarke RL, Pool W (2005) Electro-reclamation, a versatile soil remediation solution. Eng Geol 77(3-4):191–201

Lageman R, Pool W, Seffinga GA (1989) Electro-reclamation. Chem Ind 18:585–590

Lehtinen H, Saarentaus A, Rouhiainen J, Pits M, Azapagic A (2011) A review of LCA methods and tools and their suitability for SMEs. EU Project BIOCHEM, Cheshire

Lemming G, Hauschild MZ, Chambon J et al (2010) Environmental impacts of remediation of a trichloroethene-contaminated site: life cycle assessment of remediation alternatives. Environ Sci Technol 44(23):9163–9169. doi:10.1021/es102007s

Lemming G, Chambon JC, Binning PJ, Bjerg PL (2012) Is there an environmental benefit from remediation of a contaminated site? Combined assessments of the risk reduction and life cycle impact of remediation. J Environ Manage 112:392–403. doi:10.1016/j.jenvman.2012.08.002

Lesage P, Ekvall T, Deschênes L, Samson R (2007) Environmental assessment of brownfield rehabilitation using two different life cycle inventory models. Int J Life Cycle Assess 12 (7):497–513. doi:10.1065/lca2006.10.279.2

Li ZC, Yost RS, Green RE (1998) Incorporating uncertainty in a chemical leaching assessment. J Contam Hydrol 29:285–299

Li G, Guo S, Li S, Zhang L, Wang S (2012) Comparison of approaching and fixed anodes for avoiding the 'focusing' effect during electrokinetic remediation of chromium-contaminated soil. Chem Eng J 203:231–238. doi:10.1016/j.cej.2012.07.008

Lima AT, Kleingeld PJ, Heister K, Loch JPG (2011) Removal of PAHs from contaminated clayey soil by means of electro-osmosis. Sep Purif Technol 79(2):221–229. doi:10.1016/j.seppur.2011.02.021

Lima AT, Kleingeld PJ, Heister K, Loch JPG (2012a) In situ electro-osmotic cleanup of tar contaminated soil—removal of polycyclic aromatic hydrocarbons. Electrochim Acta 86:142–147

Lima AT, Ottosen LM, Heister K, Loch JPG (2012b) Assessing PAH removal from clayey soil by means of electro-osmosis and electrodialysis. Sci Total Environ 435–436:1–6

Mackay D, Shiu W-Y, Ma K-C, Lee SC (2006) Handbook of physical-chemical properties and environmental fate for organic chemicals, 2nd edn. CRC, Boca Raton

Maini G, Sharman AK, Knowles CJ, Sunderland G, Jackman SA (2000) Electrokinetic remediation of metals and organics from historically contaminated soil. J Chem Technol Biotechnol 75:657–664

Mateus EP, Zrostlíková J, da Silva MDG, Ribeiro AB, Marriott P (2010) Electrokinetic removal of creosote from treated timber waste: a comprehensive gas chromatographic view. J Appl Electrochem 40(6):1183–1193

McNab WW, Narasimhan TN (1994) Modeling reactive transport of organic compounds in groundwater using a partial redox disequilibrium approach. Water Resour Res 30:2619–2635

Meeks Y, Dean J (1990) Evaluating ground-water vulnerability to pesticides. J Water Resour Plan Manag 116:693–707

Murillo-Rivera B, Labastida I, Barron J, Oropeza-Guzman MT, Gonzalez I, Teutli-Leon MMM (2009) Influence of anolyte and catholyte composition on TPHs removal from low permeability soil by electrokinetic reclamation. Electrochim Acta 54:2119–2124
Niqui-Arroyo J-L, Ortega-Calvo J-J (2007) Integrating biodegradation and electroosmosis for the enhanced removal of polycyclic aromatic hydrocarbons from creosote-polluted soils. J Environ Qual 36:1444–1451
Nunes LM, Zhu Y-G, Stigter TY, Monteiro JP, Teixeira MR (2011) Environmental impacts on soil and groundwater at airports: origin, contaminants of concern and environmental risks. J Environ Monit 13:3026–3039
Olsen SI, Christensen FM, Hauschild M, Pedersen F, Larsen HF, Tørsløv J (2001) Life cycle impact assessment and risk assessment of chemicals—a methodological comparison. Environ Impact Assess Rev 21:385–404
Ottosen LM, Christensen IV, Rörig-Dalgård I, Jensen PE, Hansen HK (2008) Utilization of electromigration in civil and environmental engineering—processes, transport rates and matrix changes. J Environ Sci Health A 43(8):795–809. doi:10.1080/10934520801973949
Ottosen LM, Hansen HK, Laursen S, Villumsen A (1997) Electrodialytic remediation of soil polluted with copper from wood preservation industry. Environ Sci Technol 31(6):1711–1715
Owens JW (1996) LCA impact assessment categories. Int J Life Cycle Assess 1:151–158
Page CA, Diamond L (1999) Life-cycle frame work for assessment of site remediation options: case study. Environ Toxicol Chem 18:801–810
Page MM, Page CL (2002) Electroremediation of contaminated soils. J Environ Eng-ASCE 128 (3):208–219. doi:10.1061/(asce)0733-9372(2002)128:3(208)
Pamukcu S, Wittle JK (1992) Electrokinetic removal of selected heavy metals from soil. Environ Prog 11:241–250
Park J-Y, Kim S-J, Lee Y-J, Baek K, Yang J-W (2005) EK-Fenton process for removal of phenanthrene in a two-dimensional soil system. Eng Geol 77:217–224
Pazos M, Rosales E, Alcántara T, Gómez J, Sanromán MA (2010) Decontamination of soils containing PAHs by electroremediation: a review. J Hazard Mater 177:1–11
Postma D, Jakobsen R (1996) Redox zonation: equilibrium constraints on the Fe(III)/SO4-reduction interface. Geochim Cosmochim Acta 60:3169–3175
Potting J, Hauschild M (1997) Part II: spatial differentiation in life-cycle assessment via the site-dependent characterisation of environmental impact from emissions. Int J Life Cycle Assess 2:209–216
Potting J, Hauschild M, Wenzel H (1999) "Less is better" and "only above threshold": two incompatible paradigms for human toxicity in life cycle assessment? Int J Life Cycle Assess 4:16–24
Probstein RF, Hicks RE (1993) Removal of contaminants from soil by electric fields. Science 260:498–530
Rabbi MF, Clark B, Gale RJ, Ozsu-Acar E, Pardue J, Jackson A (2000) In situ TCE bioremediation study using electrokinetic cometabolite injection. Waste Manag 20:279–286
Rao PSC, Hornsby AG, Jessup RE (1985) Indices for ranking the potential for pesticide contamination of groundwater. Proc Soil Crop Sci Soc 44:1–8
Reddy K (2010) Technical challenges to in-situ remediation of polluted sites. Geotech Geol Eng 28(3):211–221. doi:10.1007/s10706-008-9235-y
Reddy KR (2013) Sustainability evaluation of electrokinetics and other remediation alternatives for a contaminated site: a case study. Paper presented at the 12th international symposium on electrokinetic remediation, Boston, 23–25 June 2013. http://nuweb9.neu.edu/erem2013/wp-content/uploads/2013/07/EREM2013_Final_Web.pdf
Ren L, Lu H, He L, Zhang Y (2014) Enhanced electrokinetic technologies with oxidization–reduction for organically-contaminated soil remediation. Chem Eng J 247:111–124. doi:10.1016/j.cej.2014.02.107

Ribeiro AB, Mateus EP, Rodríguez-Maroto J-M (2011) Removal of organic contaminants from soils by an electrokinetic process: the case of molinate and bentazone. Experimental and modeling. Sep Purif Technol 79(2):193–203

Ribeiro AB, Rodríguez-Maroto JM (2006) Electroremediation of heavy metal-contaminated soils. Processes and applications. In: Prasad MNV, Sajwan KS, Naidu R (eds) Trace elements in the environment: biogeochemistry, biotechnology and bioremediation. Taylor & Francis, Boca Raton, pp 341–368

Ribeiro AB, Rodríguez-Maroto JM, Mateus EP, Gomes H (2005) Removal of organic contaminants from soils by an electrokinetic process: the case of atrazine: experimental and modeling. Chemosphere 59(9):1229–1239. doi:10.1016/j.chemosphere.2004.11.054

Rohrs J, Ludwig G, Rahner D (2002) Electrochemically induced reactions in soils—a new approach to the in-situ remediation of contaminated soils? Part 2: remediation experiments with a natural soil containing highly chlorinated hydrocarbons. Electrochim Acta 47:1405–1414

Schwarzenbach RP, Gschwend PM, Imboden DM (2003) Environmental organic chemistry, 2nd edn. Wiley, Hoboken

Suèr P, Andersson-Sko Y (2011) Biofuel or excavation ?—life cycle assessment (LCA) of soil remediation options. Biomass Bioenergy 35:969–981

Suèr P, Nilsson-Påledal S, Norrman J (2004) LCA for site remediation: a literature review. Soil Sediment Contam 13:415–425

Sun TR (2013) Effect of pulse current on energy consumption and removal of heavy metals during electrodialytic soil remediation. PhD Thesis, Technical University of Denmark, Denmark

Therrien R, Sudicky AE (1996) Three-dimensional analysis of variably-saturated flow and solute transport in discretely-fractured porous media. J Contam Hydrol 23:1–44

Toffoletto L, Deschênes L, Samson R (2005) In LCA: case studies—using LCA to compare alternatives LCA of Ex-situ bioremediation of diesel-contaminated soil In LCA: case studies. Int J Life Cycle Assess 10:406–416

Tolle DA (1997) Special Issue: 17th SETAC Meeting 1996: LCA—selected papers lca methodology regional scaling and normalization in LCIA. Int J Life Cycle Assess 2:197–208

USEPA (2008) Life cycle assessment: principles and practice. United States Environmental Protection Agency, Washington, DC

Verones F, Bartl K, Stephan P, Jime R, Hellweg S (2012) Modeling the local biodiversity impacts of agricultural water use: case study of a wetland in the coastal arid area of Peru. Environ Sci Technol 46:4966–4974

Verones F, Pfister S, Hellweg S (2013a) Quantifying area changes of internationally important wetlands due to water consumption in LCA. Environ Sci Technol 47:9799–9807

Verones F, Saner D, Pfister S, Baisero D, Rondinini C, Hellweg S (2013b) Effects of consumptive water use on biodiversity in wetlands of international importance. Environ Sci Technol 47:12248–12257

Villanueva A, Wenzel H (2007) Paper waste—recycling, incineration or landfilling? A review of existing life cycle assessments. Waste Manag 27:S29–S46

Virkutyte J, Sillanpaa M, Latostenmaa P (2002) Electrokinetic soil remediation—critical overview. Sci Total Environ 289:97–121

Volkwein S, Hurting H-S, Klopffer W (1999) Life cycle assessment of contaminated sites remediation. Int J Food Microbiol 4:263–274

Wenzel H, Hauschild M, Alting L (1997) Environmental assessment of products—volume 1: methodology, tools and case studies in product development. Chapman & Hall, London

Yeung AT (2011) Milestone developments, myths, and future directions of electrokinetic remediation. Sep Purif Technol 79:124–132

Yuan S, Wu C, Wan J, Lu X (2008) Electromigration of cadmium in contaminated soils driven by single and multiple primary cells. J Hazard Mater 151(2–3):594–602. doi:10.1016/j.jhazmat.2007.06.029

Yuan S, Zheng Z, Chen J, Lu X (2009) Use of solar cell in electrokinetic remediation of cadmium-contaminated soil. J Hazard Mater 162(2):1583–1587
Zelm RV, Schipper AM, Rombouts M, Snepvangers J, Huijbregts MAJ (2011) Implementing groundwater extraction in life cycle impact assessment: characterization factors based on plant species richness for the Netherlands. Environ Sci Technol 45:629–635
Zhang T, Zou H, Ji M, Li X, Li L, Tang T (2014) Enhanced electrokinetic remediation of lead-contaminated soil by complexing agents and approaching anodes. Environ Sci Pollut Res 21 (4):3126–3133. doi:10.1007/s11356-013-2274-9
Zhou M, Zhu S, Liu Y, Wang X (2013) Removal of fluorine from contaminated soil by electrokinetic treatment driven by solar energy. Environ Sci Pollut Res 20:1–7. doi:10.1007/s11356-013-1595-z

Part III
Conservation of Cultural Heritage and Use in Construction Material

Chapter 11
Electro-desalination of Buildings Suffering from Salt Weathering

Lisbeth M. Ottosen and Henrik K. Hansen

11.1 Introduction

Salt weathering causes decay of monuments and buildings, and this decay is a widely recognized problem in both ancient and recent buildings made from porous materials. The crystallization of salts causes irreversible damage to many cultural objects such as wall paintings, sculptures, historic buildings, and other artworks (Pel et al. 2010). Moisture transfer processes are responsible for the transport of the salts into the porous material. The salts are transported as aqueous saline solution and the origins of the salts are numerous. They can come from the natural environment, e.g., salty groundwater, ocean aerosols, or desert dust. The salts can also originate from human activity and use of the building as de-icing salts from nearby roads, farming and animal urine or storage of salted food. In addition, the salts can originate from the construction materials themselves, e.g., in cases where unwashed sea sand is used in the mortar.

Over time the salt concentration increases on the surface of the material (efflorescence) or in the pores of the material (sub-efflorescence) as the salty water is continuously transported into the material where after the water evaporates leaving the salts. If the rate of resupply of solution to the surface is sufficient to compensate the rate of evaporation, the salts deposit on the surface. If the transport rate of the solution through the pores is too slow to compensate the evaporation, a dry zone develops beneath the surface and the crystallization occurs within the pores of the stone (Lewin 1982). Efflorescence is primarily a visual problem as it can be

L.M. Ottosen (✉)
Department of Civil Engineering, Technical University of Denmark,
Building 118, Lyngby 2800, Denmark
e-mail: lo@byg.dtu.dk

H.K. Hansen
Departamento de Ingeniería Química y Ambiental, Universidad Técnica Federico
Santa María, Avenida España 1680, Valparaíso, Chile

© Springer International Publishing Switzerland 2016

A.B. Ribeiro et al. (eds.), *Electrokinetics Across Disciplines and Continents*,
DOI 10.1007/978-3-319-20179-5_11

removed mechanically from the surface. Sub-florescence is causing salt weathering and while the removal of water-soluble salts sounds easy, nevertheless this can prove difficult in practice, particularly in the case of objects that are monumental in scale (Pel et al. 2010). It is this task which can be overcome by application of an electric DC field for removal of the solubilized salt ions by electromigration. This method is termed electro-desalination and is the topic for the present chapter. The focus is on describing the method and based on a literature survey to describe the important developments and findings.

11.2 Phase Changes for Soluble Salts in Pores

The understanding of phase changes and positions of the damaging salts is important in order to have an optimal use of electro-desalination as it is as dissolved salts which are mobile with electromigration. Crystallization of a salt within the pores of the material takes place when the saline phase reaches supersaturation and the thermodynamic characteristics of the system are adequate for the formation of nuclei, from which crystals can grow (Gomez-Heras and Fort 2007). The crystal growth in the confined pores generates a pressure that may be too large for the porous material to resist and the material is damaged. The crystal growth is usually induced by changes in ambient temperature or relative humidity. Unfavorable environmental conditions may cause repeated cycles of deliquescence–crystallization or hydration–dehydration, which lead to the rapid decay of building materials (Linnow et al. 2007). Salt weathering is thus controlled by fluctuations in temperature and moisture, where repeated oscillations in these parameters can cause recrystallization, hydration/dehydration of salts, bringing about stone surface loss in the form of, for example, granular disaggregation, scaling, and multiple flaking (McCabe et al. 2013). In some cultural heritage monuments it is the loss of surface which is of major concern, e.g., glazing from tile panels or murals in church vaults. In other cases the concern is on the whole material. Figure 11.1 shows examples of salt weathering of a sandstone sculpture where the attributes of the face are lost, a sandstone masonry where the surface is flaking, and artwork loss from a tile panel and mural in church vault.

11.3 The Principle of Electro-desalination

Electro-desalination was developed to meet limitations of the long established conservation technique poulticing, which is applied for desalination of porous materials with sub-effloresce. In general the application methodology for poulticing is relatively simple: a wet poultice material is applied to the surface of the object to be treated, and is kept in place for some period of time before being removed. At first water is transported from the poultice into the wall where it starts to dissolve

Fig. 11.1 Damage from salt weathering (**a**) Sandstone sculpture, Frederiksborg Castle, Denmark, (**b**) Sandstone masonry, Frederiksborg Castle, Denmark, (**c**) Tile panel, Coimbra, Portugal, and (**d**) Mural, Rørby Church, Denmark

the salts, then the dissolved salt ions travel in the form of an aqueous saline solution from the wall back into the poultice (Pel et al. 2010). However, due to the complex nature of salt problems within historic structures, the result of such interventions can be variable and unpredictable (Sawdy et al. 2008). The amount and depth, to which salts are mobilized during poulticing, and where they are transported to, is dependent on the interrelationship between the poultice and the stone, the drying conditions and also the initial salt distribution (Sawdy et al. 2008). Thus there is no single ideal poulticing method for extracting salts (Pel et al. 2010). In practice, one can never achieve complete desalination of a non-moveable object by use of poultice materials. Indeed, it is therefore more accurate to refer to such interventions as "salt content reduction" rather than "desalination" treatments (Pel et al. 2010). During poulticing the major transport mechanisms for the salt out from the stone into the poultice are diffusion and advection. A more controlled and efficient transport for salts out from the stone may be obtained by applying an electric potential over the infected stone as driving force as electromigration is not controlled by the pore size and thus electro-desalination will be less dependent on the material characteristic than poulticing and thus more generally applicable.

During electro-desalination electrodes are placed externally in a poultice on the surface of the salt-infected material. Figure 11.2 shows the principle. The main

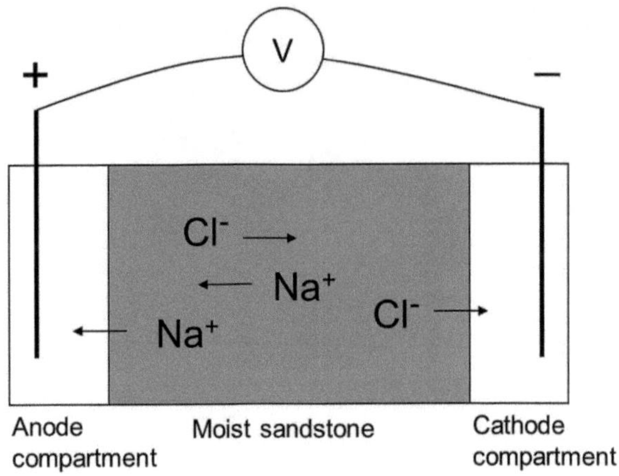

Fig. 11.2 Principle of electro-desalination

transport mechanism for the salt ions in the moist porous material is electromigration. During the process, the concentrations of ions of the damaging salts will decrease in the stone as these ions concentrate around the electrode of opposite polarity. The electrodes are placed in a poultice and the ions concentrate here. After the desalination the poultice is removed and hereby the damaging ions.

11.3.1 Neutralization of pH Changes from Electrode Reactions

At both electrodes there are pH changes due to electrolysis reactions. At the anode: $H_2O \rightarrow 2H^+ + ½O_2\ (g) + 2e^-$ and at the cathode: $2H_2O + 2e^- \rightarrow 2OH^- + H_2\ (g)$. The pH thus decreases at the anode and increases at the cathode. It is necessary to neutralize the pH changes to prevent severe pH changes in the material. Herinckx et al. (2011) underlined how important it is to avoid the acidification, as in electro-desalination experiments without pH neutralization as the experimental stones were severely damaged close to the anode. The finding was supported by (Skibsted 2013) who showed increased porosity close to the anode in a carbonate-bound sandstone in case of no pH neutralization. Also in order to obtain sufficient desalination, pH neutralization can be crucial. Kamran et al. (2012b) reported experimentally and with numerical simulations that without neutralization at the electrodes, the desalination process in bricks stopped due to formation of a sharp transition zone between the acidic and alkaline region in the brick. This zone resulted in a large electrical potential gradient due to the local depletion of ions. The experiments were conducted with constant voltage, and due to the large potential gradient in the transition zone, the electrical field diminished in the remaining brick. Consequently, the desalination stagnated and sufficient desalination could not be obtained.

In previous research it has been shown calcite-rich clay poultice can be used for neutralization of pH changes at the anode. Two types of calcite-rich clay have been tested: a calcareous clay sampled at a brick work (Ottosen and Christensen 2012) and a pure mixture of kaolinite and calcite (Rörig-Dalgaard and Ottosen 2009). In these poultices the calcite buffers the pH changes and the clay gives workability, so the poultice can have optimal contact to the surface of the object to be desalinated. The two poultices were experimentally compared (calcareous brickwork clay and a mixture of kaolin and calcite) and both efficiently neutralized the acid from electrolysis at the anode (Ottosen and Christensen 2012). At the end of the desalination pH in the stones was similar to or higher than initially. Thus the acid from electrolysis at the anode was successfully neutralized by $CaCO_3$ in the poultices. From numerical simulations and the published experimental results, the alkaline front is not buffered in a significant manner from calcite-rich poultice (Paz-Garcia et al. 2013). When using the $CaCO_3$-rich poultices, the transition zone between low and high pH is moved from the stone to the anode poultice (Paz-Garcia et al. 2012). The increased concentration of OH^- ions in the material must be counterbalanced by cations (electro-neutrality) and in Paz-Garcia et al. (2011) it is suggested from numerical-chemical simulations that these are mainly Ca^{2+} from dissolution of calcium-carbonate in the anode poultice. There may thus be precipitation of $Ca(OH)_2$ in the material. Over time $Ca(OH)_2$ react with CO_2 from air and form $CaCO_3$, but neither $Ca(OH)_2$ nor $CaCO_3$ is considered damaging, because aqueous solutions of calcium hydroxide (limewater) have been used for many centuries to protect and consolidate limestone (Clifton 1980).

Other poultices tested in combination with electro-desalination are cotton, cellulose fibers, and a mixture of calcite powder and cellulose fibers (De Clercq et al. 2014). Wet sponges were replaced regularly (Kamran et al. 2012a, b). Feijoo et al. (2013) used a combination of wet sponges and $CaCO_3$-rich poultice at the anode and kaolinite and sponge at the cathode. In Ottosen and Rörig-Dalgaard (2007) a system with electrolyte solutions and ion exchange membranes were used to avoid pH changes in the brick and the brick pH was only slightly decreased, but the system was too difficult to operate, as from the compartment next to the brick the liquid was lost either from capillary suction into the brick or leaking between the brick and electrode unit. The solid matrix is needed at the electrodes. Also Kamran et al. (2013) tested a system with ion exchange membranes at the electrodes (here with a sponge between brick and membrane), and they found that the membranes fulfilled the goal of diminishing pH changes in the brick sample, but concluded that the system was inefficient and that the Na removal was faster in a system without the membranes. The work with development of electrode units underlines the necessity for using a material at the anode which in addition to neutralizing the acid, also gives good electrical contact between metallic electrode and stone. Calcite-rich clay poultice fulfills this; however, over time the clay can dry out due to evaporation of water or electroosmosis, and thus replacement of poultice is necessary.

In addition to the pH changes also changes in redox conditions at the two electrodes occur. Oxidizing conditions are prevalent at the anode and reducing conditions at the cathode. It is unlikely that the O_2 and H_2 will diffuse through the dense water-saturated clay poultice in the time frame the poultice is used, or the thickness of the poultice can be adjusted to avoid it. Another important electrode process occurs when Cl^- reaches the surface of the anode: $2Cl^- + 2e^- \rightarrow Cl_2$. The produced Cl_2 can dissociate in water to form hypochlorite, which is an oxidizing agent. However, again diffusion through the poultice into the porous material is unlikely when using a dense poultice of a certain thickness. The redox conditions in the material to be desalinated are thus not expected to be influenced by the electro-desalination, but experimental evidence has not been published. As Cl_2 is toxic, safety precautions must be taken if the electrochemical desalination is carried out indoors. There is also a theoretical possibility for reduction of NO_3^- at the cathode, but as electromigration transports the anions in the opposite direction this reduction is considered unlikely.

11.3.2 Electromigration and Electroosmosis vs. Diffusion and Advection

The system as shown in Fig. 11.2 was tested with and without applied current for comparison in (Petersen et al. 2010). The comparison is not between the two techniques poulticing and electro-desalination as the calcite-rich poultice used was not optimized for poulticing. The investigation was focusing on the transport into the poultice with and without an applied electric field during the first 3 days of desalination. The resulting profiles in the 3.5 cm long sandstone are shown in Fig. 11.3. The concentration was decreased in both ends in the experiment without current, but not in the middle. On the contrary the concentration was decreased all through the

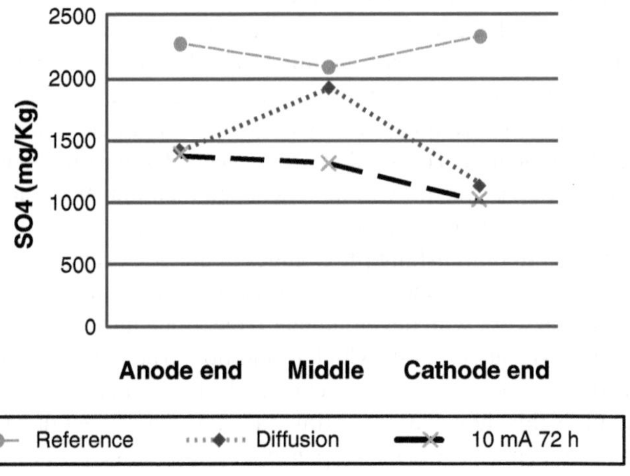

Fig. 11.3 SO_4^{2-} concentration profiles in sandstone after 3 days with carbonate-rich poultice with and without applied electric current. The reference profile shows the concentration before the experiments

stone in the experiment with applied current. The profile after application of current show highest concentration closest to the anode and lowest closest to the cathode as the anion is electromigrating from cathode towards anode and this shape of the concentration profile for anions is in consistency with other experiments, e.g., Ottosen and Christensen (2012). Diffusion of SO_4^{2-} out into the poultice was significant during these 3 days. Ottosen et al. (2014) found during desalination of tiles that diffusion of Cl^- and NO_3^- was significant in the first days, where after this effect diminished and electromigration was the only transport mechanism.

Kamran et al. (2012a) showed that for electrokinetics to exceed ion transport by diffusion a minimum level of applied voltage is necessary. Below this threshold salt transport by diffusion is dominant over electromigration. Within the time scale of their experiments, the length of the specimen used, and a setup with sponges at the electrodes, up to applied field strength of 1.20 V/cm, diffusion was dominant over electromigration. It is though important here to mention that this voltage is measured between the working electrodes, and thus the potential drop between electrode and sponges soaked in demineralized water (which must be considered considerable) was included in this value, and thus it is not a value solely related to the brick characteristics.

Electroosmotic transport of water will be determined by the material characteristics as well as on the dissolved salt content. When the concentration of dissolved ions in the pore solution is high (as in the materials decomposed by salt weathering), the electroosmotic effect will be suppressed or eliminated, as the current will be carried by both cations and anions in the bulk pore solution and not preferential from counter-ions in the electric double layer as necessary to obtain electroosmosis. This means that only after the ion content in the pores has been decreased to a low level by electromigration, electroosmosis can take place (Rörig-Dalgaard 2009; Ottosen and Rörig-Dalgaard 2009).

11.4 Electro-desalination and Material Characteristics

Electrochemical desalination of different matrices has been tested in laboratory scale on single stones representing somewhat homogeneous matrices (bricks and natural stones) and matrices with a fragile surface with other material characteristics than the main material (tiles and painted brick masonry). This chapter focuses on the different materials treated and summarizes the results published in literature.

11.4.1 Fired-Clay Bricks

Electro desalination with calcite-rich poultice at the electrodes has been tested for desalination of different types of fired-clay bricks in laboratory scale: Red Danish bricks made for restoration purposes and from old traditions (Munkesten from

Falkenløve). They are handcrafted and fired in a circular kiln as in ancient time (Rörig-Dalgaard 2009), typical modern red fired clay bricks from Denmark (Skibsted 2013), yellow Danish bricks from Wewers, where the yellow color is due to the clay used in the production (the mass of $CaCO_3$ was at least three times higher than the mass of Fe_2O_3 to have yellow bricks) (Ottosen and Rörig-Dalgaard 2009). In every case the bricks were new from the factory and had been soaked in salt solution in the laboratory. Successful desalination was obtained with every of these brick types and thus there is no limitation for the process in relation to desalination of fired-clay bricks contaminated in the laboratory. This task, desalinating fired-clay bricks with solubilized salt must also be considered the simplest case. Fired-clay bricks are considered quite inert chemically, so it is water-saturated non-reacting porous material with ions, which needs to be desalinated. A model for the electrokinetic transport phenomena based on the strongly coupled Nernst–Planck–Poisson system of equations was developed for this system (Paz-Garcia et al. 2011). The model showed high versatility, since it was able to predict results from this simple case of electrokinetic desalination of fired-clay bricks.

Electroosmosis can be significant in clays, but during firing, the clay particles sinters and thus it is not a matter of course that electroosmosis can be obtained in fired clay bricks. In Ottosen and Rörig-Dalgaard (2006) electroosmotic transport of water was found in a single red brick under application of an electric DC field, but the experimental setup was not designed for determining how dry the brick could be by electroosmosis. Bertolini et al. (2009) clearly showed that there was electroosmosis in both handcrafted and extruded single fired clay bricks and that the electroosmotic transport of water increased with the applied current. Contradictory, Kamran (2012) found that the electroosmotic pressures that are achieved by applying an electric field are insufficient to overcome the capillary pressures in order to promote drying in single fired-clay brick. The contradicting results reported on whether there electroosmosis can be obtained in bricks may be an indication that there is no simple answer, and that there can be electroosmosis in some bricks whereas not in others, depending on the clay used for the production and the sintering degree.

11.4.2 Natural Stones

Natural stones cover a variety of stone materials such as sandstone, travertine, limestone, and granite with a broad range of characteristics. They are widely employed as building and sculpture material in cultural heritage and can suffer from salt decay when exposed, but the resistance towards salt decay varies significantly with the properties of the stone. Stone durability depends heavily on both strength and pore structure properties (Benavente et al. 2004). Electro-desalination has been tested in laboratory scale for different natural stones contaminated with salt in the laboratory. To investigate the influence from the porosity on the electro-desalination NaCl or KCl removal are relevant, as these chlorides are highly soluble and the process is thus not determined by solubility in these laboratory contaminated stones. The removal of chloride has approached 100 % in different natural stones with different porosities ranging from Albero granite *3.9 %* and Rodas granite *5.9 %* (Feijoo et al. 2013), to Posta and Cotta sandstones (Ottosen et al. 2012) with porosities of *22 %* and *23 %* (Sidel 2010) and no limitations from the porosity on the high removal percentage was seen.

The resistance towards chemical attack also strongly differs between the types of natural stones. Limestones and sandstones with $CaCO_3$ as bonding materials are very sensitive, for example Nord and Tronner (1995) observed that even rain dissolve calcite and decrease the Ca concentration from samplings from two monuments built from Gotlandic sandstone. Thus it is relevant to evaluate if such poor resistance towards chemical attack also makes the stone more vulnerable to the fast change in equilibrium, which is caused by the electric field as ions electromigrate out of the stone and to pH changes, which are not fully avoided by the use of calcite-rich poultice. Skibsted (2013) investigated changes in Gotlandic sandstone ($CaCO_3$ bound) with and without buffering poultice at the electrodes in order to investigate changes in the stone due to equilibrium changes or acidification. A XRD investigation did not show changes in the mineral phases in any of the cases. In case of no buffering at the anode an increased porosity was seen in the stone in very close to the anode, and this increase was not seen in case the buffering poultice was used. So this investigation did not show the unwanted changes when using poultice, but the topic should be investigated further to make sure that there will be no unacceptable and irreversible changes. Feijoo et al. (2013) investigated possible color changes in two different granites caused by electro-desalination, and they identified slight changes, which tough were associated with traces of kaolin from the poultice rather than changes in color of the granite itself.

A possible electroosmotic effect in natural stones depends on the stone material itself. In some stones a significant electroosmotic effect can be obtained, e.g., in argillaceous sandstones (sandstones with a significant amount of silt and/or clay) (Ahmed et al. 2013). In the published investigations with electro-desalination of natural stones, electroosmosis has not been in focus, and the setups used was not designed for investigation of this topic. For example, electroosmotic suction or pressure from the poultice clay may influence the water content in the stone and hide a possible electroosmotic effect in the stone itself (Ottosen and Christensen 2012).

11.4.3 Historic Portuguese Tiles

Ceramic tiles (Azulejo) are an important part of the Portuguese cultural heritage. The tiles are vulnerable to salt decay as salts may crystallize in the interface under the glaze causing detachment as seen in Fig. 11.1c. Soluble salts are a major cause of tiles decay and, aside from human actions, are likely to be the most important cause of decay and loss of single tiles and whole panels (Mimoso et al. 2009). Electrochemical desalination of single azulejo tiles was tested in laboratory for model tiles (nineteenth century) spiked with NaCl (Ottosen et al. 2010a, b).

Successful desalination was obtained as the Cl^- concentration was lowered from 0.13 wt% Cl^- to less than 0.01 wt%. Electrochemical desalination of salt-infected tiles from Centeno Palace, Lisbon, was tested (Ottosen et al. 2010a, b, 2011), but here the charge transfer was too low to obtain full desalination; the decreases obtained were >81 % Cl^-, ~59 % NO_3^-, and ~22 % SO_4^{2-}. In Ottosen et al. (2014) the charge transfer was sufficient and the desalination successful. The initial concentrations of chloride and especially nitrate were very high in the actual tiles (around 10 g Cl^-/kg and 30 g NO_3^-/kg, respectively). Both anions were successfully removed to below 0.1 g Cl^-/kg and 0.2 g NO_3^-/kg during the electrochemical treatment.

Salts from both biscuit and interface between biscuit and glaze must be removed to have an efficient conservation technique. As received, the tiles in (Ottosen et al. 2014) had NaCl crystals under loosened glaze, whereas after electro-desalination no signs of salt crystals were found (by SEM-EDX), so in this important point, the desalination was successful. To one of the tiles in (Ottosen et al. 2011) a part of the original mortar bed was sticking to the biscuit. The electrode units were placed on the mortar and the experiment was stopped before full desalination. It was hereby seen that desalination of the mortar occurred before biscuit as little chloride and nitrate was removed from the latter, whereas the concentrations were significantly reduced in the mortar. This was most likely linked to the interface between biscuit and mortar, where poor contact and cracks with air could be the reason (the mortar was very loose), hindering the current in passing. This result indicates that loose tiles must be removed and treated separately when treating tile panels. The next step towards developing the technique for tile panels will be testing on the whole system with wall, mortar bed, and tile. It is important to remove salts from all parts. Otherwise the tiles will soon be salt infected again from salts entering from mortar and wall.

A lab experiment with electro-desalination of a small wall section with murals was reported in Rörig-Dalgaard (2009). The electrodes were placed on the opposite side of the mural as if the desalination was carried out from the top of a church vault. The experiment lasted 2 weeks with an applied current of 2.9 mA/cm. The wall had been stored indoor for a long time before the experiment, and was very dry. Thus the wall was wetted by spraying of water. The desalination efficiency was high in some parts of the wall but low in other parts, dependent on the ease of moistening of the actual part of the wall section. This underlines that moisture content in a wall is determining where the electric field will be strongest in a case where parts are rather dry.

11.5 Removal Rates and Transference Numbers

Most works on laboratory electro-desalination have been carried out with stones contaminated with a single salt; however, in real life situations a wide variety of salt mixtures can be present in the porous material. The damaging effect of such multicomponent salt solutions is often lower than in the case of single salts as the activity coefficient decreases (Cardell et al. 2008). In this chapter the removal of single salts are compared first, and following is the removal of mixed salts.

Electro-desalination (with buffering poultice at the electrodes) of bricks and natural stones contaminated with chlorides and nitrates have been shown very successful with removal of 99–100 %, e.g., Cl removal from yellow and red bricks (Ottosen et al. 2008; Skibsted et al. 2013), posta and cotta sandstones (Ottosen and Christensen 2012), and granite (Feijoo et al. 2013). Chlorides and nitrates are generally highly soluble, whereas sulfates are less soluble. Skibsted (2013) did comparable electro-desalination experiments with bricks contaminated with pure single salts: NaCl, $NaNO_3$, or Na_2SO_4. The initial concentration of Na was kept constant in these experiments. After 5 days with 2 mA applied, 99 % Cl^- and 100 % NO_3^- was removed, whereas after 8 days with the same current only 89 % SO_4^{2-} was removed. The theoretical ionic mobility for these anions are: NO_3^- $7.4 \cdot 10^{-8}$ $m^2/(s\ V)$, Cl^- 7.9×10^{-8} $m^2/(V\ s)$, and ½SO_4^{2-} 4.2×10^{-8} $m^2/(V\ s)$. The reasons for this slower removal rate for SO_4^{2-} is though a combination of lower ionic mobility and electromigration of Ca from the poultice at the anode into the brick and subsequently precipitation of $CaSO_4$. A physicochemical model for electrochemically induced reactive-transport processes was described in Paz-Garcia et al. (2013) and used for a theoretical analysis of the influence of the chemical interactions on the removal rate of the target ions. Results from the simulations supported that the lower removal efficiency of sulfates is related to the precipitation of gypsum inside the porous body.

During electro-desalination in a single salt system, the transport rate of the anion depends on the counter ion (Ottosen et al. 2009; Paz-Garcia et al. 2012). The transport rate of Cl^- was experimentally shown different in similar experiments where a brick was contaminated the pore solution contaminated with NaCl or KCl. The electro-desalination process had a similar general behavior but a slightly higher transport number for potassium with respect to sodium resulted in a slightly lower migration rate of chloride when the brick was contaminated with NaCl.

Electro-desalination in laboratory scale was tested on samples removed from salt contaminated building facades: Obernkirchen sandstone claddings from a warehouse in Copenhagen (Matyscak et al. 2014) and tiles removed from Centeno Palace, Lisbon, Portugal (Ottosen et al. 2014). In both cases the investigation included removal of Cl^-, NO_3^-, and SO_4^{2-} and in both cases the concentration of SO_4^{2-} was lowest of the three. Simultaneous removal of the three anions into the anode poultice was seen from the Obernkirchen sandstone, though the mole of SO_4^{2-} removed was much less than Cl^- and NO_3^- in every experiment. From the tiles a distinct pattern was seen of the SO_4^{2-} removal starting at the point where Cl^- and NO_3^- concentrations had been removed to a very low level in the tiles. The knowledge of the mutual influence from mixed salts on electro-desalination is a topic for future research.

Transference numbers for the target ions changes over time during electro-desalination. Very high transference numbers for Cl^- and NO_3^- have been reported in the beginning of the process: During the first day of desalination the transference numbers were 0.63 (Cl^-), 0.76 (NO_3^-), and 0.42 (SO_4^{2-}) in small bricks samples in case of single salts (Skibsted 2013); a transference number of 0.55–0.65 (Cl^-) after 1–2 weeks during desalination of a whole brick (Rörig-Dalgaard 2009) and in tiles with both Cl^- and NO_3^- the sum of transference numbers for these two was about 0.6 during the first weeks. The transference numbers for the two anions were very similar in the tiles. The sum of transference numbers for all ions (cations and anions) in a system is always 1.0 and thus these reported transference numbers for the anions are high, but it must be remembered that there is no distinction between diffusion and electromigration, and the calculation is made on the basis of both. Thus the true transference number is less. At the point where a low concentration has been reached in the porous material to be desalinated, the transference number decreases dramatically (Rörig-Dalgaard 2009; Ottosen et al. 2014).

11.6 Electrode Placement

In the conducted laboratory experiments the electrodes has been placed either on opposite sides of the porous material or at the same side. In a homogeneously contaminated sample the field is equally strong in each cross section, but when the electrodes are placed at the same side, the electric field is strongest closest to the side where the electrodes are placed, see Fig. 11.4.

With both setups successful desalination results have been obtained. In the first case, the anions are removed to a low concentration from the part closest to the cathode first and from the part closest to the anode during the last stage (e.g., Ottosen and Rörig-Dalgaard 2009). In the second case with the electrodes at the same side, the concentration decreases closest to the electrodes first and over time the desalination progresses into the depth of the material (Rörig-Dalgaard 2009).

In materials undergoing salt decay, the salt concentration is most often highest close to the surface. This means that in real cases, the salt contamination is not homogeneously distributed as in the laboratory spiked materials, with which most of the laboratory experiments has been conducted. In the work of Matyscak et al. (2014) with samples from Obernkirchen sandstone claddings from a warehouse, the uneven distribution of salts in the stone was in focus and was related to the electrode placement. Electrodes were placed on the side of the stone with the highest salt concentration or on the opposite side, and it was shown that the electrode placement influenced the electro-desalination efficiency; however, direct comparison between

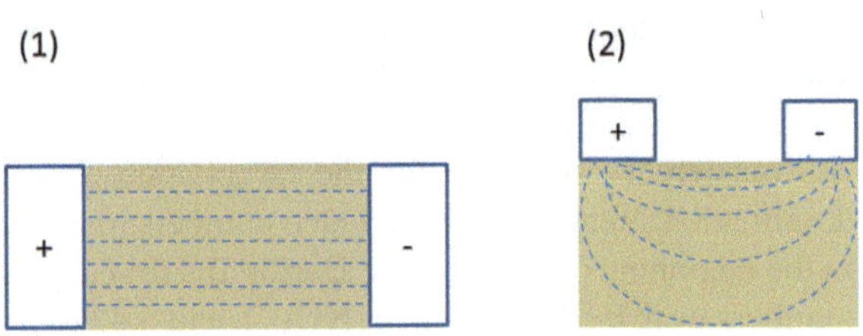

Fig. 11.4 Sketch of the electric field lines in a homogeneous stone with two different electrode placements

the experiments of the investigation was difficult due to very different concentrations of the three anions between the stones initially. Two experiments were conducted with stones from the same warehouse with electrodes placed at the same side or on opposite sides in Ottosen et al. (2014). Whether the electrodes were placed on the same side of the stone or on two opposite sides did not influence the electromigration rate of Cl^- and NO_3^-, but diffusion of the anions into the cathode poultice was seen more pronounced in the first days when the cathode was placed on the most contaminated surface, and similar diffusion also added to the removed part into the anode poultice in the very beginning of the experiment. It is fortunate that the electrodes do not have to be placed on the most polluted surface as this surface will also be the most fragile. The electrodes can be placed next to the most fragile area hindering mechanic damage from placement of the electrodes.

11.7 Pilot Scale Experiences

A pilot scale test was made on outdoor brick masonry with primitive electrode units where a bar of reinforcement steel served as metallic electrode and it was placed in brickwork clay directly on the masonry (Ottosen et al. 2007). This pilot scale test showed removal of Cl^- and NO_3^- even though the temperature was below 0 °C during most of the experiment. The test showed that the principle of electro-desalination worked on masonry, but also that the electrodes needed further development in order to maintain a sufficient electric field, as especially a good contact between clay and masonry could not be maintained over a sufficiently long period.

Electrode casings with a metallic electrode and filled with a clay poultice were developed (Ottosen et al. 2008). The electrode casings (50 cm long) were made from plastic and consisted of two parts: a box and a movable bottom (see Fig. 11.5).

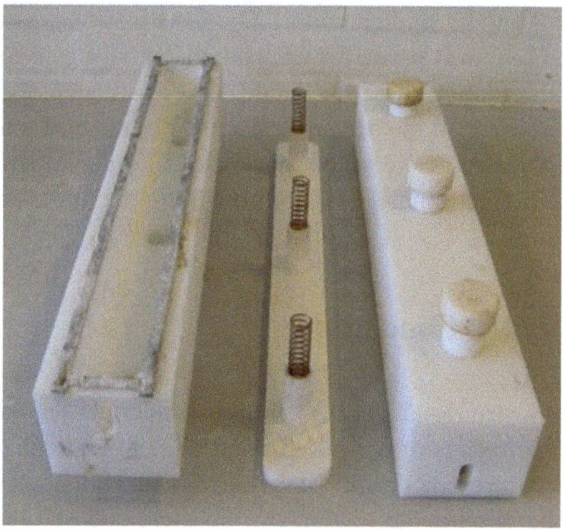

Fig. 11.5 Two electrode casings. To the left the hole in the box is shown, in which the movable bottom in the middle is to be placed with the springs down. On top of the movable bottom is first placed a metallic electrode mesh and afterwards the casing is filled with poultice

The purpose of the movable bottom was to ensure mechanical pressure between poultice and masonry during desalination, as this is crucial to passage of current. As the poultice dries slightly during the desalination process, it is shrinking and it is necessary to move the bottom closer to the masonry to maintain the pressure. Springs are aiding this, and further it is possible to tighten the springs with bottoms at the back of the casing. An inert metallic electrode mesh was placed on top of the movable bottom before the casing was filled with poultice.

The first electrode casings were black (Ottosen et al. 2008), but these black casings got very hot in sunshine and the clay tended to dry out fast. Thus the white color was chosen to meet this problem in later works (Ottosen et al. 2010a, b, 2012). Four small pilot scale tests with few electrodes mounted at walls and a larger test with 72 electrodes have been reported. In all the tests similar electrode units were used. The pilot scale tests are summarized in Table 11.1.

In Table 11.1 the transference numbers for Cl^- are given calculated on the basis of concentrations measured in the anode poultice. It is seen that the transport number was highest for masonry B, where the highest concentration was also found. The transference number in C and D are similar. The transport numbers are quite high in all cases showing that the major part of the current towards the anode was carried by chloride. So even though neither of the desalination actions finished in respect to the final concentration in the masonry, the removal progressed well and it is interesting to see if the desalination could be maintained with a high transference number until a concentration of acceptable level is obtained in the masonry.

In a large pilot scale experiment with 72 electrodes placed in two rows (see Fig. 11.6) (Ottosen et al. 2012). The test plant covered about 25 m^2 surface of a limestone wall of a historic warehouse (the same as the small pilot scale test (C), table 1). The wall was severely infected with NaCl and the average concentration of Cl^- was 0.7 wt%. The test lasted for about a year, and 3.8 kg chloride corresponding to 6.3 kg NaCl was removed during this period. The part of the wall underneath the electrodes was successfully desalinated; however, the Cl^- concentration was in the same level as initially in samples taken just between sets of anodes and cathodes. The desalination was thus not completed during the test. The removal rate for Cl^- into the anodes was constant all through the test revealing that the desalination could have continued if the test had lasted longer. The test underlined the necessity for development of a new design, which allow for shorter distance between the electrodes in order to shorten the duration of the treatment. The spent man power on mounting and replacing electrodes was extensive.

11.8 Conclusion

Electro-desalination of porous materials has been tested in laboratory scale with success, both on spiked and real salt contaminated samples. An important strength of the method is that it is not very dependent on the characteristics of the material to be desalinated. Chlorides, nitrates, and sulfates can all be removed and at high

Table 11.1 Overview of small pilot scale plants with few electrode units

	(A)	(B)	(C)	(D)
Building	Building in the countryside from 1870. Originally a farmhouse but has served different purposes	Old warehouse from 1766 from clay bricks	Old warehouse from 1748, from limestone walls	Outside Obernkirchen sandstone cladding
Salt	Main contaminant sulfates (0.68 wt% SO_4^{2-} in average). Also Cl at some spots though the average concentration was at an acceptable level 0.06 wt%	NaCl as main salt with an average concentration of 2.7 wt% Cl	The main salt was NaCl and the average concentration 0.7 wt% Cl	The main pollutant was NaCl and the initial level varied between 0.4 and 1.4 wt% Cl
Pilot scale plant	 Six electrodes in pairs (each pair with one power supply), 35 cm distance	 Four electrodes placed in a heated room on the masonry. They were alternating with distance of 40 cm and connected to one power supply	 Three electrodes (distance 40 cm) were placed outside. The central electrode was cathode and the two outer anodes	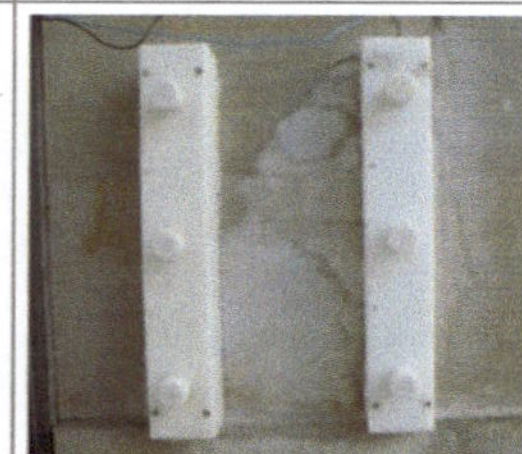 Two electrodes with a distance of 20 cm were placed outdoor
	40 mA applied to each set of electrodes for 4 month	100 mA in 5 month	50 mA in 8 month	50 mA in 5½ month

(continued)

Table 11.1 (continued)

	(A)	(B)	(C)	(D)
Current and duration				
Overall result	Average SO_4^{2-} concentration reduced to 0.46 wt%, which was not sufficient. The limiting factor was probably a low dissolution rate of the sulfates. However, the removal efficiency did not decrease during the experiment, so the process was slow, but had not slowed down	In average, 1.6 g Cl was removed per day summing up to 240 g in total. The average concentration reduced to 1.5 wt% Cl^- which is still a very high concentration, so the desalination was not finished during the time of the experiment	In average, 0.6 g Cl was removed per day summing up to 140 g in total. The average concentration reduced to 0.6 wt% Cl^- which is still very high concentration, so the desalination was not finished during the time of the experiment	The Cl^- concentration in the masonry below the cathode and between the electrodes was decreased to a sufficient level in 10 cm depth (the whole depth investigated). The concentration below the anode was decreased to a level below 0.18 wt% Cl. All together 101 g Cl was removed into the anode poultice, e.g., 0.6 g/day
Transf. number Cl		0.48	0.38	0.38
Reference	Ottosen et al. (2008)	Ottosen et al. (2010a, b)	Ottosen et al. (2010a, b)	Ottosen et al. (2014)

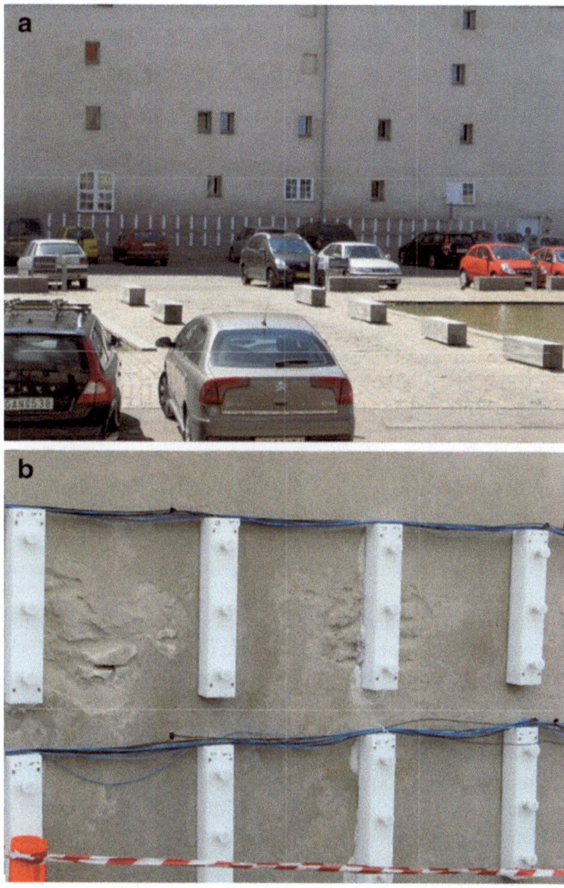

Fig. 11.6 (**a**) The test plant for electrochemical desalination with two rows of 36 electrode units and (**b**) detail with electrode units and the salt damaged masonry

transference numbers, but the removal of SO_4^{2-} is slower than the two other anions both because the theoretical ionic mobility is lower and because gypsum is formed with Ca^{2+} ions in the pore solution. Another important strength is also that the electrodes do not have to be placed directly on top of the most decayed and fragile surface with the highest salt content. The electrodes can be placed next to it and desalination of the most contaminated part is possible in this way. For electro-desalination it is important that the water content is sufficient for salts to be dissolved and not nucleated.

In pilot scale on walls the final proof of concept is still lacking, as full desalination has not been carried out yet. Different pilot scale experiments have been made, but they have all been stopped before full desalination. However, when they were stopped, the transference number for the target ion was still high, and the desalination could have progressed further.

References

Ahmed MY, Taibi S, Souli S, Fleureau J-M (2013) The Effect of pH on electro-osmotic flow in argillaceous rocks. Geotech Geol Eng 31:1335–1348

Benavente D, García del Cura MA, Fort R, Ordoñez S (2004) Durability estimation of porous building stones from pore structure and strength. Eng Geol 74:113–127

Bertolini L, Coppola L, Gastaldi M, Redaelli E (2009) Electroosmotic transport in porous construction materials and dehumidification of masonry. Constr Build Mater 23:254–263

Cardell C, Benavente D, Rodríguez-Gordillo J (2008) Weathering of limestone building material by mixed sulfate solutions. Characterization of stone microstructure, reaction products and decay forms. Mater Charact 59:1371–1385

Clifton JR (1980) Stone consolidating materials: a status report. National bureau of standards technical note 1118, Department of Commerce, Washington DC

De Clercq H, Vanhellemont Y, De Swaef V (2014) Salt extraction of limestone by means of electrophoresis : some results on type of contact material and electrode position. In: Proceedings from the international conference on salt weathering on building and stone sculptures, SWBSS2014

Feijoo J, Nóvoa XR, Rivas T, Mosquera MJ, Taboada J, Montojo C, Carrera F (2013) Granite desalination using electromigration. Influence of type of granite and saline contaminant. J Cult Herit 14:365–376

Gomez-Heras M, Fort R (2007) Patterns of halite (NaCl) crystallisation in building stone conditioned by laboratory heating regimes. Environ Geol 52:259–267

Herinckx S, Vanhellemont Y, Hendrickx R, Roels S, De Clercq H (2011) Salt removal from stone building materials using an electric field. In: Iannou I, Theodoridou M (eds) Proceedings from the international conference on salt weathering on building and stone sculptures, Limassol, Cyprus, 19–22 Oct, pp 357–364

Kamran K (2012) Electrokinetic desalination of porous building materials. Ph.D. thesis, Technical University Eindhoven

Kamran K, Pel L, Sawdy A, Huinink HP, Kopinga K (2012a) Desalination of porous building materials by electrokinetics: an NMR study. Mater Struct 45:297–308

Kamran K, van Soestbergen M, Huinink HP, Pel L (2012b) Inhibition of electrokinetic ion transport in porous materials due to potential drops induced by electrolysis. Electrochim Acta 78:229–235

Kamran K, van Soestbergen M, Oel L (2013) Electrokinetic salt removal from porous building materials using ion exchange membranes. Trans Porous Med 96:221–235

Lewin SZ (1982) The mechanism of masonry decay through crystallization, conservation of historic stone buildings and monuments. National Academy of Sciences, Washington, DC, pp 120–144

Linnow K, Juling H, Steiger M (2007) Investigation of NaCl deliquescence in porous substrate using RH-XRD. Environ Geol 52:317–327

Matyscak O, Ottosen LM, Rörig-Dalgaard I (2014) Desalination of salt damaged Obernkirchen sandstone by an applied DC field. Constr Build Mater 71:561–599

McCabe S, Smith BJ, McAlister JJ, Gomez-Heras M, McAllister D, Warke PA, Curran JM, Basheer PAM (2013) Changing climate, changing process: implications for salt transportation and weathering within building sandstones in the UK. Environ Earth Sci 69:1225–1235

Mimoso JM, Silva AS, Abreu MM, Costa DR, Goncalves TD, Coentro SX (2009) Decay of historic *azulejos* in Portugal: an assessment of research needs. In: Proceedings from international seminar conservation of glazed ceramic tiles, Lisbon, Portugal, 15–16 Apr 2009

Nord AG, Tronner K (1995) Effect of acid rain on sandstone: the Royal Palace and the Riddarholm Church, Stockholm. Water Air Soil Pollut 85:2719–2724

Ottosen LM, Christensen IV (2012) Electrokinetic desalination of sandstones for NaCl removal—test of different clay poultices at the electrodes. Electrochim Acta 86:192–202

Ottosen LM, Rörig-Dalgaard I (2006) Drying brick masonry by electro-osmosis. In: Proceedings of the 7th international masonry conference, London, 2006

Ottosen LM, Rörig-Dalgård I (2007) Electrokinetic removal of $Ca(NO_3)_2$ from bricks to avoid salt induced decay. Electrochim Acta 52(10):3454–3463

Ottosen LM, Rörig-Dalgaard I (2009) Desalination of a brick by application of an electric DC field. Mater Struct 42(7):961–971

Ottosen LM, Pedersen AJ, Rörig-Dalgaard I (2007) Salt-related problems in brick masonry and electrokinetic removal of salts. J Build Apprais 3(3):181–194

Ottosen LM, Rörig-Dalgaard I, Villumsen A (2008) Electrochemical removal of salts from masonry—experiences from pilot scale. In: Proceedings of the international conference salt weathering on buildings and stone sculptures, Copenhagen, 22–24 Oct 2008, pp 341–350

Ottosen LM, Christensen IV, Pedersen G, Paz-Garcia JM (2010) Development of electrode units for electrokinetic desalination of masonry and pilot scale tests at three locations for removal of chlorides. Book of extended abstracts from the 9th symposium on electrokinetic remediation (EREM 2010), Kaohsiung, Taiwan, 27–30 June, 2010

Ottosen LM, Ferreira CMD, Christensen IV (2010b) Electrokinetic desalination of glazed ceramic tiles. J Appl Electrochem 40:1161–1171

Ottosen LM, Ferreira C, Christensen IV (2011) Electrochemical desalination of historic Portuguese tiles, *Azulejos*, in laboratory scale. In: Iannou I, Theodoridou M (eds) Proceedings of salt weathering on buildings and stone sculptures SWBSS2011, Limassol, Cyprus, 19–22 Oct 2011, pp 349–356

Ottosen LM, Christensen LM, Rörig-Dalgaard I (2012) Electrochemical desalination of salt infected limestone masonry of a historic warehouse. In: Proceedings of structural faults and repair, Edinburg, CD-Rom. Ed MC Forde, p 15

Ottosen LM, Dias-Ferreira C, Ribeiro AB (2014) Electrochemical desalination of historic Portuguese tiles—removal of chlorides, nitrates and sulfates. J Cult Herit (in press) http://dx.doi.org/10.1016/j.culher.2014.11.003

Paz-Garcia JMP, Johannesson B, Ottosen LM, Ribeiro AB, Rodriguez-Maroto M (2011) Modeling of electrokinetic processes by finite element integration of the Nernst–Planck–Poisson system of equations. Sep Purif Technol 79(2):183–192

Paz-Garcia JMP, Johannesson B, Ottosen LM, Alshawabkeh A, Ribeiro AB, Rodroguez-Maroto M (2012) Modeling of electrokinetic desalination of Brick. Electrochim Acta 86:213–222

Paz-Garcia JMP, Johannesson B, Ottosen LM, Ribeiro AB, Rodroguez-Maroto M (2013) Simulation-based analysis of the differences in the removal rate of chlorides, nitrates and sulfates by electrokinetic desalination treatments. Electrochim Acta 89:436–444

Pel L, Sawdy A, Voronia V (2010) Physical principles and efficiency of salt extraction by poulticing. J Cult Herit 11:59–67

Petersen G, Ottosen LM, Jensen PE (2010) The possibility of using electrokinetics for desalination of sandstone with low porosity. In: Proceedings from the 8th fib international Ph.D. symposium in civil engineering, DTU, Denmark, 20–23 June 2010, pp 455–460

Rörig-Dalgaard I (2009) Preservation of masonry with electrokinetics—with focus on desalination of murals. Ph.D. thesis, Department of civil engineering, Technical University of Denmark

Rörig-Dalgaard I, Ottosen LM (2009) Method and device for removing an ionic impurity from building structures. Patent No WO/2009/124890, 15 Oct 2009

Rörig-Dalgaard I, Ottosen LM, Christensen IV (2008) Desalination of a wall section with murals by electromigration. In: Proceedings of the international conference of salt weathering on buildings and stone sculptures. Copenhagen, Oct 22–24 2008, pp 361–371

Ruedrich J, Siegesmund S (2007) Salt and ice crystallization in porous sandstones. Environ Geol 52:225–249

Sawdy A, Heritage A, Pel L (2008) A review of salt transport in porous media, assessment methods and salt reduction treatments. Proceedings from salt weathering on buildings and stone sculptures, Copenhagen, Denmark, 22–24 October 2008, pp 1–27

Sidel H (2010) The city of Dresden in the mirror of its building stones: utilization of natural stone at facades in the course of time (Chapter 8). In: Boştenaru Dan M, Přikryl R, Török A (eds) Materials, technologies and practices in historic heritage structures. Springer, Dordrecht

Skibsted G (2013) Matrix changes and side effects induced by electrokinetic treatment of porous and particulate materials. Ph.D. thesis, Department of civil engineering, Technical University of Denmark, Denmark

Skibsted G, Ottosen LM, Jensen PE. Electrochemical desalination of limestone spiked with Na_2SO_4—importance of buffering anode produced acid. Paper E in matrix changes and side effects induced by electrokinetic treatment of porous and particulate materials. Ph.D. thesis from Gry Skibsted, 2013, Technical University of Denmark

Chapter 12
Incorporation of Different Fly Ashes from MSWI as Substitute for Cement in Mortar: An Overview of the Suitability of Electrodialytic Pre-treatment

Cátia C. Magro, Paula R. Guedes, Gunvor M. Kirkelund, Pernille E. Jensen, Lisbeth M. Ottosen, and Alexandra B. Ribeiro

12.1 Background

Municipal solid waste incineration (MSWI) produces for each ton of incinerated municipal solid waste (MSW) in a mass burn unit, 15–40 kg of solid residue which requires further treatment or landfill as hazardous waste (Quina et al. 2008). Fly ash (FA) and air pollution control residue (APC) are both incineration residues (IR) which are generally characterized by high concentrations of salts, heavy metals, and organic trace-pollutants. Valorization may be an alternative, but it is important to ensure that the environment is properly protected against emissions

C.C. Magro (✉)
CENSE, Departamento de Ciências e Engenharia do Ambiente,
Faculdade de Ciências e Tecnologia, Universidade Nova de Lisboa,
Caparica 2829-516, Portugal

Department of Civil Engineering, Technical University of Denmark, B118,
Kgs. Lyngby 2800, Denmark
e-mail: c.magro@campus.fct.unl.pt

P.R. Guedes
CENSE, Departamento de Ciências e Engenharia do Ambiente,
Faculdade de Ciências e Tecnologia, Universidade Nova de Lisboa,
Caparica 2829-516, Portugal

Key Laboratory of Soil Environment and Pollution Remediation, Institute of Soil Science,
Chinese Academy of Sciences, Nanjing 210008, China

G.M. Kirkelund • P.E. Jensen • L.M. Ottosen
Department of Civil Engineering, Technical University of Denmark, B118,
Kgs. Lyngby 2800, Denmark

A.B. Ribeiro
CENSE, Departamento de Ciências e Engenharia do Ambiente,
Faculdade de Ciências e Tecnologia, Universidade Nova de Lisboa,
Caparica 2829-516, Portugal

© Springer International Publishing Switzerland 2016

A.B. Ribeiro et al. (eds.), *Electrokinetics Across Disciplines and Continents*,
DOI 10.1007/978-3-319-20179-5_12

into the earth, water, and atmosphere. Therefore, it becomes emergent to find a technique that can be both economically viable and environmental sustainable. The present review aims to understand the potential of the electrodialytic (ED) process, as a pre-treatment of the IR prior to their reuse in the production of mortars. In the first part of the chapter, different FAs from MSWI incineration are outlined together with current handling options and legislation. The second part is an outline of the reported results with electrodialytic treatment of different IR from MSWI and finally the experiences with incorporation of treated IR in mortar is reported.

12.2 Municipal Solid Waste

Municipal solid waste remains a major problem in modern societies, despite the significant efforts to prevent, reduce, reuse, and recycle it. Municipal solid waste normally contains a mixture of organic wastes, fabrics, paper, oil, rubber, plastics, metal, glass, and wood, among others (Quina et al. 2011). The waste composition, even if the technologies for management of waste are equal, will differ from country to country (Eurostat 2014).

In Europe-27 Stat Members, MSW production in 2012 was 492 kg year^{-1} *per capita*, with an annual decrease since 2003 (Eurostat 2014). Although Portugal had a *per capita* production of 453 kg MSW in 2012, the production of MSW in some countries was much higher, as in Denmark or Switzerland: 668 kg year^{-1} *per capita* and 694 kg year^{-1} *per capita*, respectively (Eurostat 2014).

Nowadays, modern systems embrace different methodologies aiming as much as possible to achieve sustainable global solutions for waste management. Life Cycle Assessment tools have been used to assess the potential environmental burdens of different waste management strategies from the environmental, energetic, and economic point of view (Quina et al. 2011; Boesch et al. 2014). These calculations have shown that landfilling, even if gas is recovered and leachate is collected and treated, should be avoided, as the resources in the waste are inefficiently utilized (Sundqvist 2005). Environmental sound alternatives include incineration, material recycling, anaerobic digestion, or composting (Quina et al. 2011).

12.2.1 Municipal Solid Waste Incineration

According to Directive 2000/76/EC of the European Parliament and Council, incineration plants *correspond to any stationary or mobile technical unit dedicated to the thermal treatment of wastes with or without recovery of the combustion heat generated. This includes the incineration by oxidation of waste as well as other thermal treatment processes such as pyrolysis or gasification in so far as the substances resulting from the treatment are subsequently incinerated.*

MSWI in waste-to-energy can be considered an environmentally friendly solution and a common alternative to landfilling, while allowing to recover a large part of the energy contained in MSW (Quina et al. 2011). MSWI presents several other advantages such as reduction of volume and weight of waste by 90 % and 70 %, respectively (Quina et al. 2011). It also destroys potential pathogens and toxic organic contaminants. Nevertheless, as other waste management techniques, MSWI also has disadvantages, as the high investment needed, operating costs and hazardous waste generated, e.g., APC, bottom ashes and FA that require safe disposal (Quina et al. 2011).

The Directive 2000/76/EC, on the incineration of wastes aims *to prevent or to limit as far as practicable negative effects on the environment, in particular pollution by emissions into air, soil, surface water, and groundwater, and the resulting risks to human health, from the incineration and co-incineration of waste.* This Directive, known as the Waste Incineration Directive, states that continuous measurements of NO_x, CO, total dust, Total Organic Carbon, HCl, HF, and SO_2 should be made, and at least twice a year for heavy metals (Cd, Tl, Hg, Sb, As, Pb, Cr, Co, Cu, Mn, Ni, V), dioxins, and furans in the exhaust gas.

Environmental legislation is becoming increasingly restrictive, and consequently industrial plants have to be constantly adapted to enhance control of gas emissions, in particular (Quina et al. 2008). Depending on the units used for post-combustion control, the resulting solid waste may have different characteristics, which means that different management strategies may be used (Quina et al. 2008).

Table 12.1 presents the various management strategies to APC and FA depending on the country, being the most common options the permanent storage in hazardous waste disposal sites. The second most common option is the treatment with hydraulic binders. This approach have as main advantages stabilize the APC or FA and be a low cost technique (Ferreira et al. 2003a). Nevertheless, the volume of waste is increased when it is mixed with cement; the cement has a high CO_2 impact which will add environmental footprint of waste incineration.

According to van der Sloot et al. (2001) positive utilization of residues from MSWI is an important goal for integrated waste management. Countries like Denmark or Germany store them in big bags until disposed of in underground sites. However, environmental problems can be related to the disposal, in short or long term, as the leaching of contaminants may occur (Hjelmar 1996b). The solid waste product from MSWI has high soluble salts content that are very difficult to stabilize. These salts removal makes the remaining inorganic residue much more manageable (van der Sloot et al. 2001).

12.2.1.1 Solid Particles Produced During Municipal Solid Waste Incineration

The solid particles produced during MSWI in mass burning units may be grouped into bottom ashes and FA. Fly ashes are defined by the International Ash Working Group IAWG: Chandler AJ et al. (1997) as *particulate matter carried over from the combustion chamber and removed from the flue gas stream prior to addition of any type of sorbent material.*

The present overview will focus on the solid residues produced during MSWI: FA and APC which may include FA and the solid material captured downstream

Table 12.1 Waste management of IR from MSWI in different countries (adapted from Quina et al. 2008)

Country	Waste management strategies	Refs.
USA	APC are mixed and disposed as "combined ash." The common option is disposal in landfills which receive only MSWI residues (monofills)	Eighmy and Kosson (1996); Sakai et al. (1996)
Denmark	APC and FA are classified as a special hazardous waste and are currently exported or stored temporarily in big bags. Significant efforts are being spent to develop treatment methods which can guarantee that APC can be landfilled in a sustainable way	Hjelmar (1996a); Hjelmar (1996b); Sorensen et al. (2001)
France	After industrial solidification and stabilization processes based on the properties of hydraulic binders, the waste is stored in confined cavities in a specific landfill (French class I and II). The high cost of this treatment is promoting the companies to search alternatives to disposal	Piantone et al. (2003)
Portugal	APC are treated with hydraulic binders (solidification/stabilization method) and landfilled in specific sites (monofills)	Quina (2005)
Switzerland	APC are solidified with a hydraulic binder and then eligible for residual material landfilling or they are disposed in disused underground salt mines. Fly ash can follow the same approach of APC or they can be subjected to acidic leaching, where 1. The washed fly ash is and deposited in landfill or 2. The metal concentrate goes to recovery of secondary Pb and Zn as co-production	Boesch et al. (2014)
Japan	The IR from MSWI are considered hazardous, and before landfill intermediate treatments must be performed, such as melting, solidification with cement, stabilization using chemical agents or extraction with acid or other solvents	Sakai (1996); Nagib and Inoue (2000); Ecke et al. (2000)

from the acid gas treatment units and before the gases are released into the atmosphere.

The composition of MSW varies over time and from country to country, due to the differences in lifestyle and waste recycling process seasons. Also, chemical and physical characteristics of the FA depend on the composition of the raw MSW, the operational conditions, the type of incinerator, and APC system design: dry/semidry (injection of lime in dry form or in a slurry to the flue gas) or wet (flue gas is subjected to a scrubber after removal of FA) (Zacco et al. 2014).

In order to select the most appropriate method of treatment or application for any residue, its main characteristics, particularly chemical properties, should be known. The major elements present in the APC are Si, Al, Fe, Ca, Mg, K, Na, and Cl (Ferreira et al. 2003b; Varela et al. 2009; Lima et al. 2010). Regarding heavy metals, Cd, Cr, Cu, Hg, Ni, Pb, and Zn are the most frequent, with Zn and Pb being generally found in the largest amounts (Quina et al. 2008). Trace quantities of very toxic organic compounds are also usually present in these residues, namely polycyclic aromatic hydrocarbons (PAH), chlorobenzenes (CB), polychlorinated biphenyls (PCB) and polychlorinated dibenzo-p-dioxins (PCDD) and furans (PCDF) (Quina et al. 2008).

Due to the high concentration of several heavy metals, the reuse of IR as a secondary material is forbidden in many countries, with the ever more stringent legislation. Therefore, taking into account the potential environmental impact of these residues, the main problems that have to be solved concern toxic heavy metals, high concentrations of soluble salts, and organic micropollutants (e.g., dioxins, furans) (Quina et al. 2008).

12.2.2 Characterization of Air Pollution Control Residues and Fly Ash

APC residues and FA characteristics have been extensively studied. Five case studies selected due to some similarities of methods used were conducted with the ash from:

i. Portuguese MSW Incinerator, *ValorSul* (Lima et al. 2012)
ii. MSWI APC after a semidry flue gas cleaning process from a Danish waste incineration plant, REFA I/S (Kirkelund et al. 2014a)
iii. (a) Wet FA—MSWI FA from the incineration plant I/S *Vestforbrænding* (Magro 2014; Kirkelund et al. 2014b)
    (b) Semidry APC—MSWI APC collected after a semidry process from I/S REFA after the injection of slaked lime and activated carbon (Magro 2014; Kirkelund et al. 2014b)
iv. MSWI FA collected from Shanghai Yuqiao Wastes Incineration Plant (presented in Sect. 13.4.2) (Shi and Kan 2009)

The obtained characteristics results can be seen in Table 12.2. Samples from ii (Kirkelund et al. 2014a) and iii (Magro 2014; Kirkelund et al. 2014b) presented an alkaline pH. The same values were obtained for the samples from Lima et al. (2012) between 12 and 12.5 of pH.

Fly ashes exhibit high solubility due to its KCl and K_2SO_4 content (Hansen et al. 2001; Lima et al. 2008). Semidry APC presented high water solubility. Once the ED process is applied, in some studies the final metal concentrations in the treated samples were higher than in the untreated ones. This is mainly due to the high water solubility of the samples: when the residues are mixed with water, soluble salts are dissolved and removed. Thus, even though part of the metals are removed by the ED treatment, the total concentration increases as a result of the high mass loss. The APC consist mainly of Al, Ca (22–30 % of the residue (Le Forestier and Libourel 1988), and Cl (Hjelmar et al. 2011; Del Valle-Zermeño et al. 2013), which is as a

Table 12.2 Resume of characteristics of IR from different authors (mean values ± standard error)

	Parameter	[a]Fly ash	[b]Semidry APC	[c]Wet fly ash	[c]Semidry APC
Chemical characteristics and morphology	pH_{H2O} (L/S = 5)	11.9 ± 0.0	–	12.46 ± 0.02	12.03 ± 0.00
	pH_{KCl} (L/S = 2.5)	–	12.2	12.42 ± 0.02	12.24 ± 0.03
	Water solubility (%)	23.2	42	20 ± 0.06	34 ± 0.06
	Loss on ignition (%)	–	0.7	0.76 ± 0.14	4.07 ± 0.04
	Carbonate content (%)	8.49 ± 0.7	–	8.37 ± 0.06	14.55 ± 0.12
	Cl content (%)	9.81	24	6.1 ± 0.16	12.8 ± 0.13
	Specific surface area ($cm^2\ g^{-1}$)	–	–	215.5 ± 2.5	284.0 ± 217.0
Macroelements	Mineral species found	Al_2O_3, $CaSO_4$, $CaCO_3$, Fe_2O_3, KCl, NaCl, SiO_2	–	$CaSO_4$, KCl, NaCl, SiO_2	$Ca(OH)_2$, $Ca(ClO)_2$, $CaCO_3$, KCl
	Al ($g\ kg^{-1}$)	–	23.45	25.46 ± 1.13	7.3 ± 0.16
	Ca ($g\ kg^{-1}$)	22.2*	–	161.1 ± 2.17	320.5 ± 4.95
Microelements	Cd ($mg\ kg^{-1}$)	83.4 ± 0.8	170	97.6 ± 0.4	88.7 ± 2.0
	Cr ($mg\ kg^{-1}$)	185 ± 6.0	93	99.0 ± 1.1	73.1 ± 0.9
	Cu ($mg\ kg^{-1}$)	586 ± 8	575	749.1 ± 7.0	551.2 ± 20.7
	Pb ($mg\ kg^{-1}$)	2462 ± 71	2200	2710 ± 300	3110 ± 110
	Zn ($mg\ kg^{-1}$)	–	14,650	22,400 ± 410	19,440 ± 370

*Value in (%)
[a]Lima et al. (2012), [b]Kirkelund et al. (2014a), [c]Magro (2014), Kirkelund et al. (2014b)

result of addition of lime during the acid gas cleaning. High chloride content is seen in MSWI APC presented in Table 12.2, and this may be the limiting factor for their reuse, as Cl can cause corrosion in reinforced steel (Wang et al. 2001).

The main mineral compounds found (Wang et al. 2001; Quina et al. 2011; Del Valle-Zermeño et al. 2013; Magro 2014) in the four residues were sylvite (KCl) and halite (NaCl), normally removed by ED treatment (Kirkelund et al. 2014a). The two FA are similar, and so are the two APC (Table 12.2). Calcium minerals such as $CaCO_3$, $CaSO_4$, or $Ca(OH)_2$ were also identified. Calcite ($CaCO_3$) and anhydrite ($CaSO_4$) were present in FA MSWI residues whereas portlandite ($Ca(OH)_2$) was only present in semidry APC (Magro 2014; Kirkelund et al. 2014b). Kirkelund et al. (2014a) performed XRD analysis in the ashes, after ED process and peaks for calcite presented higher counts which indicates that carbonate was not dissolved during ED treatment. The presence of ettringite may be important as it may immobilize Cr

(Kirkelund et al. 2014a) but, although ettringite was detected after almost all ED experiments, Cr was not stabilized. In fact, after ED Cr might have changed from Cr (III) to Cr(VI) which is hazardous and more leachable (Quina 2005).

The MSWI residues were rich in heavy metals and with higher amount of both Pb and Zn and APC are the most hazardous FA (Table 12.2).

12.2.3 Air Pollution Control Residues and Fly Ash: Applications

The solution to the waste management problems may involve incorporation of residues in material construction. The building sector uses large quantities of natural materials, thus its potential capacity to recycle waste is considerable (Rémond et al. 2002).

It is important to distinguish the goal of incorporation of IR in cement in construction sector and hazardous waste management sector (Table 12.1). Having the same physical/chemical principle (solidification/stabilization), explained more below, these two fields have very distinct aims. In the sector of waste management the goal is stabilization and storage of waste, without compromise of resources, different to the construction sector which aims to not only reduce costs, but also reduce carbon and environmental footprint, e.g., reducing the emission of CO_2.

Regardless of the goal, the incorporation of FA in concrete physically and/or chemically immobilize hazardous components initially present (Hills et al. 1993; Wiles 1996), relying on:

1. Physical—processes associated with microporosity of hydrated calcium silicate, which are capable to adsorbing ions and particles at its surface. Other hydration products, such as calcium and hydrosulfoaluminatehydroaluminate (ettringite) may also play an important role
2. Chemical—formation and precipitation of low solubility compounds (mainly metal hydroxides). The behavior of amphoteric metals are able to occur in the anionic form (e.g., Cr anions (VI))

The best approach to the solidification/stabilization (S/S) process involves initial chemical stabilization and then solidification of the waste (Zacco et al. 2014). The goal of stabilization is to turn the contaminants into less soluble and toxic forms, with or without solidification. Solidification involves the transformation of a liquid or a sludge into a solid and it may not lead to a chemical interaction between the constituent of concern and the solidifying agent. These processes reduce the mobility of the contaminants in the treated material through encapsulation, as a consequence of the reduced surface area and lower permeability (Quina 2005). The most common binders are cements or pozzolanic materials. The quantities involved are optimized as a function of the performance requirements for the final product (leaching behavior, compressive strength, setting time, etc.), which may be either solid massive (monolithic) or granular (Quina et al. 2008).

Sabbas et al. (2003) showed that, with the exception of chlorides, the immobilization of the toxic elements is possible through S/S processes. The most relevant factors for the immobilization of heavy metals are the pH, chemical speciation of metals, and redox potential. Regarding the organic compounds immobilization may occur through reactions which destroy or change their structures, or also by physical processes such as adsorption, and encapsulation (Quina 2005).

According to Quina et al. (2008) the more dynamic research area is in the field of solidification with binders, using in particular Ordinary Portland cement (OPC). At hazardous residues management level, the S/S process is the most important method for APC treatment in Europe. The process begins when water is added to cement and hydration reaction takes place. The hydration products crystallize and create a three-dimensional structure that binds together all the substances present into a hard mass. The reactions that occur are the basis of the S/S process, applied worldwide for the treatment of hazardous waste. The three-dimensional structure formed, which comprises hydration products, water, small bubbles of air, and particles of sand or stone, can also include small particles (<150 μm) (Ferreira et al. 2003a). The IR particles have a small grain size and they fill these spaces and become encapsulated inside the concrete matrix (Ferreira et al. 2003a).

The main difficulties and limitations of the methods based on cement and addition of FA are related to the following factors:

1. The release of hydrogen may occur due to the presence of metals in the residue, such as aluminum (Pera et al. 1997)
2. Presence of sulfates may lead to expand the hardening phase reactions leading to the crack formation (Collivignarelli and Sorlini 2001)
3. Soluble salts of Pb and Zinc can interfere with the compressive strength of the material (Quina 2005)
4. Chlorides are poorly retained in a solid matrix (Quina 2005)
5. The high amounts of soluble salts present can lead to solid matrix microstructure disintegration (Malone et al. 1997), which is important to predict materials behavior (Ferreira et al. 2003a; Rémond et al. 2002)

12.3 Electrodialytic Process Applied to Municipal Solid Waste Incineration Air Pollution Control Residues and Fly Ash

12.3.1 Cell Design

The electrodialytic process is a method which can be used for the removal of heavy metals from particulate waste products. Generally, the main advantage of the ED process is the ability to desorb the heavy metals into its ionic forms so they are mobile under the applied electric field (Ribeiro 1998). When applied ED treatment in APC or FA more advantages are considered. The ED process concentrates the

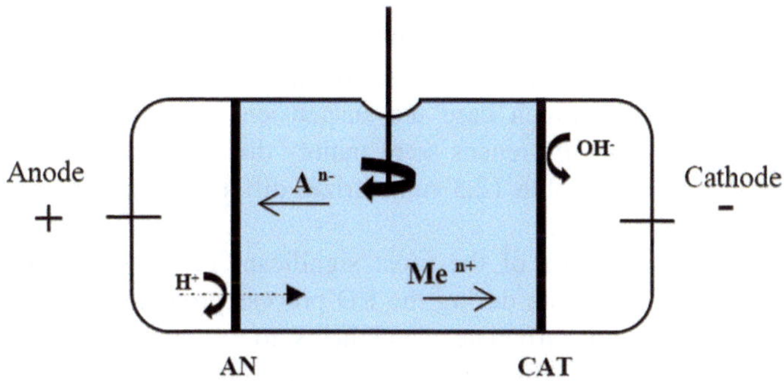

Fig. 12.1 Schematic presentation of the ED remediation principle in an experiment cell

dissolved ions in a volume of water smaller than the original volume of the residue. Thus, the contamination is concentrated rather than diluted, and such that recirculation of valuable elements may be considered more feasible. The ED process has already proven its efficiency for removal of heavy metals from soil (Ottosen 1995; Ribeiro 1998; Jensen 2005; Sun 2013) and quite a few other solid matrices with high removal efficiencies, such as harbor sediments (Nystrøm 2004; Pedersen 2014), impregnated waste wood (Christensen 2004), and different ash residues (Pedersen 2002a; Ferreira 2005; Lima 2008).

The process is based on the application of an electric potential gradient over the soil or other matrix resulting in a low level current density (Ribeiro and Rodríguez-Maroto 2006). Three main transport mechanisms take place: *electromigration*, *electroosmosis*, and/or *electrophoresis*. Diffusion is also important since concentration gradients are built up during the material transport. In case electroosmosis takes place advection of ions in the pore water can also be important. A basic scheme of a three compartment ED cell is schematized in Fig. 12.1.

Adjustments can be made to the earlier established ED cell treatment setup depending on the final objective.

12.3.2 Heavy Metals Removal from Air Pollution Control Residues and Fly Ash

Removal of heavy metals from MSWI FA using ED was first suggested by Pedersen (2002a), with a conventional ED cell where the ash was placed in the central compartment as compacted and water saturated. Difficulties during these initial experiments, like the precipitation in electrolytes, dissolution of ash and poor contact to the membranes and following increased potential drop here, poor control of the pH, and long remediation times, revealed the need to continuously mix the matrix in the cell. Pedersen (2002a) introduced a stirring rod in the central cell compartment where the ash was now suspended. After the introduction of the stirrer, several studies about ED treatment of FA from MSWI were made. Assisting agents were introduced to investigate if this could further enhance the process. Beside the assisting agents, experiments have also been conducted with the ash suspended in water, in which case the suspension is acidified during the ED treatment. The removal differences were mainly due to the types of IR and the final treated sample pH. Table 12.3 summarizes different percentages of removal obtained by several authors.

The final ash pH is one of the most significant parameters when applying ED. The pH drop that occurs during the ED process promotes the dissolution of oxides, hydroxides, and carbonates, and helps to solubilize the heavy metals (Ferreira et al. 2004a, b). After that, metals can be mobilized under the electric field and transported away from the waste, more acidification better removals rates (Ferreira et al. 2004a, b).

12.4 Materials Tests Solidification/Stabilization Process: Electrodialytic Pre-treatment

OPC is the most used material in the construction industry worldwide, it has a high level of CO_2 emissions (1 t of cement generates 1 t of CO_2) and has over the years become less attractive compared to alternative ecological new binders (Pacheco-Torgal et al. 2008). Thus, the replacement of OPC by upgraded secondary resources may be an environmentally preferable solution. Mortar is a mixture of paste (OPC and water) and sand—the aggregate. Mortar is used to hold building materials such as brick or stone together, is a substance much thicker than concrete making it ideal as a glue. The hydration is the main chemical reaction that happens in this mixture and which causes cement paste to harden and gain strength.

The quality of the paste determines the character of the material. The strength of the paste, in turn, depends on the ratio of water-to-cement. High-quality mortar is produced by lowering the water–cement ratio without sacrificing the workability of fresh mortar, allowing it to be properly placed, consolidated, and cured (ACM-PCA 2015).

When secondary resources such as upgraded IR are incorporated in substitution of cement, it is important to take into account two essential group factors: materials changes (e.g., compressive strength and workability) and environmental issues (e.g., leaching of heavy metals and chloride). These issues will be further discussed below.

12.4.1 Compressive Strength and Workability

Compressive strength is relevant to define the mortar, determining its strength capacity. According to Frutuoso (2013) a higher value gives a less deformable mortar. The compressive strength is related, among several factors, to the porosity

Table 12.3 Electrodialytic treatment of IR from MSWI reported in literature (laboratory scale)

Type of ash	Current density (mA cm^{-2})	Cell design	Liquid/solid ratio	Time of treatment (week)	Assisting agent	Removal (%)		Refs.
Fly ash: collected on the electrostatic precipitator and not exposed to any flue gas cleaning additive. From: Vestforbraending, Denmark	0.8	Stirred	6.5	2	2.5 % NH_3	Cd	100	Pedersen (2002b)
						Cr	7	
						Cu	49	
						Pb	3.5	
						Zn	42	
	0.8	Stirred	6.5	2	0.25 M Na-citrate	Cd	15	Pedersen (2002b)
						Cr	15	
						Cu	37	
						Pb	12	
						Zn	39	
	0.8	Stirred	6.5	2	0.25 M ammonium citrate in 1.25 NH_3	Cd	62	Pedersen (2002b)
						Cr	20	
						Cu	59	
						Pb	6	
						Zn	39	
	0.8	Stirred	6.5	2	Water	Pb	12	Pedersen (2002b)
						Zn	25	
	0.8	Stirred	15	2	Water	Cu	90	Pedersen (2002b)
						Pb	41	
	0.8	Stationary	0.9	10	0.5 M ammonium citrate in 2.5 NH_3	Cd	40	Pedersen et al. (2003)
						Cr	2–3	
						Cu	31	
						Pb	2.5	
						Zn	24	

(continued)

Table 12.3 (continued)

Type of ash	Current density (mA cm^{-2})	Cell design	Liquid/solid ratio	Time of treatment (week)	Assisting agent	Removal (%)		Refs.
	0.8	Stirred	6.7	3	0.25 M ammonium citrate in 1.25 NH_3	Cd	70	Pedersen et al. (2003)
	0.8	Stirred	6.5	10	Water	Cd	86	Pedersen et al. (2005)
						Cr	44	
						Cu	81	
						Pb	20	
						Zn	62	
	0.8	Stirred	6	2	Deionized water	Cu	20^b	Ottosen et al. (2006)
						Pb	$<1^b$	
	0.8	Stirred	15	2	Deionized water	Cu	90^b	Ottosen et al. (2006)
						Pb	41^b	
Fly ash: collected electrostatic precipitator before exposure to flue gas cleaning additive. From: Nuuk, Greenland	0.8	Stationary	1	10	0.5 M ammonium citrate in 2.5 % NH_3	Cd	44	Pedersen et al. (2003)
						Cr	2–3	
						Cu	18	
						Pb	8	
						Zn	73	
Air pollution control residues: collected in semidry air pollution control equipment. Complete flue gas treatment at the incineration unit—injection of ammonia for NOx removal in the boiler; addition of lime in the scrubber; collection of all the formed particulates (both electrically and non-electrically chargeable) in filters. From: Valorsul, Portugal	0.8	Stationary	0.6	3	Water	Cd	4	Ferreira et al. (2004a, b)
						Cu	5	
						Pb	5	
						Zn	4	
	0.8	Stationary	0.8	3	3 % Na-gluconate	Cd	2	Ferreira et al. (2004a, b)
						Cu	3	
						Pb	2	
						Zn	3	

	0.8	Stirred	5	2	3 % Na-gluconate	Cd	2	Ferreira et al. (2004a, b)
						Cu	0.2	
						Pb	1	
						Zn	2	
	4.0	Stationary	0.8	3	22 % Na-gluconate	Cd	5	Ferreira et al. (2004a, b)
						Cu	2	
						Pb	3	
						Zn	3	
	4.0	Stationary	0.8	3	22 % Na-gluconate	Cd	10	Ferreira et al. (2004a, b)
						Cu	5	
						Pb	6	
						Zn	6	
	0.8	Stirred	4	2	Deionized water	Cd	−17[a]	Lima et al. (2010)
						Cu	−82[a]	
						Pb	−17[a]	
	0.8	Stirred	4	2	0.25 M ammonium citrate in 1.25 ammonia	Cd	66	Lima et al. (2010)
						Cu	50	
						Pb	25	
Air pollution control residues: collected after treatment of flue gas in electrostatic filters. From: Illulissat, Greenland	1	Stirred	3.5	1	Deionized water	Al	<0.01	Kirkelund et al. (2013)
						Cd	<0.01	
						Cr	<0.1	
						Cu	<0.01	
						Pb	2	
						Zn	<0.1[b]	

(continued)

Table 12.3 (continued)

Type of ash	Current density (mA cm^{-2})	Cell design	Liquid/solid ratio	Time of treatment (week)	Assisting agent	Removal (%)		Refs.
	1	Stirred	10	2	Deionized water	Al	10	Kirkelund et al. (2013)
						Cd	60	
						Cr	6[b]	
						Cu	15[b]	
						Pb	8[b]	
						Zn	45	

[a]Negative values, due to the metal content increases after treatment: because of the dissolution of the ash
[b]Values read from graphics

of the material, which generally increases with the decreasing of the latter. The compressive strength generally increases with longer curing times. Kirkelund et al. (2014a) proved that a significant increase in compressive strength results from curing mortar during 32–56 days. However, differences in the compressive strength tests between 28 and 32 days were not statistically different (Table 12.4) and the same occurred when 5 or 15 % of cement was a replaced by the ash. So, in a future approach, 28 days of curing and tests, with an incorporation of 15 % of upgraded ash (ED pre-treatment), MSWI residue can be performed as a first step of optimization.

All the mortars containing IR showed lower compressive strength (Table 12.4) compared to the references, with one exception when replacing by 5 % untreated semidry residue (Magro 2014). The incorporation of FA gave lower compressive strength comparing with the reference. The addition of fine particles causes segmentation of large pores and increases nucleation sites for precipitation of hydration products in cement paste (Bijen 1996; Tumidajski and Chan 1996). Magro (2014) also showed that the porosity increased after the incorporation of APC in the mortars, which may explain the decreasing compressive strength.

The decreased compressive strength when a defined percentage of cement was replaced with FA or APC shows that the possible pozzolanic effect is not high enough to compensate for a 1:1 replacement with cement. Whether ED improves the quality of the FA or not in relation to the compressive strength of the mortar is ambiguous in Table 12.4. Due to this ambiguity results, statistical analysis of mortar compressive strength was carried out using GraphPad Prism 6. One-way ANOVA was achieved to understand if there are differences in the values (Table 12.1). Thus, there are no statistical differences between the two APC, contrarily with wet FA that are statistically significantly different at $\rho < 0.05$ in raw and upgraded residue and between the two FA. This analysis also proved that the incorporation of FA or APC is statistically different in this compressive strength data. Notice, the data sets are too few to make general conclusions on this point, as the results may be strongly related to both the ED treatment and the IR characteristics.

Metallic Al and sulfate in MSWI APC are regarded important factors in relation to the lower compressive strengths of mortars due to crack formation (Aubert et al. 2007). In all the results presented for upgraded APC, the amount of soluble sulfate increased with the ED treatment. Regarding Table 12.2, the wet FA had the higher concentration of Al, though the speciation of this Al is not known. So apart from loss in mortar, metallic Al or soluble sulfate can be part of the reason to strength decrease when there was incorporation of the non-upgraded residue.

Workability is also an important factor to consider when substituting cement with FA residues. Workability is one of the physical parameters of concrete which affects the strength and durability as well as the cost of labor and appearance of the finished product (Ferraris et al. 2000). Concrete is said to be workable when it is easily placed and compacted homogeneously, e.g., without bleeding or segregation. Kirkelund et al. (2014a) showed that the workability decreased when the actual MSWI APC was added. The decrease in workability is related mostly with porosity,

Table 12.4 Resume and comparison (ANOVA 1 way) of compressive strength (MPa) of the different mortars with partial replacement of cement with IR and a corresponding reference

[1]Fly ash		[2]Semidry APC			[3]Wet fly ash			[3]Semidry APC	
[#]**Ref.**	5 % Raw[**]	[#]**Ref.**	15 % Raw[*]	15 % Upgraded[*]	[#]**Ref.**	5 % Raw	5 % Upgraded	5 % Raw	5 % Upgraded
40	30[a, b, c]	**43 ± 3**	41 ± 0	40 ± 0	**47 ± 6**	38 ± 0.7[b, d]	46 ± 4[c, d]	48 ± 4[a]	35 ± 11

Raw = untreated ash, upgraded = ED treated (mean values ± standard error)

[#]Ref: Reference value correspondent to the incorporation of 0 % ash in mortars

[*]Tested after 32 days of curing

[**]Values read from graphics

[1]Lima et al. 2012, [2]Kirkelund et al. 2014a, [3]Magro 2014, Kirkelund et al. 2014b

Statistics: compressive strength statistically significantly different at $\rho < 0.05$ comparing to: [a]5 % raw FA vs. 5 % raw semi-dry APC; [b]5% raw FA vs. 5 % raw wet FA; [c]5% raw FA vs. 5 % upgraded wet FA; [d]5 % raw wet FA vs. 5 % upgraded wet FA particle size, shape, and surface characteristics of the APCs: porous particles will adsorb water in the mortar mix and reduce the workability, which should be compensated by adding water or superplasticizer (Brown et al. 2011).

12.4.2 Leaching Behavior

European Environmental Protection (2005) criteria for landfill deposition of waste and leachability thresholds of the mortars after prEN 12457-1 are used to evaluate metal leachability from the mortar with ED treated IR. Leaching from mortars cured for 28/32 days are presented in Table 12.5.

Heavy metals are considered foreign cations in a cement bed matrix and are found in considerable concentrations on ash materials. On the other hand, concrete's high pH is believed to immobilize heavy metals. However, due to FA high specific surface, some of the heavy metals are placed there and prone to leach (Pedersen 2002a).

The pH in the mortar was generally higher than in the residues and at a level where the heavy metals are expected to be stable in the matrix (Pedersen et al. 2005). Lima et al. (2010) results on FA suggest that in the presence of NaCl the leachable Cr increased compared to the reference, the same occurred in wet FA (Magro 2014). Nevertheless, these results were only seen after ED process, perhaps due to the form change from Cr(III) to Cr(VI). When comparing the amount of leached heavy metals from the APC (ash) (Kirkelund et al. 2014b) to the amount leached from the mortars with APC, it can be seen that the mortars generally incorporates the heavy metals. Leachability of Cu, Pb, and Zn was reduced or maintained compared to the reference. Whereas Ca presented higher values in the conducted studies.

There were no significant differences in compressive strength between incorporation of 5 and 15 % of ash in the results presented in this chapter (Sect. 12.4.1). Therefore, the application of concrete containing 15 % of waste incineration residues should be carefully considered. Although there are experiments in which the untreated residue gives better results (in terms of workability or compressive strength)—probably because of any inherent pozzolan effect that is destroyed during the upgrading—the use of the raw material will not be allowed, due to the high amount of soluble salts, heavy metals and the long term of leaching or problems about handling the material after demolition. There are many materials and contaminants behavior tests that must be investigated with APC, FA, or other type of ash. At the present, there are few studies about the incorporation of FA and APCs with ED pre-treatment. Chen et al. (2014) achieved promising results after the incorporation of upgraded APC in clay bricks. The study shows that, similarly with the results select in this overview, the leaching of Pb decreased when 5 % of upgraded ash residue is incorporated contrary to Cr_{total} leachability that increased after ED process.

Table 12.5 EEA landfill criteria for heavy metals leachability; experimental data obtained by prEN 12457-1 on mortar: reference; substitution of raw residues and upgraded APC and FA; resume and comparison results

EEA landfill criteria (mg L^{-1})		[a]Fly ash	[b]Fly ash		[c]Semidry APC			[d]Wet fly ash			[d]Semidry APC	
Class 1		15 % Raw	#Ref.	5 % Raw	#Ref.	15 % Raw*	15 % Upgraded*	#Ref.	5 % Raw	5 % Upgraded	5 % Raw	5 % Upgraded
Ca	–	–	–	–	**750**	1370	747	**786**	774	577	642	667
Cd	≤0.1	0.001	**0.04**	0.07	**<0.01**	<0.01	<0.01	**ND**	ND	ND	ND	ND
Cr_{total}	≤0.5	0.003	**0.07**	0.10	**0.03**	0.02	0.04	**0.04**	0.03	0.07	0.05	0.06
Cu	≤2	0.003	**0.02**	0.02	**0.02**	0.03	0.02	**0.01**	0.01	0.01	ND	0.01
Pb	≤0.5	0.01	**0.03**	0.02	**0.06**	0.20	0.07	**0.08**	0.05	0.02	0.03	0.07
Zn	≤4	0.01	–	–	**0.04**	0.10	0.04	**0.03**	0.02	0.02	0.05	0.05
pH	5.5–12	–	**12.6**	12.6	**12.6**	12.6	12.5	**12.4**	12.4	11.1	12.4	12.7

#Ref: Reference value correspondent to the incorporation of 0 % ash in mortars
*Tested after 32 days of curing; *ND*—not detected
[a]Shi and Kan (2009), [b]Lima et al. (2012), [c]Kirkelund et al. (2014a), [d]Magro (2014)

12.4.2.1 Chloride

The chloride ion promotes the oxidation of iron in the form of rust, causing not only a reduction of the section of the armature, but also an expansion due to iron oxide formation, which ultimately results in the disintegration of concrete, accelerating the corrosion process (Frutuoso 2013). Corrosion process happened, e.g., because the soluble salts present a weak interaction with cement matrices. The biggest difficulty of a S/S process is related with high amount of soluble salts content. To decrease this amount, the washing of the APC and FA before incorporation in cement may be an option (Quina 2005).

EN-206-1 European standard 206-1 (2000), Concrete – Part 1: Specification, performance, production and conformity. Brussels expresses the percentage of chloride ions by mass of cement. In Magro (2014) the chloride (%) increased after the including of 5 % ash in cement, but the values for upgraded wet FA and semidry APC showed a similarity with reference and according to the limit value in EN-206-1. In Kawamura et al. (1998) the incorporation of 20 % FA shows 2 % of chloride, showing that the chlorides are certainly present in the MSWI ash, and mainly in the form of halite and sylvite. These results show a new possibility to the improvement of the residue with ED treatment, after the S/S process.

12.5 Conclusions and Main Recommendations

This overview aimed to understand if the ED process is an appropriate pre-treatment for further reuse of IR in mortar bars. This subject is a challenging one, because this type of waste usually presents high amounts of heavy metals and salts. Contamination of soil, groundwater is one of concern associated with the disposal of MSWI.

Four parameters were resumed in this chapter: initial characteristics, compressive strength, heavy metals leachability, and chloride content. The ash characteristics are influenced by several factors related to each plant, such as the composition of the waste that is burned. Therefore, further studies should be made in order to reach more conclusions on the ashes characteristics when they are upgraded with ED treatment and incorporated in cement.

The use of ED treated IR for incorporation in cement seems a viable option if it prevents the leaching of contaminants. The results obtained till now by the group shows that the immobilization of contaminants may be achieved, and these results are very promising. Also, no statically significant differences were observed in compressive tests between the incorporation of 5 or 15 % of ash as substitute for cement in mortar, meaning that the incorporation of 15 % of ash can be the first option as it allows a high reuse of the ash. Mortars with ED treated residues show similar compressive strengths compared to the raw residue, but lower than the reference mortar. Compression tests are important to define the mortars, and must first be similar to the reference values (without residues incorporation), on the other hand, all the values are in the standard suggested by Frutuoso (2013). For the wet FA (Magro 2014) higher values were achieved when compared to the other studies (Lima et al. 2012; Kirkelund et al. 2014a).

The addition of this type of APC might present a threat for steel corrosion in reinforced concrete structures due to the highly soluble chloride content. The soluble salts have a weak interaction with cement matrices. Normally, the main problem, mostly in APC, is chloride content, and these values seem to decrease when there was an ED pre-treatment. Yet insulation measures for the building material will always be required. In fact, one of the biggest difficulties in treating this residue relates to its high content of soluble salts, which limits its application in further products. Decreasing the amount of chloride may pass through a washing process before the incorporation of this APC in cement.

The stabilization of MSWI ash may be difficult, as its characteristics may float according to the waste that is burned as well as burning conditions. As a consequence, it is hard to work out a standard for MWSI APC for concrete. In studied residues here presented pre-treatment with ED process and then incorporation in construction materials, when OPC is used, appears to be an appropriate choice.

One of the environmental benefits of this approach includes reduction of waste and the associated space for landfill, conservation of natural resources such as gypsum, limestone, and natural gas when ah is used as a replacement in cement production and reduction of greenhouse gas emissions. Still, further studies should be carried out regarding the long-term environmental impacts of the reuse of the upgraded IR in construction materials.

References

ACM-PCA American's cement manufacturers (2015). http://www.cement.org. Accessed Jan 2015

Aubert JE, Husson B, Sarramone N (2007) Utilization of municipal solid waste incineration (MSWI) fly ash in blended cement. Part 2. Mechanical strength of mortars and environmental impact. J Hazard Mater 146:12–19

Bijen J (1996) Benefits of slag and fly ash. Constr Build Mater 10(5):309–314

Boesch ME, Vadenbo C, Saner D, Huter C, Hellweg S (2014) An LCA model for waste incineration enhanced with new technologies for metal recovery and application to the case of Switzerland. Waste Manage 34(2):378–389

Brown P, Jones T, BéruBé K (2011) The internal microstructure and fibrous mineralogy of fly ash from coal-burning power stations. Environ Pollut 159:3324–3333

Chen W, Ottosen LM, Jensen PE, Kirkelund GM, Schmidt JW (2014) A comparative study on electrodialytic treated bio-ash and MSWI APC residues for use in bricks. In: Proceedings of the 5th international conference on engineering for waste and biomass valorisation, Rio de Janeiro, Brazil, 25–28 Aug 2014, pp 648–662

Christensen IV (2004) Electrodialytic remediation of CCA-treated waste wood. Ph.D. thesis, Technical University of Denmark, Denmark

Collivignarelli C, Sorlini S (2001) Optimization of industrial wastes reuse as construction materials. Waste Manage Res 19:539–544

Del Valle-Zermeño R, Formosa J, Chimenos JM, Martínez M, Fernández AI (2013) Aggregate material formulated with MSWI bottom ash and APC fly ash for use as secondary building material. Waste Manage 33:621–627

Directive 2000/76/EC of the European Parliament and the Council of 4 December 2000, on the incineration of waste

Ecke H, Sakanakura H, Matsuto T, Tanaka N, Lagerkvist A (2000) State-of-the-art treatment processes for municipal solid waste incineration residues in Japan. Waste Manage 18(1):41–51

Eighmy TT, Kosson DS (1996) U.S.A. national overview on waste management. Waste Manage 16(5/6):361–366

European Environmental Protection EEA (2005) Guidance on sampling and testing of wastes to meet landfill waste acceptance procedures

EN-206-1 European standard 206-1 (2000) Concrete—part 1: specification, performance, production and conformity, CEN, Brussels

Eurostat (2014) Municipal waste generation and treatment, by type of treatment method. Environmental statistics and accounts in Europe, Eurostat statistical books, European Commission.http://epp.eurostat.ec.europa.eu/tgm/graph.do?pcode=tsdpc240&language=en. Accessed Nov 2014

Ferraris CF, Brower L, Ozyildirim C (2000) Workability of self-compacting concrete. National Institute of Standards and Technology. In: Proceedings of the international symposium on high performance concrete, Orlando, 25–27 Sept 2000, pp 398–407

Ferreira C (2005) Removal of heavy metals from municipal solid waste incinerator fly ash by an electrodialytic process. Ph.D. thesis, Technical University of Denmark, Denmark

Ferreira C, Ribeiro AB, Ottosen LM (2003a) Possible applications of municipal solid waste fly ash. J Hazard Mater 96(2–3):201–215

Ferreira C, Ribeiro AB, Ottosen LM (2003b) Heavy metals in MSW incineration fly ashes. J Phys IV Fr 107:463–467

Ferreira C, Jensen PE, Ottosen LM, Ribeiro AB (2004a) Removal of selected heavy metals from MSW fly ash by electrodialytic process. Eng Geol 77(3–4):339–347

Ferreira C, Ribeiro AB, Ottosen LM (2004b) Treatment of MSW fly ashes using the electrodialytic remediation technique. In: Brebbia CA, Kungolos S, Popov V, Itoh H (eds) Waste manage and the environ II. WIT Press, Japan, pp 65–75

Frutuoso AR (2013) Influência de agregados provenientes de RCD e cinzas volantes não conformes em argamassas de cal aérea (Influence of aggregates from construction and demolition residues and non-compliant fly ashes in air lime mortars). Master thesis, Faculdade de Ciências e Tecnologia da Universidade Nova de Lisboa (in Portuguese)

Hansen HK, Pedersen AJ, Ottosen LM, Villumsen A (2001) Speciation and mobility of cadmium in straw and wood combustion fly ash. Chemosphere 45:123–128

Hills CD, Sollars CJ, Perry R (1993) Ordinary Portland cement based solidification of toxic wastes: the role of OPC reviewed. Cem Concr Res 23(1):196–212

Hjelmar O (1996a) Waste management in Denmark. Waste Manage 16(5/6):389–394

Hjelmar O (1996b) Disposal strategies for municipal solid waste incineration residues. J Hazard Mater 47(1–3):345–368

Hjelmar O, Johnson A, Comas R (2011) Incineration: solid residues. In: Christensen TH (ed) Solid waste technology and management, vol 1/2. Wiley, Denmark, pp 430–462

International Ash Working Group IAWG: Chandler AJ, Eighmy TT, Hartlén O, Kosson D, Sawell SE, van der Sloot H, Vehlow J (1997) Municipal solid waste incinerator residues. Stud Environ Sci 67, Elsevier Science, Amsterdam

Jensen PE (2005) Application of microbial products to promote electrodialytic remediation of heavy metal contaminated soil. Ph.D. thesis, Technical University of Denmark, Denmark

Kawamura M, Kayyali O, Haque M (1998) Effects of a fly ash on pore solution composition in calcium and sodium chloride-bearing mortars. Cem Concr Res 18:763–773

Kirkelund GM, Jensen PE, Ottosen LM (2013) Electrodialytic extraction of heavy metals from Greenlandic MSWI fly ash as a function of remediation time and L/S ratio. In ISCORD: planning for sustainable cold regions, American Society of Civil Engineers, pp 87–96

Kirkelund GM, Geiker MR, Jensen PE (2014a) Electrodialytically treated MSWI APC residue as substitute for cement in mortar. Nordic Concr Res 49:1–16
Kirkelund GM, Magro CC, Guedes PR, Jensen PE, Ribeiro AB, Ottosen LM (2014b) Electrodialytic removal of heavy metals and chloride from MSWI fly ash and APC residue in suspension—test of a new two compartment experimental cell. Electrochim Acta (in review)
Le Forestier L, Libourel G (1988) Characterization of flue gas residues from municipal solid waste combustors. Environ Sci Technol 32(15):2250–2256
Lima AT (2008) Fly ashes contaminants removal—the electrodialytic process potentialities. Ph.D. thesis, Faculdade de Ciências e Tecnologia da Universidade Nova de Lisboa, Portugal
Lima AT, Ottosen LM, Pedersen AJ, Ribeiro AB (2008) A characterization of fly ash from bio and municipal waste. Biomass Bioenerg 32(8):277–282
Lima AT, Ribeiro AB, Rodríguez-Maroto JM, Mateus EP, Castro AM, Ottosen LM (2010) Experimental and modeling of the electrodialytic and dialytic treatment of a fly ash containing Cd, Cu and Pb. J Appl Electrochem 40(9):1689–1697
Lima AT, Ottosen LM, Ribeiro AB (2012) Assessing fly ash treatment: remediation and stabilization of heavy metals. J Environ Manage 95:110–115
Magro CC (2014) Electrodialytic remediation of two types of air pollution control residues and their applicability in construction materials. M.Sc. thesis, Faculdade de Ciências e Tecnologia da Universidade Nova de Lisboa, Portugal
Malone PG, Poole TS, Wakeley LD, Burkes JP (1997) Salt related expansion reactions in Portland-cement based waste forms. J Hazard Mater 52(2–3):237–246
Nagib S, Inoue K (2000) Recovery of lead and zinc from fly ash generated from municipal incineration plants by means of acid and/or alkaline leaching. Hydrometallurgy 56(3):269–292
Nystrøm GM (2004) Electrodialytic removal of heavy metals from contaminated harbor sediments. Ph.D. thesis, Technical University of Denmark, Denmark
Ottosen LM (1995) Electrokinetic remediation—application to soil polluted from wood preservation. Ph.D. thesis, Technical University of Denmark, Denmark
Ottosen LM, Lima AT, Pedersen AJ, Ribeiro AB (2006) Electrodialytic extraction of Cu, Pb and Cl from MSWI fly ash suspended in water. J Chem Technol Biotechnol 81:553–559
Pacheco-Torgal F, Castro-Gomes JP, Jalali S (2008) Investigations on mix design of tungsten mine waste geopolymeric binder. Constr Build Mater 22:1939–1949
Pedersen AJ, (2002a) Electrodialytic removal of heavy metals from fly ashes. Ph.D. thesis, Technical University of Denmark, Denmark
Pedersen AJ (2002b) Evaluation of assisting agents for electrodialytic removal of Cd, Pb, Zn, Cu and Cd from MSWI fly ash. J Hazard Mater 95:185–198
Pedersen KB (2014) Applying multivariate analysis to developing electrodialytic remediation of harbour sediments from Arctic locations. Ph.D thesis, Arctic University of Norway, Norway
Pedersen AJ, Ottosen LM, Villumsen A (2003) Electrodialytic removal of heavy metals from different fly ash: influence of heavy metals speciation in the ashes. J Hazard Mater 100:65–78
Pedersen AJ, Ottosen LM, Villumsen A (2005) Electrodialytic removal of heavy metals from municipal solid waste incineration fly ash using ammonium citrate as assisting agent. J Hazard Mater 122:103–109
Pera J, Coutaz L, Ambroise J, Chababbet M (1997) Use of incinerator bottom ash in concrete. Cem Concr Res 27(1):1–5
Piantone P, Bodénan F, Derie R, Depelsenaire G (2003) Monitoring the stabilization of municipal solid waste incineration fly ash by phosphation: mineralogical and balance approach. Waste Manage 23(3):225–243
Quina MJ (2005) Processos de inertização e valorização de cinzas volantes—Incineração de resíduos sólidos urbanos (Inerting processes and recovery of fly ash—incineration of municipal solid waste). Ph.D. thesis, Faculdade de Ciências e Tecnologia Universidade de Coimbra, Portugal
Quina MJ, Bordado JC, Quinta-Ferreira RM (2008) Treatment and use of air pollution control residues from MSW incineration: an overview. Waste Manage 28(11):2097–2121

Quina MJ, Bordado JC, Quinta-Ferreira RM (2011) Air pollution control in municipal solid waste incinerators—the impact of air pollution on health, economy. In: Khallaf M (ed) Environ Agric Sources vol 16. Intech, Portugal, pp 331–358

Rémond S, Pimienta P, Bentz D (2002) Effects of the incorporation of municipal solid waste incineration fly ash in cement pastes and mortars: I. Experimental study. Cem Concr Res 32 (2):303–311

Ribeiro AB (1998) Use of electrodialytic remediation technique for removal of selected heavy metals and metalloids from soils. Ph.D thesis, Technical University of Denmark, Denmark

Ribeiro AB, Rodríguez-Maroto JM (2006) Electroremediation of heavy metal contaminated soils: processes and applications. In: Prasad MNV, Sajwan KS, Naidu R (eds) Trace elements in the environment: biogeochemistry, biotechnology and bioremediation, Cap 18. CRC, Boca Raton, pp 341–368

Sabbas T, Polettini A, Pomi R, Astrup T, Hjelmar O, Mostbauer P, Cappai G, Magel G, Salhofer S, Speiser C, Heuss-Assbichler S, Klein R, Lechber P (2003) Management of municipal solid waste incineration residues. Waste Manage 23(1):61–88

Sakai S (1996) Municipal solid waste management in Japan. Waste Manage 16(5/6):395–405

Sakai S, Sawell SE, Chandler AJ, Eighmy TT, Kosson DS, Vehlow J, van der Sloot HA, Hartlen J, Hjelmar O (1996) World trends in municipal solid waste management. Waste Manage 16 (5/6):341–350

Shi H-S, Kan L-L (2009) Leaching behavior of heavy metals from municipal solid wastes incineration (MSWI) fly ash used in concrete. J Hazard Mater 164:750–754

Sorensen MA, Mogensen EPB, Lundtorp K, Jensen DL, Christensen TH (2001) High temperature co-treatment of bottom ash and stabilized fly ashes from waste incineration. Waste Manage 21 (6):555–562

Sun TR (2013) Effect of pulse current on energy consumption and removal of heavy metals during electrodialytic soil remediation. Ph.D. thesis, Technical University of Denmark, Denmark

Sundqvist JO (2005) How should municipal solid waste be treated—a system study of incineration-material recycling, anaerobic digestion and composting. In: Jan-OlovSundqvist (ed) Report to Swedish Environmental Research Institute. IVL report B 1547, Stockholm

Tumidajski PJ, Chan GW (1996) Boltzmann–Matano analysis of chloride diffusion into blended cement concrete. J Mater Civ Eng 8(4):195–200

van der Sloot HA, Kosson DS, Hjelmar O (2001) Characteristics, treatment and utilization of residues from municipal waste incineration. Waste Manage 21(8):753–765

Varela A, Ribeiro AB, Monteiro O, Lima AT, Domingues H, Castelo-Branco MA (2009) Caracterização inorgânica de cinza volante de uma estação de incineração de resíduos sólidos urbanos com vista à sua eventual reciclagem (Inorganic characterization of fly ash from municipal solid waste incineration plant with the aim of their eventual recycling). Revista de Ciências Agrárias 32(1):207–215 (in Portuguese)

Wang K-S, Chiang K-Y, Lin K-L, Sun C-J (2001) Effects of a water-extraction process on heavy metals behavior in municipal solid waste incinerator fly ash. Hydrometallorgy 62:73–81

Wiles C (1996) Municipal solid waste combustion ash: state of the knowledge. J Hazard Mater 47 (1–3):325–344

Zacco A, Borgese L, Gianoncelli A, Struis RPWJ, Depero LE, Bontempi E (2014) Review of fly ash inertisation treatments and recycling. Environ Chem 12:153–175

Part IV
Modeling of the Electrokinetic Process

Chapter 13
A Coupled Reactive-Transport Model for Electrokinetic Remediation

Juan Manuel Paz-García, María Villén-Guzmán, Ana García-Rubio, Stephen Hall, Matti Ristinmaa, and César Gómez-Lahoz

13.1 Introduction

Most published works in the field of electrokinetic remediation (EKR) are experimentally based studies, reporting empirical results from a wide range of materials, contaminants, and operation conditions. These experimental results are the basis for the fundamentals of the EKR technology. However, to further develop EKR and achieve efficient field-scale treatments, it is essential to establish first a more robust theoretical framework, and second numerical models with reliable prediction capability (Alshawabkeh 2009; Yeung and Gu 2011; Reddy 2013). The aim of theoretical research and model development is not only the creation of tools that allow the prediction of experimental results, but also the increase of the overall understanding of the different processes involved, as well as to improve and optimize the experimental methods.

Research via modelling can be subdivided into, at least, two different steps. First, the theoretical researcher has to study the modelled system and, based on the experimental observation and the existing experience, identify the most significant physical–chemical processes taking place. Then, the researcher has to propose mathematical equations that describe these selected physical–chemical phenomena in a consistent manner. The set of selected physical–chemical laws and their mathematical representation is normally referred to as "theoretical model". It is worth mentioning that a good theoretical model describes all the physical–chemical

J.M. Paz-García (✉) • S. Hall • M. Ristinmaa
Division of Solid Mechanics, University of Lund, Lund, Sweden
e-mail: juanma.paz@me.com

M. Villén-Guzmán • A. García-Rubio • C. Gómez-Lahoz
Department of Chemical Engineering, University of Málaga, Málaga, Spain

© Springer International Publishing Switzerland 2016

A.B. Ribeiro et al. (eds.), *Electrokinetics Across Disciplines and Continents*,
DOI 10.1007/978-3-319-20179-5_13

processes that have relevant influence in the process, with the minimal number of equations and parameters.[1]

In a second step, the researcher has to solve (mathematically) the derived model for a specific case studied. Each case studied will be determined by its initial and boundary conditions. The solution of the mathematical problem (in the form of simulations) is appraised by comparison with experimental data. This process normally leads to the reformulation of both the theoretical model and the procedure used for the mathematical solution. A consequence of this evaluation is the repetition of the previous steps (selection of physical laws, construction of the mathematical model, and its solution) until a consistent model is obtained, in which the simulation and the experimental results are comparable.

In general, the process of developing theoretical models for the specific application on EKR processes presents the following key characteristics:

1. EKR systems are heterogeneous. The materials under study (such as soil, bricks, stones, concrete structures, sludge, wood, or fly ashes) consist of one or more porous bodies or suspensions including aqueous or liquid phases, one or more solid phases and, sometimes, gaseous phases.
2. EKR processes are multi-physical. The theoretical model should include electrochemical reactions at the electrodes, transport phenomena of chemical species driven by direct and coupled mechanisms (e.g. diffusion, electro-migration, electro-osmosis) and chemical reactions aspects (e.g. precipitation/dissolution, complexation, gas formation or electrochemical deposition).
3. EKR processes are specific for the modelled system. Aspects such as the concentration and composition of the specific contaminant and the other chemical species existing in the supporting matrix as well as the specific technique used for the remediation have to be addressed individually for each case.

Based on the list above, it is concluded that the description of a generalized model for EKR treatments is a challenging task. It is generally accepted that the most consistent theoretical model for macroscopic EKR processes consists of the combination of mass balance equations coupled with the electro-neutrality condition, where the main transport mechanisms are electro-migration for charged species and electro-osmosis for neutral (non-charged) species (Jacobs and Probstein 1996; Rodriguez-Maroto and Vereda-Alonso 2009).

Due to the complexity of the theoretical problem, its mathematical solution generally requires the use of advanced computer-assisted numerical methods. The algorithm used for obtaining the numerical solution of the theoretical model, including all the necessary assumptions to make the problem solvable, is referred to as "the numerical model". Sometimes, this term also includes the code implemented in the chosen programming language for the computer-aided solution.

[1] This means that the best model is the one with the best ratio between mathematical simplicity and physical realism.

In most cases, the implementation of a consistent and reliable numerical model is the bottleneck of model-based research processes. This is also due to the lack of commercial and user-friendly software specifically designed for EKR applications, which forces the researches to implement their own codes. There are, however, some available software packages of interest for this purpose. For example, speciation-equilibrium and batch-reaction programs, such as PHREEQC (Parkhurst and Appelo 1999); Visual Minteq (Peterson et al. 1987) or WATEQ4F (Ball and Nordstrom. 1991), among others, are useful for the characterization of the chemical system during the development of the theoretical model. These applications can be coupled with programs able to solve partial differential equations systems for a coupled reactive-transport model (Nardi et al. 2014).

In this work, we present a generalized theoretical model for EKR treatments and we propose and explain, with simple guidelines, how the numerical model was implemented for the simulations presented here. The theoretical model included many features found in existing models proposed for EKR processes, and it is explained in a very generalized manner in such a way that it can be applied for different matrices, contaminants and setups designs. The numerical model, in turn, consists of the combination of two main coupled modules: one for reactive-mass transport and another one for chemical equilibrium. The general algorithm used for the solution of the coupled model is depicted in Fig. 13.1. We assume that a subset of

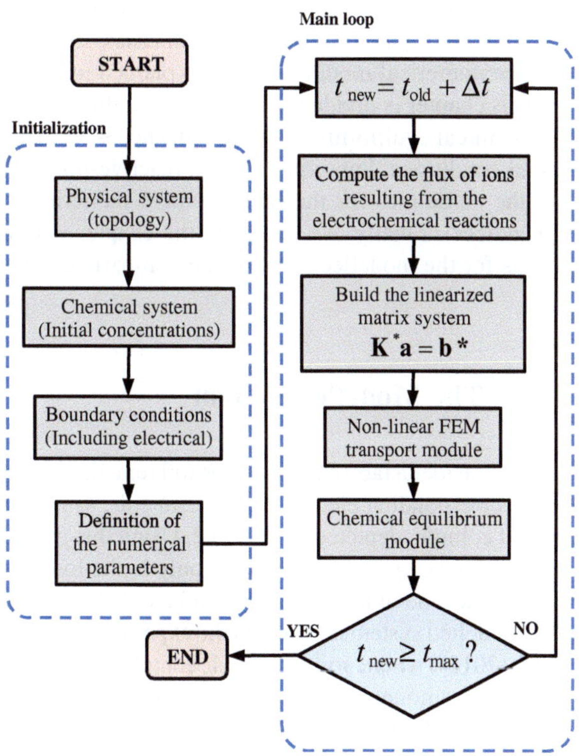

Fig. 13.1 General algorithm for the coupled reactive-transport and chemical equilibrium model. First, the physical–chemical system is defined, together with the initial and the boundary conditions depending on the case studied. The numerical integration in time is achieved by means of a finite element method for nonlinear transport, coupled with a module for re-establishing the chemical equilibrium of a set of sufficiently fast reversible reactions in the system. The algorithms for the two main modules (the one for reactive-transport and the one for chemical equilibrium) are described separately

the chemical reactions in the system are fast enough to allow for the assumption of reactive-transport under local chemical equilibrium conditions. This assumption is widely accepted for electrochemically induced reactive-transport processes and is based on the fact that the reversible equilibrium reactions involved are fast enough in both the direct (towards the products) and the reverse (towards the reactants) directions, such that the transport process can be assumed to take place in conditions close to chemical equilibrium. This assumption is valid especially for homogeneous aqueous-phase acid/base and complexation reactions. In the case of heterogeneous reactions, such as adsorption/desorption, ion-exchange or precipitation/dissolution reactions, the local chemical equilibrium assumption is in the limit of validity. However, for a large number of case studies, the assumption of local chemical equilibrium showed good agreement between experimental and simulations results. We also present guidelines for a model combining slow reactions (kinetically controlled) and fast equilibrium reactions. Other chemical aspects taken into account in the model are externally forced electrochemical reactions at the electrodes, and irreversible reactions such as gas releasing reactions, which are not included in the set of reactions that are under the local chemical equilibrium assumption.

The non-linear reactive-transport model, including the kinetically controlled, irreversible and externally forced chemical reactions, is numerically solved using a finite element method (FEM). The chemical equilibrium module, in turn, computes the chemical equilibrium concentrations of a set of chemical species from a non-equilibrium state in a time-independent manner. These two main modules are explained in separate sections. In both cases, for the sake of clarity, the theoretical and the numerical models are also explained in separated subsections.

This chapter is divided as follows: first, the modelled system is explained. Next, the chemical equilibrium is described, and, as an example, it is used to calculate the initial conditions. Subsequently, the reactive-transport model is described, including the guidelines for the finite element numerical solution (complemented with appendices). Finally, we complete the chapter with a section showing simulation results for the modelled system, and comparison with experimental results.

13.2 The Modelled System

Let us consider a lab-scale setup for soil remediation, consisting of two electrodes, in their corresponding electrode compartments, sandwiching a soil sample into a horizontal column, depicted in Fig. 13.2. Using this geometry, the transport of chemical species can be considered to occur only in the longitudinal direction of the column, i.e. towards one of the electrodes, and the system can be simplified to a 1D problem. The modelled system is based on the experimental results reported in Villen-Guzman et al. (2015), where soil with high content of carbonates and contaminated with a significant amount of lead is submitted to an acid-enhanced EKR treatment using acetic acid (HAc, where Ac^- stands for the acetate ion). The chemical system (species and chemical reactions assumed in the system) is described in Table 13.1.

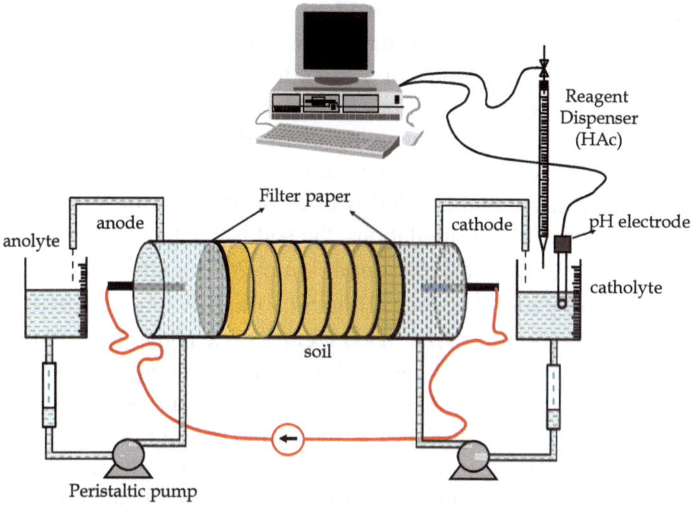

Fig. 13.2 Scheme for an acid-enhanced EKR treatment in horizontal column (20 cm length × 8 cm i.d.). Electrode compartments are well stirred and have a volume of electrolyte of 0.5 L. A constant current density of 2 mA cm^{-2} is applied to the system during 15 d. In the cathode compartment, pH < 5 is assured by dropping HAc 1 M from a burette using an automatic pH-control device

Table 13.1 Simplified chemical system used to model the case example

i	r	Species/Stoichiometry	pK_{eq}	$D_i \times 10^9$ ($m^2\,s^{-1}$)
1	-	H^+	-	9.311
2	-	OH^-	-	5.273
3	-	Na^+	-	1.134
4	-	Ca^{2+}	-	0.792
5	-	$CO_3{}^{2-}$	-	0.923
6	-	Ac^-	-	1.084
7	-	Pb^{2+}	-	0.945
8	1	$H_2O \rightleftarrows H^+ + OH^-$	14.00	100
9	2	$HCO_3{}^- \rightleftarrows H^+ + CO_3{}^{2-}$	10.33	1.185
10	3	$CaHCO_3{}^+ \rightleftarrows H^+ + Ca^{2+} + CO_3{}^{2-}$	11.6	1.185[a]
11	4	$PbHCO_3{}^+ \rightleftarrows H^+ + Pb^{2+} + CO_3{}^{2-}$	13.2	1.185[a]
12	5	$H_2CO_3 \rightleftarrows 2H^+ + CO_3{}^{2-}$	16.68	1.910
13	6	$HAc \rightleftarrows H^+ + Ac^-$	4.76	1.210
14	7	$PbAc^+ \rightleftarrows Pb^{2+} + Ac^-$	2.68	0.945[a]
15	8	$CaAc^+ \rightleftarrows Ca^{2+} + Ac^-$	1.18	0.792[a]
16	9	$CaCO_3\ (s) \rightleftarrows Ca^{2+} + CO_3{}^{2-}$	8.48	-
17	10	$2PbCO_3 \cdot Pb(OH)_2\ (s) \rightleftarrows 3Pb^{2+} + 2CO_3{}^{2-} + 2OH^-$	45.46	-

Equilibrium constants for the given stoichiometry when using molal concentrations from Parkhurst and Appelo (1999)
[a]Estimated values

To establish an external electric field in an electrolyte solution, an electric potential drop is enforced between the electrodes. For the case under study, we consider galvanostatic conditions of $2\,\mathrm{mA\,cm^{-2}}$. Due to the externally forced electric field, electrochemical reactions take place at the electrodes, i.e. oxidation at the anode, which becomes positively charged and attracts the anions; and reduction at the cathode, which becomes negative and attracts the cations.

The 1 L horizontal column is filled with a soil that is moisture-saturated with 0.1 M NaAc solution. For the simulations, the soil is assumed composed by 20 mol of inert material ($\approx$ 1.2 kg, $M = 60$ g mol^{-1}, $\rho = 2648$ g cm^{-3}), 1.2 mol of calcite ($\approx$ 120 g, $M = 100$ g mol^{-1}, $\rho = 2710$ g cm^{-3}) and 0.1 mol of hydrocerussite ($\approx$ 77.6 g, $M = 776$ g mol^{-1}, $\rho = 6800$ g cm^{-3}), leading to a soil mixture of $\approx$ 1398 g, with $\rho \approx 2.75\,\mathrm{g\,cm^{-3}}$ and solid volume fraction of $\approx$ 51 % (therefore, the initial porosity $\varepsilon_{\text{initial}} = 49\,\%$). The initial content of Ca^{2+} and Pb^{2+} in the soil is, respectively, $34.3\,\mathrm{mg\,g^{-1}}$ and $44.7\,\mathrm{mg\,g^{-1}}$ referred to mass of dry solid.

The electrode compartments in both anode and cathode consist in a column disk and an external vessel containing, each of them, a volume of electrolyte of 0.5 L in total. The electrolyte in each electrode compartment is continuously circulated between the compartment at the column containing the electrodes and the well-stirred external vessels with the pH control and sampling devices (see Fig. 13.2). Both electrode compartments are initially filled with an electrolyte resulting from dissolving 1 mmol of NaAc and 1 mmol of HAc in 1 L of water. The solution is assumed to be in equilibrium with atmospheric CO_2 (g).

The experimental results in Fig. 13.3 show that, during the treatment, the minerals containing Ca and Pb dissolve as a consequence of the reaction with the acidic front entering from the anode side of the column. Ca is gradually removed from the column and collected at the cathode compartment. Pb, however, dissolves form the anode side and precipitates again within the column in those regions where the pH is still buffered by existing carbonates. These results suggested that the remediation

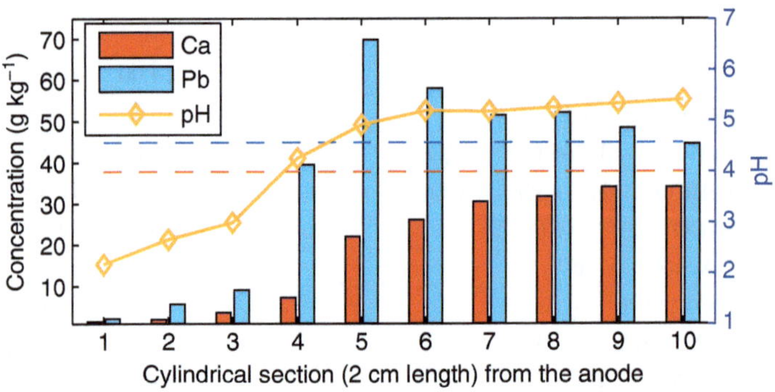

Fig. 13.3 Experimental pH and concentration of Ca and Pb referred to kg of dry soil. For the analysis, homogenized samples were taken from ten consecutive cylindrical sections (2 cm length each) along the column. Concentrations correspond to 16 days of treatment using $2\,\mathrm{mA\,cm^{-2}}$. Initial concentrations are shown with horizontal dashed lines

of this soil (with high content of carbonates of Pb and Ca) using an acid enhanced technique may require the dissolution of all the carbonates associated with Ca before dissolving those carbonates associated with Pb. Simulations results are presented in Sect. 13.5, and compared with the experimental results shown here.

13.3 Chemical Equilibrium

13.3.1 Chemical Equilibrium: Theoretical Model

The solution of the continuity equations simultaneously accounting for the chemical reaction terms is possible but very challenging. Difficulties arise due to the lack of information about the kinetic laws for the chemical reactions in each specific multi-species chemical system. Furthermore, accurate numerical solutions are difficult to obtain due to the large differences in the order of magnitude of the kinetic constants with respect to the transport terms.

For the study of the transport of ionic and non-ionic chemical species through reactive porous media, it is typically accepted that the reversible chemical reactions that take place during the process are fast (in both direct and reverse directions) compared to the rates of the transport processes involved in the system. The assumption of local chemical equilibrium allows to solve a set of fast reactions[2] using a speciation and batch reaction chemical equilibrium model. A set of non-linear algebraic equations is established from the mass balance and the mass action equations describing the chemical equilibrium (Rawlings and Ekerdt 2002; Bethke 2008; Métivier and Roussel 2012; Paz-Garcia et al. 2013, 2015, 2011).

Let i and r be, respectively, the indexes for the chemical species and the chemical reactions describing the secondary species, with $i = 1, 2, \ldots, N$ and $r = 1, 2, \ldots, S$; where N is the total number of species, $(N - S)$ is the number of master species and S is the number of chemical equilibrium reactions used to define the secondary species as function of the master species. The stoichiometric equations for each of the S reactions are defined as

$$\nu_r n_r \rightleftarrows \sum_{i=1}^{N} \nu_{i,r} n_i, \tag{13.1}$$

where the variable n_i (mol) denotes the concentration referred to an invariable volume[3]; in our case, $n_i = c_i \varepsilon \theta$ and $\nu_{i,r}$ is the stoichiometric coefficient for the ith

[2] "Fast" means here, fast enough to reach chemical equilibrium in the time interval Δt used for the numerical integration of the transient FEM.

[3] In the following, concentration in brackets are also used, as $n_i = [i]$.

species in the rth reaction,[4] where ε (–) is the porosity, θ (–) is the water content and c_i (mol m^{-3}) is the concentration referred to the volume of solvent. In this model, stoichiometric equations have to be written in the form of dissociation or dissolution reactions with only one reactant.

When N chemical species in non-equilibrium react with each other to reach the chemical equilibrium state, a mass conservation equation can be written as a function of the extent of the reaction ξ_r. The reaction extent is the extensive quantity describing the progress of a chemical reaction equal to the number of chemical transformations (McNaught and Wilkinson 1997). In order to reach the chemical equilibrium state, any reversible reaction can proceed towards the products or towards the reactants, so ξ_r can be either a positive or a negative value.

According to the previous definition, the total amount of each species at equilibrium is given by the mass balance equation along the chemical reaction path, in the form of

$$n_{i,\text{eq.}} = n_{i,\text{init.}} + \sum_{r=1}^{M} \xi_r \nu_{i,r}, \tag{13.2}$$

Let $\mathbf{x} = [\xi_1, \xi_2, \cdots, \xi_S]$ be the vector containing the reaction extents for the S reactions. The function $\Psi_r(\mathbf{x})$ is defined as a measure of the Gibbs energy of the chemical system according to the thermodynamic description of the chemical potential and it tends to zero when the system approach to the equilibrium. Thus, the function $\Psi_r(\mathbf{x})$ represents the distance to the equilibrium state for the rth reversible reaction. Using logarithms, $\Psi_r(\mathbf{x})$ can be expressed as

$$\Psi_r(\mathbf{x}) = \sum_{r=1}^{S} (\nu_{i,r} \ln a_i) - \ln K_{eq}, r, \tag{13.3}$$

where $K_{\text{eq},r}$ is the equilibrium constant for the rth reaction and a_i (–) is the chemical activity.[5]

Different theories can be used to define the activity coefficients. The Davies and Setschenow equations, for ionic and non-ionic aqueous species respectively, have been used in the simulation results presented here. For the case of the solvent and the solid species, the chemical activities are set to a fixed value of 1. Heterogeneous precipitation/dissolution reactions are included in the model. This means that the equilibrium reaction for the solid species is only included in the set of chemical reactions when the saturation index predicts the existence of the precipitated phase.

[4] For example: $[H_2CO_3] \rightleftharpoons [2H^+] + [CO_3^{2-}]$.

[5] For example: $\psi(\mathbf{x})_{H_2CO_3} = \ln a_{CO_3{}^{2-}} + 2\ln a_{H^+} - \ln K_{eq}$.

13.3.2 Chemical Equilibrium: Numerical Model

As an example of how to implement numerically the chemical equilibrium model, the initial concentration for the modelled chemical system will be calculated. The modelled chemical system consists of $N = 17$ chemical species and $S = 10$ equilibrium equations, as described in Table 13.1. Additional chemical species could have been included in the system, such as $Ca(OH)_2$, $CaOH^+$, $Pb(OH)_2$ or $NaAc^+$. The chosen chemical system is based on simulations done in PHREEQC, to select which species have significant concentrations in the range of pH < 10. As mentioned before, due to the simplicity vs. realism compromise, a good selection of the relevant chemical species and reactions in the system generally leads to a more accurate numerical solution. The stoichiometric equations are expressed in matrix notation, forming stoichiometric matrix of the chemical system **M**.

$$\mathbf{M}_{(S\times N)} = \begin{bmatrix} 1 & 1 & 0 & 0 & 0 & 0 & 0 & -1 & 0 & 0 & 0 & 0 & 0 & 0 & 0 & 0 & 0 \\ 1 & 0 & 0 & 0 & 1 & 0 & 0 & 0 & -1 & 0 & 0 & 0 & 0 & 0 & 0 & 0 & 0 \\ 1 & 0 & 0 & 1 & 1 & 0 & 0 & 0 & 0 & -1 & 0 & 0 & 0 & 0 & 0 & 0 & 0 \\ 1 & 0 & 0 & 0 & 1 & 0 & 1 & 0 & 0 & 0 & -1 & 0 & 0 & 0 & 0 & 0 & 0 \\ 2 & 0 & 0 & 0 & 1 & 0 & 0 & 0 & 0 & 0 & 0 & -1 & 0 & 0 & 0 & 0 & 0 \\ 1 & 0 & 0 & 0 & 0 & 1 & 0 & 0 & 0 & 0 & 0 & 0 & -1 & 0 & 0 & 0 & 0 \\ 0 & 0 & 0 & 0 & 0 & 1 & 1 & 0 & 0 & 0 & 0 & 0 & 0 & -1 & 0 & 0 & 0 \\ 0 & 0 & 0 & 1 & 0 & 1 & 0 & 0 & 0 & 0 & 0 & 0 & 0 & 0 & -1 & 0 & 0 \\ 0 & 0 & 0 & 1 & 1 & 0 & 0 & 0 & 0 & 0 & 0 & 0 & 0 & 0 & 0 & -1 & 0 \\ 0 & 2 & 0 & 0 & 2 & 0 & 3 & 0 & 0 & 0 & 0 & 0 & 0 & 0 & 0 & 0 & -1 \end{bmatrix}. \tag{13.4}$$

The equilibrium equations will proceed towards the direction of the products (direct) or the direction of the reactants (inverse) to reach, if possible, the chemical equilibrium. In the example, the system can be expressed in matrix notation as

$$\begin{bmatrix} n_1 \\ n_2 \\ \vdots \\ \vdots \\ n_N \end{bmatrix}_{\text{eq.}} = \begin{bmatrix} n_1 \\ n_2 \\ \vdots \\ \vdots \\ n_N \end{bmatrix}_{\text{init.}} + \mathbf{M}^{\mathrm{T}} \begin{bmatrix} x_1 \\ x_2 \\ \vdots \\ x_S \end{bmatrix}, \tag{13.5}$$

equivalent to

$$\begin{bmatrix} [\mathrm{H}^+] \\ [\mathrm{OH}^-] \\ \vdots \\ [2\mathrm{PbCO_3} \cdot \mathrm{Pb(OH)_2}] \end{bmatrix}_{\text{eq}} = \begin{bmatrix} [\mathrm{H}^+] \\ [\mathrm{OH}^-] \\ \vdots \\ [2\mathrm{PbCO_3} \cdot \mathrm{Pb(OH)_2}] \end{bmatrix}_{\text{init.}} + +\mathbf{M}^{\mathrm{T}} \begin{bmatrix} x_1 \\ x_2 \\ \vdots \\ x_{10} \end{bmatrix}, \tag{13.6}$$

and also equivalent to

$$\mathbf{n}_{\mathrm{eq}}(\mathbf{x}) = \mathbf{n}_{\mathrm{ini}} + \mathbf{M}^{\mathrm{T}}\mathbf{x}. \tag{13.7}$$

The stoichiometric system of equations is coupled with the system of equilibrium equations, written in terms of the residual function Ψ, [6] as

$$\begin{bmatrix} \psi_1(\mathbf{x}) \\ \psi_2(\mathbf{x}) \\ \vdots \\ \psi_S(\mathbf{x}) \end{bmatrix} = \mathbf{M} \begin{bmatrix} \ln a_1(\mathbf{x}) \\ \ln a_2(\mathbf{x}) \\ \vdots \\ \ln a_N(\mathbf{x}) \end{bmatrix} - \begin{bmatrix} \ln K_1 \\ \ln K_2 \\ \vdots \\ \ln K_S \end{bmatrix}, \tag{13.8}$$

equivalent to

$$\begin{bmatrix} \psi_1(\mathbf{x}) \\ \psi_2(\mathbf{x}) \\ \vdots \\ \psi_{10}(\mathbf{x}) \end{bmatrix} = \mathrm{M} \begin{bmatrix} \ln[\mathrm{H}^+] \\ \ln[\mathrm{OH}^-] \\ \ln[\mathrm{Na}^+] \\ \ln[\mathrm{Ca}^{2+}] \\ \ln[\mathrm{CO_3}^{2-}] \\ \ln[\mathrm{Pb}^{2+}] \\ \ln[\mathrm{Ac}^-] \\ 0 \\ \ln[\mathrm{HCO_3}^-] \\ \ln[\mathrm{CaHCO_3}^+] \\ \ln[\mathrm{PbHCO_3}^+] \\ \ln[\mathrm{H_2CO_3}] \\ \ln[\mathrm{HAc}] \\ \ln[\mathrm{PbAc}^+] \\ \ln[\mathrm{NaAc}] \\ \ln[\mathrm{CaAc}^+] \\ 0 \\ 0 \end{bmatrix} - \begin{bmatrix} \ln K_1 \\ \ln K_2 \\ \vdots \\ \ln K_{10} \end{bmatrix}, \tag{13.9}$$

and also equivalent to

$$\Psi(\mathbf{x}) = \mathbf{M}\,\mathbf{a}(\mathbf{x}) - \mathbf{k}. \tag{13.10}$$

[6] The reader should note that in the equilibrium equation the use of activity coefficients cancels the terms for the solids and the solvent. For the aqueous species, we approximate the chemical activities with the molal concentration.

Equations (13.7) and (13.10) form the non-linear matrix system defining the chemical equilibrium. The solution of the non-linear matrix system is achieved by means of an iterative Newton–Raphson (NR) method with line search improvement (Press et al. 1992). The NR method indicates the next iteration value for the unknowns, $\mathbf{x}_{new}$, obtained from the present value, $\mathbf{x}_{old}$, by calculating a numerical increment, $\delta\mathbf{x}$, towards the direction that decreases the residual function.

$$\mathbf{x}_{new} = \mathbf{x}_{old} + \lambda\delta\mathbf{x}. \tag{13.11}$$

The Taylor series expansion of the residual function $\Psi(\mathbf{x})$ is

$$\Psi(\mathbf{x}_{new}) = \Psi(\mathbf{x}_{old}) + \mathbf{J}\,\delta\mathbf{x} + \mathbf{O}(\delta\mathbf{x})^2, \tag{13.12}$$

where $\mathbf{O}(\delta\mathbf{x})^2$ represents the error related to terms with order greater than 2 and $\mathbf{J}$ is the Jacobian matrix.

The next step in the adopted NR method is given by taking into account that the method aims for the condition $\Psi(\mathbf{x}_{new}) = \mathbf{0}$. Ignoring the term $\mathbf{O}(\delta\mathbf{x})^2$, the increment of the variable "extent of the reactions" in the direction of the equilibrium is obtained from the solution of

$$\mathbf{J}\delta\mathbf{x} = -\Psi(\mathbf{x}_{old}). \tag{13.13}$$

A line-search with backtracking enhancement is used in the NR method. As mentioned above the vector $\delta\mathbf{x}$ indicates the direction in which the residual decreases but it does not carry information about the optimal magnitude of the increment in the extent of the reactions. If the step is too long, the system may deviate farther from the equilibrium sate, failing to minimize the residual. In addition to this, the system is restricted to the non-negative constraint for the concentrations (Baten and Szczepanski 2011). Figure 13.4 shows the algorithm used for the Newton–Raphson with line-search procedure.

As an example of the solution of the numerical model for chemical equilibrium, the initial concentration in the soil is presented in Table 13.2. The initial composition of the wet soil is calculated in two steps: first the aqueous phase is equilibrated with an infinite amount of CO_2(g) at atmospheric pressure. After that, the resulting electrolyte is equilibrated with the solid phases (calcite and hydrocerussite).[7]

[7] Note that the model predicts the changes of concentration due to changes in the amount of water in the system as, e.g., in the case of Na^+, which slightly increases its concentration even when it is inert in the chemical system considered here.

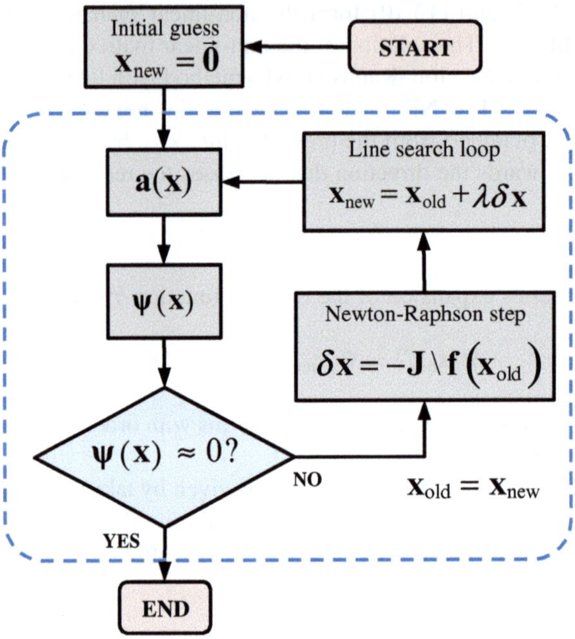

Fig. 13.4 Algorithm for the Newton–Raphson with Line-Search enhancement followed for the chemical equilibrium module. The next step in the numerical iteration is done following the direction indicated by the Newton–Raphson method, and scaled to produce an actual decrease in the overall residual, which is a measure of the distance to the equilibrium. The Line-Search routine is also adapted to assure non-negative concentrations of all species during the process. In terms of coupling with the FEM module, this algorithm is probably the one consuming more running time, as it has to be run in every node of the domain

13.4 Reactive Transport

13.4.1 Reactive Transport: Theoretical Model

Let us consider a representative volume element (RVE) in the domain under study (see Fig. 13.5). The size of the RVE depends on the physical–chemical system studied. It should be sufficiently small to be apparently discrete, but sufficiently big to contain enough particles for the continuous medium assumption. In this volume, a mass balance (or continuity equation) is defined for each of the N chemical species in the system, and coupled with the Poisson's equation,[8] that are, respectively,

[8] In most cases, the Poisson's equation is accurately approximated by the electro-neutrality condition: $\sum c_i z_i = 0$. The Poisson's equation is suggested here to present a generalized model that can also be used to describe electric double-layers formed at the interface of the electrolyte and charged surfaces, only relevant in a scale of nanometers to a few micrometers.

Table 13.2 Computing the initial composition of the pore electrolyte and the amount of solids using the chemical equilibrium model

i	r	Species	$c_{i,\text{ini}}$	$c_{i,\text{eq}}$
1	-	H^+	1×10^{-4}	7.314×10^{-7}
2	-	OH^-	1×10^{-4}	2.23×10^{-2}
3	-	Na^+	100	$100 + 4 \times 10^{-6}$
4	-	Ca^{2+}	0	0.1212
5	-	$CO_3{}^{2-}$	0	0.2055
6	-	Ac^-	100	999.31
7	-	Pb^{2+}	0	1.135×10^{-6}
8	1	H_2O	1	0.99997
9	2	$HCO_3{}^-$	0	1.172
10	3	$CaHCO_3{}^+$	0	9.642×10^{-4}
11	4	$PbHCO_3{}^+$	0	3.593×10^{-4}
12	5	H_2CO_3	1.135×10^{-2}	1.13×10^{-3}
13	6	HAc	0	2.48×10^{-2}
14	7	$PbAc^+$	0	1.98×10^{-4}
15	8	$CaAc^+$	0	0.6684
16	9	$CaHCO_3$ (s)	1.3	1.2996
17	10	$PbHCO_3$ (s)	0.1	$0.1\text{–}3.7 \times 10^{-8}$

Units are mM for aqueous species and mol for solids. For water, units are volume of water per volume of available pore

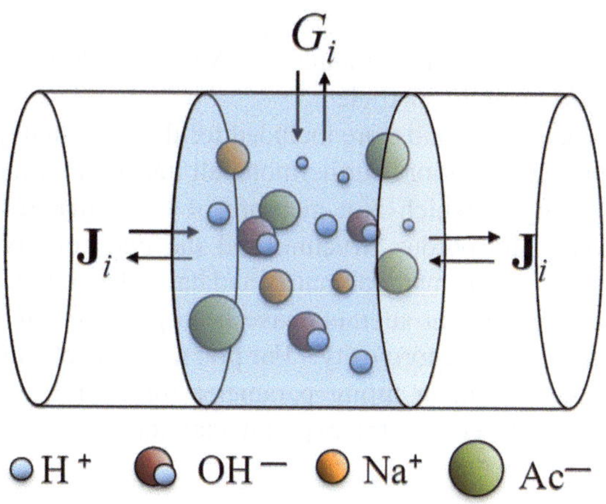

Fig. 13.5 Representative volume element of electrolyte. Concentration changes of the species i inside the volume are caused by transport through the boundaries, chemical reaction, or concentration/dilution processes due to changes of the solvent (e.g., evaporation)

$$\frac{\partial \epsilon \theta c_i}{\partial t} = -\nabla \cdot \mathbf{J}_i + G_i; (i = 1, 2, \ldots, N), \tag{13.14}$$

and

$$\sum c_i z_i = -\frac{\varepsilon}{F} \nabla^2 \phi, \tag{13.15}$$

where $i = 1, 2, \ldots, N$ is an index for the chemical species, ε (–) is the porosity, θ (–) is the water content, c_i (mol m^{-3}) is the concentration referred to the volume of solvent, $\mathbf{J}_i$ (mol m^{-2} s^{-1}) is the flux term and G_i (mol m^{-3} s^{-1}) is the chemical reaction term, F (C mol^{-1}) is the Faraday constant, z_i (–) is the ionic charge, ε (C V^{-1} m^{-1}) is the permittivity and ϕ (V) is the electric potential. This system of equations is known as Nernst–Planck–Poisson (NPP) system.

In electrochemically-induced transport of chemical species through porous media, the flux term $\mathbf{J}_i$ is a combination of different transport contributors. Here, we use an extended version of the Nernst–Planck equation accounting for diffusion, electro-migration, and advection, the latter dependent on the water transport in partially saturated porous media and electro-osmosis

$$\mathbf{J}_i = -\frac{\epsilon \theta}{\tau} \left(D_i \nabla c_i + z_i c_i \frac{F D_i}{RT} \nabla \phi + c_i D_{H_2O} \nabla \theta + c_i K_{e.o.} \nabla \phi \right), \tag{13.16}$$

where D_i (m^2 s^{-1}) is the diffusion coefficient, τ (–) is the tortuosity, R (J K^{-1} mol^{-1}) is the universal gas constant, T (K) is the absolute temperature, D_{H_2O} (m^2 s^{-1}) is the diffusivity of water (which is specific for each material matrix and strongly depends on θ in a non-linear manner (El Abd et al. 2009)), and $K_{e.o.}$ (m^2 V^{-1} s^{-1}) is the electro-osmotic coefficient.

The terms ε and θ are included for the consistency of the units of all terms in the continuity equation. Furthermore, all transport terms are assumed dependent on the tortuosity τ, which is a measure of the actual distance covered by the species through the porous structure and the connectivity of the pores. The tortuosity depends on the material matrix and has to be defined specifically for each modelled system. Different attempts have been proposed by several authors to analytically characterize the tortuosity.[9] For practical aspects, the tortuosity will be considered here as the main fitting parameter (or degree of freedom) to achieve a good agreement between the experimental and the simulation results. In the numerical

[9] In most cases, the tortuosity is studied in water saturated porous media, and the Bruggeman's equation (Chung et al. 2013) is given in terms of $\tau = \varepsilon^{\alpha}$ where, for the case of spherical particles α is $-1/2$. In our model, we propose that the decrease of the connectivity of the pores in water-unsaturated porous media affects the transport by increasing the tortuosity in the form $\tau = \varepsilon^{\alpha}\theta^{\beta}$. More detailed studies can be carried out for a better mathematical expression of this specific phenomenon, namely to include the effect of the pore size distribution and the existence of close or non-connected porosity.

implementation, the tortuosity is also used to model well-stirred compartments (by using $\tau << 1$) and to model the perm-selectivity of ion-exchange membranes (by using different τ values as a function of the ionic charge of the chemical species (Paz-Garcia et al. 2014)).

Most models for ionic transport in an aqueous phase are described without including water as a chemical species in the system. This is a good assumption in water-saturated media without advective terms.[10] As we aim here to describe a generalized model in partially saturated porous media, it is necessary to include water in the system. Taking into account that the term c_{H_2O} is a constant ($55.56\,\mathrm{mol\,m^{-3}}$), the continuity equation for water is expressed in terms of the water content θ. Combining Eqs. (13.14) and (13.16), for the reactive-transport of water and using that the diffusion and the electro-migration terms are absent,[11] one obtains

$$\frac{\partial \epsilon\theta}{\partial t} = \nabla \cdot \left(\frac{\epsilon\theta D_{H_2O}}{\tau} \nabla\theta + \frac{\epsilon\theta K_{e.o.}}{\tau} \nabla\phi \right) + G_{H_2O}, \tag{13.17}$$

Therefore, water is assumed to move through the pores as a consequence of diffusivity due to gradients of water content (Pachepsky et al. 2003; Roels et al. 2003; Nguyen et al. 2008; Johannesson et al. 2009; El Abd et al. 2009; Hall 1977) and electroosmosis (Cameselle and Reddy 2012; Delgado et al. 2007; Lyklema 2005; Paz-Garcia et al. 2014; Subires-Munoz et al. 2011).

Electro-osmotic is the flow of the pore fluid induced by the applied electric field related to the electro-migration of the non-electroneutral electric double-layer at the interface between the solid and the electrolyte. Electro-osmotic advection is essential, for example, for the EKR of soil contaminated with non-ionic species, e.g. organic contaminants. Electro-osmotic transport depends on the magnitude and sign of the zeta potential ζ (V), which is not constant as it depends on the concentration and composition of the electrolyte and the chemical properties of the solid surface (Alshawabkeh and Acar 1996; Yeung and Datla 1995). Most of the models accounting for electro-osmotic flow consider, however, a constant electro-osmotic permeability coefficient equal to the optimal experimentally observed value, which is in a short range of 10^{-8} to 10^{-9} ($\mathrm{m^2\,V^{-1}\,s^{-1}}$).

13.4.2 Reactive Transport: Numerical Model

This section focuses on the numerical model for the non-linear reactive-transport problem. Here, a Galerkin FEM is proposed for the mathematical solution of this system of non-linear partial differential equations (Ottosen and Petersson 1992; Ottosen and Ristinmaa 2005; Zienkiewicz and Taylor 2005, 2000). FEM has been already used

[10] This assumption is also consistent with the simplicity vs realism compromise.

[11] As $\nabla c_{H_2O} = 0$ and $z_{H_2O} = 0$.

before for the numerical modelling of ionic transport in porous media, as, e.g., Javadi and Al-Najjar (2007), Li and Page (2000), Lim et al. (2007), Lu et al. (2010), Paz-Garcia et al. 2011, Paz-Garcia et al. (2012), Paz-Garcia et al. (2013), and Sheng and Smith (2002). The FEM consists of the estimation of the continuous function with a piece wise approximating function defined by discrete values at nodal points. The domain is divided into a number of finite elements defined by the dimensionless nodal points. This leads to that the state variables are discretized and the system can be solved using matrix algebra. Let w denote an arbitrary weight function, and **a** and **w** columns vectors for the discrete form of the state variables a (either c_i, θ or ϕ) and w, respectively. The following standard FEM notations are defined:

$$a \approx \mathbf{N}\,\mathbf{a}_i, \frac{\partial a}{\partial x} \approx \mathbf{Ba}_i, w \approx w^{\mathrm{T}} N^{\mathrm{T}}, \frac{\partial w}{\partial x} \approx \mathbf{w}^{\mathrm{T}} B^{\mathrm{T}}, \tag{13.18}$$

where $\mathbf{N} = [1 - x/L_{\mathrm{e}}, x/L_{\mathrm{e}}]$ is a linear piecewise shape function interpolating between nodal points, $\mathbf{B} = \nabla\mathbf{N}$, L_{e} (m) is the length of finite elements, M is the total number of elements ($e = 1, 2, ..M$), and the superscript T is used to indicate the transpose of the vectors (Fig. 13.6).

Herein, we present the discretized form of the NPP system.[12] The strong formula for the Nernst–Planck continuity equation,

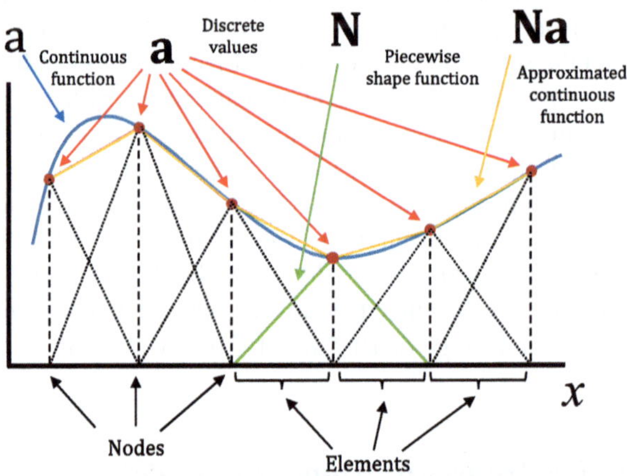

Fig. 13.6 Finite element discretization of a continuous function a into discrete node values **a** (in the example, using five elements and six nodes). The approximated continuous function **Na** is built using a Galerkin method with piecewise shape functions **N**

[12] The detailed discretization of the NPP system would be tedious. In the appendices, detailed steps to discretize the diffusion and the generation terms are presented. Following the same steps, all the terms presented below can be derived.

$$\frac{\partial \epsilon\theta c_i}{\partial t} = \frac{\partial}{\partial x}\left(\frac{\epsilon\theta D_i}{\tau}\frac{\partial c_i}{\partial x} + \frac{c_i\epsilon\theta D_w}{\tau}\frac{\partial\theta}{\partial x} + \frac{c_i\epsilon\theta K_{e.o.}}{\tau}\frac{\partial\phi}{\partial x} + \frac{z_i c_i F\epsilon\theta D_i}{\tau RT}\frac{\partial\phi}{\partial x}\right) + G_i \tag{13.19}$$

is discretized into

$$\mathbf{M}_i\frac{\partial \mathbf{a}}{\partial t} + \mathbf{D}_i\mathbf{a}_i + \mathbf{K}_i\mathbf{a}_\phi = \mathbf{b}_i, \tag{13.20}$$

where the global matrices are built from the following local matrices:

$$\mathbf{M}_{i,\text{local}} = \frac{\overline{\epsilon}\overline{\theta}L_e}{6}\frac{\partial\overline{\theta}}{\partial x}\begin{bmatrix}2 & 1\\ 1 & 2\end{bmatrix}; \tag{13.21}$$

$$\mathbf{K}_{i,\text{diffusion, local}} = \frac{\overline{\epsilon}\overline{\theta}D_i}{\overline{\tau}L_e}\begin{bmatrix}1 & -1\\ -1 & 1\end{bmatrix}; \tag{13.22}$$

$$\mathbf{K}_{i,\text{advection, local}} = \frac{\overline{\epsilon}\overline{\theta}D_w}{2\overline{\tau}}\frac{\partial\overline{\theta}}{\partial x}\begin{bmatrix}-1 & 1\\ -1 & 1\end{bmatrix}; \tag{13.23}$$

$$\mathbf{K}_{i,\text{electroosmosis, local}} = \frac{\overline{\epsilon}\overline{\theta}K_{\text{e.o.}}}{2\overline{\tau}}\frac{\partial\overline{\phi}}{\partial x}\begin{bmatrix}-1 & 1\\ -1 & 1\end{bmatrix}; \tag{13.24}$$

$$\mathbf{K}_{i,\text{local}} = \mathbf{K}_{i,\text{Electromigration, local}} = \frac{z_i\overline{c_i}\,\overline{\epsilon}\overline{\theta}FD_i}{\overline{\tau}RTL_e}\frac{\partial\overline{\theta}}{\partial x}\begin{bmatrix}1 & -1\\ -1 & 1\end{bmatrix}. \tag{13.25}$$

In Eq. (13.20), the following simplification has been made[13]:

$$\mathbf{D}_{i,\text{local}} = \mathbf{K}_{\text{diffusion, local}} + \mathbf{K}_{i,\text{advection, local}} + \mathbf{K}_{i,\text{electroosmosis, local}}; \tag{13.26}$$

[13] During the discretization of the advective and the electro-osmotic terms, it was decided to use the concentration of the aqueous chemical species as the state variable, and the others (θ, $\nabla\theta$ and $\nabla\phi$) as variables to be recalculated in the iterative non-linear solution. However, for the case of the water equation, it was linearized on θ and $\nabla\theta$. This is not shown in this text, and left as an exercise for the reader. The procedure causes the existence of non-symmetric advective matrices.

$$\mathbf{b}_i = \mathbf{f}_i + \mathbf{g}_i. \tag{13.27}$$

In turn, the Poisson's equation in continuous form,

$$\sum_{i=1}^{N} c_i z_i + \frac{\epsilon}{\epsilon \theta F} \frac{\partial^2 \phi}{\partial x^2} = 0, \tag{13.28}$$

is discretized into

$$\sum_{i=1}^{N} \mathbf{E}_i \mathbf{a}_i + K_\phi \mathbf{a}_\phi + \mathbf{f}_\phi = 0, \tag{13.29}$$

where the global matrices are built from the following local matrices

$$\mathbf{E}_{i,\text{local}} = \frac{z_i L_e}{6} \begin{bmatrix} 2 & 1 \\ 1 & 2 \end{bmatrix}; \tag{13.30}$$

$$\mathbf{K}_{\phi,\text{local}} = \frac{\epsilon}{\overline{\epsilon \theta} F L_e} \begin{bmatrix} 1 & -1 \\ -1 & 1 \end{bmatrix}. \tag{13.31}$$

Putting together the $N + 1$ equations of the discrete NPP system, a coupled matrix system of equations is obtained, in the form:

$$\mathbf{M} \frac{\partial \mathbf{a}}{\partial t} + \mathbf{K}\, a + \mathbf{b} = 0, \tag{13.32}$$

where the absence of subscripts stands for global matrices and vectors, i.e., defined for the entire coupled system. Equation (13.32) comes from:

$$\underbrace{\begin{bmatrix} \mathbf{M}_1 & 0 & \cdots & 0 \\ 0 & \ddots & \ddots & \vdots \\ \vdots & \ddots & \mathbf{M}_N & 0 \\ 0 & \cdots & 0 & 0 \end{bmatrix}}_{\mathbf{M}} \frac{\partial}{\partial t} \begin{bmatrix} \mathbf{a}_1 \\ \vdots \\ \mathbf{a}_N \\ \mathbf{a}_\phi \end{bmatrix} + \underbrace{\begin{bmatrix} \mathbf{D}_1 & 0 & \cdots & \mathbf{K}_1 \\ 0 & \ddots & \ddots & \vdots \\ \vdots & \ddots & \mathbf{D}_N & \mathbf{K}_N \\ \mathbf{E}_1 & \cdots & \mathbf{E}_N & \mathbf{K}_\phi \end{bmatrix}}_{\mathbf{K}} \begin{bmatrix} \mathbf{a}_1 \\ \vdots \\ \mathbf{a}_N \\ \mathbf{a}_\phi \end{bmatrix} + \begin{bmatrix} \mathbf{b}_1 \\ \vdots \\ \mathbf{b}_N \\ \mathbf{b}_\phi \end{bmatrix} = 0. \tag{13.33}$$

The discretization in the time, based on a truly implicit and unconditionally stable scheme, is

$$\frac{\partial \mathbf{a}}{\partial t} = \frac{\mathbf{a}(t_0 + \Delta t) - \mathbf{a}(t_0)}{\Delta t}, \tag{13.34}$$

where t_0 is the present time and $t_0 + \Delta t$ is the next time step in the numerical integration. Accordingly, Eq. (13.32) can be rearranged to the following

$$\underbrace{\left(\frac{\mathbf{M}}{\Delta t} + \mathbf{K}\right)}_{\mathbf{K}^*} \mathbf{a}(t_0 + \Delta t) = \underbrace{\frac{\mathbf{C}}{\Delta t}\mathbf{a}(t_0) - \mathbf{b}}_{\mathbf{b}^*}, \tag{13.35}$$

where Eq. (13.35) is solved taking into account the boundary conditions (Ibrahimbegovic 2009; Ottosen and Ristinmaa 2005; Ristinmaa et al. 2011).

As the reader should have noted, all terms in the previous system of equations are non-linear, and they depend on more than one state variable. During the discretization process proposed previously, some state variables have been included into temporary constant matrices or vectors (indicated with an overlined bar), in order to create a linearized matrix system of equations that can be easily solved by means of matrix algebra. For example, in the electro-migration local matrix described in Eq. (13.25) the values of ϵ, θ, and c_i are estimated from the known values (denoted as $\overline{\epsilon}, \overline{\theta}$, and $\overline{c_i}$), and the linearized equation is solved in terms of the gradient of electric potential $\partial\phi/\partial x$. The solution obtained while solving the linearized matrix system of equations $\mathbf{K}^*\mathbf{a}(t_0 + \Delta t) = \mathbf{b}^*$ is, therefore, an approximated solution that can be improved with an algorithm to correct and minimize the error induced in the linearization process. The procedure is carried out using the iterative method depicted in Fig. 13.7.

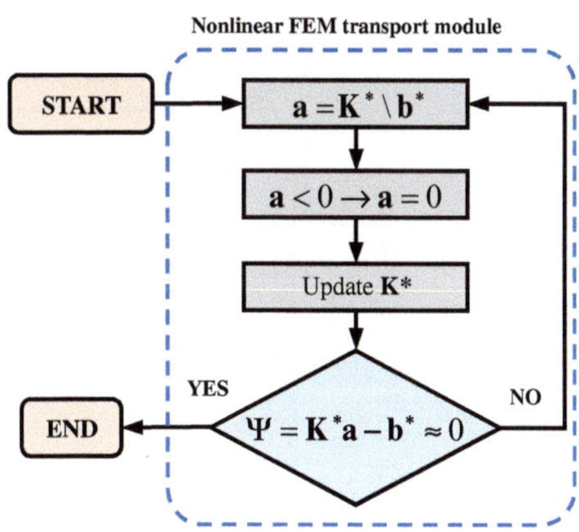

Fig. 13.7 Subalgorithm for minimizing the error in the solution of the non-linear finite elements reactive-transport model. A residual function Ψ is defined using the updated value of the global matrix $\mathbf{K}^*$. In each iteration step, spurious negative concentrations are cancelled. The algorithm can be further enhanced with a Newton–Raphson method for non-linear systems to calculate a better approximation of the discrete function

13.5 Simulation Results

In this section simulation results based on the modelled system described in Sect. 13.2 are shown and compared to the experimental values in Fig. 13.3, from Villen-Guzman et al. (2015).

Figure 13.8 shows the pH values in the column obtained in the simulations in 1, 5, and 15 days. The intermediate profiles (1 and 5 days) from the simulations are shown to illustrate the tendency of the process in time. Figure 13.8 shows that the pH control is properly modelled, keeping during the entire process the pH at the cathode equal to or lower than 5 by neutralization with HAc. The pH at the anode, which was not controlled, decreases to very acidic values of ≈ 2 due to the water oxidation reaction at the electrode. The comparison of the experimental and the simulation results in the 16th day of treatment indicates that the model predicts fairly well the experiments. During the treatment, the soil has pH ≈ 5, which corresponds to the equilibrium pH of $CaCO_3$ in acidic media. At the anode end of the column, pH ≈ 2 is obtained, as a consequence of the total dissolution of $CaCO_3$, as shown in Fig. 13.9.

In both, the experimental and the simulation results, it can be seen that the dissolution of the Pb-containing mineral (hydrocerussite in the chemical system modelled here) takes place almost exclusively when the $CaCO_3$ is totally dissolved and the pH of the pore solution decreases to values below 5 (see Fig. 13.10). However, the Pb that migrates towards the cathode and reaches regions in the soil with pH values of ≈ 5 precipitates again. Accordingly, in order to remove Pb from the entire column using this technique, a long-term treatment would be necessary producing the total dissolution of the $CaCO_3$.

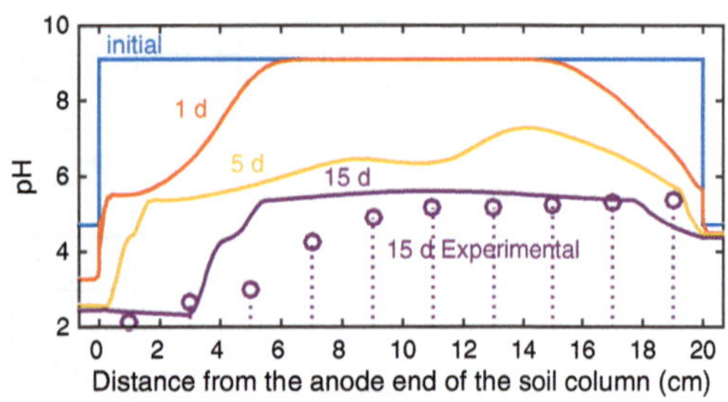

Fig. 13.8 Simulation results for the pH profile along the column in 1d, 5d, and 15d, as well as the experimental results for 15d for comparison

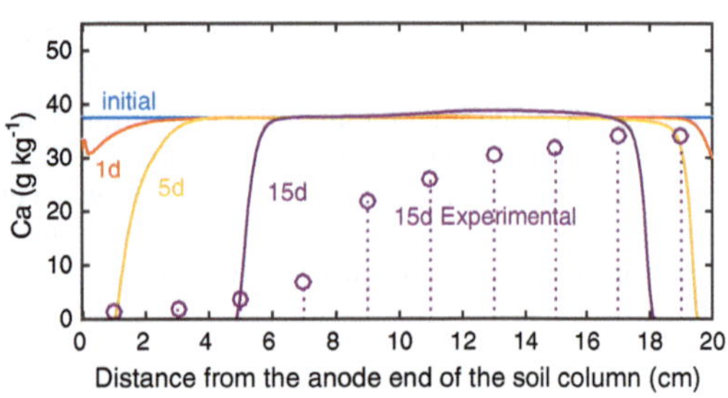

Fig. 13.9 Simulation results for the total amount of Ca (accounting for form $CaCO_3$, $CaHCO^+$, $CaAc^+$ and Ca^{2+}) along the column in 1d, 5d, and 15d, as well as the experimental results for 15d for comparison

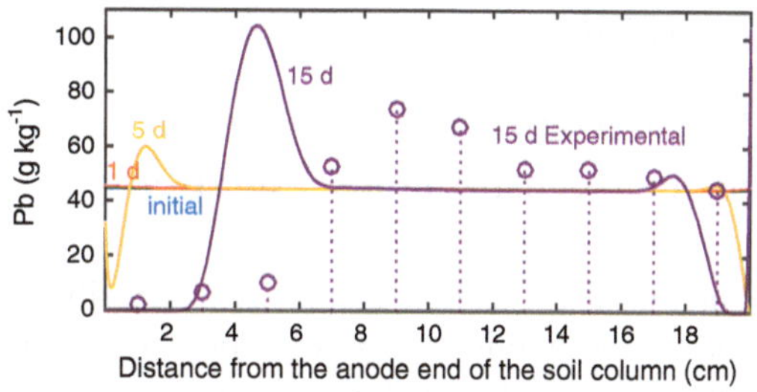

Fig. 13.10 Simulation results for the total amount of Pb (accounting for form $1PbCO_3 \cdot Pb(OH)_2$, $PbHCO^+$, $PbAc^+$ and Pb^{2+}) along the column in 1d, 5d, and 15d, as well as the experimental results for 15 d for comparison

Comparison between the simulation and the experimental results for the migration of Ca and Pb during the EKR treatment show that the model predicts fairly well the behaviour of the system. The simulation results show sharper profiles than those obtained experimentally. These sharp profiles are probably consequence of the assumption of local chemical equilibrium adopted in the model, and indicate that the kinetics of the dissolution and precipitation of the solids may play an important role in the process, as mentioned in Villen-Guzman et al. (2015). Despite the sharp profiles, the model predicts with remarkable accuracy the experimental results, taking into account the uncertainties due to the simplified chemical system. A precise modelled chemical system based on a detailed characterization of the mineral phases present in the soil, and the consideration of chemical kinetics aspects could increase the predictive capability of the model.

13.6 Conclusions

In this chapter a theoretical and numerical was presented for its application on EKR treatments. The model is based on the assumption of local chemical equilibrium. The model is described in a generalized manner, so it could be easily adapted to different geometries, solid matrices, contaminants, and operation conditions.

Simulation results are presented for the remediation of a Pb-contaminated soil in horizontal column applying $2\,A\,cm^{-2}$ during 15 days, and using an acid-enhanced method consisting of buffering the cathode $pH < 5$ with acetic acid. The numerical model predicts fairly well the experimental results. The model can therefore be used as a predictive tool to forecast the results and to design and optimize experimental procedures.

Results suggest that for the case of the precipitation/dissolution reaction of the minerals in the system, the assumption of local chemical equilibrium may not be accurate. This encourages the authors to further develop the model to include kinetically controlled reactions as a future work.

Acknowledgements Paz-Garcia acknowledges the financial support from the International Campus of Excellence (ICE) Andalucia Tech. Villen-Guzman acknowledges the FPU grant obtained from the Spanish Ministry of Education.

Appendix

Discretization of the Diffusion and the Generation Terms of the Nernst–Planck Equation

For the discretization of the term describing the diffusive transport, the strong formula is multiplied by the arbitrary weight function w to construct the weak formula, and integrated over the length of the element, i.e., $0 < x < L_e$, where L_e is the distance between two consecutive nodes limiting the eth one-dimensional element. The integration-by-parts technique is used leading to

$$\int_0^{L_e} w\frac{\partial}{\partial x}\left(\epsilon\theta D_i\frac{\partial c_i}{\partial x}\right)dx = \left[w\epsilon\theta D_i\frac{\partial c_i}{\partial x}\right]_0^{L_e} - \int_0^{L_e}\frac{\partial w}{\partial x}\epsilon\theta D_i\frac{\partial c_i}{\partial x}dx, \tag{13.36}$$

which, by substitution of the FEM terminology described in Eq. (13.20), becomes

$$\int_0^{L_e} w\frac{\partial}{\partial x}\left(\epsilon\theta D_i\frac{\partial c_i}{\partial x}\right)dx = \left[w^T N^T\epsilon\theta\quad D_i\frac{\partial c_i}{\partial x}\right]_0^{L_e} - \int_0^{L_e}\mathbf{w}^T B^T\epsilon\theta D_i\mathbf{B}\mathbf{a}_i dx. \tag{13.37}$$

The arbitrary function $\mathbf{w}^T$, common in all terms, cancels. We can define a nonlinear diffusion matrix $\mathbf{K}_{i,\text{diffusion,local}}$[14] for the right-hand term, and a boundary vector term of Eq. (13.37) as

$$\mathbf{K}_{i,\text{diffusion, local}} = \int_0^{L_e} B^T\epsilon\theta D_i\mathbf{B}dx = \frac{\overline{\epsilon}\overline{\theta}D_i}{L_e}\begin{bmatrix}1 & -1\\ -1 & 1\end{bmatrix}; \tag{13.38}$$

$$\mathbf{f}_{i,\text{diffusion, local}} = \left[\mathbf{N}^T\epsilon\theta\quad D_i\frac{\partial c_i}{\partial x}\right]_0^{L_e} = J_{i,\text{diffusion, local}}\begin{bmatrix}1\\ -1\end{bmatrix}, \tag{13.39}$$

where $\overline{\epsilon}$ and $\overline{\theta}$ are the known average value in the finite element. This means that the diffusion term is linearized to depend only on the concentration at the nodal points.

The local matrices, defined for each element, are subsequently assembled over the entire domain. This procedure involves the overlapping of the matrices to form sparse tridiagonal matrices defining the entire system. The assembled global matrix for the diffusion term is

[14] Subscript "local" indicates that the matrix is defined for one finite element.

$$\mathbf{K}_{i,\mathrm{diffusion}} = \begin{bmatrix} k_1 & -k_1 & 0 & \cdots & 0 \\ -K_1 & K_1 + k_2 & \ddots & \ddots & \vdots \\ 0 & \ddots & \ddots & -k_{M-1} & 0 \\ \vdots & \ddots & -k_{M-1} & k_{M-1} + k_M & -k_M \\ 0 & \cdots & 0 & -k_M & k_M \end{bmatrix}, \tag{13.40}$$

where

$$k_e = \frac{\overline{\varepsilon}\overline{\theta} D_i}{L_e} \quad ; \quad (e = 1, 2, \ldots, M) \tag{13.41}$$

is the mean value of the element e.

The assembled boundary vector for the diffusion term is

$$\mathbf{f}_{i,\mathrm{diffusion}} = \begin{bmatrix} J_{i,\mathrm{diff}} \\ 0 \\ \vdots \\ 0 \\ -J_{i,\mathrm{diff}} \end{bmatrix}, \tag{13.42}$$

which is non-zero only at the boundaries of the system.

Following the same steps than before, the generation term is multiplied by the weight function, integrated over the domain of the element, and substituted with the FEM terminology, leading to

$$\mathbf{g}_{i,\ \mathrm{local}} = \int_0^{L_e} N^T G_i dx = \frac{G_i L_e}{2} \begin{bmatrix} 1 \\ 1 \end{bmatrix}, \tag{13.43}$$

which, once assembled, becomes

$$\mathbf{g}_i = \frac{1}{2} \begin{bmatrix} k_1 \\ k_1 + k_2 \\ \vdots \\ k_{M-1} + k_M \\ k_M \end{bmatrix}, \tag{13.44}$$

where

$$k_e = \overline{G}_{i,e} L_e \quad ; \quad (e = 1, 2, \ldots, M)\,. \tag{13.45}$$

Electrical Conditions: Constant Current vs. Constant Voltage Drop

In electrochemically induced transport, the electric field is applied between the electrodes to produce electrochemical reactions that are the driving force for the overall transport process. Normally, the electric field is applied using either a constant current density or a constant voltage difference between electrodes.

Under conditions of constant current density, the flux of ions that enter the system can be easily related to the applied electrical current. Let us assume that the solid matrix is a perfect electrical insulator and that the only significant electrochemical reactions taking place are the electrolysis of water,[15] namely

$$\frac{1}{2}H_2\mathrm{O} \rightarrow \mathrm{H}^+ + \frac{1}{4}\mathrm{O}_2(\mathrm{g}) + e^-, \tag{13.46}$$

at the anode, and

$$\mathrm{H}_2\mathrm{O} + e^- \rightarrow OH^- + \frac{1}{2}\mathrm{H}_2(\mathrm{g}), \tag{13.47}$$

at the cathode.

The flux of ions at the boundaries is combined with the Faraday's law to set the boundary conditions in the form of incoming fluxes of ions (H^+ and OH^- at the anode and cathode, respectively), and outgoing fluxes of water molecules as a consequence of the electrolysis of water.

$$\mathbf{f}_{\mathrm{H}^+} = \frac{I}{F}\begin{bmatrix} 1 \\ 0 \\ \vdots \\ \vdots \\ 0 \end{bmatrix}; \mathbf{f}_{\mathrm{OH}^-} = \frac{I}{F}\begin{bmatrix} 0 \\ \vdots \\ \vdots \\ 0 \\ 1 \end{bmatrix}; \mathbf{f}_{H_2O} = \frac{18 \times 10^{-6} I}{F}\begin{bmatrix} -1/2 \\ 0 \\ \vdots \\ 0 \\ -1 \end{bmatrix}. \tag{13.48}$$

where I is the applied current density, normally in the order of $\approx 10\,\mathrm{A\,m^{-2}}$, and the factor 18×10^{-6} comes for converting from units of c_i to units of θ.

The cell voltage (the voltage that a multimeter would measure between the electrodes) is a combination of different contributors; namely, the voltage drops related to the electrochemical reactions and the voltage drops related to the ohmic losses in the different sections of the cell (See Fig. 13.11).

The electrochemical potential for the electrolysis of water, without taking into account the overpotential, would be

[15] Although competitive reactions may occur depending on the system under study.

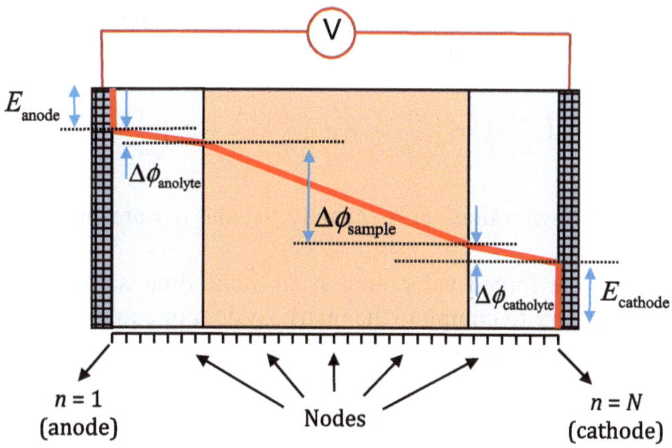

Fig. 13.11 Different contributors affecting the electric potential drop between electrodes. The voltage drop between electrodes is a combination of the electrochemical potential E, and the ohmic potential drop in the electrolyte. Over-potential due to different phenomena may affect the cell voltage

$$E_{\text{anode}} = 1.23 - 0.0592\,\text{p}H + \frac{0.0592}{4}\log_{10}p_{O_2}, \tag{13.49}$$

for the oxidation reaction at the anode, and

$$E_{\text{canode}} = -0.0592\,\text{p}H - \frac{0.0592}{2}\log_{10}p_{H_2}, \tag{13.50}$$

for the reduction reaction at the cathode.

The electrochemical potential, $E = E_{\text{anode}} - E_{\text{cathode}}$, will be in the range of 1.5–3 V, depending on the pH at the electrolyte chambers and the partial pressure of the gases released.

The FEM presented here can be adapted to work under operation conditions of constant voltage drop between the electrodes $\Delta\phi_{\text{cell}}$. In these conditions, in each time step of the numerical integration, the ohmic potential drop, obtained from the established minus the calculated electrochemical potentials, $\Delta\phi_{\Omega} = \Delta\phi_{\text{cell}} - E$, is set as boundary conditions in the system. In particular, as the absolute value of the electric potential is meaningless, an absolute potential of 0 V is set to the node corresponding to the cathode and an electric potential of the calculated $\Delta\phi_{\Omega}$ is set to the anode. The solution of the matrix system of equations constricted to boundary conditions is made by means of a matrix substitution procedure. For example, let us consider a matrix system with four equations in the form:

$$\begin{pmatrix} k_{11} & k_{12} & k_{13} & k_{14} \\ k_{21} & k_{22} & k_{23} & k_{24} \\ k_{31} & k_{32} & k_{33} & k_{34} \\ k_{41} & k_{42} & k_{43} & k_{44} \end{pmatrix} \begin{pmatrix} a_1 \\ a_2 \\ a_3 \\ a_4 \end{pmatrix} = \begin{pmatrix} b_1 \\ b_2 \\ b_3 \\ b_4 \end{pmatrix}. \tag{13.51}$$

If the values a_1 and a_4 are given, e.g. $\Delta\phi_\Omega$ and 0, they do not have to be computed in the solution of the matrix system of equation, which is simplified to

$$\begin{pmatrix} k_{22} & k_{24} \\ k_{32} & k_{34} \end{pmatrix}\begin{pmatrix} a_2 \\ a_3 \end{pmatrix} = \begin{pmatrix} b_2 \\ b_3 \end{pmatrix} + a_1\begin{pmatrix} k_{32} \\ k_{33} \end{pmatrix} + a_4\begin{pmatrix} k_{32} \\ k_{33} \end{pmatrix}; \tag{13.52}$$

where only the unknown values of $\mathbf{a}$ (namely, a_2 and a_3) are present at the left-hand side.

In order to keep the forced value and, at the same time, satisfy the continuity condition, it is necessary to complete the matrix system of equations with a vector, $\mathbf{r}$, denoted as reactive vector, as

$$\mathbf{K}^\star\ \mathbf{a} = b^\star + \mathbf{r}. \tag{13.53}$$

In the FEM, $\mathbf{r}$ is the flux necessary to keep the corresponding forced value in any specific node with a boundary condition. Consequently, $\mathbf{r}$ will be zero in all elements except for those with a predefined forced value. In the example, it would be

$$\begin{pmatrix} k_{11} & k_{12} & k_{13} & k_{14} \\ k_{21} & k_{22} & k_{23} & k_{24} \\ k_{31} & k_{32} & k_{33} & k_{34} \\ k_{41} & k_{42} & k_{34} & k_{44} \end{pmatrix}\begin{pmatrix} a_1 \\ a_2 \\ a_3 \\ a_4 \end{pmatrix} = \begin{pmatrix} b_1 \\ b_2 \\ b_3 \\ b_4 \end{pmatrix} + \begin{pmatrix} r_1 \\ 0 \\ 0 \\ r_4 \end{pmatrix}, \tag{13.54}$$

where $\mathbf{r}$ is calculated from

$$\mathbf{r} = \mathbf{K}^\star\ \mathbf{a} - \mathbf{b}^\star. \tag{13.55}$$

References

Alshawabkeh AN (2009) Electrokinetic soil remediation: challenges and opportunities. Sep Sci Technol 44(10):2171–2187

Alshawabkeh AN, Acar YB (1996) Electrokinetic remediation. II: Theoretical model. J Geotech Eng 122(3):186–196

Ball JW, Nordstrom DK (1991) User's manual for WATEQ4F, with revised thermodynamic data base and test cases for calculating speciation of major, trace, and redox elements in natural waters. US Geological Survey, Denver

Bethke CM (2008) Geochemical and biogeochemical reaction modeling, 2nd edn. Cambridge University Press, New York

Cameselle C, Reddy KR (2012) Development and enhancement of electro-osmotic flow for the removal of contaminants from soils. Electrochim Acta 86:10–22

Chung DW, Ebner M, Ely DR, Wook V, Edvin Garcia R (2013) Validity of bruggeman relation for porous electrodes. Model Simul Mater Sci Eng 21:074009–074025

Delgado AV, Gonzalez-Caballero F, Hunter RJ, Koopal LK, Lyklema J (2007) Measurement and interpretation of electrokinetic phenomena. J Colloid Interface Sci 309(2):194–224

El Abd A, Czachor A, Milczarek J (2009) Neutron radiography determination of water diffusivity in fired clay brick. Appl Radiat Isot 67(4):556–559

Hall C (1977) Water movement in porous building materials. I. Unsaturated flow theory and its applications. Build Environ 12:117–125

Ibrahimbegovic A (2009) Nonlinear solid mechanics: theoretical formulations and finite element solution methods. Springer, Dordrecht

Jacobs RA, Probstein RF (1996) Two-dimensional modeling of electroremediation. AIChE J 42 (6):1685–1696

Javadi A, Al-Najjar M (2007) Finite element modeling of contaminant transport in soils including the effect of chemical reactions. J Hazard Mater 143(3):690–701

Johannesson B, Hosokawa Y, Yamada K (2009) Numerical calculations of the effect of moisture content and moisture flow on ionic multi-species diffusion in the pore solution of porous materials. Comput Struct 87:39–46

Li LY, Page CL (2000) Finite element modelling of chloride removal from concrete by an electrochemical method. Corros Sci 42(12):2145–2165

Lim J, Whitcomb J, Boyd J, Varghese J (2007) Transient finite element analysis of electric double layer using Nernst-Planck-Poisson equations with a modified stern layer. J Colloid Interface Sci 305:159–174

Lu B., Holst MJ, Andrew McCammon J, Zhou YC (2010) Poisson-Nernst-Planck equations for simulating biomolecular diffusion-reaction processes. I: Finite element solutions. J Comput Phys 229(19):6979–6994

Lyklema J (2005) Fundamentals of interface and colloid science. Academic, New York

McNaught AD, Wilkinson A (1997) IUPAC. Compendium of chemical terminology (the "Gold Book"), 2nd edn. Blackwell Scientific Publications, Oxford

Métivier L, Hervé Roussel H (2012) Accounting robustly for instantaneous chemical equilibriums in reactive transport: a numerical method and its application to liquid–liquid extraction modeling. Comput Chem Eng 45:50–61

Nardi A, Idiart A, Trinchero P, Manuel de Vries L, Molinero J (2014) Interface comsol-phreeqc (icp), an efficient numerical framework for the solution of coupled multiphysics and geochemistry. Comput Geosci 69:10–21

Nguyen TQ, Petkovic J, Dangla P, and Baroghel-Bouny V (2008) Modelling of coupled ion and moisture transport in porous building materials. Constr Build Mater 22:2185–2195

Ottosen NS, Petersson H (1992) Introduction to the finite element method. Prentice-Hall International, Trowbridge, Wiltshire

Ottosen NS, Ristinmaa M (2005) The mechanics of constitutive modeling. Elsevier Science, Amsterdam

Pachepsky Y, Timlin D, Rawls W (2003) Generalized Richards' equation to simulate water transport in unsaturated soils. J Hydrol 272:3–13

Parkhurst DL, Appelo CAJ (1999) User's guide to PHREEQC (version 2) - a computer program for speciation, batch-reaction, one-dimensional transport, and inverse geochemical calculations. U.S. Department of the Interior, Water-Resources Investigations Reports (99-4259)

Paz-Garcia JM, Johannesson B, Ottosen LM, Ribeiro AB, Rodriguez-Maroto JM (2011a) Modeling of electrokinetic processes by finite element integration of the Nernst-Planck-Poisson system of equations. Sep Purif Technol 79(2):183–192

Paz-Garcia JM, Johannesson B, Ottosen LM, Rodriguez-Maroto JM, Ribeiro AB (2011b) Modelling of electrokinetic processes in civil and environmental engineering applications. In: Topping BHV, Tsompanakis Y (eds) Proceedings of the thirteenth international conference on civil, structural and environmental engineering computing. Civil-Comp Press, Stirlingshire, Paper 128. doi: 10.4203/ccp.96.128

Paz-Garcia JM, Johannesson B, Ottosen LM, Alshawabkeh AN, Ribeiro AB, Rodríguez-Maroto JM (2012) Modeling of electrokinetic desalination of bricks. Electrochim Acta 86:213–222

Paz-Garcia JM, Johannesson B, Ottosen LM, Ribeiro AB, Rodriguez-Maroto JM (2013a) Computing multi-species chemical equilibrium with an algorithm based on the reaction extents. Comput Chem Eng 58:135–143

Paz-Garcia JM, Johannesson B, Ottosen LM, Ribeiro AB, Rodriguez-Maroto JM (2013b) Simulation-based analysis of the differences in the removal rate of chlorides, nitrates and sulfates by electrokinetic desalination treatments. Electrochim Acta (89):436–444

Paz-Garcia JM, Johannesson B, Ottosen LM, Ribeiro AB, Rodriguez-Maroto JM (2014a) Modeling of electric double-layers including chemical reaction effects. Electrochim Acta 150:263–268

Paz-Garcia JM, Schaetzle O, Biesheuvel PM, Hamelers HVM (2014b) Energy from CO_2 using capacitive electrodes – theoretical outline and calculation of open circuit voltage. J Colloid Interface Sci 418:200–207

Paz-Garcia JM, Dykstra JE, Biesheuvel PM, Hamelers HVM (2015) Energy from CO_2 using capacitive electrodes—a model for energy extraction cycles. J Colloid Interface Sci 442:103–109

Peterson SR, Hostetler CJ, Deutsch WJ, Cowan CE (1987) Minteq user's manual. Technical report, Pacific Northwest Lab., Richland, WA (USA); Nuclear Regulatory Commission, Washington, DC (USA). Division of Waste Management

Press WH, Teukolsky SA, Vetterling WT, Flannery BP (1992) Root finding and nonlinear sets of equations, Chap. 9 In Numerical recipes in C: the art of scientific computing. Cambridge University Press, Cambridge

Rawlings JB, Ekerdt JG (2002) Chemical reactor analysis and design fundamentals. Nob Hill Publishing, Madison

Reddy KR (2013) Electrokinetic remediation of soils at complex contaminated sites: technology status, challenges, and opportunities. In: Manassero M et al. (ed) Coupled phenomena in environmental geotechnics. Taylor and Francis Group, London

Ristinmaa M, Ottosen NS, Johannesson B (2011) Mixture theory for a thermoelasto-plastic porous solid considering fluid flow and internal mass exchange. Int J Eng Sci 49(11):1185–1203

Rodriguez-Maroto JM, Vereda-Alonso C (2009) Electrokinetic modeling of heavy metals. In Reddy KR, Camesselle C (ed) Electrochemical remediation technologies for polluted soils, sediments and groundwater, Chap 25, Wiley Online Library, New York, pp. 537–562

Roels S, Carmeliet J, Hens H (2003) Modelling unsaturated moisture transport in heterogeneous limestone. Transp Porous Media 52(3):333–350

Sheng D, Smith DW (2002) 2D finite element analysis of multicomponent contaminant transport through soils. Int J Geom 2(1):113–134

Subires-Munoz JD, Garcia-Rubio A, Vereda-Alonso C, Gomez-Lahoz C, Rodriguez-Maroto JM, Garcia-Herruzo F, Paz-Garcia JM (2011) Feasibility study of the use of different extractant agents in the remediation of a mercury contaminated soil from almaden. Sep Purif Technol 79:151–156

van Baten J, Szczepanski R (2011) A thermodynamic equilibrium reactor model as a cape-open unit operation. Comput Chem Eng 35:1251–1256

Villen-Guzman M, Paz-Garcia JM, Rodriguez-Maroto JM, Garcia-Herruzo F, Amaya-Santos G, Gomez-Lahoz C, Vereda-Alonso C (2015a) Scaling-up the acid-enhanced electrokinetic remediation of a real contaminated soil. Electrochim Acta. http://dx.doi.org/10.1016/j.electacta.2015.02.067

Villen-Guzman M, Paz-Garcia JM, Amaya-Santos G, Rodriguez-Maroto JM, Vereda-Alonso C, Gomez-Lahoz C (2015b) Effects of the buffering capacity of the soil on the mobilization of heavy metals. Equilibrium and kinetics. Chemosphere 131:78–84

Yeung AT, Datla S (1995) Fundamental formulation of electrokinetic extraction of contaminants from soil. Can Geotech J 32(4):569–583

Yeung AT, Gu YY (2011) A review on techniques to enhance electrochemical remediation of contaminated soils. J Hazard Mater 195:11–29. doi: 10.1016/j.jhazmat.2011.08.047

Zienkiewicz OC, Taylor RL (2000) The finite element method, The Basis, 5 edn, vol 1 Butterworth-Heinemann, Oxford

Zienkiewicz OC, Taylor RL (2005) The finite element method for solid and structural mechanics, 6th edn, vol 2. Butterworth-Heinemann, Elsevier, Oxford

Chapter 14
Electrokinetics and Zero Valent Iron Nanoparticles: Experimental and Modeling of the Transport in Different Porous Media

Helena I. Gomes, José M. Rodríguez-Maroto, Alexandra B. Ribeiro, Sibel Pamukcu, and Celia Dias-Ferreira

14.1 Introduction

The use of granular zero valent iron (ZVI) in permeable reactive barriers (PRB) allowed groundwater remediation for more than 20 years (USEPA 2011a), targeting both organic and inorganic contaminants. In 1996, Lehigh University researchers developed a method to synthesize zero valent iron nanoparticles (nZVI) using sodium borohydride as reductant (Wang and Zhang 1997; Zhang et al. 1998).

H.I. Gomes (✉)
CENSE, Departamento de Ciências e Engenharia do Ambiente, Faculdade de Ciências e Tecnologia, Universidade Nova de Lisboa, Caparica 2829-516, Portugal

CERNAS—Research Center for Natural Resources, Environment and Society, Escola Superior Agraria de Coimbra, Instituto Politecnico de Coimbra, Bencanta, Coimbra 3045-601, Portugal

Department of Civil Engineering, Technical University of Denmark, Lyngby, Denmark
e-mail: hrg@campus.fct.unl.pt

J.M. Rodríguez-Maroto
Department of Chemical Engineering, University of Málaga, Campus de Teatinos, Málaga 29071, Spain

A.B. Ribeiro
CENSE, Departamento de Ciências e Engenharia do Ambiente, Faculdade de Ciências e Tecnologia, Universidade Nova de Lisboa, Caparica 2829-516, Portugal

S. Pamukcu
Department of Civil and Environmental Engineering, Lehigh University, Bethlehem, PA 18015, USA

C. Dias-Ferreira
CERNAS—Research Center for Natural Resources, Environment and Society, Escola Superior Agraria de Coimbra, Instituto Politecnico de Coimbra, Bencanta, Coimbra 3045-601, Portugal

© Springer International Publishing Switzerland 2016

A.B. Ribeiro et al. (eds.), *Electrokinetics Across Disciplines and Continents*,
DOI 10.1007/978-3-319-20179-5_14

Since then, these nanoparticles were used with several modifications for the remediation of different organic and inorganic contaminants in groundwaters and soils (Gomes 2014). The use of these nanomaterials in pilot and full-scale applications in the last decade is expressive. In November 2011, there were 36 projects at pilot and full scale (11 were Superfund sites) in the United States (USEPA 2011b). There were also 15 pilot tests in Europe (Mueller et al. 2012) and full-scale applications in Italy, Germany, Czech Republic, and Slovakia (Rejeski et al. 2014). Most of them target soil and groundwater contamination with volatile organic compounds (perchloroethylene—PCE, trichloroethylene—TCE, and corresponding daughter products, perchlorate, polychlorinated biphenyls (PCB), and other organochlorines, as well as diesel products), in sandy or silty sandy soils. There are also applications in glacial till soils and unconsolidated sediments (USEPA 2011b).

Iron nanoparticles have some particular advantages when compared to the granular ZVI, such as high reactivity, lower degradation times, generation of less toxic intermediate products, and also the possibility of injection as aqueous slurries (Gomes 2014). However, the mobility of nZVI in the subsurface is frequently less than a few meters, as several field applications show (Elliott and Zhang 2001; Quinn et al. 2005; Henn and Waddill 2006; Zhang and Elliott 2006; Su et al. 2012), ranging from 1 m (Kocur et al. 2014) to 6–10 m (Zhang and Elliott 2006). This is due to Brownian motion, the density of iron, long-range magnetic attractive forces, and groundwater ionic strength, which increase the aggregation of nZVI (Phenrat et al. 2007). There are numerous studies on the transport of iron nanoparticles, mostly in column tests with sand (Hydutsky et al. 2007; Kanel et al. 2007; Phenrat et al. 2009a; Raychoudhury et al. 2010; Saleh et al. 2008), glass beads (Jiemvarangkul et al. 2011; Kanel et al. 2007; Lin et al. 2010; Saleh et al. 2008), and model soils (He et al. 2007; Schrick et al. 2004; Yang et al. 2007). Studies using high concentrations representative of field applications show that, although stabilized nanoparticles are more mobile than bare nZVI (Jiemvarangkul et al. 2011; Kanel et al. 2007; Phenrat et al. 2009b; Raychoudhury et al. 2010), aggregation remains an important process. The particle size distribution and Fe^0 content of nZVI as well as groundwater ionic strength and composition (Lin et al. 2010; Saleh et al. 2008) can affect aggregation. The affinity with soil minerals, resulting in nZVI deposition onto the porous matrix (Tosco et al. 2014) also limits their mobility. The main factors that influence nZVI adsorption onto soil and aquifer materials are: (1) surface chemistry of soil and the nanoparticles; (2) groundwater chemistry (ionic strength, pH, organic matter content); (3) hydrodynamic conditions (pore size, porosity, flow velocity, and degree of mixing and turbulence) (Noubactep et al. 2012).

This chapter reports on the application of the electrokinetic process to enhance the nZVI transport. Integrating both technologies, the role of direct electric current would be to increase the nZVI transport into the soil for in situ transformation, and subsequent destruction of the contaminants, instead of aiming at the removal of the contaminants. The discussion is focused on (a) the experimental data obtained in the different porosity media and (b) numerical model and prediction of the nZVI assisted transport.

14.2 Coupling Electrokinetics and Iron Nanoparticles

The use of treatment trains for the remediation of contaminated soils minimizes the cost of achieving risk-based end points for regulatory compliance or liability reduction (Rao et al. 2001). The general principle of treatment trains is the use of a combination of techniques, simultaneously or in succession, to improve treatment performance in a quicker, more efficient, and cost-effective way. Electrokinetic remediation and nZVI were used in conjunction, both to enhance the nZVI transport in low permeability fine-grain soils, and to degrade organic contaminants (Table 14.1).

A direct comparison of the previous studies on nZVI-enhanced transport with direct current (Table 14.1) is inconclusive. The studies use different experimental conditions, such as experimental setups; soils or other solid media used; types of iron nanoparticles; injection places (i.e., directly in the soil, anode, or cathode compartments); value and duration of the voltage gradients applied. In general, electrophoretic transport of the particles was predominant in sandy soil (Yang et al. 2008; Jones et al. 2010; Chowdhury et al. 2012), while electroosmotic transport was more important in kaolin clay and loamy sand soil (Yang et al. 2007; Pamukcu et al. 2008).

14.3 Case Study

The experimental data were published previously by Gomes et al. (2013, 2014), where the experimental conditions and setup are described in detail. In the experiments, the transport of nZVI took place in the domain of a layer of porous solid (kaolin and/or glass beads) saturated with an electrolyte (Table 14.2).

The experiments were performed in a modified commercial electrophoretic cell (Econo-Submarine Gel Unit, model SGE-020) as shown in Fig. 14.1. The cell is a rectangular translucent box 10 cm height, 40 cm long, and 23 cm width, with a central square (20 cm $\times$ 20 cm) sample tray and a lid that covers the whole cell (Gomes et al. 2013, 2014b). Two liquid chambers on either side of the sample tray hold the anolyte and the catholyte, and platinum working electrodes (Fig. 14.1). In all experiments, both chambers were filled with the same electrolyte solution (NaCl 0.001 M; volume of 650 mL each) that was also used to saturate the porous specimen. The level of the solutions in the side compartments was held slightly below the specimen surface, thus preventing preferential transport of nZVI through a liquid pool over it. Compressed fiberglass wool pads, saturated and immersed in the electrolyte solution, helped transport the migrating ions from the solution into the specimen, and vice versa (Gomes et al. 2013, 2014b). The test media (Table 14.2) used had different porosity and surface reactivity, ranging from glass

Table 14.1 Coupling of electrokinetic remediation and zero valent iron nanoparticles in bench-scale experiments for enhanced transport and remediation (updated after Gomes et al. 2012a)

Matrix	Target contaminant	Duration of test(s)	Electrolyte	Voltage gradient (V cm^{-1})	% Contaminant removal best results	Reference
White Georgia kaolin clay	–	46 h	0.2 M NaCl	0.25	–	Pamukcu et al. (2008)
Loamy sand	–	6 days	Simulated groundwater	1.0	–	Yang et al. (2007)
Spiked soil	KNO_3	6 days	Simulated groundwater	1.0	99	Yang et al. (2008)
Spiked kaolin	PCP	427 h; 960 h	Deionized water	1.0	55	Reddy and Karri (2008)
Spiked silica sand	PCE	100 h	0.01 M Na_2CO_3	1.0	76	Chen et al. (2010)
40/60 or 100/200 sands	–	10 days	0.007 or 0.02 M NaCl	0.55; 1.30	–	Jones et al. (2010)
Spiked sandy soil	TCE	10 days	Simulated groundwater	1.0	70	Yang and Chang (2011)
Coarse and fine sand	–	48 h	Simulated groundwater	0.25; 0.5	–	Chowdhury et al. (2012)
Spiked turf	PCP	8 days; 14 days	0.025 M Na_2SO_4 and 0.025 M Na_2CO_3	6.0	70	Yuan et al. (2012)
Kaolin clay	Cr(IV)	24 h	0.01 M NaCl	0.25	62	Gomes et al. (2012b)
Contaminated soil	PCB	14 days	0.01 M $NaNO_3$	1.0 → 2.0	20	Fan et al. (2013)
Boom clay	–	160 h; 360 h	0.01 M NaCl	4.0	–	Rosales et al. (2014)
Spiked soils (sandy and sandy loam)	Molinate	6 days	0.01 M $NaNO_3$	~5	90	Gomes et al. (2014c)
Contaminated soil	PCB	5 days	0.01 M NaCl	1.0	76	Gomes et al. (2014a)

Table 14.2 Experimental setup and model parameters using 0.001 M NaCl as electrolyte

Test number	Layer thickness (mm)	Matrix	Porosity	Moisture content (%)	D^*_{nZVI} $(\text{cm}^2 s^{-1})$	U^*_{nZVI} $(\text{cm}^2 V^{-1} s^{-1})$
Solid layer length, L: 20 cm						
Solid layer area, A: 10 or 4 cm^2						
Number of volume elements used to represented solid column, N: 20						
Electrode compartment volume, V_0 and V_{N+1}: 650 cm^3						
Faradaic efficiency $\eta = 1$						
Δt: 5 s; *Total duration of the experiments:* 48 h						
Initial mass of nZVI injected in the system: 0.008 g (2 mL of 4 g L^{-1} solution)						
Electric potential: 5 V (experiments 5–8)						
Diffusion control tests						
1	5	100 % Kaolin	0.65	60	5.9×10^{-5}	-1.7×10^{-4}
2	2	50 % Glass beads and 50 % kaolin	0.57	30	4.6×10^{-5}	-1.5×10^{-4}
3	2	75 % Glass beads and 25 % kaolin	0.35	30	3.9×10^{-5}	-0.9×10^{-4}
4	2	100 % Glass beads	0.20	20	1.8×10^{-5}	-0.5×10^{-4}
Enhanced transport tests						
5	5	100 % Kaolin	0.65	60	1.6×10^{-5}	-1.1×10^{-4}
6	2	50 % Glass beads and 50 % kaolin	0.57	30	2.8×10^{-5}	2.4×10^{-5}
7	2	75 % Glass beads and 25 % kaolin	0.35	30	1.4×10^{-5}	2.1×10^{-6}
8	2	100 % Glass beads	0.20	20	4.6×10^{-6}	8×10^{-6}

beads (with particle diameter < 1 mm, previously sieved) to white Georgia kaolinite clay (>2 μm). The polyacrylic acid stabilized nanoparticles (PAA-nZVI) suspensions were freshly prepared before each experiment according to the method used by Kanel et al. (2008) and had a concentration of 4 g L^{-1}. The particle size distribution of the nanoparticles had a mean particle diameter value of 63 nm and the median size was 60.2 nm, based on a count of 420 particles in TEM images (Gomes et al. 2013, 2014b). Two sets of control experiments were conducted for each mixture under the same conditions, one without direct current but with PAA-nZVI, and another with current but without PAA-nZVI. In the experiments with current, a constant potential was applied for 48 h. During the experiments, the cell was kept in a dark location to prevent iron photo-oxidation. The nanoparticle suspension was delivered in the media using a syringe to inject 2 mL through a tube, which allowed the suspension disperse into a pre-cut shallow channel between the auxiliary electrodes E2 and E3 (Fig. 14.1).

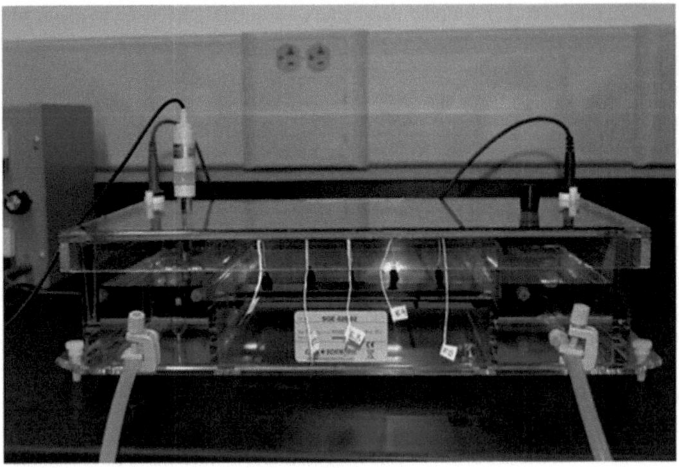

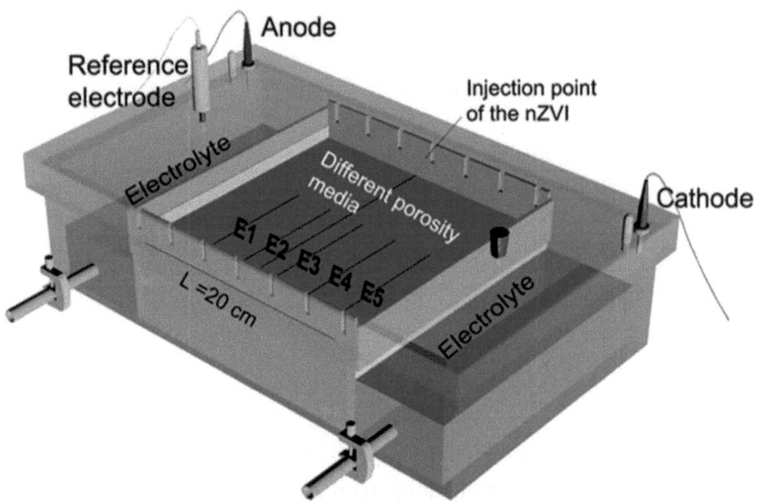

Fig. 14.1 Modified electrophoretic cell used in the experiments (Gomes et al. 2013, 2014b). E1 to E5 are auxiliary platinum electrodes and represent the locations of the solid samples for iron quantification

14.4 Theoretical Formulation of the Model

Until now, the available analytical models of the nanoparticle transport in the literature included only the electrophoretic effect that mostly takes place in sands (Jones et al. 2010; Chowdhury et al. 2012). This generalized physicochemical model has been developed to describe the electrically induced transport of nZVI particles through different types of media with varying porosity and surface reactivity (Gomes et al. 2015). The model is more detailed, including the fundamental processes, and its numerical solution offers a reliable prediction of the nZVI transport.

The model operates in two steps: first the kinetic process is simulated by integrating forward in time the one-dimensional transport equations and the electrochemical reactions at the electrodes; and second the chemical equilibriums are reestablished before the next step of integration. This is done because chemical equilibriums are considered instantaneous when compared with the transport time.

The mass conservation equation for i^{th} species in a jth volume element, including electrochemical reactions, is described by:

$$V_j\left(\frac{dc_{ij}}{dt}\right) = \left(N_{i,\,j-1} + N_{i,\,j}\right)A + R_iV_j \tag{14.1}$$

where V_j is volume of water in j^{th} cell (cm^3), c_{ij} is the concentration of ith species (ions and nZVI) in the aqueous phase of the jth volume element (mol cm^{-3}), t is the time, $N_{i,\,j-1}$ and $N_{i,\,j}$ the mass flux of ith species from $(j-1)^{th}$ into j^{th} element volume and from j^{th} into $(j+1)^{th}$ volume element (mol cm^2 s^{-1}), A cross-sectional area of the domain (cm^2), and R_i the reaction rate for i species. With respect to the chemical reactions, only the chemical equilibria and the electrochemical reactions at the electrodes are considered (Gomes et al. 2015).

A coupled system of Nernst–Planck equations is the foundation of the transport model, accounting for the mass balance of the ionic species in a fluid medium when diffusion and electromigration are considered. In the case of charged nZVI (i.e., the nanoparticles are stabilized with polyacrylic acid—PAA, which gives them a negative charge), diffusion and electrophoretic terms have to be considered. The electroosmotic flow is always included.

Therefore, the flux of any chemical species or charged particles I from a jth volume element of the system can be expressed as:

$$N_i = -D_i^*\nabla c_i - U_i^*c_i\nabla\phi - k_ec_i\nabla\phi \tag{14.2}$$

where (subindex j is omitted), c_i is the molar concentration, D_i^* is the effective diffusion coefficient, $\nabla\phi$ is the electrical potential, k_e is the electroosmotic permeability coefficient, and U_i^* is the effective electrophoretic mobility for nZVI-charged particles or effective ionic mobility, estimated by the Einstein–Nernst relation for ions (Newman 1991):

$$U_i^* = \frac{D_i^*z_iF}{RT} \tag{14.3}$$

where R is the ideal gas constant, F is the Faraday constant, z_i is the ionic charge of the species, and T is the temperature (K), assuming a constant room temperature of 25 °C (Gomes et al. 2015). The value of the electroosmotic permeability usually is in a very tight range of 10^{-5} to 10^{-4} cm^2 s^{-1} V^{-1} (Mitchell 1993). Electroosmotic permeability and mobility can be combined for a new effective mobility in the porous medium, U_i^{**}:

$$U_i^{**} = U_i^{*} + k_{\mathrm{e}} \tag{14.4}$$

The mass balance equations for nZVI and the ionic species are integrated over the one-dimensional region limited by the electrodes compartments to obtain the concentration profile for a given set of experimental conditions (Gomes et al. 2015). Due to the negative charge of polyacrylic acid-coated nZVI, the sign of electrophoretic term is negative, whereas the electroosmotic term is positive, resulting in a reduced effective mobility.

The principal chemical reactions that need to be considered are the electrochemical reduction and oxidation of water at the electrodes, so the generation term rate is not included in the continuity equation for the porous specimen. The Nernst equation is used to calculate the redox potential for each electrochemical half-reaction, as:

$$E = E^0 - \frac{RT}{\upsilon F} \ln Q \tag{14.5}$$

where Q is the reaction quotient, defined as the product of the activities of the chemical species to the power of their stoichiometric coefficients, for nonequilibrium conditions. In the special case that the reaction is at equilibrium, the reaction quotient is equal to the equilibrium constant at 25 °C (Gomes et al. 2015). υ in this case is the stoichiometric coefficient of the electrons in the redox equation, i.e., the number of electrons exchanged during the oxidation or reaction process. E (V) is the redox potential in the reduction sense, and E^0 (V) is the standard redox potential, which is measured under standard conditions which are 25 °C, 1 M concentration of each ion participating in the reaction, a partial pressure of 1 atm is assigned for each gas that is part of the reaction and metals in their pure state (Chang and Overby 2011).

The cations Na^+ are attracted to the cathode, but the redox potential of alkali metals is too high to be competitive to that of water. Consequently, it seems reasonable to assume that only water reduction is taking place at the cathode. Thus, the only electrochemical half-reaction at the cathode is:

$$2H_2O + 2e^- \rightarrow 2OH^-(aq) + H_2(gas) \quad E^0 = -0.828\,V \tag{14.6}$$

On the other hand, anions are attracted to the anode, where oxidation reactions occur. However, in this case, only water oxidation is expected. Therefore, the possible half-reaction at the anode is given by (14.7):

$$2H_2O \rightarrow 4H^+(aq) + O_2(gas) + 4e^- \quad E^0 = -1.229\,V \tag{14.7}$$

The electrochemical reactions were included in the mass balance equations of anode and cathode compartments as given in (14.8) and (14.9):

$$V_0\left(\frac{dc_{10}}{dt}\right)_{\mathrm{ER}}=\frac{I}{F}\eta \tag{14.8}$$

$$V_{N+1}\left(\frac{dc_{2N+1}}{dt}\right)_{\mathrm{ER}}=\frac{1}{F}\eta \tag{14.9}$$

where V_0 and V_{N+1} are the volumes of electrolyte in the anode and cathode compartments, c_{10} and c_{2N+1}, H^+ and OH^- concentrations generated there by electrochemical reactions, I is the current intensity, F, the Faraday's constant, and η the Faradaic efficiency (Gomes et al. 2015).

At each time step, after completing the transport calculations, the concentration corresponding to the chemical equilibrium of every species is calculated from the last value obtained from the transport. Therefore, in every volume element is solved a system of nonlinear equations comprising the mass balances, the electrical neutrality condition, and the equilibrium mass action equations. The extremely rapid reactions between protons and hydroxyls to form water and reverse are considered. The chemical equilibrium of water is:

$$\mathrm{H^+ + OH^- \leftrightarrows H_2O} \quad K_{\mathrm{w}}=[\mathrm{H^+}][\mathrm{OH^-}]=10^{-14} \tag{14.10}$$

In the experiments using NaCl (0.001 M) as electrolyte, as no equilibrium process affects Cl^- and Na^+, the conservation equations for them are trivial.

14.5 Transport of the Iron Nanoparticles

The experimental results showed that, in general, higher concentrations of iron across the test bed (in samples over the auxiliary electrodes E1–E5) were measured when a direct current was applied (Fig. 14.2), indicating that the nZVI transport was enhanced in comparison with diffusion (Gomes et al. 2013). The model results reproduce in a satisfactory way the nZVI concentration profiles, as shown by the lines in Fig. 14.2 (Gomes et al. 2015). In some cases, a substantial fraction of the nZVI tends to aggregate when the concentration is high relative to the available pore volume, becoming immobile. Considering a ratio between nZVI out of the injection point and the injected nZVI, in experiments 2 and 4 only about 19 % and 8 % of the injected nZVI remained mobile over the experiment, respectively (Gomes et al. 2015).

In all the experiments with glass beads, there was a well-defined peak of concentration at E3 (i.e., practically the injection point). This was probably due to the nZVI aggregation, or to the fast corrosion of the iron nanoparticles, or to both

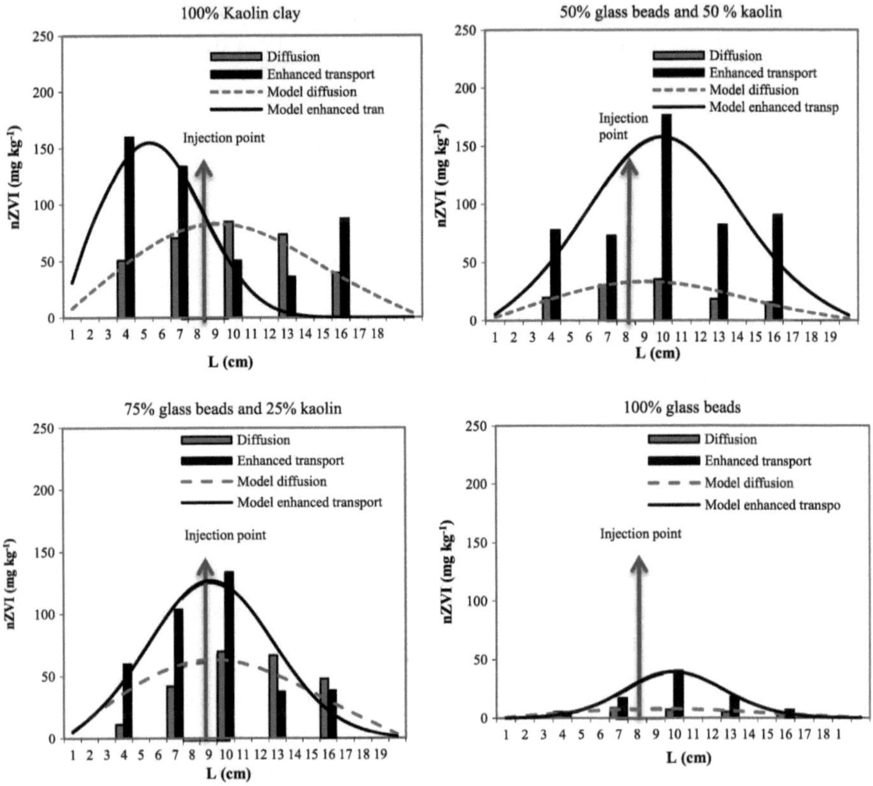

Fig. 14.2 Iron concentrations across the electrophoretic cell in the experiments compared with the model results for the porous media tested (modified from Gomes et al. 2013 and Gomes et al. 2015)

phenomena. It has been previously shown that at high particle concentrations (1–6 g L^{-1}) nZVI have a greater tendency for agglomeration (Phenrat et al. 2009a). When nZVI aggregate and form clusters larger than the soil pores, their transport becomes restricted (Reddy et al. 2011). Changes to the nZVI mobility can be due to volumetric expansion with corrosion (14.11 and 14.12):

$$2Fe^0_{(s)} + 4H^+_{(aq)} + O_{2\ (aq)} \rightarrow 2Fe^{2+}_{(aq)} + 2H_2O_{(l)} \quad (14.11)$$

$$Fe^0_{(s)} + 2H_2O_{(l)} \rightarrow Fe^{2+}_{(aq)} + H_{2(g)} + 2OH^-_{(aq)} \quad (14.12)$$

The volume of corrosion products (Fe hydroxide or oxide) is larger than that of the original metal (Fe^0) and these products are likely to contribute to porosity loss, promoting simultaneously particle agglomeration (Noubactep et al. 2012).

PAA-nZVI did not move into the aqueous phase in the electrode chambers, except for the cathode chamber in the enhanced transport tests with 100 % kaolin

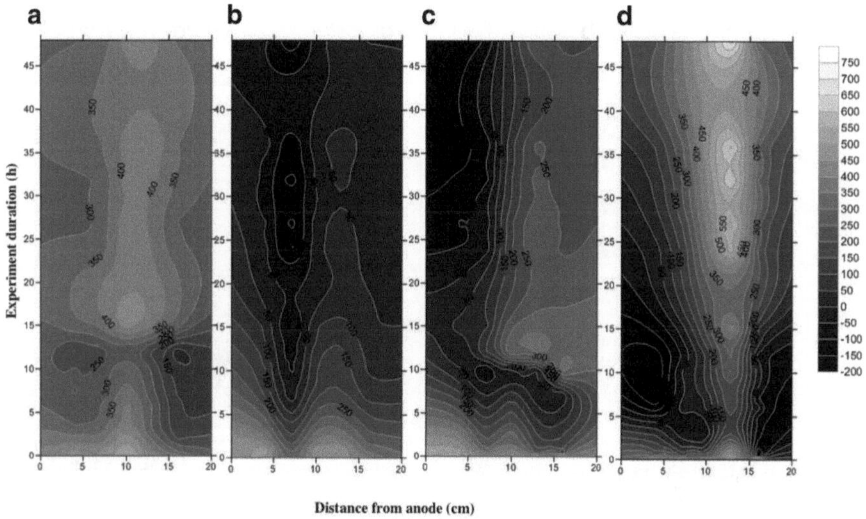

Fig. 14.3 Variation of the oxidation-reduction potential across the electrophoretic cell (*horizontal axis*), during the experiments (*vertical axis*), with the different porous media (Gomes et al. 2013). (**a**) 100 % kaolin; (**b**) 50 % kaolin and 50 % glass beads; (**c**) 25 % kaolin and 75 % glass beads; (**d**) 100 % glass beads

(final concentration of 0.43 mg L^{-1}) and 100 % glass beads (0.74 mg L^{-1} in the anode compartment and 0.09 mg L^{-1} in the cathode) (Gomes et al. 2013). It appears that the EOF and electrophoresis compete, resulting in the prolonged presence of iron in the pores and potential capture on the clay surfaces. Other experimental results showed that there was greater deposition of nZVI onto clay minerals compared to similar sized silica fines due to charge heterogeneity on clay mineral surfaces (Kim et al. 2012).

The pH and ORP conditions measured during the experiments were favorable to the nZVI oxidation. Also, the initial conditions were not advantageous to the mobility of PAA-nZVI because low pH (such as the kaolin pH = 4.97) increases both the deposition of nZVI in clay and its aggregation (Kim et al. 2012). Basically, the presence of clay appear to push the ORP up to more oxidizing conditions, while over time the ORP values decreased, favoring reducing conditions in all tests (Fig. 14.3). In the experiments with more percentage of glass beads, and in particular in the experiment 4, there is an accentuated drop in the ORP values in the electrodes E1, E2, and E3, due to the injection and fast electrophoretic transport of PAA-nZVI in the absence of clay (Gomes et al. 2013). The system attains equilibrium with more or less uniform and constant ORP after 15 h. The typical profile of electrokinetic treatments of a pH front increasing from the anode to the cathode was not observed in the media in the experiments (Fig. 14.4), which can be attributed to the small values of current density applied, or to the absence of the physical conditions for fast transport of H^+ and OH^- from the electrode compartments into the media (Gomes et al. 2013). Only when using 100 % glass beads, the

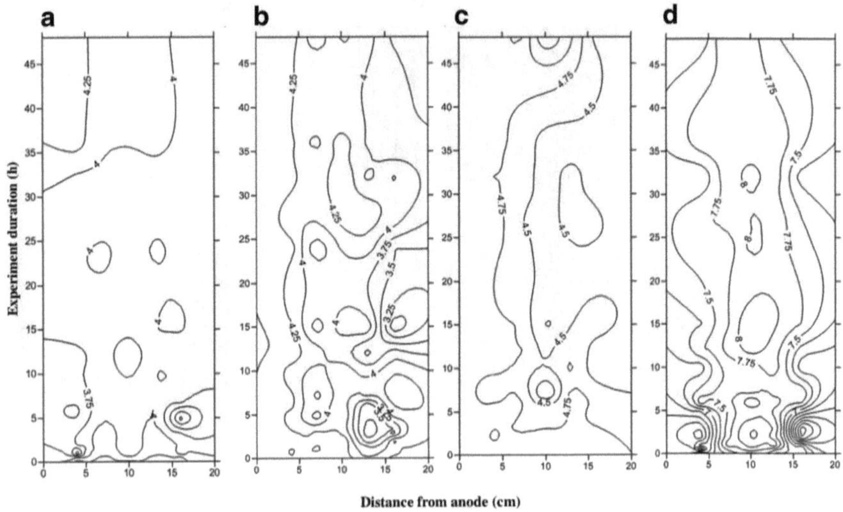

Fig. 14.4 Variation of the pH across the electrophoretic cell (*horizontal axis*), during the experiments (*vertical axis*) in the different porous media (Gomes et al. 2013). (**a**) 100 % kaolin; (**b**) 50 % kaolin and 50 % glass beads; (**c**) 25 % kaolin and 75 % glass beads; (**d**) 100 % glass beads

effect on pH from the injection of the PAA-nZVI was noticed, particularly in E2, E3, and E4, in which pH values higher than 8 were measured.

The model predicts very low effective mobility values in the porous medium (U_i^{**}, Table 14.2) as a result of the opposing transport directions between the electroosmotic advection and the electrophoretic migration of the negatively charged nanoparticles (Gomes et al. 2015). This effect manifests itself as higher concentrations close to the injection point in most of the experiments. Thus, if the nanoparticles could be stabilized with a surface modifier to give them a positive charge, the nZVI-effective mobility could potentially be increased. However, the probability of the positively charged particles be attracted to the soil particle surfaces, particularly clays, could increase. Also, the use of stabilizers without charge could enhance the electroosmotic transport of the iron nanoparticles.

The transport model has still some limitations. The surface charge of PAA will be changed according to the surrounding pH because of protonation or deprotonation. Therefore, the transport is highly dependent on the surrounding pH. The pH evolution in the different porosity media was included in the model when pH decreases it is expected both electrophoretic mobility and electroosmotic coefficients to decrease too. Both effects counterbalance inside the global mobility U^{**}, this is why we have used only an average value U^{**} along the experiments, because additional experimental information would be necessary. Further research is needed to allow the collection of experimental data of the effects of pH on electrophoretic and electroosmotic mobilities. Then, it will be possible to include the pH effect on U^{**} in a more complex and realistic model. Other issues such as the striping of the polymer from the nanoparticle surface and the aging of the nanoparticles still need further research in order to improve the model.

14.6 Conclusions

The use of direct current to enhance the nanoparticles transport also prevents or hinders their aggregation, increasing their mobility, even using high concentrations, typical of field applications. However, the opposing directions of electrophoretic transport and the electroosmotic advection still produces limited nZVI transport. Further improvement could be achieved by using neutrally charged nanoparticles that could be transported by electroosmotic advection. The use of positively charged nanoparticles is not recommended, due to the attraction to the surface of soil particles (which are negatively charged). The use of polymers that increase the repulsion between nanoparticles (inhibiting aggregation) and between nanoparticles and surfaces (inhibiting attachment) aims at enhancing the dispersibility and transport of nZVI through the porous medium. However, despite the use of polyacrylic acid to modify the surface, an important aggregation of nZVI is observed in the experiments when the particles are allowed to diffuse into a porous medium from an injection point. The higher the nZVI concentration is in the matrix, the higher the aggregation; therefore, low concentrations suspensions must be used for successful field application.

A detailed physicochemical model that included advective, diffusive and electrophoretic transport, electrolyte-specific electrochemical reactions, as well as mass balances, was developed and solved. The model predictions agreed well with the experimental findings of the nZVI distribution in various porous systems, as well as the pH evolution in the anolyte and catholyte in the enhanced transport and diffusion tests. However, further research and experimental data are needed to incorporate in the model the striping, protonation or deprotonation of the nanoparticles coating, and the aging of the nanoparticles.

Acknowledgments This work has been funded by the research grant SFRH/BD/76070/2011, by project PTDC/AGR-AAM/101643/2008 NanoDC under Portuguese National funds through "Fundação para a Ciência e a Tecnologia" and by FP7-PEOPLE-IRSES-2010-269289-ELECTROACROSS. The Department of Civil and Environmental Engineering at Lehigh University is acknowledged for the funding of equipment development, testing, and analysis of the nZVI transport experiments.

References

Chang R, Overby J (2011) General chemistry—the essential concepts. McGraw-Hill, New York

Chen S-S, Huang Y-C, Kuo T-Y (2010) The remediation of perchloroethylene contaminated groundwater by nanoscale iron reactive barrier integrated with surfactant and electrokinetics. Ground Water Monit Rem 30(4):90–98

Chowdhury AIA, O'Carroll DM, Xu Y, Sleep BE (2012) Electrophoresis enhanced transport of nano-scale zero valent iron. Adv Water Resour 40:71–82. doi:10.1016/j.advwatres.2012.01.014

Elliott DW, Zhang W (2001) Field assessment of nanoscale bimetallic particles for groundwater treatment. Environ Sci Technol 35:4922–4926

Fan G, Cang L, Qin W, Zhou C, Gomes HI, Zhou D (2013) Surfactants-enhanced electrokinetic transport of xanthan gum stabilized nano Pd/Fe for the remediation of PCBs contaminated soils. Sep Purif Technol 114:64–72. doi:10.1016/j.seppur.2013.04.030

Gomes HI (2014) Coupling electrokinetics and iron nanoparticles for the remediation of contaminated soils. Ph.D. Dissertation. Faculdade de Ciencias e Tecnologia, Universidade Nova de Lisboa, Portugal

Gomes HI, Dias-Ferreira C, Ribeiro AB (2012a) Electrokinetic remediation of organochlorines in soil: enhancement techniques and integration with other remediation technologies. Chemosphere 87(10):1077–1090. doi:10.1016/j.chemosphere.2012.02.037

Gomes HI, Dias-Ferreira C, Ribeiro AB, Pamukcu S (2012b) Electrokinetic enhanced transport of zero valent iron nanoparticles for chromium (VI) reduction in soils. Chem Eng Trans 28:139–144. ISBN:978-188-95608-95619-95608, ISSN:91974-99791, doi:95610.93303/CET1228024

Gomes HI, Dias-Ferreira C, Ribeiro AB, Pamukcu S (2013) Enhanced transport and transformation of zerovalent nanoiron in clay using direct electric current. Water Air Soil Pollut 224 (12):1–12. doi:10.1007/s11270-013-1710-2

Gomes HI, Dias-Ferreira C, Ottosen LM, Ribeiro AB (2014a) Electrodialytic remediation of polychlorinated biphenyls contaminated soil with iron nanoparticles and two different surfactants. J Colloid Interface Sci 433:189–195. doi:10.1016/j.jcis.2014.07.022

Gomes HI, Dias-Ferreira C, Ribeiro AB, Pamukcu S (2014b) Influence of electrolyte and voltage on the direct current enhanced transport of iron nanoparticles in clay. Chemosphere 99:171–179. doi:10.1016/j.chemosphere.2013.10.065

Gomes HI, Fan G, Mateus EP, Dias-Ferreira C, Ribeiro AB (2014c) Assessment of combined electro–nanoremediation of molinate contaminated soil. Sci Total Environ 493:178–184. doi:10.1016/j.scitotenv.2014.05.112

Gomes HI, Rodríguez-Maroto JM, Ribeiro AB, Pamukcu S, Dias-Ferreira C (2015) Numerical prediction of diffusion and electric field-induced iron nanoparticle transport. Electrochim Acta (in press). doi:10.1016/j.electacta.2014.11.157

He F, Zhao D, Liu J, Roberts CB (2007) Stabilization of Fe-Pd nanoparticles with sodium carboxymethyl cellulose for enhanced transport and dechlorination of trichloroethylene in soil and groundwater. Ind Eng Chem Res 46:29–34

Henn KW, Waddill DW (2006) Utilization of nanoscale zero-valent iron for source remediation—a case study. Remediation 16(2):57–77

Hydutsky BW, Mack EJ, Beckerman BB, Skluzacek JM, Mallouk TE (2007) Optimization of nano- and microiron transport through sand columns using polyelectrolyte mixtures. Environ Sci Technol 41(18):6418–6424. doi:10.1021/es0704075

Jiemvarangkul P, Zhang WX, Lien HL (2011) Enhanced transport of polyelectrolyte stabilized nanoscale zero-valent iron (nZVI) in porous media. Chem Eng J 170(2–3):482–491. doi:10.1016/j.cej.2011.02.065

Jones EH, Reynolds DA, Wood AL, Thomas DG (2010) Use of electrophoresis for transporting nano-iron in porous media. Ground Water 49(2):172–183. doi:10.1111/j.1745-6584.2010.00718.x

Kanel S, Nepal D, Manning B, Choi H (2007) Transport of surface-modified iron nanoparticle in porous media and application to arsenic(III) remediation. J Nanoparticle Res 9(5):725–735. doi:10.1007/s11051-007-9225-7

Kanel SR, Goswami RR, Clement TP, Barnett MO, Zhao D (2008) Two dimensional transport characteristics of surface stabilized zero-valent iron nanoparticles in porous media. Environ Sci Technol 42:896–900

Kim H-J, Phenrat T, Tilton RD, Lowry GV (2012) Effect of kaolinite, silica fines and pH on transport of polymer-modified zero valent iron nano-particles in heterogeneous porous media. J Colloid Interface Sci 370(1):1–10. doi:10.1016/j.jcis.2011.12.059

Kocur CM, Chowdhury AI, Sakulchaicharoen N, Boparai HK, Weber KP, Sharma P, Krol MM, Austrins LM, Peace C, Sleep BE, O'Carroll DM (2014) Characterization of nZVI mobility in a field scale test. Environ Sci Technol 48(5):2862–2869. doi:10.1021/es4044209

Lin Y-H, Tseng H-H, Wey M-Y, Lin M-D (2010) Characteristics of two types of stabilized nano zero-valent iron and transport in porous media. Sci Total Environ 408(10):2260–2267. doi:10.1016/j.scitotenv.2010.01.039

Mitchell JK (1993) Fundamentals of soil behavior, 2nd edn. Wiley, New York

Mueller NC, Jr B, Bruns J, Černík M, Rissing P, Rickerby D, Nowack B (2012) Application of nanoscale zero valent iron (NZVI) for groundwater remediation in Europe. Environ Sci Pollut Res 19(2):550–558. doi:10.1007/s11356-011-0576-3

Newman J (1991) Electrochemical Systems. Prentice Hall, Englewood Cliffs

Noubactep C, Caré S, Crane R (2012) Nanoscale metallic iron for environmental remediation: prospects and limitations. Water Air Soil Pollut 22(3):1363–1382. doi:10.1007/s11270-011-0951-1

Pamukcu S, Hannum L, Wittle JK (2008) Delivery and activation of nano-iron by DC electric field. J Environ Sci Health A 43(8):934–944

Phenrat T, Saleh N, Sirk K, Tilton RD, Lowry GV (2007) Aggregation and sedimentation of aqueous nanoscale zerovalent iron dispersions. Environ Sci Technol 41:284–290

Phenrat T, Kim H-J, Fagerlund F, Illangasekare T, Tilton RD, Lowry GV (2009a) Particle size distribution, concentration, and magnetic attraction affect transport of polymer-modified Fe(0) nanoparticles in sand columns. Environ Sci Technol 43:5079–5085

Phenrat T, Liu Y, Tilton RD, Lowry GV (2009b) Adsorbed polyelectrolyte coatings decrease Fe^0 nanoparticle reactivity with TCE in water: conceptual model and mechanisms. Environ Sci Technol 43:1507–1514

Quinn J, Geiger C, Clausen C, Brooks K, Coon C, O'Hara S, Krug T, Major D, Yoon W-S, Gavaskar A, Holdsworth T (2005) Field demonstration of DNAPL dehalogenation using emulsified zero-valent iron. Environ Sci Technol 39(5):1309–1318. doi:10.1021/es0490018

Rao PSC, Jawitz JW, Enfield CG, Falta RW, Annable MD, Wood AL (2001) Technology integration for contaminated site remediation: clean-up goals and performance criteria. In: Groundwater quality: natural and enhanced restoration of groundwater pollutions. Proceedings of the groundwater quality 2001 conference, Sheffield, IAHS publications no. 275, pp 571–578

Raychoudhury T, Naja G, Ghoshal S (2010) Assessment of transport of two polyelectrolyte-stabilized zero-valent iron nanoparticles in porous media. J Contam Hydrol 118 (3–4):143–151. doi:10.1016/j.jconhyd.2010.09.005

Reddy KR, Karri MR (2008) Electrokinetic delivery of nanoiron amended with surfactant and cosolvent in contaminated soil. In: Proceedings of the international conference on waste engineering and management, Hong Kong, May 2008

Reddy KR, Darko-Kagy K, Cameselle C (2011) Electrokinetic-enhanced transport of lactate-modified nanoscale iron particles for degradation of dinitrotoluene in clayey soils. Sep Purif Technol 79(2):230–237

Rejeski D, Kuiken T, Polischuk P, Pauwels E (2014) Nanoremediation map. Project on emerging nanotechnologies. http://www.nanotechproject.org/inventories/remediation_map/. Accessed 5 Mar 2014

Rosales E, Loch JPG, Dias-Ferreira C (2014) Electro-osmotic transport of nano zero-valent iron in Boom Clay. Electrochim Acta 127:27–33. doi:10.1016/j.electacta.2014.01.164

Saleh N, Kim H-J, Phenrat T, Matyjaszewski K, Tilton RD, Lowry GV (2008) Ionic strength and composition affect the mobility of surface-modified Fe0 nanoparticles in water-saturated sand columns. Environ Sci Technol 42(9):3349–3355. doi:10.1021/es071936b

Schrick B, Hydutsky BW, Blough JL, Mallouk TE (2004) Delivery vehicles for zerovalent metal nanoparticles in soil and groundwater. Chem Mater 16(11):2187–2193

Su C, Puls RW, Krug TA, Watling MT, O'Hara SK, Quinn JW, Ruiz NE (2012) A two and half-year-performance evaluation of a field test on treatment of source zone tetrachloroethene and its chlorinated daughter products using emulsified zero valent iron nanoparticles. Water Res 46(16):5071–5084. doi:10.1016/j.watres.2012.06.051

Tosco T, Petrangeli Papini M, Cruz Viggi C, Sethi R (2014) Nanoscale zerovalent iron particles for groundwater remediation: a review. J Clean Prod 77:10–21. doi:10.1016/j.jclepro.2013.12.026

USEPA (2011a) Permeable reactive barriers, permeable treatment zones, and application of zero-valent iron: overview, technology innovation and field services division. USEPA, Washington, DC

USEPA (2011b) Fact sheet on selected sites using or testing nanoparticles for remediation. United States Environmental Protection Agency

Wang C-B, Zhang W-x (1997) Synthesizing nanoscale iron particles for rapid and complete dechlorination of TCE and PCBs. Environ Sci Technol 31(7):2154–2156

Yang GCC, Chang Y-I (2011) Integration of emulsified nanoiron injection with the electrokinetic process for remediation of trichloroethylene in saturated soil. Sep Purif Technol 79:278–284

Yang GCC, Tu H-C, Hung C-H (2007) Stability of nanoiron slurries and their transport in the subsurface environment. Sep Purif Technol 58:166–172

Yang GCC, Hung C-H, Tu H-C (2008) Electrokinetically enhanced removal and degradation of nitrate in the subsurface using nanosized Pd/Fe slurry. J Environ Sci Health A 43(8):945–951

Yuan S, Long H, Xie W, Liao P, Tong M (2012) Electrokinetic transport of CMC-stabilized Pd/Fe nanoparticles for the remediation of PCP-contaminated soil. Geoderma 185–186:18–25. doi:10.1016/j.geoderma.2012.03.028

Zhang W, Elliott DW (2006) Applications of iron nanoparticles for groundwater remediation. Remediat J 16(2):7–21

Zhang W, Wang C-B, Lien H-L (1998) Treatment of chlorinated organic contaminants with nanoscale bimetallic particles. Catal Today 40:387–395

Chapter 15
Feasibility Study of the Electrokinetic Remediation of a Mercury-Polluted Soil

Ana García-Rubio, María Villén-Guzmán, Francisco García-Herruzo, José M. Rodríguez-Maroto, Carlos Vereda-Alonso, César Gómez-Lahoz, and Juan Manuel Paz-García

15.1 Introduction

It is widely accepted that the electrokinetic remediation (EKR) requires the use of some enhancements to avoid the effects induced by the electrolysis reactions that frequently occur at the electrodes. In general, the main objectives of these enhancing techniques are to solubilize contaminants in soil keeping them in mobile states and to control the pH within a range of values that improve the remediation performance. One of the most typical enhancements used in electrochemical remediation is based on reagent addition.

Regarding the selection of the enhancement reagent, some interesting papers about the addition of different acids to the cathode have been published (Reddy and Chinthamreddy 2004; Al-Shahrani and Roberts 2005; Pazos et al. 2009; Villen-Guzman et al. 2014). Reddy and Chinthamreddy (2004) studied different purging solutions for the removal of Cr (VI), Ni (II), and Cd (II) from glacial till. The results highlighted the importance of an adequate selection of the enhancing reagent for heavy metals removal, which depends on the contaminant characteristics and the soil composition. Also, it should be taken into account that the in situ electrokinetic process involves a significant quantity of enhancing solution and, therefore this reagent should be nontoxic, effective, and economical. With respect to the acid-enhanced EKR, it is known that weak acids present some advantages such as: (1) organic acids are environmentally safe and biodegradable, (2) possess certain buffer capacities, (3) can behave as complexing agents, and (4) induce small

A. García-Rubio • M. Villén-Guzmán (✉) • F. García-Herruzo • J.M. Rodríguez-Maroto
C. Vereda-Alonso • C. Gómez-Lahoz
Chemical Engineering Department, University of Málaga, Málaga 29071, Spain
e-mail: mvillen@uma.es

J.M. Paz-García
Division of Solid Mechanics, University of Lund, Lund, Sweden

© Springer International Publishing Switzerland 2016
A.B. Ribeiro et al. (eds.), *Electrokinetics Across Disciplines and Continents*,
DOI 10.1007/978-3-319-20179-5_15

increases in the soil conductivity (Pazos et al. 2009; Yeung and Gu 2011). Regarding comparisons between strong and weak acids, Villen-Guzman et al. (2014) carried out a study of the results obtained for a real Pb-contaminated soil treated by EKR enhanced with a strong and a weak acid, nitric and acetic acid respectively, as neutralizing reagents. This study showed that the removal yield of Pb was larger for the acetic acid even when the electric charge circulated was larger for the nitric acid.

One of the main difficulties for a generalized use of the EKR technique is the lack of available tools to make reliable feasibility studies. The sequential extraction procedures (SEPs) is the most promising method for the prediction of the amount of contaminants that will be recovered, the environmental risks remaining after the remediation and the time and economical cost recovery for each particular contaminated site. The most used SEPs for heavy metals in soils are the Tessier (Tessier et al. 1979) and the BCR (Rauret et al. 2000). Several researches have used these techniques to obtain information about the mobility of the toxic metals after the contamination ageing (Liang et al. 2014a) or after a remediation process (Kirkelund et al. 2010; García-Rubio et al. 2011; Subirés-Muñoz et al. 2011; Liang et al. 2014b).

Usually, the feasibility of the electrokinetic design and operation requires the studies at laboratory scale together with mathematical models and analytical procedures. However, as it is reported by some authors, the data obtained at this scale cannot always be extrapolated and used for field applications. Although there are some reports regarding the scaling-up processes (Acar and Alshawabkeh 1996), it is clear that it is necessary to continue the study of this subject for a better understanding. Gent et al. (2004) studied the removal of chromium and cadmium from soil at laboratory and field conditions, concluding that the process was more efficient in the field due to the lower energy levels for higher extraction rates. In this study, the voltage drop at the electrolytes versus the drop across the electrodes was suggested as the main reason for these differences (Gent et al. 2004). Lopez-Vizcaino et al. presented the results for scaling-up of EKR for soils contaminated with phenanthrene and also found out important discrepancies between the two scales (López-Vizcaíno et al. 2014).

This chapter is focused on the study of electroremediation of heavy metals from a real soil. Some interesting aspects are discussed such as, the selection of an adequate enhancing agent, the implementation of a mathematical model for a better understanding of the experimental results, and the differences between two operating scales of EKR.

15.2 Case Study

The Almaden mining district, located in south-central Spain, constitutes one of the major concentrations of mercury on Earth. Before mining, it is estimated to contain about a third of the known global mercury resources (250,000 t) (Hernández

Fig. 15.1 Contaminated area

et al. 1999; García Herruzo et al. 2010). Exploitation of this deposit and the mining-metallurgic activity has lasted over 2000 years (from S.II b.C. to 2003). Today, once the exploitation and processing of ore has been stopped, this area could be regarded as one of the most mercury-polluted places on the planet (Higueras et al. 2006). As could be expected, the soils of the Almaden district are highly contaminated, with some zones displaying very high values of mercury concentrations, well above 1000 mg kg^{-1} (Higueras et al. 2003; Martínez-Coronado et al. 2011). Moreover, preliminary studies (Higueras et al. 2006) have revealed high levels of mercury absorption by plants. Higueras et al. (2012) measured the concentration of mercury in the leaves of the olive trees from Almaden area (Fig. 15.1), and this is above from olive trees from uncontaminated areas. Additionally, slightly high values in the hair from inhabitants of Almaden were found (Díez et al. 2011). These facts highlight a certain health risk and the need for soil remediation.

The geological dispersion is probably due to erosion of minerals such as cinnabar. Taking into account the low solubility of HgS (10^{-54} mol dm^{-3}) (Palmieri et al. 2006), it is predictable that this Hg is mainly associated with the less mobile fractions of the speciation analysis (Fernández-Martínez et al. 2006; Subirés-Muñoz et al. 2011). Hg with origin in anthropogenic dispersion is generated by the mining activity (transport and stockpiling, and metallurgical activities), and will probably be associated with the semi-mobile fractions (Fernández-Martínez et al. 2006).

Nowadays, it is well known that the risk assessment of heavy metals depends on the mobility and bioavailability of these metals, and not only on the total concentration (Tack and Verloo 1995; Biester and Nehrke 1997; Filgueiras et al. 2002; Issaro et al. 2009). In turn, the distribution of the metals depends on the association form in the solid phase to which they are bound (Ure et al. 1993; Filgueiras

Table 15.1 Properties of the soil sample

Moisture		<1 %	Concentration (mg kg^{-1})	
Bulk density (g cm^{-3})		1.13	SO_4^{2-}	92 ± 6
Porosity (%)		36	CO_3^{2-}	$68{,}000 \pm 4000$
Solid density (g cm^{-3})		2.24	Hg	6100 ± 800
Hydraulic conductivity (cm s^{-1})		2.1×10^{-6}	Fe	$45{,}900 \pm 1700$
Organic matter (%)		0.90	Zn	96 ± 10
pH (1/2.5 w/w)		8.5	Cu	31 ± 8
Particle size-distribution (%)	Sand	64.6	Pb	29 ± 5
	Silt	30	Cd	0.60
	Clay	5.4	As	17
CEC (meq/100 g)		14.0 ± 1.2	Ca	68,000
			Mg	7000

et al. 2002). The target of the most used fractionation analyses is not to identify nor quantify one or more individual chemical species (Templeton et al. 2000; Issaro et al. 2009), but rather a distribution of the total content of the element among different fractions according to the behavior of the toxic element upon the addition of an specific chemical agents (Tack and Verloo 1995; Filgueiras et al. 2002).

The soil samples used for this study were collected from a zone close to the metallurgical plant of Almadenejos in the Almaden mining district. Table 15.1 shows the properties of the soil sample.

The results indicate that this soil has low organic matter content (<1 %), is slightly basic, with a high buffer capacity and a medium-low cation exchange capacity (similar to the one of a kaolinite). The hydraulic conductivity value is lower than those corresponding to a sandy-loam soil (according to International Soil Science Classification), and therefore the EKR is probably more suitable than other soil remediation techniques, such as pump and treat or flushing.

As can be seen, there are very large concentrations of Fe, ($45{,}900 \pm 1700$) mg kg^{-1}, and of Hg, (6100 ± 800) mg kg^{-1}, whereas other metals of environmental relevance are Zn (96 mg kg^{-1}) and Cu (31 mg kg^{-1}). As was mentioned previously, besides the total concentration the mobility is also important for the risk assessment. In order to classify the chemical metal into different fractions according to their difficulty to be mobilized, the BCR fractionation was performed. As can be seen in Fig. 15.2, except for mercury, almost no metal is present in the weak acid soluble fraction probably because these metals have already been washed down by the rainwater. Around 40 % of mercury is present in the reducible fraction, probably associated with Fe/Mn oxides, and another 40 % is present in the oxidizable fraction, probably as sulfides, whereas the remainder is found under the residual form. Thus, an important environmental risk can be associated with the presence of this large amount of Hg, since significant amounts are present in the most mobile fractions.

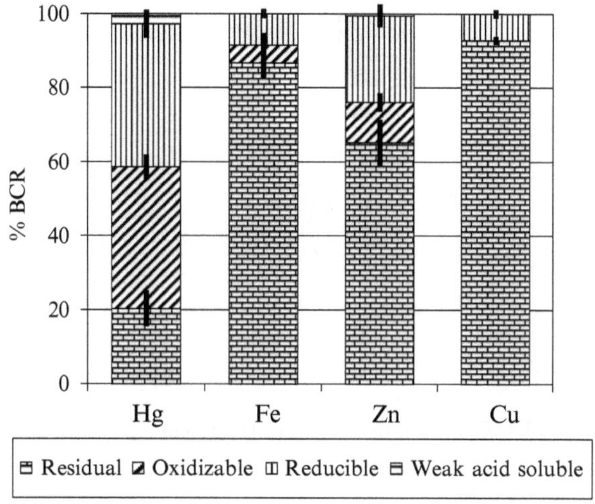

Fig. 15.2 Speciation of metals present in the original soil by BCR sequential extraction procedure

15.3 Electroremediation Experiments

Electrokinetic experiments, with two different enhancing reagents, were carried out at different scales to study the effectiveness of the scaling-up. It should be highlighted that all the assays were performed at a constant current density. Thus, the divergences that are usually obtained when using a constant voltage drop were avoided. Most authors working this way have found that when the assays are carried out at a constant electric potential drop, the current intensity could drop or increase depending on the soil conductivity which is changing during the experiments. The reliability of these constant-current experiments is always checked by the comparison of the results obtained for duplicated experiments.

Previous to the electrokinetic experiments, a kinetic study in batch reactor was performed to determine both the adequate extractant agent and its optimal concentration (Subirés-Muñoz et al. 2011). The best results to remove mercury from the soil were obtained for iodide. When the soil presents enough hydraulic conductivity, a possible in situ technique is the soil flushing (Roote 1997; Environmental Management and Inc 1997; García-Delgado et al. 1998). We have also explored the use of different extracting solutions for the soil remediation using soil flushing, and found that, when iodide was the extracting reagent, 35 % of the total mercury was removed from the soil by this technique (Subirés-Muñoz et al. 2011).

The experimental system used for the EKR essays is presented schematically in Fig. 15.3. Details of the experimental device are shown elsewhere (García-Rubio et al. 2011).

In both scales, all the mercury removed was recovered at the anode (30–35 % of mercury from initial soil); this indicates that the formation of negatively charged complexes (HgI_4^{2-} and HgI_3^-) is the main dissolved species of mercury. The possible mercury-iodide complex formation equilibria are:

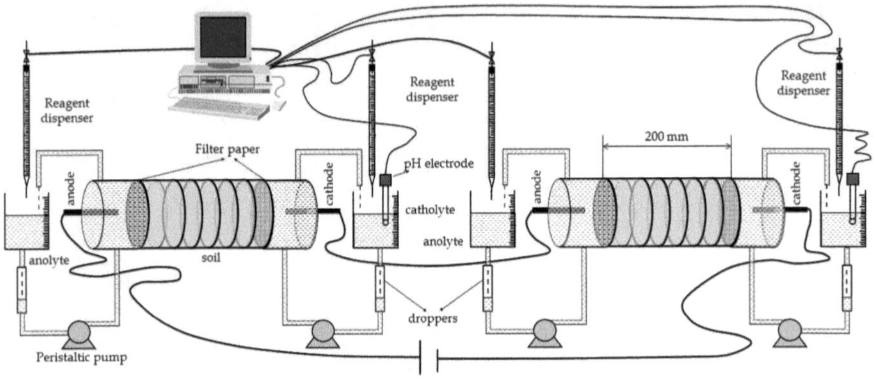

Fig. 15.3 Outline of the experimental system

$Hg^{2+} + I^- \rightleftarrows HgI^+$	**log** $K_1 = 12.87$
$Hg^{2+} + 2I^- \rightleftarrows HgI_2(aq)$	**log** $K_2 = 23.82$
$Hg^{2+} + 3I^- \rightleftarrows HgI_3^-$	**log** $K_3 = 27.60$
$Hg^{2+} + 4I^- \rightleftarrows HgI_4^{2-}$	**log** $K_4 = 29.83$

According to the formation constants, HgI_4^{2-} and HgI_3^- complexes would predominate; therefore, Hg transport by electromigration will take place mainly for these species.

For both scales, the BCR results before and after soil treatment show two main issues: first, the most important mercury removal takes place for the reducible fraction, and second, the weak acid soluble fraction of mercury increases with respect to the initial soil (Figs. 15.4 and 15.5).

The same results were observed when the soil was treated by flushing. It entails an increase of the environmental risk because the WAS fraction is the most mobile. This fraction is easier to be treated by another remediation technique, such as acid-enhanced EKR. Although, the kinetic studies showed that no mercury was extracted by the nitric acid from the initial soil, after iodide-enhanced EKR, it is possible to recover the metal present in the WAS fraction using this acid after iodide-enhanced EKR, that will be denominated AAI EKR.

Figure 15.6 shows the percentage of mercury in each fraction of the BCR in soil before and after AAI EKR experiments. In this figure, it can be observed that further removals can be obtained with this technique, with an important redistribution of the BCR fractions, that indicates that an important abatement of the environmental risks are achieved with respect to the single iodide-enhanced EKR.

In the laboratory scale with nitric acid experiments, most of the mercury recovered (25 % relative to the mercury content after treatment with iodide) was

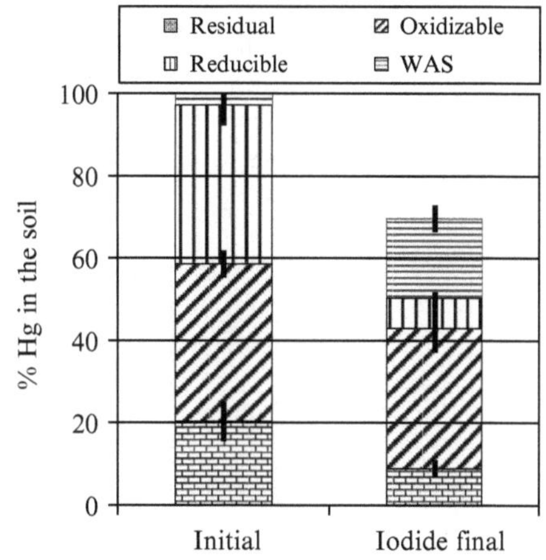

Fig. 15.4 Comparison of the BCR distribution of Hg in initial soil and soil after EKR laboratory scale with iodide

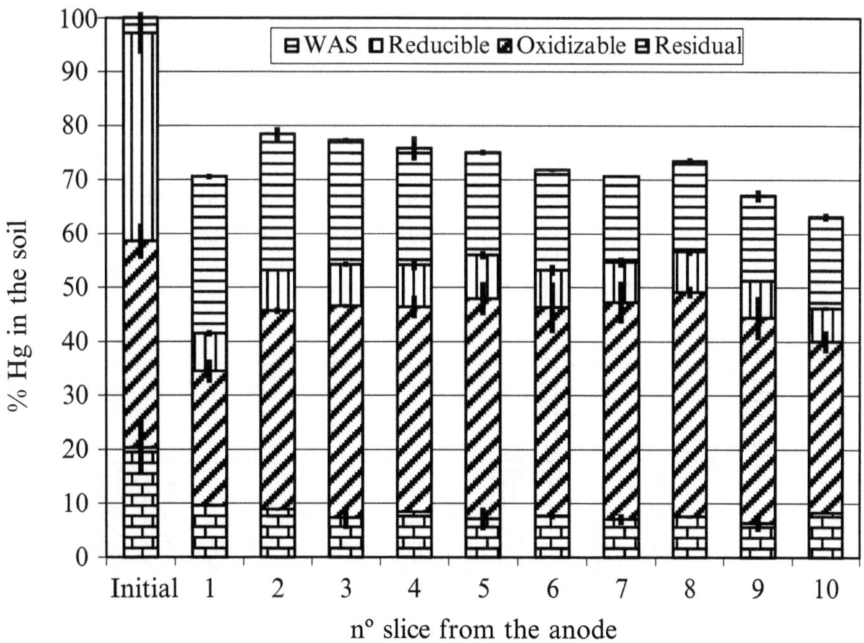

Fig. 15.5 Comparison of the BCR distribution of Hg in initial soil and soil after EKR semi-pilot scale with iodide

collected at the anode. Therefore, also for this acid-enhanced EKR, the iodide still present in the soil introduced during the iodide-enhanced EKR formed negatively charged complexes with mercury.

In the semi-pilot scale experiments, the percentage of mercury extracted was not more than 5 %, however, after this treatment, a change in the distribution of

Fig. 15.6 Comparison of the BCR distribution of Hg in initial soil, soil after EKR laboratory scale with iodide and acid after iodide (AAI) experiments

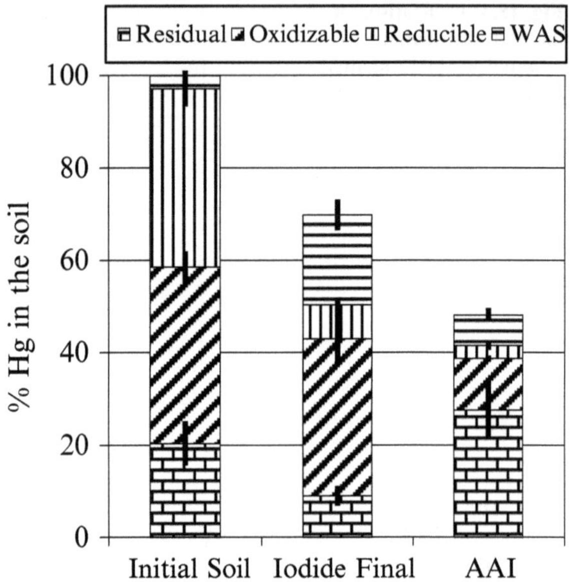

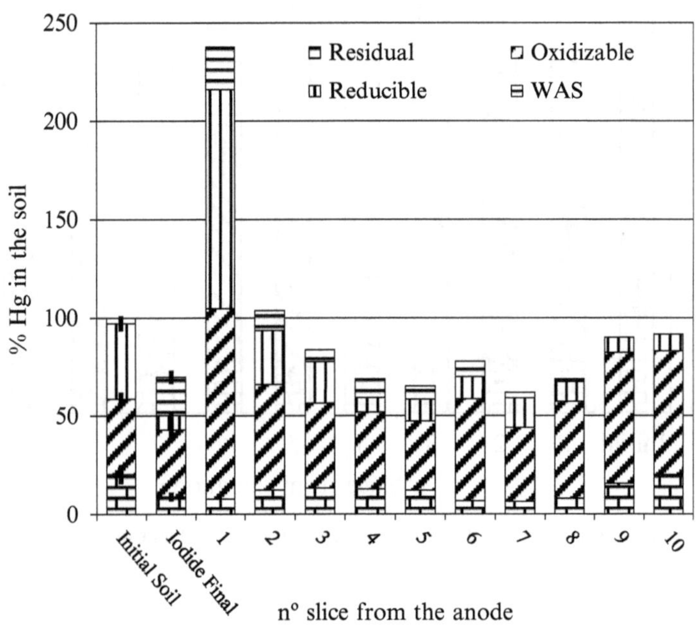

Fig. 15.7 Distribution of Hg in the soil for AAI EKR semi-pilot scale experiments

mercury in soil was observed: the mercury associated with the more mobile fraction decreased (Fig. 15.7). In addition, there was an accumulation of mercury in the soil section adjacent to the anode and a decrease of its concentration in the central sections; this confirms that the iodide present in the soil continued to form negatively charged complexes with mercury.

As can be seen from the different results, the conclusions about the phenomena taking place during the experiments are similar for both scales. In order to obtain further information about the scaling-up, Villen-Guzman et al. (2015) presented a simple model for the prediction of the energy requirements of different scales for a lead contaminated site, obtaining satisfactory results. In that study, it was defined a specific-energy parameter based on the amount of metal mobilized and the maximum that can be removed, which resulted to be scale independent. When this concept was applied to the case of Almaden soil, also similar values of specific energy were obtained for the two scales: $E_{Hg} = 590 \pm 130$ kJ/kg and 390 ± 90 kJ/kg for the lab and semi-pilot scales, respectively. These results indicated this specific-energy depends on the target metal to be recovered, and it is quite similar for mercury and the experimental conditions of the two scales essayed.

The specific-energy requirements for the Hg-contaminated site were much lower than those obtained for the Pb-contaminated site. These differences could be due to the fact that, for the Pb-contaminated case, a large fraction of the electric current is dedicated to the transport of protons, which present a large concentration, as expected for an acid-enhanced EKR. Moreover, an important amount of protons are dedicated to the dissolution and mobilization of other alkaline species different from the target contaminant. Instead, for the iodide-enhanced remediation, the H^+ and OH^- ions generated at the electrode reactions are neutralized and consequently, a larger fraction of the current is dedicated to the transport of iodide and the iodide complexes of mercury, making this kind of remediation more energy efficient.

15.4 A Tool for Predictions: Mathematical Model

A mathematical model has been developed to predict the results that would be obtained in future essays or field scale, following the one described by C. Vereda (Reddy and Cameselle 2009). The transport equation, when the porosity and water content in the soil remains constant during the electrokinetic treatment is given by:

$$\frac{dC_k}{dt} = -\nabla J_k + R_k$$

where J_k, C_k, and R_k are the ionic flux, the concentration, and the generation, respectively, per volume unit of the kth ion.

The ionic flux is due to the electromigration, electroosmotic phenomena, and diffusion. No advection is included because in laboratory tests advection was avoided. Therefore, assuming that for the transport processes only the first two phenomena are significant, the ion flux is given by:

$$J_k = -z_k C_k u_k^* \Phi_e - k_e C_k \Phi_e$$

where the terms on the right correspond, respectively, to the flux due to electromigration and the electroosmotic flux. Here, z_k is the charge of the kth ion, u_k^* is its electrochemical mobility, k_e is the electroosmotic permeability of the soil, and Φ_e is the electric potential gradient.

The analytical simultaneous solution of these equations is difficult to obtain; therefore, it is obtained numerically. To achieve this, the model is divided into two parts: one for the equations of transport of ions (∇J_k) and the second part for the equations for equilibria of the reactions (R_k) (Wilson et al. 1995).

(1) *Transport equations*—The variation of the concentration of each ion, due to electromigration and electroosmotic flow, in each one of the sections into which the soil column is divided, is:

$$\Delta V_W \frac{dC_{ik}}{dt} = QM_{ik} - QM_{i-1k} + Ak^*_{osm}\frac{\Delta V}{L}(C_{i-1k} - C_{ik}) \quad \text{for } k = 1, 2, 3, \ldots$$

where ΔV_W is the volume of water in the ith cell, A is the area of the cross section of the column, k^*_{osm} is the electroosmotic effective permeability, ΔV is the electrical potential drop through the soil, and C_{ik} is the concentration of the kth ion in the ith cell. QM_{ik} is the molar flow due to the electromigration of the kth ion between the cells i and $(i+1)$:

$$QM_{ik} = \frac{-I}{z_k F}\frac{\lambda_k[C_{ik}S(z_k) + C_{i+1k}S(-z_k)]}{K_i} \quad \text{for } 0 \le i < N$$

where F is the Faraday constant (96,500 C/mol e^-). The second fraction is the transport number for the kth ion between the cells i and $i+1$ (t_k), where the denominator is the conductivity K_i at the boundary between this two cells.

The boundary conditions in this study are given in the compartments of the electrodes. At the cathode, water reduction takes place; therefore, the generation of hydroxyl ions must be taken into account in the mass balance in this compartment ($i = N+1$).

Cathode $\rightarrow i = N+1$

$$V_{N+1}\frac{dC_{N+1k}}{dt} = -QM_{Nk} + Ak^*_{osm}\frac{\Delta V}{L}C_{Nk} \quad \text{for } k \neq OH^-$$

$$V_{N+1}\frac{dC_{N+1k}}{dt} = -QM_{Nk} + Ak^*_{osm}\frac{\Delta V}{L}C_{Nk} + \frac{I}{F}(1 - nr) \quad \text{for} k = OH^-$$

$$V_{N+1}\frac{dC_{N+1k}}{dt} = -QM_{Nk} + Ak^*_{osm}\frac{\Delta V}{L}C_{Nk} + nr\frac{I}{z_k F} \quad \text{for } k = \text{ anions of the acid used in the enhanced EKR}$$

where *I*/*F* is the generation of hydroxyl ions, and *nr* is the ratio of hydroxyl ions neutralized.

Analogously, at the anode, the oxidation of water takes place, and therefore the generation of protons should be taken into account in the mass balance in that compartment ($i = 0$).

Anode $\rightarrow i = 0$

$$V_0 \frac{dC_{0k}}{dt} = QM_{0k} + Ak^*_{osm} \frac{\Delta V}{L}(-C_{0k}) \qquad \text{for } k \neq H^+$$

$$V_0 \frac{dC_{0k}}{dt} = QM_{0k} + Ak^*_{osm} \frac{\Delta V}{L}(-C_{0k}) + \frac{I}{F}(1 - nr') \qquad \text{for } k = H^+$$

$$V_0 \frac{dC_{0k}}{dt} = QM_{0k} + Ak^*_{osm} \frac{\Delta V}{L}(-C_{0k}) + nr' \frac{I}{z_k F} \qquad \text{for } k = \text{cations of the base used in the enhanced EKR}$$

where *I*/*F* is the generation of protons, and *nr*' is the ratio of protons neutralized.

Finally, concentrations in the next increment of time ($t = t + \Delta t$) are calculated as follows:

$$C_{ik}(t + \Delta t) = C_{ik}(t) + \frac{dC_{ik}}{dt}\Big|_t \Delta t$$

(2) *Chemical Equilibria*—The simplest model considers that mercury is present in the soil as mercury hydroxide ($Hg(OH)_2$). Also the presence of calcium carbonate, which acts as a buffer, is included because the soil does not undergo changes in the pH values after electrokinetic treatment. Therefore, it is taken into account the balance of water, formation of complex equilibria, the equilibrium carbonate-bicarbonate (since the pH value is above 9 the bicarbonate-carbonic equilibrium is neglected) and the balance of the possible precipitation reactions (calcium carbonate, mercury hydroxide, and mercury iodide). All these equations can be arranged in a matrix for the solution through the numerical Newton–Raphson method (Press et al. 1992). The arrangement of variables and equations are shown in Fig. 15.8.

This system is resolved as follows:

(a) Concentrations in the aqueous phase of the C species are calculated from the concentrations of the master species (X_a, independent variables), the stoichiometric coefficients (E_a), and the equilibrium constants (K_a).

$$\log(C_a) = \log K_a + E_a \cdot \log(X_a)$$

(b) Mass balances are obtained when the product stoichiometric coefficients by the corresponding concentration in aqueous phase is added in columns. Also the independent variables of each precipitate must be multiplied by its stoichiometric coefficient. Instead of the mass balance for protons, the system is defined by the

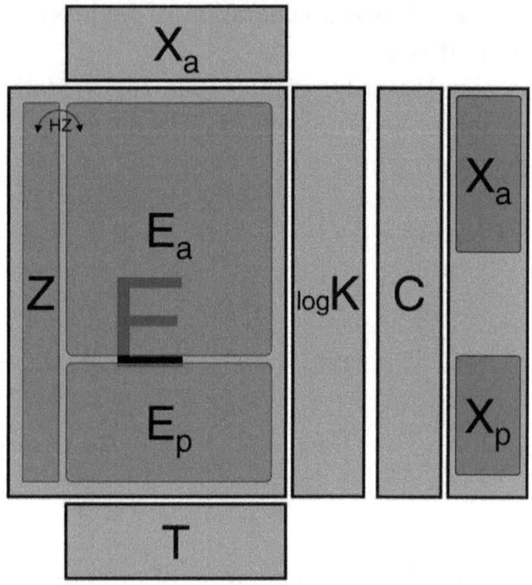

Fig. 15.8 Outline of the arrangement of equations and variables of the model

introduction of the electroneutrality condition, which provides a more stable algorithm. The Newton–Raphson method has to reset the residue of each balance by changing of the independent variables X. In matrix form, it is

$$E_{aHZ}^{T} \cdot C_{a} + E_{pHZ}^{T} \cdot X_{p} - T = 0 = f_{vec}$$

where E_{aHZ}^{T} and E_{pHZ}^{T} are the transposed matrixes of E_{aHZ} and E_{pHZ}, E_{aHZ} and E_{pHZ} are E_{a} and E_{p} changing protons column with the electrical charge column.

(c) Finally, the saturation indexes of each precipitated species are calculated. When a precipitated species is present, the corresponding value of saturation index is appended to the matrix of the residues for the Newton–Raphson method. The sum over the row of the precipitated species of the product of the stoichiometric coefficients corresponding to the precipitates ($\boldsymbol{E_p}$) and the logarithm of the corresponding values of the concentrations of the "master species" minus the logarithm of the corresponding solubility product constant. The saturation coefficient must be zero if the precipitate is present and negative if it is not.

$$\mathrm{SI} = E_{p} \times \log(X_{a}) - \log K_{p} = f_{vec} \begin{cases} = 0 & \text{if there is precipitate} \\ < 0 & \text{if there isn't precipitate} \end{cases}$$

This mathematical model has been done to predict the results of semi-pilot experiments. The initial Hg concentration to model is about 42 mmol Hg L^{-1}. This concentration is obtained from the 31.4 % of the initial mercury in the soil, treating it as mercury mobile in the soil because this is the recovered mercury in laboratory tests. The Fig. 15.9. shows the mercury recovered experimentally (dots) and using the model (curve) versus time. As can be seen, the model predicts satisfactorily the experimental results.

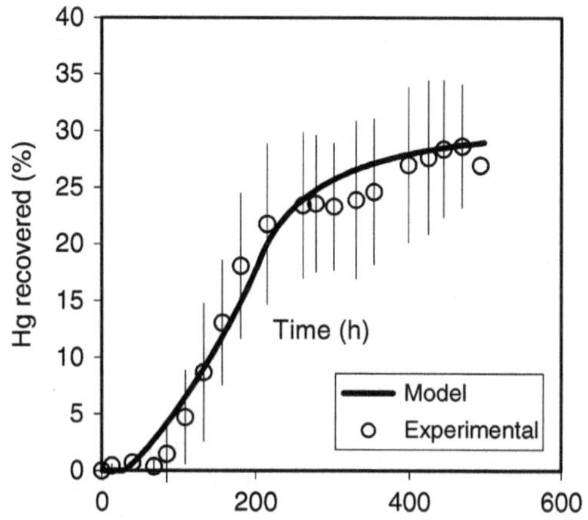

Fig. 15.9 Hg recovered versus time

Also, the results obtained from lab essays, such as acid and base consumption at the electrodes, the variation of the potential and the concentration of mercury in the sections of the soil column when the test is ended, were very similar to the model results.

15.5 Conclusions

The results are similar for laboratory and semi-pilot scale. In both cases, for those experiments using acid after iodide-enhanced experiments (AAI EKR), 35 % of mercury in the soil was recovered. An important redistribution of the remaining mercury among the different BCR fractions is also observed, which indicates that the AAI achieve an important decrease of the mobile mercury with respect to the use of the iodide-enhanced EKR alone. The scaling-up was also studied applying a simple model for the prediction of the energy requirements. The specific-energy parameter defined is based on the amount of metal mobilized and the maximum recoverable. This amount was obtained from a sequential extraction procedure (BCR). Similar values of this specific-energy were obtained for the two scales studied, so it was found to be scale independent. Therefore, it can be used for the scaling-up calculations.

In this study, the importance of using simple analytical characterization techniques together with mathematical models were highlighted. These tools are useful for the prediction of the amount of metals recovered, the energy consumption at a full-scale EKR, and the risk abatement achieved.

Acknowledgements Authors acknowledge the financial support provided by the Spanish Ministry of Innovation and the FEDER fund of the EU through the Research Project ERHMES, CTM2012-16824, and the UE project Electroacross IRSES-GA-2010 269289. Villen-Guzman also acknowledges the FPU grant obtained from the Spanish Ministry of Education. Paz-Garcia acknowledges the financial support from the International Campus of Excellence (ICE) Andalucía Tech.

References

Acar YB, Alshawabkeh AN (1996) Electrokinetic remediation. I: pilot-scale tests with lead-spiked kaolinite. J Geotech Eng 122:173–185

Al-Shahrani SS, Roberts EPL (2005) Electrokinetic removal of caesium from kaolin. J Hazard Mater 122:91–101. doi:10.1016/j.jhazmat.2005.03.018

Biester H, Nehrke G (1997) Quantification of mercury in soils and sediments—acid digestion versus pyrolysis. Fresenius J Anal Chem 358:446–452

Díez S, Esbrí JM, Tobias A, Higueras P, Martínez-Coronado A (2011) Determinants of exposure to mercury in hair from inhabitants of the largest mercury mine in the world. Chemosphere 84:571–577. doi:10.1016/j.chemosphere.2011.03.065

PCR Environmental Management, Inc (1997) Recent developments for in situ treatment of metal contaminated soils

Fernández-Martínez R, Loredo J, Ordóñez A, Rucandio MI (2006) Physicochemical characterization and mercury speciation of particle-size soil fractions from an abandoned mining area in Mieres, Asturias (Spain). Environ Pollut 142:217–226. doi:10.1016/j.envpol.2005.10.034

Filgueiras AV, Lavilla I, Bendicho C (2002) Chemical sequential extraction for metal partitioning in environmental solid samples. J Environ Monit 4:823–857. doi:10.1039/b207574c

García Herruzo F, García Rubio A, Gómez Lahoz C, Vereda Alonso C, Rodríguez Maroto JM (2010) Mercury: actual situation, problems and solutions. Ing Quimica Spain 42:84–91

García-Delgado RA, Rodríguez-Maroto JM, Gómez-Lahoz C, Vereda-Alonso C, García-Herruzo F (1998) Soil flushing with EDTA solutions: a model for channeled flow. Sep Sci Technol 33:867–886

García-Rubio A, Rodríguez-Maroto JM, Gómez-Lahoz C, García-Herruzo F, Vereda-Alonso C (2011) Electrokinetic remediation: the use of mercury speciation for feasibility studies applied to a contaminated soil from Almadén. Electrochim Acta 56:9303–9310

Gent DB, Bricka RM, Alshawabkeh AN, Larson SL, Fabian G, Granade S (2004) Bench- and field-scale evaluation of chromium and cadmium extraction by electrokinetics. J Hazard Mater 110:53–62. doi:10.1016/j.jhazmat.2004.02.036

Hernández A, Jébrak M, Higueras P, Oyarzun R, Morata D, Munhá J (1999) The Almaden mercury mining district, Spain. Miner Deposita 34:539–548. doi:10.1007/s001260050219

Higueras P, Oyarzun R, Biester H, Lillo J, Lorenzo S (2003) A first insight into mercury distribution and speciation in soils from the Almadén mining district, Spain. J Geochem Explor 80:95–104. doi:10.1016/S0375-6742(03)00185-7

Higueras P, Oyarzun R, Lillo J, Sánchez-Hernández JC, Molina JA, Esbrí JM, Lorenzo S (2006) The Almadén district (Spain): anatomy of one of the world's largest Hg-contaminated sites. Sci Total Environ 356:112–124. doi:10.1016/j.scitotenv.2005.04.042

Higueras P, Amorós JA, Esbrí JM, García-Navarro FJ, Pérez de los Reyes C, Moreno G (2012) Time and space variations in mercury and other trace element contents in olive tree leaves from the Almadén Hg-mining district. J Geochem Explor 123:143–151. doi: 10.1016/j.gexplo.2012.04.012

Issaro N, Abi-Ghanem C, Bermond A (2009) Fractionation studies of mercury in soils and sediments: a review of the chemical reagents used for mercury extraction. Anal Chim Acta 631:1–12. doi:10.1016/j.aca.2008.10.020

Kirkelund GM, Ottosen LM, Villumsen A (2010) Investigations of Cu, Pb and Zn partitioning by sequential extraction in harbour sediments after electrodialytic remediation. Chemosphere 79:997–1002. doi:10.1016/j.chemosphere.2010.03.015

Liang S, Guan D-X, Ren J-H, Zhang M, Luo J, Ma LQ (2014a) Effect of aging on arsenic and lead fractionation and availability in soils: coupling sequential extractions with diffusive gradients in thin-films technique. J Hazard Mater 273:272–279. doi:10.1016/j.jhazmat.2014.03.024

Liang Y, Cao X, Zhao L, Arellano E (2014b) Biochar- and phosphate-induced immobilization of heavy metals in contaminated soil and water: implication on simultaneous remediation of contaminated soil and groundwater. Environ Sci Pollut Res 21:4665–4674. doi:10.1007/s11356-013-2423-1

López-Vizcaíno R, Alonso J, Cañizares P, León MJ, Navarro V, Rodrigo MA, Sáez C (2014) Electroremediation of a natural soil polluted with phenanthrene in a pilot plant. J Hazard Mater 265:142–150. doi:10.1016/j.jhazmat.2013.11.048

Martínez-Coronado A, Oyarzun R, Esbrí JM, Llanos W, Higueras P (2011) Sampling high to extremely high Hg concentrations at the Cerco de Almadenejos, Almadén mining district (Spain): the old metallurgical precinct (1794 to 1861AD) and surrounding areas. J Geochem Explor 109:70–77. doi:10.1016/j.gexplo.2010.04.007

Palmieri HEL, Nalini HA Jr, Leonel LV, Windmöller CC, Santos RC, de Brito W (2006) Quantification and speciation of mercury in soils from the Tripuí Ecological Station, Minas Gerais, Brazil. Sci Total Environ 368:69–78. doi:10.1016/j.scitotenv.2005.09.085

Pazos M, Alcántara MT, Cameselle C, Sanromán MA (2009) Evaluation of electrokinetic technique for industrial waste decontamination. Sep Sci Technol 44:2304–2321. doi:10.1080/01496390902979867

Press WH, Teukolsky SA, Vetterling WT, Flannery BP (1992) Numerical recipes in C: the art of scientific computing. Cambridge University Press, New York

Rauret G, Lopez-Sanchez J-F, Sahuquillo A, Barahona E, Lachica M, Ure AM, Davidson CM, Gomez A, Luck D, Bacon J, Yli-Halla M, Muntau H, Quevauviller P (2000) Application of a modified BCR sequential extraction (three-step) procedure for the determination of extractable trace metal contents in a sewage sludge amended soil reference material (CRM 483), complemented by a three-year stability study of acetic acid and EDTA extractable metal content. J Environ Monit 2:228–233. doi:10.1039/b001496f

Reddy KR, Cameselle C (2009) Electrochemical remediation technologies for polluted soils, sediments and groundwater. Wiley, New Jersey

Reddy KR, Chinthamreddy S (2004) Enhanced electrokinetic remediation of heavy metals in glacial till soils using different electrolyte solutions. J Environ Eng 130:442–455

Roote DS (1997) In situ flushing. Ground-water Remediation Technologies Analysis Center, Pittsburgh

Subirés-Muñoz JD, García-Rubio A, Vereda-Alonso C, Gómez-Lahoz C, Rodríguez-Maroto JM, García-Herruzo F, Paz-García JM (2011) Feasibility study of the use of different extractant agents in the remediation of a mercury contaminated soil from Almaden. Sep Purif Technol 79:151–156. doi:10.1016/j.seppur.2011.01.032

Tack FMG, Verloo MG (1995) Chemical speciation and fractionation in soil and sediment heavy metal analysis: a review. Int J Environ Anal Chem 59:225–238. doi:10.1080/03067319508041330

Templeton DM, Ariese F, Cornelis R, Danielsson L-G, Muntau H, Van Leeuwen HP, Łobiński R (2000) Guidelines for terms related to chemical speciation and fractionation of elements. Definitions, structural aspects, and methodological approaches (IUPAC recommendations 2000). Pure Appl Chem 72:1453–1470

Tessier A, Campbell PGC, Blsson M (1979) Sequential extraction procedure for the speciation of particulate trace metals. Anal Chem 51:844–851

Ure AM, Quevauviller P, Muntau H, Griepink B (1993) Speciation of heavy metals in soils and sediments. An account of the improvement and harmonization of extraction techniques undertaken under the auspices of the BCR of the Commission of the European Communities. Int J Environ Anal Chem 51:135–151. doi:10.1080/03067319308027619

Villen-Guzman M, Paz-Garcia JM, Rodriguez-Maroto JM, Gomez-Lahoz C, Garcia-Herruzo F (2014) Acid enhanced electrokinetic remediation of a contaminated soil using constant current

density: strong vs. weak acid. Sep Sci Technol 49:1461–1468. doi:10.1080/01496395.2014.898306
Villen-Guzman M, Paz-Garcia JM, Rodriguez-Maroto JM, Garcia-Herruzo F, Amaya-Santos G, Gomez-Lahoz C, Vereda-Alonso C (2015) Scaling-up the acid-enhanced electrokinetic remediation of a real contaminated soil. Electrochim Acta. doi:10.1016/j.electacta.2015.02.067
Wilson DJ, Rodriguez-Maroto JM, Gomez-Lahoz C (1995) Electrokinetic remediation. II. Amphoteric metals and enhancement with a weak acid. Sep Sci Technol 30:3111–3128
Yeung AT, Gu Y-Y (2011) A review on techniques to enhance electrochemical remediation of contaminated soils. J Hazard Mater 195:11–29. doi:10.1016/j.jhazmat.2011.08.047

Part V
Coupling Electrokinetic Process with Other Technologies to Enhance Performance and Sustainability

Chapter 16
Phytoremediation Coupled to Electrochemical Process for Arsenic Removal from Soil

Paula R. Guedes, Nazaré Couto, Alexandra B. Ribeiro, and Dong-Mei Zhou

16.1 Introduction

16.1.1 Background

Mining industries produces tons of wastes and tailings that needs to be disposed of or, alternatively, gets deposited at the surface. This represents pollution sources that may directly affect loss of cultivated land, forest or grazing land, and the overall loss of production (Wong 2003) and indirectly promote water pollution and siltation of rivers that will eventually lead to the loss of biodiversity, amenity, and economic wealth (Bradshaw 1993). One intrinsic problem of mining contamination is that pollution may persist for hundreds of years after the cessation of mining operations (Nagajyoti et al. 2010).

Arsenic (As) is ubiquitous in the environment and highly toxic to all life forms. It is a crystalline "metalloid," a natural element with features intermediate between metals and nonmetals, occurs naturally as an element, ranks as the 20th most occurring trace element in the earth's crust, 14th in seawater, and 12th in the human body (Mandal and Suzuki 2002). Soil deterioration by As contamination

---

P.R. Guedes (✉) • N. Couto
CENSE, Departamento de Ciências e Engenharia do Ambiente, Faculdade de Ciências e Tecnologia, Universidade Nova de Lisboa, Caparica 2829-516, Portugal

Key Laboratory of Soil Environment and Pollution Remediation, Institute of Soil Science, Chinese Academy of Sciences, Nanjing 210008, China
e-mail: p.guedes@campus.fct.unl.pt

A.B. Ribeiro
CENSE, Departamento de Ciências e Engenharia do Ambiente, Faculdade de Ciências e Tecnologia, Universidade Nova de Lisboa, Caparica 2829-516, Portugal

D.-M. Zhou
Key Laboratory of Soil Environment and Pollution Remediation, Institute of Soil Science, Chinese Academy of Sciences, Nanjing 210008, China

© Springer International Publishing Switzerland 2016
A.B. Ribeiro et al. (eds.), *Electrokinetics Across Disciplines and Continents*,
DOI 10.1007/978-3-319-20179-5_16

may result in high exposure for human body through soil ingestion and the food chain. Therefore, there is the need to remediate soil contaminated by As.

EK process is based on the application of a low intensity DC or AC current aiming to improve the mobilization of contaminants. When coupled to phytoremediation, the electric field may enhance the removal of the contaminants by the plant by increasing its bioavailability (enhancement of solubilization and/or selective mobilization) by desorption and transport of metals or metalloids, thus facilitating phytoextraction process. Also, plant simultaneously rehabilitates soil properties changed by the presence of EK, recovery of soil properties, and improvement of its structure.

This chapter aims to give an overview of both and highlight some research perspectives on EK remediation of As alone and coupled with other technologies, specially focusing on EK-enhanced phytoremediation.

16.1.2 Soil Arsenic Contamination

Some metals or metalloids represent severe danger for health (Vamerali et al. 2010). Arsenic (As) is cited as the most hazardous substance by the USA Agency for Toxic Substance and Disease Registry (2007) and is considered one of the priority pollutants by the US EPA and the EU (Filella et al. 2002). Arsenic is toxic and bioaccumulative with harmful environmental effects even when in low levels and when ingested for prolonged periods causes extensive health problems, class-one carcinogen (National Research Council 2001), leading to ultimate untimely death.

Environmental contamination by As has been reported worldwide (Singh et al. 2015). The sources of As include both natural (i.e., through dissolution of As compounds adsorbed onto pyrite ores into the water by geochemical factors) and anthropogenic (i.e., through use of pesticides, animal manures and phosphate fertilizers, semi-conductor industries, mining and smelting, industrial processes, coal combustion, timber preservatives, etc.) (Mondal et al. 2006; Bundschuh et al. 2011).

According to the U.S. Environmental Protection Agency, the permissible limit of arsenic in soil is 24 mg/kg. Arsenic background concentration can significantly differ depending on soil conditions but its average concentration in European topsoil is estimated at 7.0 mg/kg (Stafilov et al. 2010).

One of the most important anthropogenic sources of As in the environment is the mining and refining of metals with grinding and concentrating ores, and disposal of tailings. Therefore, special attention has been paid to the environmental management of As-bearing mining sites throughout the world (Kwon et al. 2012).

China has large amount of As reserves. The known arsenic reserves were reported to be 3977 kt, and 2796 kt preserved reserves, of which 87.1 % existed in paragenetic or associated ores up to the end of 2003 (Xiao et al. 2008). In 2011, China was the top producer of white arsenic with almost 50 % world share, followed by Chile, Peru, and Morocco (U.S. Geological Survey 2012). The sum of arsenic reserves located in Guangxi, Yunnan, and Hunan Provinces represents

Table 16.1 Common As chemical species found in natural systems (adapted from Wilson et al. 2010)

Specie	Chemical formula
Arsenopyrite	FeAsS
Orpiment	As_2S_3
Realgar	AsS
Arsenic acid	$AsO(OH)_3$ (or H_3AsO_4)
Dihydrogenarsenate	$AsO_2(OH)_2^-$ (or $H_2AsO_4^-$)
Monohydrogenarsenate	AsO_3OH^{2-} (or $HAsO_4^{2-}$)
Arsenous acid	$As(OH)_3$ (or H_3AsO_3)
Arsenite	$AsO(OH)_2^-$ (or $H_2AsO_3^-$)
	$HAsO_3^{2-}$
Arsine	AsH_3
Trimethylarsine	$As(CH_3)_3$
Monomethylarsonic acid (MMAA)	$CH_3AsO(OH)_2$
Dimethylarsinic acid (DMAA)	$(CH_3)_2As(O)OH$
Dimethylarsine	$(CH_3)_3As$

more than 60 % of the total arsenic reserves explored in China. As reported, up to 87 % of reserves are believed to be present in sulfide ores which are paragenetic or associated with transition metals ores.

Different species of As present in the biosphere can be seen in Table 16.1. Arsenic occurs as a major constituent in more than 200 minerals, including elemental As, arsenides, sulfides, oxides, and arsenates (Smedley and Kinniburgh 2002). However, most of those arsenic-containing minerals are rare in nature and generally found as sulfides associated with Au, Cu, Pb, Zn, Sn, Ni, and Co in ore zones. The most dominant minerals existing in environment are arsenopyrite (FeAsS), realgar (AsS), and orpiment (As_2S_3). In terms of speciation in natural environment, arsenic is present as As(V) (arsenate) in oxidizing environment, while As(III) (arsenite) is the dominate form in reducing environments (Smedley and Kinniburgh 2002). As(V), an analog of phosphate, has moderate toxicity as it is a potent inhibitor of oxidative phosphorylation. Arsenate is less mobile and therefore less bioavailable as it more readily gets attached to inorganic surfaces, prominently Fe(III) (Akter et al. 2005). It can be present as various protonated forms depending on pH: H_3AsO_4, $H_2AsO_4^-$, $HAsO_4^{2-}$, and AsO_4^{3-}. As(III) has the highest toxicity as it has high affinity to bind with sulfhydryl groups of proteins disrupting enzymes involved in cellular metabolism. It is more mobile, and quickly comes into aqueous solvents which makes it more bioavailable (Akter et al. 2005).

Arsenite was found under long-term anaerobic environment only (Smedley and Kinniburgh 2002). Above 80 % of arsenic exists as arsenate rather than as arsenite in the contaminated soil. Conventional off-site remediation technologies of soil replacement, solidification/stabilization, and acid washing have been used to remediate As-contaminated soils (Tokunaga and Hakuta 2002) which were unsuccessful due to high cost in labor and operation for ex situ technologies. Consequently, a need exists for an in situ effective technique to remediate arsenic contamination that heavily impacts the subsurface environment. Besides, most arsenic still remains in the environment and there is always a risk of leaching caused by changes in the environmental condition (Wang et al. 2007). As a consequence, in situ technologies for As removal for soil are highly desired.

16.1.3 Electrokinetic Remediation

Electrokinetic (EK) remediation is a promising innovative technology for decontamination in low permeability media (Table 16.2). EK technique is based on the action of an electric field generated between inserted electrodes in the medium by applying a direct current or a constant voltage. Removal of contaminants by EK is accomplished by the mechanisms of electrolysis of water, electroosmosis, electrophoresis, and electromigration (Acar and Alshawabkeh 1993; Acar et al. 1993). During EK process, the contaminant migration in soil is simultaneously controlled by mechanisms of sorption/desorption, precipitation, and dissolution (Yeung 2006). Factors affecting contaminant extractability from soils in EK process include soil type (Saichek and Reddy 2003), contaminant type and concentration (O'Connor et al. 2003), zeta potential of soil (Yeung and Hsu 2005), electrode spacing (Alshawabkeh et al. 1999) and enhancement techniques (Reddy and Chinthamreddy 2003a; Sawada et al. 2003) which were discussed and reviewed in detail by Yeung (2006).

EK process has already been demonstrated to be successful and cost-effective in removing a wide variety of heavy metals in many bench- and field-scale studies

Table 16.2 Advantages and limitations of electrokinetic remediation (adapted from Virkutyte et al. 2002)

Advantages	Limitations
In situ and ex situ technology	Limited by the solubility of the contaminant and the desorption of contaminants from the soil matrix
Simultaneously treats inorganic and organic compounds in porous media	Acidic conditions and corrosion of the anode may create difficulties in in situ remediation
Use of the pH shift produced by the electrolysis of the water to effectively desorb contaminating ions	Precipitation of species close to the electrode is an impediment to the process
Potentially effective in both saturated and unsaturated soils	Necessity to apply enhancing solution
Effective method for inducing movement of water, ions, and colloids through fine-grained materials	The application of higher voltages to the soil, process efficiency decreases due to the increased temperature
Competitive in cost and remediation effectiveness compared	Removal efficiency is significantly reduced if soil contains carbonates and hematite, as well as large rocks or grave

(Acar and Alshawabkeh 1993; Virkutyte et al. 2002). The process has also been applied to different matrices for the remediation of both organic and inorganic contamination (Ribeiro et al. 2000; Ribeiro et al. 2005; Ottosen et al. 2007; Wang et al. 2007; Lima et al. 2008; Alcántara et al. 2010; Mateus et al. 2010; Ribeiro et al. 2011; Guedes et al. 2015).

Mention must be made that EK process is mainly to remove and concentrate pollutants in a small portion of soil or in reservoir solution. Being an optimum technology is highly needed to enhance EK remediation efficiency (Yuan and Chiang 2007).

Various enhancement techniques, e.g., careful management of the pH within the reservoir (Reddy and Chinthamreddy 2003b; Saichek and Reddy 2003; Zhou et al. 2004), addition of hydroxyl ion membranes, chelating agent, and enhancing reagent in the cathode reservoir (Reddy et al. 2004; Kim et al. 2005b; Yuan and Chiang 2008) were proposed to improve effectiveness in EK process. Other parameters like the potential gradient and processing time also affect EK performance. Still, EK depends on the characteristics of soil texture and contaminants.

As(V) dissolves under alkaline conditions but not under acidic conditions, whereas As(III) behaves in the opposite way (Alam and Tokunaga 2006). Therefore, depending on the dominant As species, both acids and bases can be used to mobilize As (Lee et al. 2007; Baek et al. 2009; Yang et al. 2009). Catholyte conditioning with 0.1 M nitric acid showed removal efficiency for As of 62 % after 28 days, at 4 mA/cm of current density (Baek et al. 2009). When the dominant species is As(V) (under oxidized conditions), NaOH extraction usually gives good results. The impact of NaOH on As mobility is based not only on the ability of OH^- to displace As oxyanions but also on the weaker re-adsorption of anions under alkaline conditions (Jackson and Miller 2000; Jang et al. 2007). NaOH has been used for selective extraction of anions, which is an advantage when mobilization of toxic heavy metal cations is not desired, for example, extraction of As from mine tailings that contain residuals of Pb and Zn (Yang et al. 2009). For soil and other materials that are co-contaminated with cations and As, more acidic extractants or sequential extractions have been employed (Ko et al. 2005; Isosaari and Sillanpää 2012). One drawback with NaOH extraction is that it seems to be effective only for the removal of adsorbed As fractions whereas ammonium oxalate, oxalic acid, or their mixture (Tamm's reagent) are more effective for mobilizing As associated with Fe (oxy)hydroxides and arsenates (Daus et al. 1998; Ahn et al. 2005).

Ethylenediaminetetraacetic acid (EDTA) is the most commonly used chelate because of its strong chelating ability for different heavy metals (Chaiyaraksa and Sriwiriyanuphap 2004). Citric acid is a weak acid, which would acidify the soil, making the As(V) easy to desorb from soil and it can form mononuclear, binuclear, or polynuclear and bi-, tri-, and multidentate complexes with metallic ions (Bassi et al. 2000). It is also relatively inexpensive, rather easy to handle, and has a comparatively low affinity for alkaline earth metals.

An enhanced electrokinetic remediation process for removal of arsenic, presented as As(V) form, from spiked soil has been investigated with groundwater and chemical reagents of cetylpyridinium chloride (a cationic surfactant), EDTA, and citric acid under potential gradient of 2.0–3.3 V/cm for 5 days treatment (Yuan and Chiang 2008). The removal efficiency of As(V) in EK-EDTA system was better than that in other two EK systems. As potential gradient increased from 2.0 V/cm to 3.0 V/cm, the removal efficiency of As(V) was increased from 35.4 to 44.8 % in EK-EDTA system. It showed that the arsenic removal could be enhanced by selecting suitable chemical reagent and increasing potential gradient. The intense electroosmotic flow towards the cathode caused a significant retardation of electromigration of arsenic towards the anode. The quantity of As(V) collected in anode reservoir was 1.4–2.5 times greater than that in cathode reservoir for all EK systems. The obtained results implied that As(V) removal was directly related to the electromigration rather than electroosmosis mechanism in EK systems (Yuan and Chiang 2008).

The similarity of phosphate with arsenate anions makes it feasible for displacing adsorbed As. Isosaari and Sillanpää (2012) showed that oxalate and phosphate solutions might also serve as mobility-enhancing agents and EK experiments conducted with mine tailings showed that in the section near the anode, As removal was lower with oxalate when compared with phosphate. But, due to the accumulation of As in the middle section, the respective overall removals in the entire tailings material were only 6 and 12 %. Potassium phosphate was also the most efficient in extracting arsenic from kaolinite, probably due to anion exchange of arsenic species by phosphate whereas sodium hydroxide seemed to be the most efficient in removing arsenic from the tailing-soil (Kim et al. 2005a). This result may be explained by the fact that the sodium hydroxide increased the soil pH and accelerated ionic migration of the species through the desorption of arsenic species as well as the dissolution of arsenic-bearing minerals (Kim et al. 2005a). Using different soil samples, Tao et al. (2006) also determined that phosphate extracted more As (11–32 %) than similar solutions (10 mM, pH 5.5) of oxalate (2 %). Contrary to the previous studies, Anawar et al. (2008) found that, using mine soils, phosphate extracted less As (6 %) than ammonium oxalate (41 %).

The efficiency of an extraction solution depends on factors such as the mineralogy of As and other elements, source of As and aging processes in the contaminated soil (Beeston et al. 2008). In fact, the association of elements with different mineral fractions and their dissolution has a higher impact on the transportation of different elements in mine tailings than EK mobility of soluble ions (Isosaari and Sillanpää 2010). The chemical speciation under the pH–Eh conditions prevailing between the electrodes also affects the mobility of As: oxidation of sulfides may release As but reduction of secondary Fe minerals is a faster release mechanism. One particular problem is that uncharged species with poor EK mobility, such as arsenite (H_3AsO_3), might be formed if Eh is not high enough or if pH is below 2 (Masscheleyn et al. 1991; Wang and Zhao 2009).

16.1.4 Coupling EK Process with Other Technologies

There are several technologies that can be coupled with EK trying to enhance and improve the soil remediation.

One alternative is a passive method employing permeable reactive barriers (PRBs), which represents an innovative and cost-effective technology for in situ treatment of groundwater contamination (Blowes et al. 2000). A PRB is an engineered zone of reactive materials placed in an aquifer that allows to intercept the pollution plume carried within the aquifer by retaining or degrading the contaminants (Yuan and Chiang 2007). Elemental iron was commonly applied in remediation of chlorinated organic compounds by dechlorination reaction (Su and Puls 2001; Yuan and Chiang 2008) and remediation of arsenic by promoting sorption on to the iron surface (Farrell et al. 2001; Zhang et al. 2004; Bednar et al. 2005). PRB combined with EK process was reported to enhance EK remediation performance. The common PRB materials applied included ion exchange membrane for metals removal (Kim et al. 2005a) and Fe(0)/FeOOH for As(V) removal (Yuan and Chiang 2008).

Yuan and Chiang (2007) carried batch tests with EK/PRB media of Fe(0) and FeOOH under a potential gradient of 2 V/cm for 5 days to evaluate the removal mechanisms of As in soil matrix. Arsenic enhancement of 1.6–2.2 times was achieved when the PRB system was installed in an EK system. A better performance of As removal was shown in EK/FeOOH systems, than in EK/Fe(0) systems, when PRB was located in the middle of EK cell. The results obtained by Yuan and Chiang (2007) indicated that the removal of As in EK/PRB systems was much more contributed by surface adsorption/precipitation on PRB media than by EK process. Among the electrical removal mechanisms, electromigration was predominant than electroosmotic flow. Surface adsorption and precipitation were, respectively, the principal removal mechanism under acid environment, e.g., near anode side, and under basic environment, e.g., near cathode side.

An enhanced EK/PRB of carbon nanotube coated with cobalt (CNT-Co) has been investigated for As(V) removal from soil under potential gradient of 2.0 V/cm for 5 days treatment. The results obtained by Yuan et al. (2009) showed that removal efficiency of As(V) was greater than 70 % in EK/CNT-Co system with EDTA as processing fluid, which was enhanced by a factor of 2.2 compared to EK system and EK/CNT systems. A better removal performance in EK/CNT-Co system was attributed to higher sorption of As(V) onto CNT-Co than onto CNT. Removal of As(V) in EK/CNT-Co system was mainly contributed by surface sorption on CNT-Co rather than by EK process. The surface characteristics of CNT-Co, which was qualified by SEM coupled with EDS, clearly confirmed that arsenic was adsorbed on the passive layer surface. Among EK processes, As(V) removal was dominated by electroosmosis flow and electromigration in EK/CNT-Co system with groundwater and EDTA as processing fluid. An investigation with sequential extraction revealed that As(V) associated with soils was considerably shifted from strong binding forms, i.e., Fe-Mn oxide, organic, and residual, to weak binding forms, i.e., exchange and carbonate, after EK/CNT-Co treatment (Yuan et al. 2009).

16.2 Phytotechnologies

Phytoremediation is a cost-effective natural option to upgrade contaminated lands by different mechanisms (Ali et al. 2013). The time to perform a remediation scheme, geological and climatic determinants or toxicity may be a constraint to the success of this technique (Table 16.3). Metals may be removed by phytoextraction. In this process, plants uptake the contaminant by roots followed by its translocation and accumulation in the aboveground biomass. This biomass may be further harvested and further disposed or reused for environmental applications (e.g., biomass production, phytomining). The level of accumulation depends on plant tissue (shoot or root) (Qi et al. 2011) and can be season dependent (Murciego et al. 2007). The accumulation levels are also related with soil type, bioavailability, and plant species (Qi et al. 2011).

In the specific case, mining areas can be rehabilitated by plants with potential to accumulate metals and/or metalloids. Plant species spontaneously growing in mining areas contaminated with As naturally accumulate this metalloid, mainly when in available form (Jana et al. 2012; Levresse et al. 2012; Pérez-Sirvent et al. 2012; Vaculík et al. 2013). The mechanism of metal tolerance and detoxification

Table 16.3 Advantages and limitations of phytoremediation (adapted from Gomes 2012; Favas et al. 2014)

Advantages	Limitations
In situ and passive technique	Limited to shallow soils or where contamination is localized to the surface (<5 m) and plants are selective in metal remediation
Solar-driven and low cost	Limited practical experience and therefore not accepted by many regulatory agencies
High acceptance by the public	There is little knowledge of farming, genetics, reproduction and diseases of phytoremediating plants
Reduced environmental impact, contributes to the landscape improvement, provides wildlife for animal life	Can only be applied depending heavy metals concentrations in the soil (can be toxic and lethal to plants) and plants may not adapt to climatic and environmental conditions at contaminated sites
Reduction of soil erosion, dispersal of dust and contaminants by wind	Toxicity and bioavailability of degradation products remain largely unknown
Reduction of surface runoff, leaching, and mobilization of contaminants in soil	Treatment slower than the traditional physicochemical techniques
Easy harvesting of the plants	Contamination may spread through the food chain if accumulator plants are ingested by animals
The harvested biomass can be economically valuable	If the plants release compounds to increase the mobility of the metals, these can be leached into groundwater
Plant process more easily controlled than those of microorganisms	The area to be decontaminated must be large enough to allow application of cultivation techniques

enables some plant species to survive, grow, and reproduce in contaminated sites (Pratas et al. 2005).

The *Pteris vittata* (Chinese brake fern) was found to be resistant to As, having the capability of hyperaccumulating large amounts of As in its fronds (Ma et al. 2001), in which the contaminants are picked up by the roots of plants and transported to their aboveground parts, and then removed together with the crops (phytostabilization, phytoextraction, and phytovolatilization). The brake fern can accumulate between 1442 and 7526 mg/kg As in fronds from contaminated soils, and up to 27,000 mg/kg As in its fronds in hydroponics culture. The As hyperaccumulation capacity has also been demonstrated in other plants (Visoottiviseth et al. 2002; Zhao et al. 2002; Meharg 2003).

Besides phytoremediation, phytostabilization methods using plants can also be applied for long-term remediation of As. This method limits uptake and excludes mobilization of As. The major benefit of phytostabilization is that the vegetative biomass above ground is not contaminated with As, thus reduces the risk of As transfer through food chains (Madejón et al. 2002).

16.3 Electrokinetic-Enhanced Phytoremediation

One of the key aspects of innovation in the remediation field refers to the development of combined treatments and operational methods that contribute to improve the mobilization and the transport of the contaminants out of the matrix. Due to its mobilization potential, EK process can be considered an integrated tool for contaminants removal, alone and coupled with phytoremediation. The coupling of EK process with phytoremediation (also known as EK-assisted phytoremediation) is an innovative technique that deserves a deeper knowledge to enlarge the scope of EK application (Fig. 16.1).

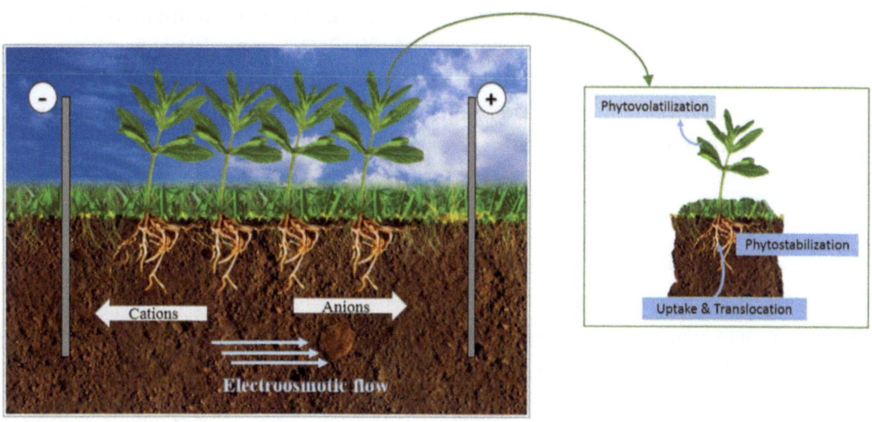

Fig. 16.1 Schematic representation of in situ electrokinetic enhanced phytoremediation

Phytoremediation coupled with EK process was reported to evaluate metals uptake by different plant species (O'Connor et al. 2003; Cang et al. 2011; Putra et al. 2013). With the coupled technique a redistribution of soil metals between electrode compartments and changes (normally increase) of plant uptake compared with the absence of an electric field. Soil pH also changes, normally, decreasing in the anode compartment (O'Connor et al. 2003; Cang et al. 2011) which may lead to the activation of soil heavy metals in the anode region (Cang et al. 2011). The applied voltage may be determinant in the efficiency of the process (Cang et al. 2011). Literature (Aboughalma et al. 2008; Bi et al. 2011) reports the use of DC and alternate field (AC) in EK-assisted phytoremediation and, in general, concludes that AC field does not promote considerable pH variation or metal redistribution, as happens with DC field. The root and shoot development may change as a reflex of the application of electric field from a slightly inhibition in the anode compartment due to soil acidification and movement of nutrient cations (e.g., Ca) (O'Connor et al. 2003) to positive or negative effect depending on the applied voltage (Cang et al. 2011).

On one hand, EK process affects soil chemistry, most commonly the acidification of the soil (when a DC field is applied), and possible disappearance of most of the natural microflora due to the toxic effect of the acidic pH, but, on the other hand, plant growth favors increased enzymatic and microbial activity (Cameselle et al. 2013). Cang et al. (2012) carried out a study with the purpose of understanding the effect of electric current on physicochemical soil properties and enzymatic and microbial activity and reported differences between the three soil compartments (anode, central compartment, and cathode). The NO_3^-, NH_4^+, available K and P increased compared with their initial soil concentration. Basal soil respiration and microbial biomass carbon significantly increased near the anode and cathode. All tested enzymatic parameters (urease, invertase, and phosphatase) were inhibited by DC field. DC field was the main factor affecting the soil properties but plant growth counteracted to same extent its impact on soil properties.

The efficiency and extent of phytoextraction is dependent on contaminant bioavailability for uptake, speciation, and efficiency of each plant species regarding metal uptake (ability to intercept, absorb, and accumulate contaminants) (Saifullah et al. 2009; Bech et al. 2012; Bhargava et al. 2012; Ali et al. 2013) as plants have specific responses to metal/metalloid tolerance, accumulation mechanisms, and genetic and environmental factors, e.g., as light and temperature in greenhouse conditions (Antoniadis et al. 2006).

16.4 Case Study

The study was conducted using a sandy clay soil collected from a mine area located in southern Hunan Province, China. The soil contained 66 mg/kg of As (Couto et al. 2015). Desorption tests were performed in the soil with and without phosphate amendment and higher extractions were found when no amendment was added (pH 13.5, 98 % for As).

Table 16.4 Arsenic remediation after application of electrokinetic remediation, phytoremediation, and coupled technology (Couto et al. 2015)

Technology	Chemical agent[a]	Plant	Treatment time (days)	Final soil pH[b]	Total removal (μg/g or μg/pot[d])
Electrokinetic[c] (stationary cell)	–	–	90 (20 V)	4.2–10.3	0.37
	PO_4^{3-}	–		6.8–11.0	4.01
	NaOH	–		4.0–8.9	0.67
	PO_4^{3-}+NaOH	–		10.6–11.2	11.41
Phytoremediation	–	Indian Mustard	15	Approx. 6.5[e]	79.3
		Ryegrass		Approx. 6.5[e]	86.9
	PO_4^{3-}	Indian Mustard		Approx. 6.5[e]	289.7
		Ryegrass		Approx. 6.5[e]	96.4
Electrokinetic-Enhanced Phytoremediation	–	Indian Mustard	15 (20 V 8 h *per* day)	6.2–7.2	176
		Ryegrass		6.1–6.9	96.7
	PO_4^{3-}	Indian Mustard		5.9–7.5	335.6
		Ryegrass		6.3–7.2	143.7

[a]PO_4^{3-} (0.5 M) was added to the soil prior to the experiment and NaOH was added to the anolyte to keep its pH above 12
[b]Initial soil pH was 6.3
[c]Removal values were calculated by the sum of As concentration found in the anolyte and catholyte divided by the initial concentration
[d]Total As and Sb levels (calculated using metalloids concentration and biomass values of each plant tissue) in ryegrass and Indian mustard after different treatments (mg/pot)
[e]Minimum and maximum values between 6.4 and 6.7

The group tested the feasibility of anolyte conditioning with NaOH and/or phosphate soil amendment on EK remediation of a mine soil contaminated with As. Controlling soil pH during EK was beneficial for increasing contaminant solubilization and migration from the soil region adjacent to the anode. Still, after 90 days, the highest removal rate was achieved for the experiment where both anolyte conditioning and phosphate amendment were performed, ca. 17 % of As was removed from the soil (Table 16.4). The main removal mechanism was electromigration towards the anode. In this study, lower remediation efficiencies were obtained comparing with literature. Better efficiencies for As removal were achieved for a contaminated soil through anolyte conditioning with NaOH, up to 43 % by Ryu et al. (2011) and with EDTA, 45 % of As(V) for a spiked soil (Yuan and Chiang 2008). It must be noticed that when remediation schemes are applied to spiked matrices, they usually present higher remediation although they are less "environmentally" reproducible. The electrolyte conditioning has the advantage of not only increasing the removal of As but also decreasing energy consumption (Baek et al. 2009). Another option to improve remediation efficiencies is to couple EK with other technologies.

Phytoremediation alone and coupled with electrokinetic was also tested for the remediation of the contaminated soil. Two plants were used, Indian mustard (*Brassica juncea*) and ryegrass (*Lolium perenne*). Indian mustard accumulated more metalloids than ryegrass (approximately 65 % more than ryegrass) (Couto et al. 2015) but the total uptake may be counteracted by higher biomass of ryegrass that is higher than those from I. mustard (Table 16.4; Couto et al. 2015). Phosphorus amendment may also help improving metalloids uptake and this was of particular importance for I. mustard.

Compared with plant alone, phosphorus and EK process increased As (25 and 48 %) and Sb (25 and 30 %) uptake for ryegrass and I. mustard, respectively.

An integrated assessment was also carried out to study the effect of DC in the heavy metals bioavailability after the EK treatment in the stationary cell as well as the electric current effect on plants biomass, enzymatic activities, and available soil nutrients (Couto et al. 2015).

After a two-step extraction, to simulate the mobile and mobilizable fractions, the initial soil contained 43 % of the As in a mobile form (most available to biota and most easily leached to groundwater) indicating that 28 mg of As *per* kg of soil can be potentially released into the environment in the mine area where the soil was sampled. Comparing to the initial soil, the application of EK decreased value of the mobilizable As fraction, but no changes in the mobile form was observed.

Through the evaluation of the biomass changes, it was possible to assess that I. mustard belowground biomass was positively affected by phosphate and EK. Together, phosphate amendment and EK increased biomass in both plants (Couto et al. 2015). The slight pH changes did not negatively affect biomass in both species. Regarding soil nutrients, potassium electromigrated towards cathode compartment in the treatment with combined technologies and available nitrate was redistributed between soil compartments in the presence of EK, with or without phosphate amendment. Between plant species, there was a similar pattern of nutrient distribution suggesting that, for the tested conditions, the used plants did not directly affect the nutrients pattern but factors such as the application of a DC electric field did it. Urease slightly decreased in the presence of EK but increased in the presence of ryegrass with or without phosphate amendment and also increased in the presence of I. mustard. But, for example, the activity of neutral phosphatase did not significantly change between the applied treatments (Couto et al. 2015).

16.5 Conclusions and Final Remarks

There are several approaches that can be applied aiming soil remediation/rehabilitation and both EK and phytoremediation seem to be viable option. Hybrid technologies may be a reliable option to achieve better remediation ratios. One option is to apply EK-assisted phytoremediation, a combination of technologies that may be considered to upgrade mine-contaminated areas. Nevertheless, in both cases, the efficiency of both techniques is dependent on time constraints.

Gathering fundamentals and research on applied EK-assisted phytoremediation will allow deeper knowledge on the application of this hybrid technology that assembles advantages from both processes, in the design of a remediation scheme. As so, more tests may be performed to achieve optimum conditions and better efficiencies for remediation of mine contaminated areas. In this scope, parameters such as life cycle costs, energy consumption, availability, resource use, and pollution may also be taken into account in order to make the process economically viable.

Acknowledgements Financial support was provided by FP7-PEOPLE-2010-IRSES-269289-*ELECTROACROSS—Electrokinetics across disciplines and continents*: *an integrated approach to finding new strategies for sustainable development* and PTDC/ECM/111860/2009—*Electrokinetic treatment of sewage sludge* and *membrane concentrate*: *Phosphorus recovery and dewatering* and National Natural Science Foundation of China (21177135). RIARTAS-Red Iberoamericana de Aprovechamiento de Residuos Industriales para el Tratamiento de Suelos y Aguas Contaminadas, Programa Iberoamericano de Ciencia y Tecnología para el Desarrollo (Cyted). N. Couto acknowledges Fundação para a Ciência e a Tecnologia for her Post-Doc fellowship (SFRH/BPD/81122/2011).

References

Aboughalma H, Bi R, Schlaak M (2008) Electrokinetic enhancement on phytoremediation in Zn, Pb, Cu and Cd contaminated soil using potato plants. J Environ Sci Health A Tox Hazard Subst Environ Eng 43:926–933

Acar YB, Alshawabkeh AN (1993) Principles of electrokinetic remediation. Environ Sci Technol 27:2638–2647

Acar YB, Alshawabkeh AN, Gale RJ (1993) Fundamentals of extracting species from soils by electrokinetics. Waste Manage 13:141–151

Agency for Toxic Substances and Disease Registry (2007) Toxicological profile for arsenic. U.S. Department of Health and Human Services, Public Health Service

Ahn J, Park Y, Kim J-Y, Kim K-W (2005) Mineralogical and geochemical characterization of arsenic in an abandoned mine tailings of Korea. Environ Geochem Health 27:147–157

Akter KF, Owens G, Davey DE, Naidu R (2005) Arsenic speciation and toxicity in biological systems. Rev Environ Contam Toxicol 184:97–149

Alam MGM, Tokunaga S (2006) Chemical extraction of arsenic from contaminated soil. J Environ Sci Health A Tox Hazard Subst Environ Eng 41:631–643

Alcántara MT, Gómez J, Pazos M, Sanromán MA (2010) Electrokinetic remediation of PAH mixtures from kaolin. J Hazard Mater 179:1156–1160

Ali HH, Khan E, Sajad MA (2013) Phytoremediation of heavy metals—concepts and applications. Chemosphere 91:869–881

Alshawabkeh A, Yeung A, Bricka M (1999) Practical aspects of in-situ electrokinetic extraction. J Environ Eng 125:27–35

Anawar HM, Garcia-Sanchez A, Santa Regina I (2008) Evaluation of various chemical extraction methods to estimate plant-available arsenic in mine soils. Chemosphere 70:1459–1467

Antoniadis V, Tsadilas C, Samaras V, Sgouras J (2006) Availability of heavy metals applied to soil through sewage sludge. In: Prasad MNV, Sajwan KS, Naidu R (eds) Trace elements in the environment: biogeochemistry, biotechnology and bioremediation. Taylor and Francis, Florida

Baek K, Kim D-H, Park S-W, Ryu B-G, Bajargal T, Yang J-S (2009) Electrolyte conditioning-enhanced electrokinetic remediation of arsenic-contaminated mine tailing. J Hazard Mater 161:457–462

Bassi R, Prasher SO, Simpson BK (2000) Extraction of metals from a contaminated sandy soil using citric acid. Environ Prog 19:275–282

Bech J, Corrales I, Tume P, Barceló J, Duran P, Roca N, Poschenrieder C (2012) Accumulation of antimony and other potentially toxic elements in plants around a former antimony mine located in the Ribes Valley (eastern Pyrenees). J Geochem Explor 113:100–105

Bednar AJ, Garbarino JR, Ranville JF, Wildeman TR (2005) Effects of iron on arsenic speciation and redox chemistry in acid mine water. J Geochem Explor 85:55–62

Beeston MP, van Elteren JT, Šlejkovec Z, Glass HJ (2008) Migration of arsenic from old tailings ponds—a case study on the King Edward Mine, Cornwall, UK. Environ Res 108:28–34

Bhargava A, Carmona FF, Bhargava M, Srivastava S (2012) Approaches for enhanced phytoextraction of heavy metals. J Environ Manage 105:103–120

Bi R, Schlaak M, Siefert E, Lord R, Connolly H (2011) Influence of electrical fields (AC and DC) on phytoremediation of metal polluted soils with rapeseed (Brassica napus) and tobacco (Nicotiana tabacum). Chemosphere 83:318–326

Blowes DW, Ptacek CJ, Benner SG, McRae CWT, Bennett TA, Puls RW (2000) Treatment of inorganic contaminants using permeable reactive barriers. J Contam Hydrol 45:123–137

Bradshaw AD (1993) Understanding the fundamentals of succession. In: Miles J, Walton DH (eds) Primary succession on land. Blackwell, Oxford

Bundschuh J, Bhattacharya P, Sracek O, Mellano MF, Ramírez AE, Storniolo AR, Martín RA, Cortés J, Litter MI, Jean J-S (2011) Arsenic removal from groundwater of the Chaco-Pampean plain (Argentina) using natural geological materials as adsorbents. J Environ Sci Health A Tox Hazard Subst Environ Eng 46:1297–1310

Cameselle C, Chirakkara RA, Reddy KR (2013) Electrokinetic-enhanced phytoremediation of soils: status and opportunities. Chemosphere 93:626–636

Cang L, Wang QY, Zhou DM, Xu H (2011) Effects of electrokinetic-assisted phytoremediation of a multiple-metal contaminated soil on soil metal bioavailability and uptake by Indian mustard. Sep Purif Technol 79:246–253

Cang L, Zhou DM, Wang QY, Fan GP (2012) Impact of electrokinetic-assisted phytoremediation of heavy metal contaminated soil on its physicochemical properties, enzymatic and microbial activities. Electrochim Acta 86:41–48

Chaiyaraksa C, Sriwiriyanuphap N (2004) Batch washing of cadmium from soil and sludge by a mixture of $Na_2S_2O_5$ and Na_2EDTA. Chemosphere 56:1129–1135

National Research Council (2001) Arsenic in drinking water 2001 update. National Academy Press, Washington, DC

Couto N, Guedes P, Zhou DM, Ribeiro A (2015) Integrated perspectives of a greenhouse study to upgrade an antimony and arsenic mine soil—potential of enhanced phytotechnologies. Chem Eng J 262:563–570

Daus B, Weiß H, Wennrich R (1998) Arsenic speciation in iron hydroxide precipitates. Talanta 46:867–873

Farrell J, Wang J, O'Day P, Conklin M (2001) Electrochemical and spectroscopic study of arsenate removal from water using zero-valent iron media. Environ Sci Technol 35:2026–2032

Favas PJC, Pratas J, Varun M, D'Souza R, Paul M (2014) Chapter 17: Phytoremediation of Soils Contaminated with Metals and Metalloids at Mining Areas: Potential of Native Flora. In: Soriano, MCH (ed) Environmental Risk Assessment of Soil Contamination. In Tech

Filella M, Belzile N, Chen Y-W (2002) Antimony in the environment: a review focused on natural waters: I. Occurrence. Earth Sci Rev 57:125–176

Gomes HI (2012) Phytoremediation for bioenergy: challenges and opportunities. Environ Technol Rev 1:59–66

Guedes P, Magro C, Couto N, Mosca A, Mateus EP, Ribeiro A (2015) Potential of the electrodialytic process for emerging organic contaminants remediation and phosphorus separation from sewage sludge. Electrochim Acta (in press). http://dx.doi.org/10.1016/j.electacta.2015.03.167

Isosaari P, Sillanpää M (2010) Electromigration of arsenic and co-existing metals in mine tailings. Chemosphere 81:1155–1158

Isosaari P, Sillanpää M (2012) Effects of oxalate and phosphate on electrokinetic removal of arsenic from mine tailings. Sep Purif Technol 86:26–34

Jackson BP, Miller WP (2000) Effectiveness of phosphate and hydroxide for desorption of arsenic and selenium species from iron oxides. Soil Sci Soc Am J 64:1616–1622

Jana U, Chassany V, Bertrand G, Castrec-Rouelle M, Aubry E, Boudsocq S, Laffray D, Repellin A (2012) Analysis of arsenic and antimony distribution within plants growing at an old mine site in Ouche (Cantal, France) and identification of species suitable for site revegetation. J Environ Manage 110:188–193

Jang M, Hwang JS, Choi SI (2007) Sequential soil washing techniques using hydrochloric acid and sodium hydroxide for remediating arsenic-contaminated soils in abandoned iron-ore mines. Chemosphere 66:8–17

Kim S-O, Kim W-S, Kim K-W (2005a) Evaluation of electrokinetic remediation of arsenic-contaminated soils. Environ Geochem Health 27:443–453

Kim W-S, Kim S-O, Kim K-W (2005b) Enhanced electrokinetic extraction of heavy metals from soils assisted by ion exchange membranes. J Hazard Mater 118:93–102

Ko I, Chang Y-Y, Lee C-H, Kim K-W (2005) Assessment of pilot-scale acid washing of soil contaminated with As, Zn and Ni using the BCR three-step sequential extraction. J Hazard Mater 127:1–13

Kwon JC, Lee J-S, Jung MC (2012) Arsenic contamination in agricultural soils surrounding mining sites in relation to geology and mineralization types. Appl Geochem 27:1020–1026

Lee M, Paik I, Do W, Kim I, Lee Y, Lee S (2007) Soil washing of As-contaminated stream sediments in the vicinity of an abandoned mine in Korea. Environ Geochem Health 29:319–329

Levresse G, Lopez G, Tritlla J, López EC, Chavez AC, Salvador EM, Soler A, Corbella M, Sandoval LGH, Corona-Esquivel R (2012) Phytoavailability of antimony and heavy metals in arid regions: the case of the Wadley Sb district (San Luis, Potosí, Mexico). Sci Total Environ 427–428:115–125

Lima AT, Ottosen LM, Pedersen AJ, Ribeiro AB (2008) Characterization of fly ash from bio and municipal waste. Biomass Bioenerg 32:277–282

Ma LQ, Komar KM, Tu C, Zhang W, Cai Y, Kennelley ED (2001) A fern that hyperaccumulates arsenic. Nature 409:579

Madejón P, Murillo JM, Marañón T, Cabrera F, López R (2002) Bioaccumulation of As, Cd, Cu, Fe and Pb in wild grasses affected by the Aznalcóllar mine spill (SW Spain). Sci Total Environ 290:105–120

Mandal BK, Suzuki KT (2002) Arsenic round the world: a review. Talanta 58:201–235

Masscheleyn PH, Delaune RD, Patrick WH (1991) Effect of redox potential and pH on arsenic speciation and solubility in a contaminated soil. Environ Sci Technol 25:1414–1419

Mateus E, Zrostlíková J, Gomes da Silva M, Ribeiro A, Marriott P (2010) Electrokinetic removal of creosote from treated timber waste: a comprehensive gas chromatographic view. J Appl Electrochem 40:1183–1193

Meharg AA (2003) Variation in arsenic accumulation—hyperaccumulation in ferns and their allies. New Phytol 157:25–31

Mondal P, Majumder CB, Mohanty B (2006) Laboratory based approaches for arsenic remediation from contaminated water: recent developments. J Hazard Mater 137:464–479

Murciego AM, Sánchez AG, González MAR, Gil EP, Gordillo CT, Fernández JC, Triguero TB (2007) Antimony distribution and mobility in topsoils and plants (Cytisus striatus, Cistus ladanifer and Dittrichia viscosa) from polluted Sb-mining areas in Extremadura (Spain). Environ Pollut 145:15–21

Nagajyoti PC, Lee KD, Sreekanth TVM (2010) Heavy metals, occurrence and toxicity for plants: a review. Environ Chem Lett 8:199–216
O'Connor CS, Lepp NW, Edwards R, Sunderland G (2003) The combined Use of electrokinetic remediation and phytoremediation to decontaminate metal-polluted soils: a laboratory-scale feasibility study. Environ Monit Assess 84:141–158
Ottosen LM, Pedersen AJ, Hansen HK, Ribeiro AB (2007) Screening the possibility for removing cadmium and other heavy metals from wastewater sludge and bio-ashes by an electrodialytic method. Electrochim Acta 52:3420–3426
Pérez-Sirvent C, Martínez-Sánchez MJ, Martínez-López S, Bech J, Bolan N (2012) Distribution and bioaccumulation of arsenic and antimony in Dittrichia viscosa growing in mining-affected semiarid soils in southeast Spain. J Geochem Explor 123:128–135
Pratas J, Prasad MNV, Freitas H, Conde L (2005) Plants growing in abandoned mines of Portugal are useful for biogeochemical exploration of arsenic, antimony, tungsten and mine reclamation. J Geochem Explor 85:99–107
Putra RS, Ohkawa Y, Tanaka S (2013) Application of EAPR system on the removal of lead from sandy soil and uptake by Kentucky bluegrass (Poa pratensis L.). Sep Purif Technol 102:34–42
Qi C, Wu F, Deng Q, Liu G, Mo C, Liu B, Zhu J (2011) Distribution and accumulation of antimony in plants in the super-large Sb deposit areas, China. Microchem J 97:44–51
Reddy K, Chinthamreddy S (2003a) Sequentially enhanced electrokinetic remediation of heavy metals in low buffering clayey soils. J Geotech Geoenviron 129:263–277
Reddy KR, Chinthamreddy S (2003b) Effects of initial form of chromium on electrokinetic remediation in clays. Adv Environ Res 7:353–365
Reddy K, Danda S, Saichek R (2004) Complicating factors of using ethylenediamine tetraacetic acid to enhance electrokinetic remediation of multiple heavy metals in clayey soils. J Environ Eng 130:1357–1366
Ribeiro AB, Mateus EP, Ottosen LM, Bech-Nielsen G (2000) Electrodialytic removal of Cu, Cr, and As from chromated copper arsenate-treated timber waste. Environ Sci Technol 34:784–788
Ribeiro AB, Rodriguez-Maroto JM, Mateus EP, Gomes H (2005) Removal of organic contaminants from soils by an electrokinetic process: the case of atrazine: experimental and modeling. Chemosphere 59:1229–1239
Ribeiro AB, Mateus EP, Rodríguez-Maroto J-M (2011) Removal of organic contaminants from soils by an electrokinetic process: the case of molinate and bentazone. Experimental and modeling. Sep Purif Technol 79:193–203
Ryu B-G, Park G-Y, Yang J-W, Baek K (2011) Electrolyte conditioning for electrokinetic remediation of As, Cu, and Pb-contaminated soil. Sep Purif Technol 79:170–176
Saichek RE, Reddy KR (2003) Effect of pH control at the anode for the electrokinetic removal of phenanthrene from kaolin soil. Chemosphere 51:273–287
Saifullah ME, Qadir M, de Caritat P, Tack FMG, Du Laing G, Zia MH (2009) EDTA-assisted Pb phytoextraction. Chemosphere 74:1279–1291
Sawada A, Tanaka S, Fukushima M, Tatsumi K (2003) Electrokinetic remediation of clayey soils containing copper(II)-oxinate using humic acid as a surfactant. J Hazard Mater 96:145–154
Singh R, Singh S, Parihar P, Singh VP, Prasad SM (2015) Arsenic contamination, consequences and remediation techniques: a review. Ecotox Environ Safe 11:247–270
Smedley PL, Kinniburgh DG (2002) A review of the source, behaviour and distribution of arsenic in natural waters. Appl Geochem 17:517–568
Stafilov T, Šajn R, Pančevski Z, Boev B, Frontasyeva MV, Strelkova LP (2010) Heavy metal contamination of topsoils around a lead and zinc smelter in the Republic of Macedonia. J Hazard Mater 175:896–914
Su C, Puls RW (2001) Arsenate and arsenite removal by zerovalent iron: effects of phosphate, silicate, carbonate, borate, sulfate, chromate, molybdate, and nitrate, relative to chloride. Environ Sci Technol 35:4562–4568
Tao Y, Zhang S, Jian W, Yuan C, Shan X-Q (2006) Effects of oxalate and phosphate on the release of arsenic from contaminated soils and arsenic accumulation in wheat. Chemosphere 65:1281–1287

Tokunaga S, Hakuta T (2002) Acid washing and stabilization of an artificial arsenic-contaminated soil. Chemosphere 46:31–38

U.S. Geological Survey (2012) Arsenic. In: Brooks WE (ed) Mineral commodity summaries 2012. U.S. Geological Survey, Reston, pp 20–21

Vaculík M, Jurkovič Ľ, Matejkovič P, Molnárová M, Lux A (2013) Potential risk of arsenic and antimony accumulation by medicinal plants naturally growing on old mining sites. Water Air Soil Pollut 224:1–16

Vamerali T, Bandiera M, Mosca G (2010) Field crops for phytoremediation of metal-contaminated land. A review. Environ Chem Lett 8:1–17

Virkutyte J, Sillanpää M, Latostenmaa P (2002) Electrokinetic soil remediation—critical overview. Sci Total Environ 289:97–121

Visoottiviseth P, Francesconi K, Sridokchan W (2002) The potential of Thai indigenous plant species for the phytoremediation of arsenic contaminated land. Environ Pollut 118:453–461

Wang S, Zhao X (2009) On the potential of biological treatment for arsenic contaminated soils and groundwater. J Environ Manage 90:2367–2376

Wang J-Y, Huang X-J, Kao JCM, Stabnikova O (2007) Simultaneous removal of organic contaminants and heavy metals from kaolin using an upward electrokinetic soil remediation process. J Hazard Mater 144:292–299

Wilson SC, Lockwood PV, Ashley PM, Tighe M (2010) The chemistry and behaviour of antimony in the soil environment with comparisons to arsenic: a critical review. Environ Pollut 158:1169–1181

Wong MH (2003) Ecological restoration of mine degraded soils, with emphasis on metal contaminated soils. Chemosphere 50:775–780

Xiao XY, Chen TB, Liao XY, Wu B, Yan XL, Zhai LM, Xie H, Wang LX (2008) Regional distribution of arsenic contained minerals and arsenic pollution in China. Geograph Res 27:201–212

Yang J-S, Lee JY, Baek K, Kwon T-S, Choi J (2009) Extraction behavior of As, Pb, and Zn from mine tailings with acid and base solutions. J Hazard Mater 171:443–451

Yeung AT (2006) Contaminant extractability by electrokinetics. Environ Eng Sci 23(1):202–224

Yeung A, Hsu C (2005) Electrokinetic remediation of cadmium-contaminated clay. J Environ Eng 131:298–304

Yuan C, Chiang T-S (2007) The mechanisms of arsenic removal from soil by electrokinetic process coupled with iron permeable reaction barrier. Chemosphere 67:1533–1542

Yuan C, Chiang T-S (2008) Enhancement of electrokinetic remediation of arsenic spiked soil by chemical reagents. J Hazard Mater 152:309–315

Yuan C, Hung C-H, Chen K-C (2009) Electrokinetic remediation of arsenate spiked soil assisted by CNT-Co barrier—the effect of barrier position and processing fluid. J Hazard Mater 171:563–570

Zhang W, Singh P, Paling E, Delides S (2004) Arsenic removal from contaminated water by natural iron ores. Mineral Eng 17:517–524

Zhao FJ, Dunham SJ, McGrath SP (2002) Arsenic hyperaccumulation by different fern species. New Phytol 156:27–31

Zhou DM, Deng CF, Cang L (2004) Electrokinetic remediation of a Cu contaminated red soil by conditioning catholyte pH with different enhancing chemical reagents. Chemosphere 56:265–273

Chapter 17
Nanoremediation Coupled to Electrokinetics for PCB Removal from Soil

Helena I. Gomes, Guangping Fan, Lisbeth M. Ottosen, Celia Dias-Ferreira, and Alexandra B. Ribeiro

17.1 Introduction

Polychlorinated biphenyls (PCB) are synthetic organic compounds classified as Persistent Organic Pollutants (POP). Despite their use has been banned in the majority of countries, PCB can still be found in the environment, even in remote areas where they were never manufactured or used (Rosalinda et al. 2013). These pollutants represent an important environmental problem due to their persistence and to the lack of a cost-effective and sustainable remediation technology for contaminated soils and sediments (Gomes et al. 2013a).

H.I. Gomes (✉)
CENSE, Departamento de Ciências e Engenharia do Ambiente, Faculdade de Ciências e Tecnologia, Universidade Nova de Lisboa, Caparica 2829-516, Portugal

CERNAS—Research Center for Natural Resources, Environment and Society, Escola Superior Agraria de Coimbra, Instituto Politecnico de Coimbra, Bencanta, Coimbra 3045-601, Portugal

Department of Civil Engineering, Technical University of Denmark, Brovej, Building 118, Kongens Lyngby 2800, Denmark
e-mail: hrg@campus.fct.unl.pt

G. Fan
Key Laboratory of Soil Environment and Pollution Remediation, Institute of Soil Science, Chinese Academy of Sciences (ISSCAS), East Beijing Road, Nanjing 210008, China

L.M. Ottosen
Department of Civil Engineering, Technical University of Denmark, Brovej, Building 118, Kongens Lyngby 2800, Denmark

C. Dias-Ferreira
CERNAS—Research Center for Natural Resources, Environment and Society, Escola Superior Agraria de Coimbra, Instituto Politecnico de Coimbra, Bencanta, Coimbra 3045-601, Portugal

A.B. Ribeiro
CENSE, Departamento de Ciências e Engenharia do Ambiente, Faculdade de Ciências e Tecnologia, Universidade Nova de Lisboa, Caparica 2829-516, Portugal

© Springer International Publishing Switzerland 2016

A.B. Ribeiro et al. (eds.), *Electrokinetics Across Disciplines and Continents*,
DOI 10.1007/978-3-319-20179-5_17

The solutions commonly adopted for PCB-contaminated soils and sediments are "dig and dump" and "dig and incinerate," despite the development and testing of innovative technologies at lab scale (Gomes et al. 2013a). In the late 1990s, the emergence of nanotechnologies and especially the use of zero valent iron nanoparticles (nZVI) were considered encouraging for organochlorines, such as PCB, in contaminated groundwaters and soils (Wang and Zhang 1997; Lowry and Johnson 2004; He et al. 2010). Zero valent iron nanoparticles, as a strong reductant, can promote reductive dechlorination through a multi-step removal of chlorine atoms in different pathways, which can occur in parallel or sequentially. In field applications, nZVI can be easily injected in the aquifers through injection wells, or incorporated to topsoil to adsorb or degrade pollutants (Crane and Scott 2012). However, results in aquifers show that nZVI have limited mobility, ranging from 1 m (Kocur et al. 2014) to 6–10 m (Zhang and Elliott 2006). One of the methods tested to enhance nZVI mobility was the use of low-level direct current (DC) (Gomes et al. 2013b; Pamukcu et al. 2008; Jones et al. 2010; Yang et al. 2007), using the same principles of electrokinetic remediation (EKR). Combining both technologies, the role of the electric current would be to enhance the nZVI transport into the soil for *in situ* transformation, and subsequent destruction of the contaminants would occur.

The main objective of this chapter is to provide an overview of the experimental work combining electro- and nanoremediation of PCB-contaminated soils. It is organized as follows: after a general introduction to the chemical properties of PCB and their use, the existing technologies for PCB-contaminated soils and sediments are briefly mentioned, detailing the use of EKR and iron nanoparticles, and mainly in the simultaneous use of both technologies targeting PCB in polluted soils.

17.2 Polychlorinated Biphenyls

The classification of PCBs as persistent organic pollutants (POP) arises from the United Nations (UN) Stockholm Convention in 2001, due to accumulated evidence of the potential toxicity of these chemicals based on in vitro or in vivo assays (Valentín et al. 2013). These pollutants are synthetic organic compounds resistant to environmental degradation through chemical, biological, and photolytic processes (Ritter et al. 1997), which make them persist in the environment, transport through long-range distances and reach remote areas where they have never been used, bioaccumulate in human and animal tissue, and biomagnificate in food chains (Koblizkova et al. 2009; Nizzetto et al. 2010; Donaldson et al. 2010). Soil plays an important role in the global fate and distribution of POP, as an efficient sink for these chemicals due to its large retention capacity for hydrophobic compounds (Holoubek et al. 2009). Soil contamination with POP is especially alarming.

PCBs are a group of synthetic aromatic compounds that comprise two benzene rings connected at the C-1 carbon, with the general formula $C_{12}H_{10-x}Cl_x$. Each ring

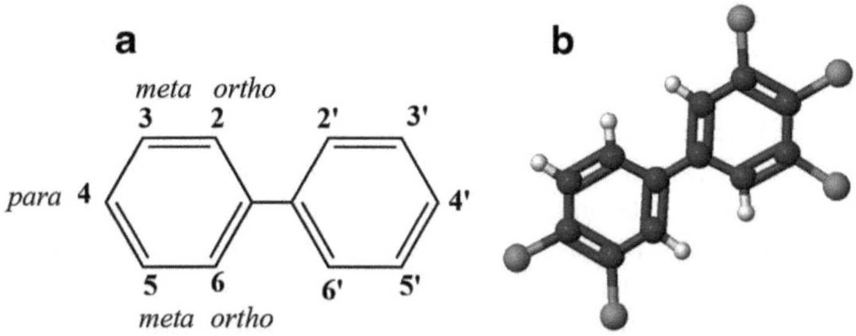

Fig. 17.1 Chemical structure of PCB. (**a**) The possible positions of chlorine atoms on the benzene rings are denoted by numbers assigned to the carbon atoms. (**b**) Chemical structure of 3,3′,4,4′,5-pentachlorobiphenyl (PCB126)

can have up to five chlorines in the *ortho*, *meta*, or *para* positions (Fig. 17.1), ranging from mono- to decachlorobiphenyl. There is a total of 209 structural arrangements (congeners) differing in chlorine number and position and exhibiting unique chemical properties. Ballschmiter and Zell (BZ), in 1980, introduced a system in which congeners were arranged in ascending numerical order based on the number of chlorine atoms and their position on the biphenyl structure. The BZ system was recognized by the International Union of Pure and Applied Chemistry (IUPAC) and is the generally accepted notation (Maervoet et al. 2003).

Manufactured by subjecting biphenyl to chlorine, commercial PCB typically came in a mixture of many congeners, classified by the amount of chlorine by weight present. For instance, Aroclor 1260, marketed by Monsanto from 1930 to 1977, contained 60 % wt. of chlorine (Yak et al. 1999). Other commercial mixtures of different congeners were Clophen (Bayer, Germany), Kanechlor (Kanegafuchi, Japan), Santotherm (Mitsubishi, Japan), Phenoclor and Pyralene (Prodolec, France) (Breivik et al. 2002a; Diaz-Ferrero et al. 1997). PCB were used for industrial application in closed systems: lubricants (industrial oils), dielectric fluids in electrical equipment such as transformers, capacitors, heat transfer, and hydraulic systems (Borja et al. 2005; Erickson and Kaley 2011; Kas et al. 1997; UNEP 2002). They were also employed in open uses, such as pesticide extenders, sealants, carbonless copy paper, lubricants, paints, adhesives, plastics, flame retardants, and dedusting agents on roads (ATSDR 2000; Diamond et al. 2010; Kohler et al. 2005; Priha et al. 2005). Other uses, in partially open systems, include: heat transfer fluids; hydraulic fluid in lifting equipment, trucks and high-pressure pumps; vacuum pumps; voltage regulators; liquid filled electrical cables; and liquid filled circuit breakers (ATSDR 2000; Eisler and Belisle 1996; Erickson and Kaley 2011; Furukawa and Fujihara 2008). Recently, PCB congeners (namely PCB11 or 3,3′-dichlorobiphenyl) were detected in azo and phthalocyanine pigments, commonly used in paint and also in inks, textiles, paper, cosmetics, leather, plastics, food, and other materials (Hu and Hornbuckle 2010; Rodenburg et al. 2010), as by-products (inadvertently produced) of industrial synthetic process of paint pigments (Hu and Hornbuckle 2010).

PCB pose a real human health threat through numerous exposure pathways (Mikszewski 2004). Of the 209 different types of PCBs, 13 exhibit a dioxin-like toxicity (UN 2001). The Toxic Equivalency Factor (TEF) defined by the World Health Organization (WHO) for PCB varies between 10^{-5} for 2,3′,4,4′,5,5′-hexachlorobiphenyl and 0.1 for 3,3′,4,4′,5-pentachlorobiphenyl. The non-dioxin-like PCB and their metabolites do not interact substantially with the aryl hydrocarbon receptor and may act through different pathways than the dioxin-like chemicals, so their effects are not accounted for in TEF (CDC 2009). Adverse effects associated to the exposure of PCB comprise damage to the immune system, liver, skin irritation (acne), reproductive system, gastrointestinal tract, and thyroid gland (ATSDR 2011; Faroon et al. 2003; Xing et al. 2009). Research has also shown that PCB can cause severe neurological problems in children, including impairment of cognitive and motor abilities, and can be transmitted from mother to child during breastfeeding (Faroon et al. 2003). PCB are listed as probable human carcinogens (UN 2001). Occurrences of PCB toxic effects in invertebrates, fish, birds, and other animals are also well documented (Eisler 1986; Eisler and Belisle 1996).

In the 1970s, several countries limited PCB use due to the concerns on human toxicity and later, in 1985, the European Community heavily restricted the use and marketing of PCB. Nevertheless, PCB production was estimated around 1.2 and 2 million t, with some of the most detailed data indicating a total global production of approximately 1.3 million t between 1929 and 1993 (Breivik et al. 2007). Of this cumulative global production, previsions point to 440–92,000 t emitted into the environment (Breivik et al. 2002b, 2007). According to Ockenden et al. (2003), most of the PCB manufactured/used (perhaps >70 %) have not yet entered the environmental pool because they are still associated with diffusive source materials. Around 10 % of the total produced PCB (Schmidt 2010) is accumulated in the geosphere because of their low volatility, low solubility in water, and high affinity for particulates, both biotic and abiotic (Alder et al. 1993; Yak et al. 1999; Schmidt 2010).

The extent of PCB contamination in soils and sediments worldwide is unknown. In the USA, 25 % of the Superfund Sites (total number of 433 sites) have PCB-contaminated soils or sediments (USEPA 2014), whereas in Canada there are 338 sites according to the Federal Contaminated Sites Inventory (TBS-SCT 2014). In European countries, an estimate points towards 272,000 contaminated sites with chlorinated hydrocarbons, but a total quantification of PCB-contaminated sites is missing (Marc van Liedekerke et al. 2014). Additionally to the local contamination near industrial sites, an inventory on atmospheric deposition in background surface soil estimates a PCB global soil burden of 21,000 t (Meijer et al. 2003).

Under the UN Stockholm Convention, the parties (currently 179) have to eliminate the use of PCB in equipment (e.g., transformers and capacitors) by 2025 and to ensure the environmentally sound management of PCB waste (including contaminated soils) by 2028 (UN 2001).

17.3 Technologies for PCB-Contaminated Soils and Sediments

The state of the art of the technologies for in situ and ex situ remediation of PCB-contaminated soils and sediments was recently reviewed, including laboratory, pilot, and full-scale case studies (Gomes et al. 2013a). The main emergent remediation technologies were described and their current status was evaluated, assessing the main advantages and also potential obstacles to their full-scale application (Gomes et al. 2013a). A classification was also developed, considering in situ and ex situ methods, differencing biological, physical, chemical, and thermal methods, and also natural attenuation and combined technologies (Fig. 17.2). Although there are promising results in bench-scale studies, most technologies are still in the initial stage of development. Further research and scale-up are needed for the progress of cost-effective and sustainable alternatives for PCB remediation.

After this review, recent studies include the electrokinetically enhanced persulfate oxidation of PCB in contaminated soils with (Fan et al. 2014) and without surfactant (Yukselen-Aksoy and Reddy 2012); the phytoremediation by

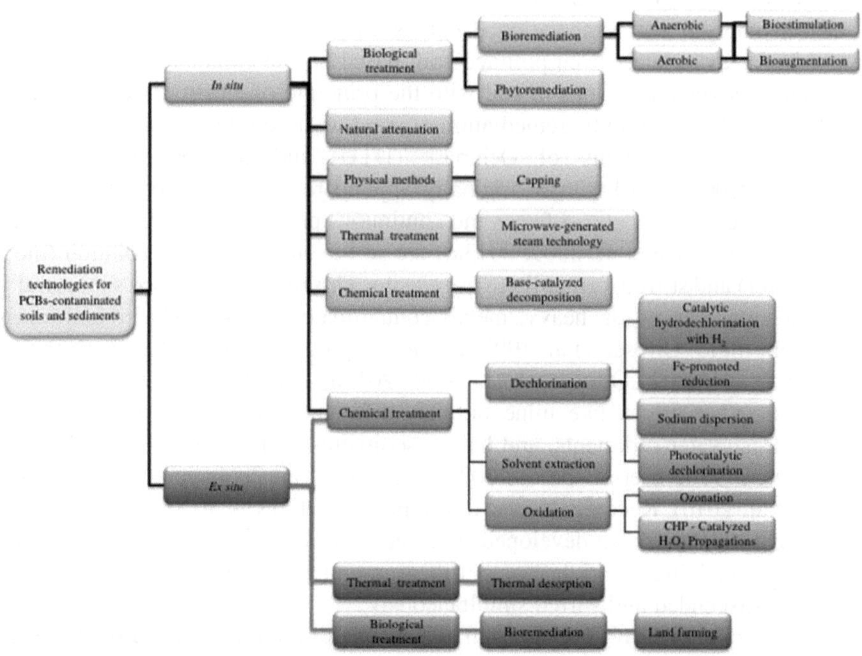

Fig. 17.2 Classification of remediation technologies for remediation of PCB-contaminated soils and sediments (Gomes et al. 2013a)

alfalfa, tall fescue, single, and mixed plants cultivation (Li et al. 2013); and the use of biosurfactant on combined chemical–biological treatment (Viisimaa et al. 2013). In none of those studies was clearly demonstrated the method efficiency for full-scale implementation.

17.3.1 Electroremediation

EKR, also called electrokinetics, electroremediation, or electroreclamation, is a solid technology that has been successfully used since the late 1980s to treat contaminated soils, especially low permeability soils (Lageman et al. 1989; Pamukcu and Wittle 1992; Acar and Alshawabkeh 1993; Probstein and Hicks 1993; Ottosen et al. 1997). Electrokinetics can be defined as the application or induction of a low-level direct current on the order of mA cm^{-2} of cross-sectional area between the electrodes or an electric potential difference about few V cm^{-1} across a soil mass containing fluid or a high fluid content slurry/suspension, causing or caused by the motion of electricity, charged soil, and/or fluid particles (Sun 2013). In this method, the current acts as the "cleaning agent," generating transport processes (as electroosmosis, electromigration, and electrophoresis) and electrochemical reactions (electrolysis and electrodeposition) (Acar and Alshawabkeh 1993).

Electrodialytic soil remediation (EDR) is a remediation method developed for removing heavy metals from polluted soil, based on the combination of the electrokinetic movement of ions in soil with the principle of electrodialysis (Ottosen et al. 1997). Electrodialytic remediation of solid waste products started in 1992 at the Technical University of Denmark (DTU) and was patented in 1995 (PCT/DK95/00209). EDR was originally applied to moist and consolidated soil for *in situ* treatment. Later, a faster and continuous process was developed (Jensen et al. 2007; Ottosen et al. 2009, 2013a): the soil is suspended in a solution (most often water) and stirred, primarily to use ex situ. This method was used successfully for the remediation of heavy metals-contaminated soils, both in soil mass containing fluid (Ottosen et al. 1997) and in a high fluid content slurry/suspension (Jensen et al. 2007; Ottosen et al. 2009, 2013a), for the cleanup of different contaminated matrices like mine tailings, different types of fly ashes, sewage sludge, freshwater sediments, and harbor sediments (Ribeiro et al. 2000; Hansen et al. 2005; Ottosen et al. 2006, 2007; Kirkelund et al. 2009; Jensen et al. 2010; Pazos et al. 2010). Recently, a new development in EDR is the two-compartment electrodialytic setup also developed (Ottosen et al. 2013b) at DTU, in which the anode is placed directly in the compartment in which the soil or the contaminated matrix is suspended and stirred simultaneously.

The removal of heavy metals from soils is one of the most studied processes in electroremediation (Virkutyte et al. 2002). Despite the success at bench scale, there are limited full-scale applications for soil remediation. In the last years, numerous research teams developed methods to increase electrokinetics removal efficiency, such as pH control (Lee and Yang 2000; Saichek and Reddy 2003a), enhancement solutions (Ottosen et al. 2005), desorbing agents (Nystroem et al. 2006), surfactants (Saichek and Reddy 2003b), pulse current (Hansen and Rojo 2007; Sun and Ottosen 2012), and alternating current (Rojo et al. 2012).

Gomes et al. (2012) made a literature review of the experimental work carried out to improve organochlorines electroremediation from contaminated soils, both with enhancement solutions and using other remediation technologies simultaneously. When this review was made, EK had never been used to extract PCB from soils. Since then, a study on spiked soils reported an 87 % removal efficiency for PCB with electrokinetics (Istrate et al. 2013). Still this paper fails to explain this removal because no quantification of PCB in the anolyte and the catholyte was made, nor degradation processes were identified or proposed. Recently, two studies tested the electrokinetically enhanced persulfate oxidation of PCB in spiked soils (Yukselen-Aksoy and Reddy 2012) and in contaminated soils using also a surfactant (Fan et al. 2014). In the first study, the highest level of PCB oxidation, 78 %, was achieved in spiked kaolin with temperature-activated persulfate in 7 days (Yukselen-Aksoy and Reddy 2012). These results could not be replicated in the glacial till soil (with only 14 % oxidation), probably due to the high buffering capacity, nonhomogeneous mineral content, and high organic content (Yukselen-Aksoy and Reddy 2012). In the second study, the surfactant used was Igepal CA720 and zero valent iron (added in the cathode reservoir) was the persulfate activator (Fan et al. 2014). The highest PCB degradation rate was 38 % in the treatment without activator. The authors found that the use of iron as persulfate activator was ineffective as it consumed most of persulfate and inhibited its transport into the cell.

17.3.2 Nanoremediation

Zero valent iron nanoparticles were considered a promising alternative for PCB degradation in aqueous solutions (Wang and Zhang 1997; Lowry and Johnson 2004; He et al. 2010) because they can promote reductive dechlorination (Fig. 17.3). However, a 95 % dechlorination in historically contaminated soils was

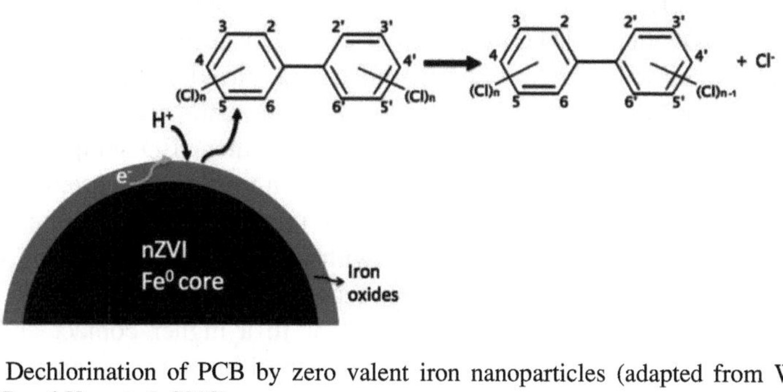

Fig. 17.3 Dechlorination of PCB by zero valent iron nanoparticles (adapted from Wang and Zhang 1997 and Yan et al. 2013)

only achieved at high temperatures (300 °C) (Varanasi et al. 2007). Another recent study reported 81.5 % and 53.4 % PCB removal from soils from an e-waste recycling area in batch tests with 12 days duration, using the Fe/Pd bimetallic nanoparticles, compared with 67.4 and 48.3 % using bare nZVI (Chen et al. 2014). Until now, the efficiency of PCB dechlorination in soils and soil slurries is limited when compared with aqueous solutions, due to competing reactions occurring and also to the strong PCB adsorption to soil organic matter.

17.3.3 Nanoremediation Coupled to Electrokinetics

The idea of combining nanoremediation and electrokinetics of PCB-contaminated soil was presented in 2011 (Gomes et al. 2011) and, in 2013, Fan et al. (2013) report experimental work on the topic, in which a contaminated soil from a capacitor landfill (100 g), in Zhejiang, China, was diluted with a clean soil collected from a farmland (2400 g) in Nanjing, China, that was previously spiked with PCB28 (Fan et al. 2013). The sum of PCBs in the final soil was 46.3 mg kg^{-1}. The four nanoremediation and electrokinetic experiments had 14 days duration; the voltage gradient was 1 V cm^{-1} initially and was doubled at day 7, the electrolyte was 10^{-2} M $NaNO_3$ and different stabilizers for bimetallic Fe/Pd nanoparticles containing xanthan gum and surfactants were tested (Exp-01. 1 g L^{-1} xanthan gum; Exp-02. 1 g L^{-1} xanthan gum + 3 % sodium dodecyl benzene sulfonate—SDBS; Exp-03. 1 g L^{-1} xanthan gum + 3% Brij35 and Exp-04. 1 g L^{-1} xanthan gum + 3 g L^{-1} rhamnolipid). In EK test, 20 g L^{-1} of stabilized nano Fe/Pd was injected in the compartment located 3 cm from the anode chamber daily. After the experiments, the removal efficiencies were 17.1 %, 11.0 %, 20.4 %, and 11.7 %, respectively. Brij35-xanthan gum stabilized nanoparticles obtained the highest PCBs removal efficiency due to its higher reactivity and higher electroosmotic flow toward cathode (Fan et al. 2013). This limited removal is due to the strong adsorption of PCB to soils. The other surfactants tested (SDBS and rhamnolipid) showed even lower PCB removal in the EK experiments (Fig. 17.4).

The combined electro-nano remediation of PCB-contaminated soils was also tested in two different experimental setups—both developed at the Technical University of Denmark (DTU) (Gomes et al. 2015). In the electrodialytic (ED) - two-compartment cell, the soil is suspended and stirred simultaneously with the addition of nZVI vs. a conventional three-compartment electrokinetic (EK) cylindrical cell (Fig. 17.5). The two-compartment ED setup showed PCBs removal percentages of 83 % with and 29 % without direct current (Fig. 17.6). These removal percentages are higher than that in the previous studies (Fan et al. 2013) and also than in batch tests without current (Chen et al. 2014). The suspension and stirring of the soil can enhance the PCB dechlorination by nZVI, due to an increase in desorption from soil or to a higher contact and reaction between nZVI and PCB, or to both. In the traditional three-compartment EK setup, the iron has to be transported across the compacted saturated soil to reach

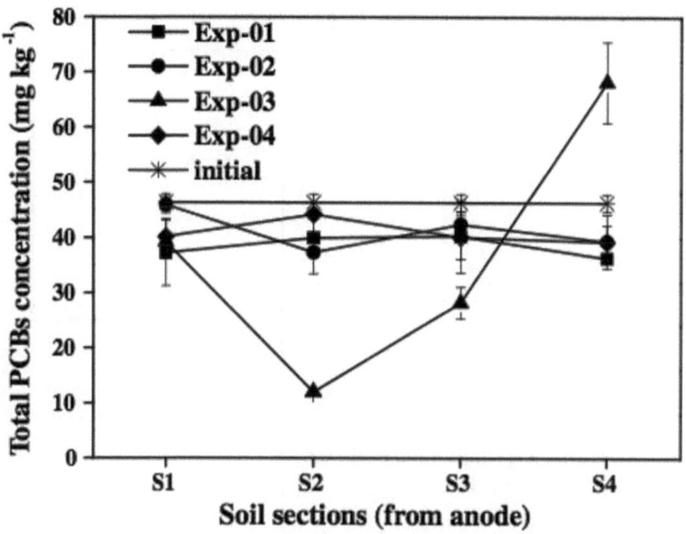

Fig. 17.4 PCB concentrations in soil sections after the different treatments applied (Fan et al. 2013)

the contaminants. Previous studies show that even a low proportion of carbonate minerals may cause an increase in the deposition of PAA-nZVI particles and aggregates due to the lower negative surface charge (Laumann et al. 2013). As the soil used in the experiments has high carbonate content (18 %), the limited dechlorination (12–58 %) observed is probably related to the iron precipitation with the carbonates (Gomes 2014).

Experiments with the EK setup (Fig. 17.5b) were conducted with different durations to assess if longer times would increase the PCB dechlorination. Comparing a 10 day experiment (A) with a 45 days experiment (D), the PCB removal has a small increase (27 % vs. 36 %) (Fig. 17.6). Although the removal percentages were higher than in previous studies (Fan et al. 2013), their values were not encouraging for a scale-up of the process (pilot and full scale) for the remediation of PCB-contaminated soils and sediments.

Direct current can be used to enhance nZVI transport in different porous matrices or model soils (Gomes et al. 2013b, 2014b) but, in the new ED setup, the contact between the nanoparticles and the contaminated soil is ensured by the stirring, so the current might not be needed for the PCBs dechlorination. However, an experiment with direct current (Exp. F) had a higher PCBs removal rate (83 %) than a similar experiment without current (Exp. E) (29 %) (Fig. 17.6). Of major importance to this was the high pH and buffer capacity of the soil tested (Fig. 17.6). In the experiment without current (Exp. E) the soil suspension with nZVI kept a constant alkaline pH, which promotes the passivation of the iron nanoparticles.

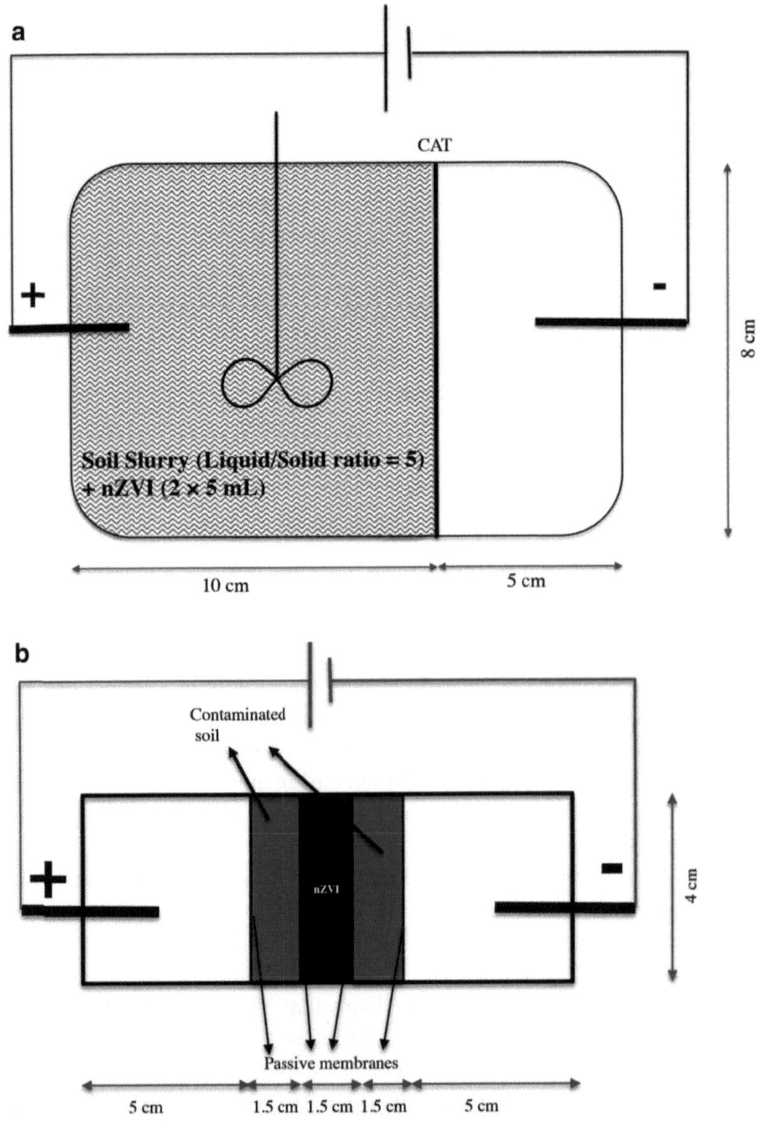

Fig. 17.5 Schematic representation of the experimental setups used in the experiments: (**a**) setup A—new electrodialytic cell (CAT—cation exchange membrane); (**b**) electrokinetic cell (setup B)

In the experiment with current (Exp. F) water electrolysis produces H^+ in the anode, thus lowering the pH in the soil suspension (pH range between 2.9 and 10.2). A slightly acidic pH (4.90–5.10) increases the dechlorination rate of PCB by nZVI and nZVI/Pd (Wang et al. 2012).

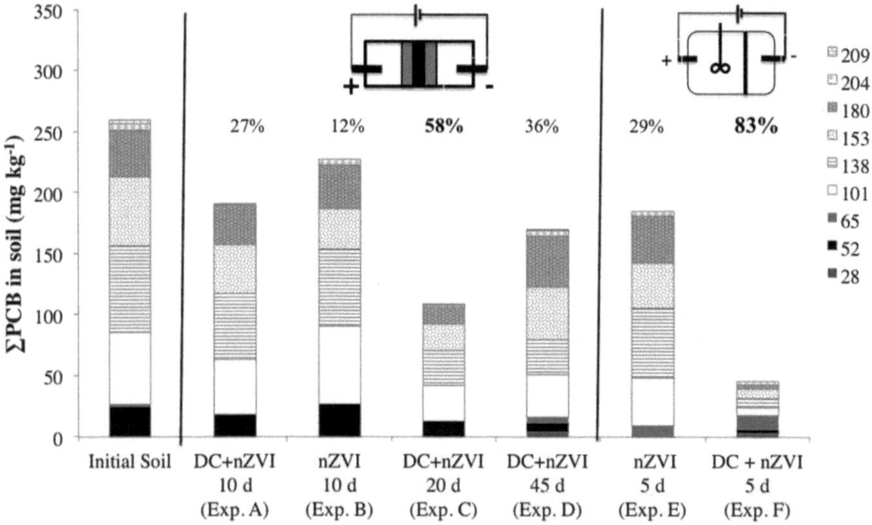

Fig. 17.6 Average concentration of PCB congeners (PCB28, 52, 65, 101, 138, 153, 180, 204 and 209) in soil before and after the experiments, using the three-compartment cell and the two-compartment cell. Percentages on the top of each column represent PCB removal regarding the sum of congeners analyzed in the initial soil (Gomes 2014; Gomes et al. 2015)

Two different surfactants (saponin and Tween 80) were also tested to enhance PCBs desorption and removal efficiency from a contaminated soil in the two-compartment ED setup (Gomes et al. 2014a). The soil was stirred in a slurry with 1 % surfactant, 10 mL of nZVI commercial suspension, and a voltage gradient of 1 V cm^{-1}. The results showed that, in the tested conditions, saponin allowed to obtain higher PCB removal from soil when compared with Tween 80 (Gomes et al. 2014a). The most efficient removal (76 %) was obtained with saponin with current and iron nanoparticles, while the lowest removal was obtained with Tween 80 and nZVI (8 % removal), without application of a direct current. This removal is consistent with previous studies that showed that Tween 80 was one of the surfactants with the least efficient degradation of 1-(2-chloro-benzoyl)-3-(4-chlorophenyl) urea by nZVI, when compared with Triton X-100, Tween 20, sodium dodecyl sulfonate (SDS), and cetyltrimethylammonium bromide (CTAB), probably due to the effect of the hydrophobic chain length (Zhou and Lin 2013). Comparing the energy consumption of the experiments, calculated according to Sun et al. (2012), Tween 80 has higher energy consumption when compared with saponin (Gomes et al. 2014a). Added to the low removal percentages, this also contributes to show that Tween 80 is not a suitable surfactant to use with PCB, despite the good results obtained for PAH with this method (Lima et al. 2012).

Two different soils (Table 17.1), historically contaminated with PCB, were used in experiments using the two-compartment ED setup (Gomes 2014). Soil 1 was provided by a hazardous waste operator in Portugal and is a mixture of

Table 17.1 Physical and chemical characteristics of the soils (Gomes 2014)

Parameter	Soil 1	Soil 2
Particles distribution (%)		
Coarse sand (200 < Ø < 2000 μm)	19.1	3.2
Fine sand (20 < Ø < 200 μm)	67.3	69.6
Silt (2 < Ø < 20 μm)	12.7	23.6
Clay (Ø < 2 μm)	0.9	3.6
Textural classification	Loamy sand	Silt loam
pH (H_2O)	12.2	8.20
Conductivity (mS cm^{-1})	18.76	0.221
Exchangeable cations ($cmol_{(c)}$ kg^{-1})		
Ca^{2+}	83.75	259.14
Mg^{2+}	3.2	9.75
K^+	26.88	7.36
Na^+	9.37	8.34
Sum of exchangeable cations ($cmol_{(c)}$ kg^{-1})	123.2	284.59
Calcium carbonate (%)	18.0	1.3
Organic matter (%)	16.46	0.57
PCBs[a] (μg kg^{-1})	258 ± 24	156 ± 2
Metals[b] (mg kg^{-1})		
Al	20,980 ± 590	4952 ± 71
As	9 ± 2	0.6 ± 0.97
Cd	0.7 ± 0.1	0.4 ± 0.04
Cr	52 ± 3	2.5 ± 0.04
Cu	142 ± 95	10 ± 0.3
Fe	13,162 ± 301	6773 ± 97
Ni	32 ± 1	6 ± 0.3
Pb	45 ± 3	25 ± 0.9
Zn	2155 ± 40	135 ± 0.1

[a]Sum of PCB 52, 65, 101, 138, 153, 180, 204, and 209
[b]Acid digestion with HNO_3 according to the Danish Standard DS259

contaminated soils from industrial sites with transformers oils spills. Soil 2 is a surface soil sampled in a decommissioned school in Region Hovedstaden (Capital Region of Denmark), and the PCB resulted from the weathering of the windows joint sealants used in 1955–1977 (Jensen 2009).

The results of PCB removal were different according to soil type. Soil 1 has the higher PCB removal without saponin, only with direct current, stirring and nZVI (83 %). In soil 2, the highest removal was obtained only with nZVI and saponin. The highest removal percentage in Soil 1 can be related to the pH of the soil slurry during the experiments. The initial soil pH and carbonate content allowed to maintaining the pH between 6 and 7 for about half the time of the experiment. In the experiments with Soil 2, the pH values turn acidic faster (~30 h after the beginning of the experiment), and this contributes to the corrosion of the iron nanoparticles. This is also consistent with the higher PCB removal obtained in

Soil 2 without the direct current, and consequently, without acidification of the soil slurry. The pH values in soil 2 reached 2.91–3.35, corresponding to less favorable conditions to PCB dechlorination due to the corrosion of zero valent iron. In all the experiments where nZVI was added, higher pH values were measured in the suspension. Also, in the experiments without direct current the pH values showed very little variation.

17.3.4 Overview of the Main Results

Table 17.2 summarizes the experimental conditions and main results obtained for the PCB electroremediation with iron nanoparticles. The results show that the soil characteristics are critical and affect the reaction between nZVI and the PCB, as well as the iron transport under electric fields. The use of surfactants as an enhancement method of electrokinetic coupled with nZVI needs to be carefully assessed, to minimize the inhibitory effect of the surfactants on iron nanoparticles reactivity. Combining surfactant and nZVI originated less successful remediation than combining surfactant and electrodialytic remediation. The use of surfactants as enhancements may be a poor choice and needs to be previously assessed.

17.4 Conclusions

The characteristics that led to a widespread use of PCB, in various industrial and domestic applications, do present a challenge for the remediation of contaminated soils and sediments. Following PCB entrance into the soil environment, they rapidly adsorb to mineral and organic matter (solid phases). The ability to desorb these contaminants determines, in most cases, the effectiveness of the remediation technologies. Currently, there is no cost-effective alternative to landfilling and incineration of PCB-contaminated soils. Also, there is no single, portable technology that is applicable to both ex situ and in situ remediation of PCB in contaminated soils and sediments. Each case is unique and several factors must be considered. The success of the treatment is dependent on proper selection, design, and adjustment of the remediation technology, based on the congeners present, soils properties and the system performance. The combined use of remediation technologies and the so-called treatment trains is a promising approach for persistent contaminants.

The use of nanoremediation in conjunction with electrokinetics showed limited results in the traditional electrokinetic setup, even with bimetallic Fe/Pd nanoparticles. However, the new electrodialytic setup allowed PCB dechlorination from contaminated soil ex situ at a higher percentage, in a shorter time, with lower nZVI consumption, and with the use of half of the voltage gradient when compared with the traditional electrokinetic setup. In addition, there is no need to treat and

Table 17.2 Summary of the experimental conditions and results of the electroremediation of PCB-contaminated soils with iron nanoparticles

Soil	Experimental setup	Duration of test(s)	Electrolyte	Voltage gradient (V cm^{-1})	Observations	% PCB removal best results	References
PCB-contaminated soil from a capacitor landfill (100 g), in Zhejiang, China, diluted with a clean soil collected from a farmland (2400 g) in Nanjing, China	EK	14 days	0.01 M $NaNO_3$	$1.0 \rightarrow 2.0$	Stabilized bi-metallic Fe/Pd nanoparticles	20	Fan et al. (2013)
Mixture of contaminated soils from industrial sites with transformers oils spills (Portugal)—S1	ED	5 days	0.01 M NaCl	1.0	2 surfactants: saponin and Tween 80	76	Gomes et al. (2014a)
S1	EK and ED	5 days (ED) 10, 20, 45 days (EK)	0.01 M NaCl	1.0	Comparison of the two setups	87	Gomes (2014)
S1 and S2—surface soil sampled in a decommissioned school in Denmark with PCB from the weathering of the windows joint sealants	ED	5 days	0.01 M NaCl	1.0	Comparison between the two soils	87 (S1) 48 (S2)	Gomes (2014)

dispose of the anolyte. Still, additional testing with different soils, repeated application of the technique on the same material (also with different duration experiments), testing of enhancement methods and further optimization and scale-up of the process are needed to prove the versatility of the electrodialytic setup.

Acknowledgment This work has been funded by the research grant SFRH/BD/76070/2011, by project PTDC/AGR-AAM/101643/2008 NanoDC under Portuguese National funds through "Fundação para a Ciência e a Tecnologia," and by FP7-PEOPLE-IRSES-2010-269289-ELECTROACROSS.

References

Acar YB, Alshawabkeh AN (1993) Principles of electrokinetic remediation. Environ Sci Technol 27(13):2638–2647

Alder AC, Haggblom MM, Oppenheimer SR, Young LY (1993) Reductive dechlorination of polychlorinated biphenyls in anaerobic sediments. Environ Sci Technol 27:530–538

ATSDR (2000) Toxicological profile for polychlorinated biphenyls (PCBs). Agency for Toxic Substances and Disease Registry/U.S. Department of Health and Human Services, Atlanta

ATSDR (2011) Toxicological profiles. Agency for Toxic Substances and Disease Registry/U.S. Department of Health and Human Services. http://www.atsdr.cdc.gov/toxprofiles/index.asp. Accessed 13 Jul 2011

Borja J, Taleon DM, Auresenia J, Gallardo S (2005) Polychlorinated biphenyls and their biodegradation. Process Biochem 40:1999–2013

Breivik K, Sweetman A, Pacyna JM, Jones KC (2002a) Towards a global historical emission inventory for selected PCB congeners—a mass balance approach 1. Global production and consumption. Sci Total Environ 290:181–198

Breivik K, Sweetman A, Pacyna JM, Jones KC (2002b) Towards a global historical emission inventory for selected PCB congeners—a mass balance approach 2. Emissions Total Sci Environ 290:199–224

Breivik K, Sweetman A, Pacyna JM, Jones KC (2007) Towards a global historical emission inventory for selected PCB congeners—a mass balance approach 3. An update. Sci Total Environ 377:296–307

CDC (2009) Fourth National Report on Human Exposure to Environmental Chemicals. Department of Health and Human Services/Centers for Disease Control and Prevention, Atlanta

Chen X, Yao X, Yu C, Su X, Shen C, Chen C, Huang R, Xu X (2014) Hydrodechlorination of polychlorinated biphenyls in contaminated soil from an e-waste recycling area, using nanoscale zerovalent iron and Pd/Fe bimetallic nanoparticles. Environ Sci Pollut Res 27:1–10. doi:10.1007/s11356-013-2089-8

Crane RA, Scott TB (2012) Nanoscale zero-valent iron: future prospects for an emerging water treatment technology. J Hazard Mater 211–212:112–125. doi:10.1016/j.jhazmat.2011.11.073

Diamond ML, Melymuk L, Csiszar SA, Robson M (2010) Estimation of PCB stocks, emissions, and urban fate: will our policies reduce concentrations and exposure? Environ Sci Technol 44:2777–2783

Diaz-Ferrero J, Rodriguez-Larena MC, Comellas L, Jimhnez B (1997) Bioanalytical methods applied to endocrine disrupting polychlorinated biphenyls, polychlorinated dibenzo-p-dioxins and polychlorinated dibenzofurans. A review trends. Anal Chem 16(10):563–573

Donaldson SG, Oostdam JV, Tikhonov C, Feeley M, Armstrong B, Ayotte P, Boucher O, Bowers W, Chan L, Dallaire F, Dallaire R, Dewailly É, Edwards J, Egeland GM, Fontaine J, Furgal C, Leech T, Loring E, Muckle G, Nancarrow T, Pereg D, Plusquellec P, Potyrala M,

Receveur O, Sheare RG (2010) Environmental contaminants and human health in the Canadian Arctic. Sci Total Environ 408:5165–5234
Eisler R (1986) Planar PCB hazards to fish, wildlife, and invertebrates: a synoptic review. Patuxent Wildlife Research Center. U.S., Fish and Wildlife Service, Laurel
Eisler R, Belisle AA (1996) Planar PCB hazards to fish, wildlife, and invertebrates: a synoptic review. Patuxent Wildlife Research Center. U.S., National Biological Service, Laurel
Erickson MD, Kaley RG (2011) Applications of polychlorinated biphenyls. Environ Sci Pollut Res 18:135–151
Fan G, Cang L, Qin W, Zhou C, Gomes HI, Zhou D (2013) Surfactants-enhanced electrokinetic transport of xanthan gum stabilized nano Pd/Fe for the remediation of PCBs contaminated soils. Sep Purif Technol 114:64–72, http://dx.doi.org/10.1016/j.seppur.2013.04.030
Fan G, Cang L, Fang G, Zhou D (2014) Surfactant and oxidant enhanced electrokinetic remediation of a PCBs polluted soil. Sep Purif Technol 123:106–113, http://dx.doi.org/10.1016/j.seppur.2013.12.035
Faroon OM, Keith LS, Smith- C, Simon, Rosa CTD (2003) Polychlorinated biphenyls: human health aspects. In: World Health Organization. Report published under the joint sponsorship of the United Nations Environment Programme, the International Labour Organization, and the World Health Organization, and produced within the framework of the Inter-Organization Programme for the Sound Management of Chemicals, Geneva, Switzerland
Furukawa K, Fujihara H (2008) Microbial degradation of polychlorinated biphenyls: biochemical and molecular features. J Biosci Bioeng 105(5):433–449
Gomes HI (2014) Coupling electrokinetics and iron nanoparticles for the remediation of contaminated soils. Ph.D. Dissertation, Faculdade de Ciencias e Tecnologia, Universidade Nova de Lisboa, Portugal
Gomes H, Dias-Ferreira C, Ribeiro AB, Loch G, Ottosen LM (2011) A new approach to soil remediation: coupling nanotechnology with electrically induced particle transport (Electrokinetics). In: Castro F, Vilarinho C, Carvalho J (eds) Book of proceedings of the 1st international conference WASTES: solutions, Treatments and opportunities. CVR—Centro para a valorização de Resíduos, Guimarães, Portugal, pp 732–737. ISBN 978-989-97429-1-8
Gomes HI, Dias-Ferreira C, Ribeiro AB (2012) Electrokinetic remediation of organochlorines in soil: enhancement techniques and integration with other remediation technologies. Chemosphere 87(10):1077–1090. doi:10.1016/j.chemosphere.2012.02.037
Gomes HI, Dias-Ferreira C, Ribeiro AB (2013a) Overview of *in situ* and *ex situ* remediation technologies for PCB-contaminated soils and sediments and obstacles for full-scale application. Sci Total Environ 445–446:237–260, http://dx.doi.org/10.1016/j.scitotenv.2012.11.098
Gomes HI, Dias-Ferreira C, Ribeiro AB, Pamukcu S (2013b) Enhanced transport and transformation of zerovalent nanoiron in clay using direct electric current. Water Air Soil Poll 224 (12):1–12. doi:10.1007/s11270-013-1710-2
Gomes HI, Dias-Ferreira C, Ottosen LM, Ribeiro AB (2014a) Electrodialytic remediation of polychlorinated biphenyls contaminated soil with iron nanoparticles and two different surfactants. J Colloid Interf Sci 433:189–195, http://dx.doi.org/10.1016/j.jcis.2014.07.022
Gomes HI, Dias-Ferreira C, Ribeiro AB, Pamukcu S (2014b) Influence of electrolyte and voltage on the direct current enhanced transport of iron nanoparticles in clay. Chemosphere 99:171–179, http://dx.doi.org/10.1016/j.chemosphere.2013.10.065
Gomes HI, Dias-Ferreira C, Ottosen LM, Ribeiro AB (2015) Treatment of a suspension of PCB contaminated soil using iron nanoparticles and electric current. J Environ Manage 151:550–555, http://dx.doi.org/10.1016/j.jenvman.2015.01.015
Hansen HK, Rojo A (2007) Testing pulsed electric fields in electroremediation of copper mine tailings. Electrochim Acta 52(10):3399–3405, http://dx.doi.org/10.1016/j.electacta.2006.07.064
Hansen HK, Rojo A, Ottosen LM (2005) Electrodialytic remediation of copper mine tailings. J Hazard Mater 117(2–3):179–183, http://dx.doi.org/10.1016/j.jhazmat.2004.09.014
He F, Zhao D, Paul C (2010) Field assessment of carboxymethyl cellulose stabilized iron nanoparticles for in situ destruction of chlorinated solvents in source zones. Water Res 44:2360–2370

Holoubek I, Dušek L, Sáňka M, Hofman J, Čupr P, Jarkovský J, Zbíral J, Klánová J (2009) Soil burdens of persistent organic pollutants—their levels, fate and risk. Part I variation of concentration ranges according to different soil uses and locations. Environ Pollut 157 (12):3207–3217, http://dx.doi.org/10.1016/j.envpol.2009.05.031

Hu D, Hornbuckle KC (2010) Inadvertent polychlorinated biphenyls in commercial paint pigments. Environ Sci Technol 44(8):2822–2827. doi:10.1021/es902413k

Istrate I, Cocarta D, Neamtu S, Cirlioru T (2013) The treatment of PCB polluted soil—the approach based on the application of electrochemical treatment. Water Air Soil Pollut 224 (4):1–14. doi:10.1007/s11270-013-1516-2

Jensen SF (2009) PCB in Soil. The contamination of PCB in selected locations around Roskilde and Copenhagen. Roskilde University, Denmark

Jensen PE, Ottosen LM, Ferreira C (2007) Electrodialytic remediation of soil fines (<63 μm) in suspension—influence of current strength and L/S. Electrochim Acta 52(10):3412–3419, http://dx.doi.org/10.1016/j.electacta.2006.03.116

Jensen PE, Ferreira CMD, Hansen HK, Rype J-U, Ottosen LM, Villumsen A (2010) Electroremediation of air pollution control residues in a continuous reactor. J Appl Electrochem 40:1173–1181. doi:10.1007/s10800-010-0090-1

Jones EH, Reynolds DA, Wood AL, Thomas DG (2010) Use of electrophoresis for transporting nano-iron in porous media. Ground Water 49(2):172–183. doi:10.1111/j.1745-6584.2010.00718.x

Kas J, Burkhard J, Demnerová K, Kost'ál J, Macek T, Macková M, Pazlarová J (1997) Perspectives in biodegradation of alkanes and PCBs. Pure Appl Chem 69(11):2357–2369

Kirkelund GM, Ottosen LM, Villumsen A (2009) Electrodialytic remediation of harbour sediment in suspension—evaluation of effects induced by changes in stirring velocity and current density on heavy metal removal and pH. J Hazard Mater 169(1–3):685–690, http://dx.doi.org/10.1016/j.jhazmat.2009.03.149

Koblizkova M, Ruzicková P, Cupr P, Komprda J, Holoubek I, Klánová J (2009) Soil burdens of persistent organic pollutants: their levels, fate and risks. Part IV, quantification of volatilization fluxes of organochlorine pesticides and polychlorinated biphenyls from contaminated soil surfaces. Environ Sci Technol 43:3588–3595

Kocur CM, Chowdhury AI, Sakulchaicharoen N, Boparai HK, Weber KP, Sharma P, Krol MM, Austrins LM, Peace C, Sleep BE, O'Carroll DM (2014) Characterization of nZVI mobility in a field scale test. Environ Sci Technol 48(5):2862–2869. doi:10.1021/es4044209

Kohler M, Tremp J, Zennegg M, Seiler C, Minder-Kohler S, Beck M, Lienemann P, Wegmann L, Schmid P (2005) Joint sealants: an overlooked diffuse source of polychlorinated biphenyls in buildings. Environ Sci Technol 39:1967–1973

Lageman R, Pool W, Seffinga GA (1989) Electro-reclamation. Chem Ind 18:585–590

Laumann S, Micić V, Lowry GV, Hofmann T (2013) Carbonate minerals in porous media decrease mobility of polyacrylic acid modified zero-valent iron nanoparticles used for groundwater remediation. Environ Pollut 179:53–60, http://dx.doi.org/10.1016/j.envpol.2013.04.004

Lee H-H, Yang J-W (2000) A new method to control electrolytes pH by circulation system in electrokinetic soil remediation. J Hazard Mater B 77:227–240

Li Y, Liang F, Zhu Y, Wang F (2013) Phytoremediation of a PCB-contaminated soil by alfalfa and tall fescue single and mixed plants cultivation. J Soil Sediment 13(5):925–931. doi:10.1007/s11368-012-0618-6

Lima AT, Ottosen LM, Heister K, Loch JPG (2012) Assessing PAH removal from clayey soil by means of electro-osmosis and electrodialysis. Sci Total Environ 435–436:1–6, http://dx.doi.org/10.1016/j.scitotenv.2012.07.010

Lowry G, Johnson K (2004) Congener-specific dechlorination of dissolved PCBs by microscale and nanoscale zerovalent iron in a water/methanol solution. Environ Sci Technol 38:5208–5216

Maervoet J, Covaci A, Schepens P, Sandau CD, Letcher RJ (2003) A reassessment of the nomenclature of polychlorinated biphenyl (PCB) metabolites. Environ Health Perspect 112 (3):291–294

Meijer SN, Ockenden WA, Sweetman A, Breivik K, Grimalt JO, Jones KC (2003) Global distribution and budget of PCBs and HCB in background surface soils: implications for sources and environmental processes. Environ Sci Technol 37:667–672

Mikszewski A (2004) Emerging technologies for the *in situ* remediation of PCB-contaminated soils and sediments: bioremediation and nanoscale zero-valent iron. U.S. Environmental Protection Agency, Office of Solid Waste and Emergency Response. Office of Superfund Remediation and Technology Innovation Program, Washington, DC

Nizzetto L, Macleod M, Borgå K, Cabrerizo A, Dachs J, Guardo AD, Ghirardello D, Hansen KM, Jarvis A, Lindroth A, Ludwig B, Monteith D, Perlinger JA, Scheringer M, Schwendenmann L, Semple KT, Wick LY, Zhang G, Jones KC (2010) Past, present, and future controls on levels of persistent organic pollutants in the global environment. Environ Sci Technol 44 (17):6526–6531. doi:10.1021/es100178f

Nystroem GM, Pedersen AJ, Ottosen LM, Villumsen A (2006) The use of desorbing agents in electrodialytic remediation of harbour sediment. Sci Total Environ 357(1–3):25–37

Ockenden WA, Breivik K, Meijer SN, Steinnes E, Sweetman AJ, Jones KC (2003) The global re-cycling of persistent organic pollutants is strongly retarded by soils. Environ Pollut 121 (1):75–80, http://dx.doi.org/10.1016/S0269-7491(02)00204-X

Ottosen LM, Hansen HK, Laursen S, Villumsen A (1997) Electrodialytic remediation of soil polluted with copper from wood preservation industry. Environ Sci Technol 31(6):1711–1715

Ottosen LM, Pedersen AJ, Ribeiro AB, Hansen HK (2005) Case study on the strategy and application of enhancement solutions to improve remediation of soils contaminated with Cu, Pb and Zn by means of electrodialysis. Eng Geol 77(3–4):317–329, http://dx.doi.org/10.1016/j.enggeo.2004.07.021

Ottosen LM, Lima AT, Pedersen AJ, Ribeiro AB (2006) Electrodialytic extraction of Cu, Pb and Cl from municipal solid waste incineration fly ash suspended in water. J Chem Technol Biotechnol 81(4):553–559. doi:10.1002/jctb.1424

Ottosen LM, Pedersen AJ, Hansen HK, Ribeiro AB (2007) Screening the possibility for removing cadmium and other heavy metals from wastewater sludge and bio-ashes by an electrodialytic method. Electrochim Acta 52(10):3420–3426, http://dx.doi.org/10.1016/j.electacta.2006.06.048

Ottosen LM, Jensen PE, Hansen HK, Ribeiro A, Allard B (2009) Electrodialytic remediation of soil slurry–removal of Cu, Cr, and As. Sep Sci Technol 44(10):2245–2268. doi:10.1080/01496390902979651

Ottosen L, Jensen P, Kirkelund G, Hansen H (2013a) Electrodialytic remediation of different heavy metal-polluted soils in suspension. Water Air Soil Pollut 224(12):1–10. doi:10.1007/s11270-013-1707-x

Ottosen LM, Jensen PE, Kirkelund GM, Ebbers B (2013b) Electrodialytic separation of heavy metals from particulate material. Patent application EPC 13183278:4–1352

Pamukcu S, Wittle JK (1992) Electrokinetic removal of selected heavy metals from soil. Environ Progress 11:241–250

Pamukcu S, Hannum L, Wittle JK (2008) Delivery and activation of nano-iron by DC electric field. J Environ Sci Health A 43(8):934–944

Pazos M, Kirkelund GM, Ottosen LM (2010) Electrodialytic treatment for metal removal from sewage sludge ash from fluidized bed combustion. J Hazard Mater 176(1):1073–1078

Priha E, Hellman S, Sorvari J (2005) PCB contamination from polysulphide sealants in residential areas—exposure and risk assessment. Chemosphere 59(4):537–543. doi:10.1016/j.chemosphere.2005.01.010

Probstein RF, Hicks RE (1993) Removal of contaminants from soil by electric fields. Science 260:498–530

Ribeiro AB, Mateus EP, Ottosen LM, Bech-Nielsen G (2000) Electrodialytic removal of Cu, Cr and As from chromated copper arsenate-treated timber waste. Environ Sci Technol 34:784–788

Ritter L, Solomon KR, Forget J, Stemeroff M, O'Leary C (1997) Persistent organic pollutants. An Assessment Report on: DDT-Aldrin-Dieldrin-Endrin-Chlordane, Heptachlor-Hexachlorobenzene,

Mirex-Toxaphene, Polychlorinated Biphenyls, Dioxins and Furans. The International Programme on Chemical Safety (IPCS) within the framework of the Inter-Organization Programme for the Sound Management of Chemicals (IOMC)

Rodenburg LA, Guo J, Du S, Cavallo GJ (2010) Evidence for unique and ubiquitous environmental sources of 3,3′-Dichlorobiphenyl (PCB 11). Environ Sci Technol 44(8):2816–2821. doi:10.1021/es901155h

Rojo A, Hansen HK, Cubillos M (2012) Electrokinetic remediation using pulsed sinusoidal electric field. Electrochim Acta 86:124–129, http://dx.doi.org/10.1016/j.electacta.2012.04.070

Rosalinda G, Jordi D, Luca N, Rainer L, Kevin CJ (2013) Atmospheric transport, cycling and dynamics of polychlorinated biphenyls (PCBs) from source regions to remote oceanic areas. In: Occurrence, fate and impact of atmospheric pollutants on environmental and human health, vol 1149. ACS Symposium Series, vol 1149. American Chemical Society, pp 3–18. doi:10.1021/bk-2013-1149.ch001

Saichek RE, Reddy KR (2003a) Effect of pH control at the anode for the electrokinetic removal of phenanthrene from kaolin soil. Chemosphere 21:273–287

Saichek RE, Reddy KR (2003b) Effects of system variables on surfactant enhanced electrokinetic removal of polycyclic aromatic hydrocarbons from clayey soils. Environ Technol 24 (4):503–515

Schmidt C (2010) How PCBs are like grasshoppers. Environ Sci Technol 44(8):2752

Sun TR (2013) Effect of pulse current on energy consumption and removal of heavy metals during electrodialytic soil remediation. Ph.D. Dissertation. Technical University of Denmark, Denmark

Sun TR, Ottosen LM (2012) Effects of pulse current on energy consumption and removal of heavy metals during electrodialytic soil remediation. Electrochim Acta 86:28–35, http://dx.doi.org/10.1016/j.electacta.2012.04.033

Sun TR, Ottosen LM, Jensen PE, Kirkelund GM (2012) Electrodialytic remediation of suspended soil—comparison of two different soil fractions. J Hazard Mater 203–204:229–235, http://dx.doi.org/10.1016/j.jhazmat.2011.12.006

TBS-SCT (2014) Contaminants & Media. http://www.tbs-sct.gc.ca/fcsi-rscf/cm-eng.aspx?clear=1. Accessed 27 Feb 2014

UN (2001) Stockholm convention on persistent organic pollutants. http://chm.pops.int/Convention/tabid/54/language/en-GB/Default.aspx. 2014

UNEP (2002) PCB transformers and capacitors from management to reclassification and disposal. UNEP Chemicals, United Nations Environmental Programme, Geneva, Switzerland

USEPA (2014) Search superfund site information. http://cumulis.epa.gov/supercpad/cursites/srchsites.cfm. Accessed 27 Feb 2014

Valentín L, Nousiainen A, Mikkonen A (2013) Introduction to organic contaminants in soil: concepts and risks. In: Vicent T, Caminal G, Eljarrat E, Barceló D (eds) Emerging organic contaminants in sludges: analysis, fate and biological treatment. Springer, Berlin. doi:10.1007/698_2012_208

Marc van Liedekerke GP, Sabine Rabl-Berger, Mark Kibblewhite, Geertrui Louwagie (2014) Progress in the management of Contaminated Sites in Europe. Report EUR 26376 EN. Institute for Environment and Sustainability. Joint Research Center. European Commission, Luxembourg. http://dx.doi.org/10.1016/j.seppur.2013.12.035

Varanasi P, Fullana A, Sidhu S (2007) Remediation of PCB contaminated soils using iron nanoparticles. Chemosphere 66:1031–1038

Viisimaa M, Karpenko O, Novikov V, Trapido M, Goi A (2013) Influence of biosurfactant on combined chemical-biological treatment of PCB-contaminated soil. Chem Eng J 220:352–359, http://dx.doi.org/10.1016/j.cej.2013.01.041

Virkutyte J, Sillanpaa M, Latostenmaa P (2002) Electrokinetic soil remediation—critical overview. Sci Total Environ 289:97–121

Wang C-B, Zhang W (1997) Synthesizing nanoscale iron particles for rapid and complete dechlorination of TCE and PCBs. Environ Sci Technol 31(7):2154–2156

Wang Y, Zhou D, Wang Y, Wang L, Cang L (2012) Automatic pH control system enhances the dechlorination of 2,4,4′-trichlorobiphenyl and extracted PCBs from contaminated soil by nanoscale Fe^0 and Pd/Fe^0. Environ Sci Pollut Res 19(2):448–457. doi:10.1007/s11356-011-0587-0

Xing GH, Chan JKY, Leung AOW, Wu SC, Wong MH (2009) Environmental impact and human exposure to PCBs in Guiyu, an electronic waste recycling site in China. Environ Int 35:76–82

Yak HK, Wenclawiak BW, Cheng IF, Doyle JG, Wai CM (1999) Reductive dechlorination of polychlorinated biphenyls by zerovalent iron in subcritical water. Environ Sci Technol 33 (8):1307–1310

Yan W, Lien H-L, Koel BE, Zhang W (2013) Iron nanoparticles for environmental clean-up: recent developments and future outlook. Environ Sci Proc Imp 15:63–77

Yang GCC, Tu H-C, Hung CH (2007) Stability of nanoiron slurries and their transport in the subsurface environment. Sep Purif Technol 58:166–172

Yukselen-Aksoy Y, Reddy KR (2012) Effect of soil composition on electrokinetically enhanced persulfate oxidation of polychlorobiphenyls. Electrochim Acta 86:164–169. doi:10.1016/j.electacta.2012.03.049

Zhang W, Elliott DW (2006) Applications of iron nanoparticles for groundwater remediation. Remediation J 16(2):7–21

Zhou Q, Lin H (2013) Influence of surfactants on degradation of 1-(2-Chlorobenzoyl)-3-(4-chlorophenyl) urea by nanoscale zerovalent iron. Clean 41(2):128–133. doi:10.1002/clen.201100650

Chapter 18
Removal of Pharmaceutical and Personal Care Products in Aquatic Plant-Based Systems

Ana R. Ferreira, Nazaré Couto, Paula R. Guedes, Eduardo P. Mateus, and Alexandra B. Ribeiro

18.1 Introduction

18.1.1 Pharmaceutical and Personal Care Products Contamination

A set of organic contaminants, especially pharmaceuticals and personal care products (PPCPs), are daily discharged into the aquatic system, due to its worldwide consumption (Barceló and Petrovic 2007). These compounds are an extraordinary diverse group of chemicals used in veterinary medicine, agricultural practices, human health and cosmetic care (Barceló and Poschenrieder 2003). The use of PPCPs is growing, as they are used not only for treatment but also for prevention of illnesses. Therefore, new pharmacologically active substances are being constantly developed with unknown fates and effects on the environment. These reveal pharmaceuticals as continuous emerging pollutants (Cunningham et al. 2006).

One of the major sources of PPCPs in the aquatic environment is the effluent discharge from wastewater treatment plants (WWTPs), where PPCPs are not completely removed during wastewater treatment (Onesios et al. 2009). Several studies have presented evidence that some pharmaceuticals are not efficiently removed in the municipal WWTPs and are, thus, discharged as contaminants into the receiving water bodies

A.R. Ferreira (✉) • E.P. Mateus • A.B. Ribeiro
CENSE, Departamento de Ciências e Engenharia do Ambiente, Faculdade de Ciências e Tecnologia, Universidade Nova de Lisboa, Campus de Caparica, Caparica 2829-516, Portugal
e-mail: arl.ferreira@campus.fct.unl.pt

N. Couto • P.R. Guedes
CENSE, Departamento de Ciências e Engenharia do Ambiente, Faculdade de Ciências e Tecnologia, Universidade Nova de Lisboa, Campus de Caparica, Caparica 2829-516, Portugal

Key Laboratory of Soil Environment and Pollution Remediation, Institute of Soil Science, Chinese Academy of Sciences, Nanjing 210008, China

© Springer International Publishing Switzerland 2016

A.B. Ribeiro et al. (eds.), *Electrokinetics Across Disciplines and Continents*,
DOI 10.1007/978-3-319-20179-5_18

(Aga 2008; Fent et al. 2006; Heberer 2002). The WWTPs receive wastewaters that contain a lot of different trace polluting compounds, for which conventional treatment technologies have not been specifically designed (Verlicchi et al. 2012).

The PPCPs removal rates vary depending on the wastewater treatment technology used and the compound selected. In fact, removal efficiencies of some biodegradable pharmaceuticals such as triclosan (TCS) in WWTPs may typically be quite high (in some cases, up to 90 %). Still, this is not sufficient to satisfactorily reduce contamination as the remaining amounts of pharmaceutical after the treatment may still be too high for discharge in receiving water bodies (Aga 2008; Fent et al. 2006; Heberer 2002). This is because PPCPs are referred to as "pseudo-persistent" contaminants (i.e. high transformation/removal rates are compensated by their continuous introduction into the environment) and often have the same type of physico-chemical behaviour as other harmful xenobiotics (persistence in order to avoid the substance from becoming inactive before having a curing effect, and lipophilicity in order to be able to pass through membranes) (Zhang et al. 2013a).

The widespread uses of PPCPs such as caffeine (CAFF), oxybenzone (HMB) and TCS, coupled with their physico-chemical characteristics, and their generally inefficient removal from WWTPs, may cause their spread into the sewage system from where they can be potentially released for the aquatic environment (WFD-UKTAG 2009; Aga 2008; Fent et al. 2006; Heberer 2002).

18.1.1.1 Caffeine

1,3,7-Trimethylpurine, CAFF, is a purine alkaloid and one of the most widely consumed drugs in the world (Buerge et al. 2003). Buerge et al. (2003) demonstrated that CAFF was a suitable anthropogenic marker of domestic sewage contamination in surface waters due to its distinctive sewage origin, and ubiquitous and elevated consumption pattern in the human population.

18.1.1.2 Oxybenzone

2-Hydroxy-4-methoxybenzophenone, HMB, occurs naturally in flower pigments and is synthesized for use in sunscreens, as a UV stabilizer in various cosmetic products, and in plastic surface coatings and polymers (Magi et al. 2012).

18.1.1.3 Triclosan

5-Chloro-2-(2,4-dichlorophenoxy)phenol, TCS, is an ingredient added to many consumer products to reduce or prevent bacterial contamination. It may be found in products such as clothing, kitchenware, furniture and toys. It also may be added to antibacterial soaps and body washes, toothpastes and some cosmetics (Ying and Kookana 2007).

18.1.1.4 PPCPs in the Aquatic Environment

Several studies have reported the frequency and concentration of various PPCPs in the aquatic environment, with CAFF and TCS among the most frequently detected compounds (Fent et al. 2006). A study performed by Blair et al. (2013) showed the presence of 54 PPCPs in surface waters and sediments in a distance of 3.2 km from one WWTP. In total, 32 PPCPs were detected in Michigan Lake, 30 detected in sediment samples, and many of them still detected more than 3.2 km away from the WWTP. CAFF and TCS were two of the four PPCPs with a frequency higher than 50 % in all sampling sites in Lake Michigan: CAFF (97.6 %) and TCS (71.4 %). Similarly, Kolpin et al. (2002) compiled information about the occurrence of PPCPs and other organic contaminants in wastewater and water resources. CAFF was detected at a rate of 70 % and TCS of 60 %.

Table 18.1 shows some values reported in the literature for the occurrence of CAFF, HMB and TCS in the aquatic environment.

Despite their low concentrations in the aquatic environment, the ecotoxicological effects of pharmaceuticals are unpredictable because of the large number of compounds possibly present in that medium and due to the fact that they were designed as biologically active molecules (Fent et al. 2006). It is known that the increasing ecotoxicological impacts on organisms, either in the aquatic or in the

Table 18.1 Concentrations of caffeine, oxybenzone and triclosan detected in different waters

Contaminant	Type of water	Concentration ($\mu g\ L^{-1}$)	References
Caffeine (CAFF)	Influent	7–73	Buerge et al. (2003)
	Effluent	0.15–11.4	Aga (2008)
		6.0	Kolpin et al. (2002)
		0.03–9.5	Buerge et al. (2003)
	Superficial water	0.11–1.39	Blair et al. (2013)
		0.01–0.25	Buerge et al. (2003)
	Groundwater	0.23	
Oxybenzone (HMB)	Effluent	0.081–0.61	Aga (2008)
		3×10^{-5}–3.6×10^{-4}	Loraine and Pettigrove (2006)
	Superficial water	0.11	Aga (2008)
Triclosan (TCS)	Influent	0.38–3.36	Aga (2008)
	Effluent	0.106–0.321	Aga (2008)
		0.02–0.43	Ying and Kookana (2007)
		1–10	Lindström et al. (2002)
	Superficial water	0.07	Kolpin et al. (2002)
		0.01–0.06	Blair et al. (2013), Aufiero et al. (2012)
		1.4×10^{-3}–0.07	Lindström et al. (2002)
	Groundwater	$<4 \times 10^{-4}$	Lindström et al. (2002)

terrestrial environment, may include the development of antimicrobial resistance, decrease in plankton diversity and inhibition of growth of human embryonic cells (Klamerth et al. 2010; Reinhold et al. 2010; Fent et al. 2006). So, there is an increasing need of environmental friendly alternatives for PPCPs removal from wastewater for searching for higher removal efficiencies at reasonable costs.

18.1.2 Physical and Chemical Characteristics

Fate and effects of PPCPs depend on their structural characteristics and physico-chemical properties (Susarla et al. 2002).

18.1.2.1 Octanol–Water Partition Coefficient

The octanol–water partition coefficient (K_{ow}), which is the ratio of the concentration of unionized compound between octanol and water, has been adopted as the standard measure of hydrophobicity of a chemical compound (Reinhold et al. 2010; Burken and Schnoor 1998).

A high Log K_{ow} corresponds to high hydrophobicity. Extremely hydrophobic molecules Log $K_{ow} > 3$, like HMB and TCS, are tightly bound to soil organic matter and do not dissolve in the soil pore water (Pilon-Smits 2005).

Organics with moderate to high water solubility (Log $K_{ow} < 3$) will be able to migrate in the soil pore water to an extent that is inversely correlated with their Log K_{ow} (Pilon-Smits 2005).

Hydrophilic compounds (Log $K_{ow} < 1$) as CAFF (Log $K_{ow} = -0.77$) would not be significantly absorbed by plants (Burken and Schnoor 1998). For the author Pilon-Smits (2005) organic compounds with $0.5 <$ Log $K_{ow} < 3$ have adequate properties to move through cell membranes and the plants transpiration stream, thereby being easily taken up by the plants. However, recent studies have shown otherwise. Dettenmaier et al. (2009) and Zhang et al. (2013b) have shown that highly soluble hydrophilic organic compounds are more likely to be absorbed by plant roots and translocated into the tissues. These data led to a new model of plant root uptake potential, which states that compounds with Log $K_{ow} < 0.5$ present the greatest potential for root uptake and translocation.

18.1.2.2 Organic Carbon Partition Coefficient

The organic carbon partition coefficient (K_{oc}) is defined as the ratio of the mass of contaminants adsorbed by unit weight of organic carbon in the soil, with the concentration in solution. The K_{oc} values give an indication of the mobility of organic compounds in the soil. Low values of Log K_{oc} are related to mobile compounds like CAFF unlike the HMB and TCS (Grathwohl 1990 in Zhang et al. 2013a).

18.1.2.3 Henry Constant

Another property of chemical compounds that affects movement in soils is volatility. The volatility is expressed by Henry's constant (Hi) derived from the vapour pressure of the compounds and water solubility (in mg L^{-1}). Pollutants with Hi $> 10^{-4}$ tend to move in the air spaces between soil particles, whereas pollutants with Hi $< 10^{-6}$ move predominantly in water. If Hi is between 10^{-4} and 10^{-6}, compounds are mobile in both air and water (Pilon-Smits 2005).

Table 18.2 shows some of the most relevant physical and chemical properties of these contaminants.

18.2 Phytoremediation

Environmental pollution by organic compounds such as PPCPs is a global problem. Therefore, the development of efficient remediation technologies that cause minimal damage to the environment is of great importance (Abhilash et al. 2009).

The use of plants to remediate contaminated systems is not a new technique and started to be developed in the 1950s. The term phytoremediation appeared in 1980 but the expansion in this area began only in the 1990s (McCutcheon and Schnoor 2003; Cunningham and Berti 1993). Phytoremediation is based on the use of plants for in situ environmental clean-up (Pereira et al. 2007; Pilon-Smits 2005; Newman and Reynolds 2004; Meagher 2000; Salt et al. 1998).

The fact that phytoremediation is a solar-driven technology, and thus more sustainable, usually carried out in situ contributes to its cost-effectiveness. It allows the reduction exposure of the polluted substrate to humans, wildlife and the environment. Phytoremediation also enjoys popularity with the general public as a "green clean" (Pilon-Smits 2005).

Furthermore, phytoremediation offers other benefits at contaminated sites. It increases the soil microbial activity, stabilizes soil reducing erosion and protects the soil of direct sunlight.

Phytoremediation has advantages but also limitations. The time needed for successful phytoremediation is an important factor that needs to be taken into consideration for practical applications. Phytoremediation is also highly related to the growth and plant metabolism (which in turn is dependent on genetic and environmental factors) that, in combination with soil characteristics and environmental conditions, strongly influence the rate of the remediation process (Pilon-Smits 2005; McCutcheon and Schnoor 2003; Susarla et al. 2002).

The soil properties, toxicity level and climate should allow plant growth. Phytoremediation may also be limited by root depth and bioavailability of pollutants (Susarla et al. 2002; Dietz and Schnoor 2001).

Table 18.2 Physico-chemical properties of caffeine, oxybenzone and triclosan

Compounds	Category	Chemical structure	Molecular weight[a] (g mol^{-1})	Solubility[b] (mg L^{-1} at 25 °C)	Log K_{ow}[a–c]	Log K_{oc}[b,d]	Hi[b,d] (Pa m^3 mol^{-1} at 25 °C)
Caffeine	Stimulant	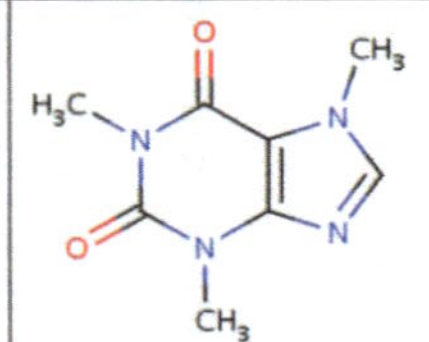	194.19	2.16×10^4	−0.07	1	3.58×10^{-11}
Oxybenzone	Fragrance ingredient	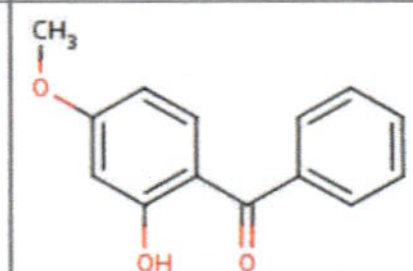	228.24	69	3.8	2.7	1.50×10^{-8}
Triclosan	Antiseptic	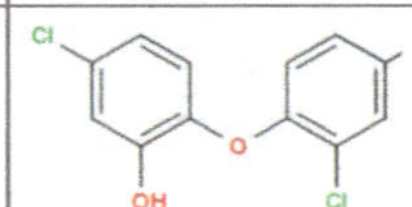	289.55	10	4.76	3.96	2.4×10^{-7}

[a]Sigma-Aldrich (2014)
[b]Pubchem (2014)
[c]Zhang et al. (2013a)
[d]WFD-UKTAG (2009)

Table 18.3 SWOT analysis

Strengths	Weaknesses
– In situ	– Influenced by the weather and climate conditions of the site
– Solar driven	– Treatment time (generally, longer than physico-chemical methods)
– Conserves natural resources	– Limited by root action
– Low cost	
– Implementation with little or no environmental disturbances	
– Environmental benefits, e.g. controlling soil erosion, carbon sequestration and habitat	
– Application over large areas where other technologies are not viable	
– Versatility to treat a wide variety of hazardous materials	
– Possibility tore-use the soil	
Opportunities	*Threats*
– High public acceptance	– High concentrations of contaminants can be toxic to plants and animals that consume the plants
– Possibility of combination with other techniques	– Contaminants may be mobilized into the ground water
– Aesthetically attractive	– Toxicity and bioavailability of biodegradation products is sometimes unknown
– The biomass produced can be economically valued in raw material for various activities	

Although phytoremediation has been primarily applied for the removal of inorganic pollutants in the soil, this technology has proven to be effective in the treatment of organic pollutants such as chlorinated solvents, petroleum hydrocarbons including polycyclic aromatic hydrocarbons (PAHs), pesticides and herbicides, radioactive chemicals, industrial organic waste and explosives (Couto et al. 2011; Pilon-Smits 2005; Newman and Reynolds 2004; Williams 2002; Meagher 2000; Salt et al. 1998).

Table 18.3 presents a SWOT (strengths, weaknesses, opportunities and threats) analysis of the technique here discussed.

18.2.1 Phytoremediation Processes

Phytoremediation is an efficient clean-up technology for a variety of contaminants that can be achieved by plant roots and subjected to various mechanisms: immobilized, sequestered, degraded or metabolized in place, either inside or outside the plant, depending on the type of contaminant (Cameselle et al. 2013; Susarla et al. 2002; Dietz and Schnoor 2001). These processes are illustrated in Fig. 18.1.

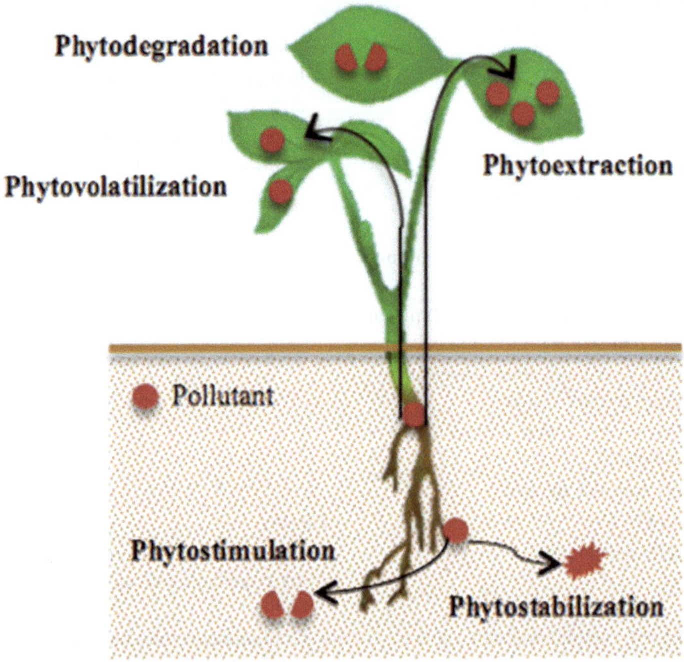

Fig. 18.1 Possible fates of contaminants during phytoremediation (Adapted from Pilon-Smits 2005)

The term phytostabilization refers to the use of plants to stabilize pollutants in soil, either simply by preventing erosion, leaching, or runoff, or by converting pollutants to less bioavailable forms (e.g. via precipitation in the rhizosphere). Plants can also be used to extract pollutants and accumulate them in their tissues, followed by harvesting of the (above ground) plant material. This technology is called phytoextraction. The phytostimulation or rhizodegradation is the degradation of organic contaminants near the root mass by bacteria and fungi, which activity is stimulated by the root exudes and enzymes released by the plants (USEPA 2000). Plants can also degrade organic pollutants directly via their own enzymatic activities, a process called phytodegradation. After uptake in plant tissue, certain pollutants can leave the plant in volatile form, this is called phytovolatilization. These various phytoremediation technologies are not isolated and may occur simultaneously; for instance, accumulation, stabilization and volatilization can occur together. Because the processes involved in phytoremediation occur naturally, vegetated polluted sites have a tendency to clean themselves up without human interference (Pilon-Smits 2005).

Some of the factors affecting chemical uptake and distribution within living plants include: (1) physical and chemical properties of the compound (e.g. water solubility, vapour pressure, molecular weight and octanol–water partition coefficient); (2) environmental characteristics (e.g. temperature, pH, organic matter and soil moisture content); (3) plant characteristics (e.g. root system, type of exudates) (Pilon-Smits 2005).

18.2.2 Wetlands

Wetlands encompass a broad range of ecosystems ranging from submerged coastal grass beds to salt marshes, swamp forests and boggy meadows. In the present chapter, the term "wetlands" refers to transition zones between terrestrial and aquatic systems with soils water saturated or covered by shallow water along with characteristic wetland plant species.

Surface waters, like salt marsh areas, are considered sinks for pollutants due to the continuous discharge of treated effluents from the WWTPs. This is of major concern as wetlands provide an ecologically important environment and numerous benefits for humans (Boorman 1999). These important ecosystem services are often at risk due to the anthropogenic inputs from upstream catchments as well as from metropolitan areas and industries located at the vicinity of those areas. The importance of salt marsh ecosystems and their capability for degrading pollutants makes imperative the pursuit of remediation strategies to favour the recovery of coastal sediments. Therefore, the application of adapted strategies that cause the minimum environmental and ecological impact is needed to restore these contaminated areas.

Reboreda and Caçador (2007b) say that retention of suspended particles by marsh vegetation acts as a trap for many contaminants reducing the availability of the same in the ecosystem.

Assessing the phytoremediation potential of wetlands is complex due to variable conditions of hydrology, soils/sediment types, plant species diversity, growing season and water chemistry. The remediation and restoration of contaminated sites requires a detailed understanding of how contaminants and plants behave in a particular ecosystem, before remedial activities could be carried out (Punshon et al. 2003).

18.2.3 Constructed Wetlands

There are two types of wetlands, the natural occurring ones and those that have been purposely constructed. Constructed wetlands (CWs) are defined as engineered wetlands that utilize natural processes involving wetland vegetation, soil and their associated microbial assemblages to assist, at least partially, in treating wastewater or other polluted water sources (Helt et al. 2012).

Sustainable wastewater treatment technologies such as CWs have been receiving increasing attention due to their ability to remove a large variety of pollutants (Matamoros et al. 2008; Williams 2002).

With the unique advantages of low-cost, simple operation/maintenance, and environmental friendliness, aquatic plant-based systems have a long history for treatment of all kinds of wastewater (Zhang et al. 2013a; Vymazal 2005).

These technologies can be used for primary, secondary and tertiary treatment of municipal or domestic wastewaters, storm water, agricultural and industrial wastewaters (such as landfill leachate, petrochemicals, food wastes, pulp and paper) and mining, usually combined with an adequate pretreatment (Mackova et al. 2006; Kadlec et al. 2000).

Once in wetlands, contaminants are subjected to a variety of removal mechanisms and processes including microbial degradation, photodegradation, plant uptake and physical interactions (Matamoros et al. 2005). However, to date there are only a few reports on the ability of CW systems to reduce concentrations of pharmaceutical compounds and very few studies on the use of the physico-chemical properties of these compounds for predicting their fate in CWs (Matamoros et al. 2008). CWs have been often viewed as a "black box", with only influent and effluent concentrations measured to gauge performance and without detailed studies on the actual fate of the compounds or their removal pathways. Additionally, the specific roles played by microbial degradation, photodegradation and plant uptake with aquatic plants to remediate PPCPs (Zhang et al. 2013a).

CWs have shown higher efficiencies in removing pharmaceuticals from wastewaters compared with conventional wastewater treatment processes with plants having an important role in the removal of some of those compounds (Carvalho et al. 2012; Matamoros et al 2012; Dordio et al. 2011; Matamoros et al. 2007). For example, Dordio et al. (2010) achieved a removal of 96 % in selected PPCPs: ibuprofen, carbamazepine and clofibric acid and Matamoros et al. (2012) reported between 83 and 96 % removal of CAFF in systems with macrophytes beds (opposing to the ratio reported in the absence of plants: 0–30 % removal).

The choice of substrate in CWs is of major importance as it supports living organisms and provides storage for many contaminants. The substrates can be natural, such as gravel, sand and organic materials including compost and waste material. Its permeability affects the wastewater flow through the CW, and is where chemical and biological transformations occur, either by microorganisms or plants (Korkusuz 2005).

Light expanded clay aggregates (LECA) are thus being increasingly considered as alternative low-cost adsorbents for organic pollutants (Dordio et al. 2007). LECA is an artificially modified natural material that is produced by subjecting clay materials to a high temperature treatment, causing injected CO_2 to expand within the clay aggregates and thus creating a highly porous, lightweight material LECA is mainly used for construction purposes but over the last years it is also being used for different types of water and wastewater treatment processes such as filtering and in CWs (Dordio et al. 2009a, b; Brix et al. 2001). Previous studies have shown that LECA are able to remove by sorption PPCPs from water and wastewater and adequate for the development of plants and microorganisms (Dordio et al. 2007, 2009a, b, 2010), while other more commonly used media such granitic gravel showed almost no capacity to sorb it (Matamoros et al. 2008; Matamoros et al. 2005).

18.2.4 Macrophytes Plant Species: Remediation Potential

For phytoremediation to be successful it is necessary that the selected plant species exhibit a deep and dense root system, ability to grow in various environments, considerable growth rate and biomass production (especially for metals) and resistance to pests and diseases (Douglas and Rats 2007; Pilon-Smits 2005).

Some macrophyte species of plants have these characteristics. Macrophytes play a vital role in healthy ecosystems because they serve as primary producers of oxygen through photosynthesis, provide a substrate for algae and shelter for many invertebrates, aid in nutrient cycling to and from the sediments, and help stabilize river and stream banks. Furthermore, these species have a variable capacity for nutrient uptake and heavy metals accumulation also affecting the functioning and structure of bacterial communities involved in the removal of contaminants (Ruiz-Rueda et al. 2009).

Tagus estuary is mainly colonized by macrophyte species *Halimione portulacoides* and *Spartina maritima*, occupying different areas of the estuary (Duarte et al. 2007). The complex systems that constitute salt marsh sediments may be strongly influenced by plant activity. The species *S. maritima* is distributed in the lower areas, while *H. portulacoides* colonizes the central areas. This fact increases the differences between the chemical characteristics of the sediment, and the different times of submersion (Reboreda and Caçador 2007a). The study performed by Caçador et al. (2000) showed that in this estuary the sediments between the *S. maritima* roots concentrate more metals than the sediments between *H. portulacoides* roots. In plant tissues there was an opposite trend. Thus, it was concluded that the use of *S. maritima* and *H. portulacoides* for the phytoremediation of marshes contaminated with metals may be applied in two different perspectives: (1) phytostabilization of metals in rizhosediments or (2) phytoextraction through accumulation in the tissues of the aerial parts of the plant, followed by plant harvesting.

Previous studies have also demonstrated that macrophyte species have a positive role in the removal of organic contaminants like PPCPs (Zarate et al. 2012; Hijosa-Valsero et al. 2010a).

Hijosa-Valsero et al. (2010b) reported that *Phragmites australis* was more efficient than *Typha angustifolia* for the removal of ibuprofen and CAFF. Stevens et al. (2009) evidenced that different plant species may accumulate TCS, methyl-triclosan (MTCS) and triclocarban (TCC). To date, there is little information regarding the ability of different macrophyte species to bioaccumulate pharmaceuticals and which aquatic plant species may maximize the pharmaceutical elimination rates (Zarate et al. 2012).

According to literature, the choice of plants is also an important issue, since the metabolic activity in the plant rhizosphere releases oxygen, which helps nitrification process through direct absorption of nutrients. In turn, the access and availability of nutrients affect plant response in terms of growth and resource allocation, a fact that will influence the removal efficiency in salt marsh areas (Zhang et al. 2013a). The species of plant *S. maritima* presents a significant rhizome structure, with adventitious roots borne in it, and *H. portulacoides* presents a fibrous and dense root system.

Even within the same plant species, the biomass can influence the process of removing the contaminants. Carvalho et al. (2012) reports that a higher percentage of biomass corresponds to a higher removal percentage. In this sequence, we can say that the remediation values for both the *H. portulacoides* as for *S. maritima* could be higher between July and September or January and March (when there are the largest underground biomass values) in relation to time of the study, October and November. Although the several reported studies the performance between different macrophyte species for PPCPs removal has not been well demonstrated (Zhang et al. 2013a).

18.3 Case Studies

In this chapter, is also given an overview of the research performed by the team. The first case study is based on the simulation of salt marsh aiming to understand the behaviour of the selected PPCPs (CAFF, HMB and TCS) in natural environments and to assess the potential of *S. maritima* and *H. portulacoides* to remediate the target contaminants. The second case study is with the same plant species but searching for its potential on constructed wetlands and the use of LECA as support media for the CW.

18.3.1 Experimental Procedure

Case study 1: Two sets of experiments with *S. maritima* and *H. portulacoides* were carried out during 10 days in parallel with different media: *set I* consisted of only sediment elutriate (E) and *set II* consisted of sediment mixed with the respective elutriate (ES).

Elutriate (E) is a simpler medium which might mimic hydroponic plant exposure to contaminants for phytoremediation purposes and would facilitate the interpretation of results. Experiments carried out with sediment soaked in the respective elutriate (ES) allowed simulation of nutrients and contaminants exchange among plants, solution and sediment, as in the natural environment.

The plants, water and colonized sediment were collected in October of 2013 at low tide from a salt marsh, located in the Tagus Estuary, Portugal (38°36′59.39″N; 9°02′33.41″W).

Case study 2: A microcosm CWs system was prepared with LECA, deionized water (H_2O) and planted with *S. maritima* were used to evaluate their ability to remove CAFF, HMB and TCS. The *S. maritima* plant species was chosen as it presented a good performance in the previous work.

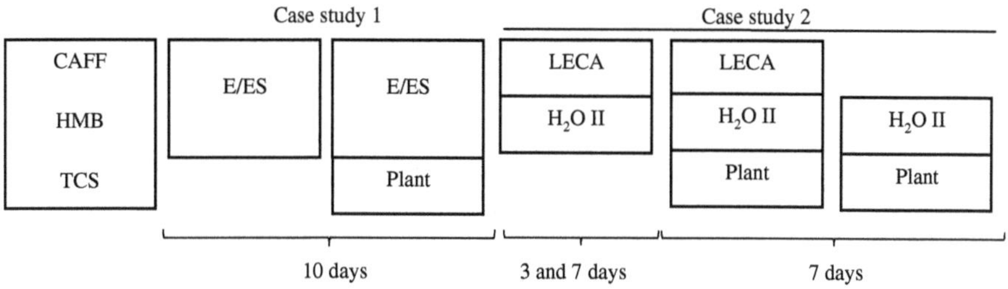

Fig. 18.2 Experimental design of the work with each experimental treatment carried out in triplicate

The sorption of LECA was evaluated during 3 and 7 days. The mass of contaminants found in the different matrices at the end of the experiment was compared with the mass added at the beginning of the experiment in order to obtain removal percentages in the liquid and solid phases.

Experimental design of the assay is shown in Fig. 18.2.

18.3.2 General Procedure

In both case studies, all microcosms were wrapped in aluminium foil to simulate a real system (no light penetration at substrate level, preventing photodegradation of the compounds).

Groups of plant species were homogeneously distributed by different treatment and exposed to the media spiked with 1 mg L^{-1}. Control experiments without plants and plants without contamination (to assess the vigour of the plants) were also carried out in parallel. At the end of the experiments, plant stress indicators were obtained by evaluating the levels of chlorophylls *a*, *b* and carotenoids in plant leaves, according to the procedure of Lichtenthaler and Wellburn (1983).

Gas chromatography with flame ionization detection (GC-FID) and high performance liquid chromatography (HPLC) were used to quantify the levels of different contaminants in the studied matrices.

18.3.3 Case Study 1: Potential of Two Salt Marsh Plants for PPCPs Remediation in Contaminated Salt Marsh Areas

The percentage of each contaminant present in different phases, liquid phase and sediment, at the end of the experiments are presented in Table 18.4 (Couto et al. (Submitted)).

Table 18.4 Percentage of contaminant at end of assay in elutriate (E) and elutriate soaked in sediment (ES) for control and species of plants *S. maritima* and *H. portulacoides* (mean ± SD, $n = 3$) (Couto et al. (Submitted))

	Control			*S. maritima*			*H. portulacoides*		
		ES			ES			ES	
Contaminant	E	Liquid phase	Sediment	E	Liquid phase	Sediment	E	Liquid phase	Sediment
CAFF	100 ± 3	83 ± 4	18 ± 6	100 ± 4	49 ± 4	25 ± 3	107 ± 3	56 ± 1	20 ± 6
HMB	78 ± 3	18 ± 0	72 ± 22	35 ± 4	12 ± 1	31 ± 9	43 ± 4	18 ± 0	31 ± 9
TCS	75 ± 3	4 ± 0	56 ± 17	10 ± 4	3 ± 0.1	25 ± 8	15 ± 4	4 ± 0	25 ± 8

The results show that the contaminants have different behaviours which might be explained by physical and chemical characteristics. Hydrophobic compounds like HMB and TCS showed that can be adsorbed onto the organic matter present in the granular medium (sediment) and become more recalcitrant to biodegradation resulting in high accumulation on the medium. Whereas, hydrophilic compounds (Log $K_{ow} < 1$), as CAFF, are removed by different processes like biodegradation and plant uptake (Zhang et al. 2013b, c; García et al. 2010; Buerge et al. 2003).

The authors Burken and Schnoor (1998) reported that hydrophilic compounds would not be significantly absorbed by plants. This fact is coincident with the results obtained in the presented study (Couto et al. (Submitted)), where in the E experiment, only a small percentage of CAFF was remediated in the presence of plants. However, the presence of sediment (ES treatments) and plants appear to have stimulated CAFF remediation due to possible interactions between them (Couto et al. (Submitted)). It is not expected that CAFF adsorbs to the sediment (Log $K_{ow} = -0.07$ and Log $K_{oc} = 1$). Instead it should solubilize (solubility of 2.16×10^4 mg L^{-1}) being present in the liquid phase. It is not expected that photodegradation has a significant influence on the degradation of this compound as it does not absorb light at wavelengths exceeding 290 nm (PubChem 2014).

The contaminants HMB and TCS showed a different behaviour from CAFF. In E assay, HMB and TCS remediation seem to be enhanced in the presence of plants, when the experiment was conducted in the elutriate (Couto et al. (Submitted)). These results can be explained by the characteristics of the compounds (e.g. Log K_{ow}). It is expected that these contaminants may adsorb to plant roots but it is not expected to be absorbed (Sect. 18.1.2).

When the experiments were carried out with elutriate and sediment (ES) the unvegetated sediment (control) presented higher concentrations of HMB and TCS, compared to the vegetated sediment (Couto et al. (Submitted)). The contaminants may have adsorbed to the plant root and sediment where it may have been biodegraded. This behaviour is explained by the physico-chemical characteristics of the compound: high hydrophobicity (Log $K_{ow} > 3$) and low solubility.

These results suggest that the presence of plants may have stimulated the bio/rhizoremediation of contaminants. The "rhizosphere effect" is caused by physical (e.g. gas exchange) and chemical (exudates) processes resulting from the interaction between plant roots and sediments (McCully 1999). This favourable habitat for microorganisms in the root zone of plants can lead to increased degradation of contaminants by specific microorganisms.

As for abiotic factors, the contaminants HMB and TCS are susceptible to suffer photodegradation (PubChem 2014). The HMB and TCS control in E assay showed 3 % and 25 %, respectively, enhanced by the photodegradation remediation. The other abiotic factors were considered low or even negligible because the adsorption of compounds to the walls of the containers was about 0.02 % and the volatilization is not expected to be an important fate process given an estimated Henry's Law.

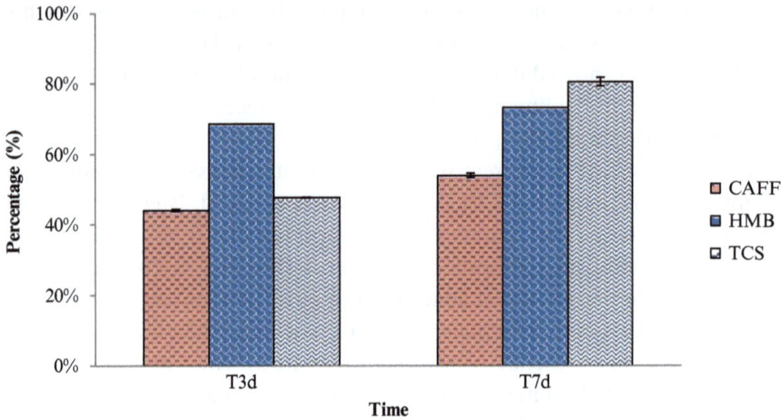

Fig. 18.3 Concentration (mg L^{-1}) of CAFF, HMB and TCS, after 3 days and 7 days for assay with LECA (mean ± SD, $n = 3$) (Adapted from Ferreira et al. (Submitted))

18.3.4 Case Study 2: Pharmaceuticals Removal Efficiency by the Microcosm Constructed Wetlands

The removal of PPCPs like CAFF, HMB and TCS in CWs can involve a diverse and complex set of physical, chemical and biological processes. Optimization of CWs for removal contaminants can be achieved by careful selection of its components, such as the plant species used and the materials which compose the support matrix. To date, only a few studies have been completed on the specific roles played by microbial degradation, photodegradation and plant uptake in aquatic plant-based systems used to remediate pharmaceutical compounds (Hijosa-Valsero et al. 2010a; Matamoros et al. 2005).

18.3.4.1 Unplanted Beds

Figure 18.3 resumes of the removal percentage of each contaminant after 3 and 7 days of treatment relative to the initial concentration used in the study (1 mg L^{-1}). The results show that considerably high removals (44 % for CAFF, 69 % HMB and 48 % TCS) were attained after 3 days of treatment (Ferreira et al. (Submitted)).

After 3 days of experiment, the removal effectiveness started to decrease for the more soluble compound because the absorption capacity of LECA decreases (Ferreira et al. (Submitted)). This may be explained by high water solubility and low hydrophobicity of compounds, as CAFF, which induces its presence in the liquid phase instead of entrapped in LECA. However, for less soluble contaminant, as TCS, the removal continues to increase substantially over time, achieving 80 % after 7 days of treatment (more 32 % than the third day) (Ferreira et al. (Submitted)).

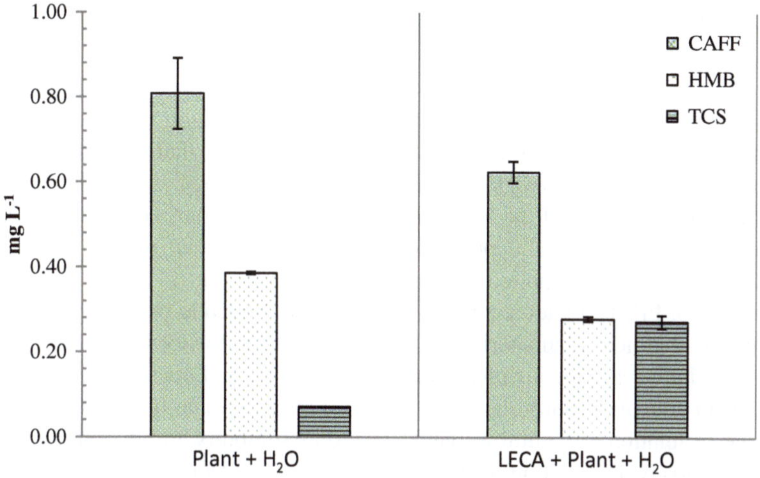

Fig. 18.4 Pharmaceutical concentrations (mg L^{-1}) detected in solution after 7 days for assays only planted with *S. maritima* and planted beds (mean ± SD, $n = 3$) (Adapted from Ferreira et al. (Submitted))

Sorption processes associated with the presence of LECA may be of major importance to achieve this efficient removal, especially for hydrophobic compounds. The previous work (Sect. 18.3.3) has shown that hydrophobic compounds like HMB and TCS have properties (e.g. Log K_{ow} and Log K_{oc}) that allow their adsorption onto the granular medium.

18.3.4.2 Planted Beds

The results for the assay with plants and LECA are shown in Fig. 18.4 (Ferreira et al. (Submitted)).

The assay results with only plant suggest that plants can have an important role in the remediation of less soluble compound with Log $K_{ow} > 3$ (HMB and TCS). The action of plants as an enhancing remediation agent is connected to the root surface, where in the absence of solid surface, the contaminants adsorbed only the roots of plants. These results are coincident with those obtained in the previous case study (Sect. 18.2.3), where the plant also enhanced the removal of HMB and specially TCS.

In addition, the presence of LECA, unlike the sediment, did not enhance TCS removal. This can be justified by the absence of microorganisms in LECA that would naturally occur in the sediment providing bio/rhizoremediation processes.

18.4 Conclusions

Due to contamination of PPCPs in aquatic systems is relevant demand for efficient, economic and causing as little disturbance in the remediation process.

The results of phytoremediation experiments show that salt marsh plants, *H. portulacoides* and *S. maritime*, have potential to remediate CAFF, HMB and TCS. The performance of the two plants studied was similar. However, its characteristics (morphological and physiological) may be a determining factor in the choice of species to be applied in future remediation work.

The fate of PPCPs in aquatic plant-based system proved to be strongly influenced by physico-chemical characteristics of them, which influenced the distribution of substances between aqueous phase and solid surfaces, and consequently affects the transport in aqueous environment and regulates their ultimate fate.

In the simulation of natural wetlands, due to its solubility and lipophilicity, CAFF remained mostly in the liquid phase and was poorly remediate by plants during 10 days. However, the interaction between sediment and plant seemed to stimulate CAFF remediation. The compounds with Log $K_{ow} > 3$, HMB and TCS, undergo different remediation mechanisms/environmental partitioning: photodegradation, bio/rhizoremediation, adsorption to sediment and roots.

In the microcosms CWs, an important role in the removal processes was played by LECA due to contaminants sorption to this solid matrix. The first period seems to be most important stage of the whole sorption process. CAFF was observed to be extensively absorbed by the LECA matrix in 3 days, although for HMB and TCS, removal has to be attributed to adsorption.

Future research work is necessary to complement and substantiate the information provided by the present studies.

Techniques such as bioaugmentation and biostimulation may be considered initially in the laboratory and then in field scale. Furthermore, coupling the phytoremediation with other process such as the electrokinetic process may increase the availability of organic contaminants and to stimulate the microbial community.

Additionally, more field investigations of the specific fate of a wider spectrum of PPCPs in aquatic plant-based systems with wastewater are necessary, as well as, performance of full-scale systems should be assessed. Further studies are needed in order to better understand the mechanisms involved in PPCPs removal at the plant level.

References

Abhilash PC, Jamil S, Singh N (2009) Transgenic plants for enhanced biodegradation and phytoremediation of organic xenobiotics. Biotechnol Adv 27:474–488

Aga A (2008) Fate of pharmaceuticals in the environment and in water treatment systems. CRC, Boca Raton

Aufiero M, Butler C, Jaser J (2012) An analysis of methods for detecting triclosan and removal of triclosan from water using activated carbon and zeolites. Major Qualifying Project completed in partial fulfillment of the Bachelor of Science Degree at Worcester Polytechnic Institute, Worcester

Barceló D, Petrovic M (2007) Pharmaceuticals and personal care products (PPCPs) in the environment. Anal Bioanal Chem 4:1141–1142

Barceló J, Poschenrieder C (2003) Phytoremediation: principles and perspectives. Institut d'Estudis Catalans, Barcelona. Contrib Sci 2:333–344

Blair BD, Crago JP, Hedman CJ, Klaper RD (2013) Pharmaceuticals and personal care products found in the Great Lakes above concentrations of environmental concern. Chemosphere 93:2116–2123

Boorman LA (1999) Salt marshes—present functioning and future change. Mangrove Salt Marshes 3:227–241

Brix H, Arias CA, Del Bubba M (2001) Media selection for sustainable phosphorus removal in subsurface flow constructed wetlands. Water Sci Technol 44:47–54

Buerge IJ, Poiger T, Müller MD, Buser HR (2003) Caffeine, an anthropogenic marker for wastewater contamination of surface waters. Environ Sci Technol 37:691–700

Burken JG, Schnoor JL (1998) Predictive relationships for uptake of organic contaminants by hybrid poplar trees. Environ Sci Technol 32:3379–3385

Caçador I, Vale C, Catarino F (2000) Seasonal variation of Zn, Pb, Cu and Cd concentrations in the root-sediment system of *Spartina maritima* and *Halimione portulacoides* from Tagus estuary salt marshes. Mar Environ Res 49:279–290

Cameselle C, Chirakkara K, Reddy K (2013) Electrokinetic-enhanced phytoremediation of soils: status and opportunities. Chemosphere 93:626–636

Carvalho PN, Basto MCP, Almeida CMR (2012) Potential of *Phragmites australis* for the removal of veterinary pharmaceuticals from aquatic media. Bioresour Technol 116:497–501

Couto N, Ferreira AR, Guedes P, Mateus E, Ribeiro AB (Submitted). Partition and remediation potential of caffeine, oxybenzone and triclosan in estuarine environment.

Couto MNPFS, Basto MCP, Vasconcelos MTSD (2011) Suitability of different salt marsh plants for petroleum hydrocarbons remediation. Chemosphere 84:1052–1057

Cunningham SD, Berti WR (1993) Remediation of contaminated soils with green plants: an overview. Tissue Culture Association. In Vitro Cell. Dev Biol 29:207–212

Cunningham VL, Buzby M, Hutchinson T, Mastrocco F, Parke N, Roden N (2006) Effects of human pharmaceuticals on aquatic life: next steps. Environ Sci Technol 40:3456–3462

Dettenmaier EM, Doucette WJ, Bugbee B (2009) Chemical hydrophobicity and uptake by plant roots. Environ Sci Technol 43:324–329

Dietz AC, Schnoor JL (2001) Advances in phytoremediation. Environ Health Perspect 109:163–168

Dordio AV, Teimao J, Ramalho I, Palace Carvalho AJ, Estevao Candeias AJ (2007) Selection of a support matrix for the removal of some phenoxyacetic compounds in constructed wetlands systems. Sci Total Environ 380:237–246

Dordio A, Pinto J, Dias CB, Pinto AP, Carvalho AJP, Teixeira DM (2009a) Atenolol removal in microcosm constructed wetlands. Int J Environ Anal Chem 89:835–848

Dordio AV, Estêvão Candeias AJ, Pinto AP, Teixeira da Costa C, Carvalho AJP (2009b) Preliminary media screening for application in the removal of clofibric acid, carbamazepine and ibuprofen by SSF-constructed wetlands. Ecol Eng 35:290–302

Dordio AV, Carvalho AJP, Teixeira D, Dias CB, Pinto AP (2010) Removal of pharmaceuticals in microcosm constructed wetlands sing Typha spp. and LECA. Bioresour Technol 101:886–892

Dordio AV, Gonçalves P, Texeira D, Candeias AJ, Castanheiro JE, Pinto AP, Carvalho AJP (2011) Pharmaceuticals sorption behaviour in granulated cork for the selection of a support matrix for a constructed wetlands system. Int J Environ Anal Chem 91:615–631

Douglas H, Rats A (2007) Fitorremediação: Considerações Gerais e Características de Utilização. Silva Lus 15:103–117 (in Portuguese)

Duarte B, Delgado M, Caçador I (2007) The role of citric acid in cadmium and nickel uptake and translocation, in *Halimione portulacoides*. Chemosphere 69:836–840

Fent K, Weston AA, Caminada D (2006) Ecotoxicology of human pharmaceuticals. Aquat Toxicol 76:122–159

Ferreira AR, Couto N, Ribeiro A (Submitted) Removal of pharmaceuticals and personal care products in microcosm constructed wetland with *Spartina maritima* and LECA

García J, Rousseau DPL, Morató J, Lesage E, Matamoros V, Bayona JM (2010) Contaminant removal processes in subsurface-flow constructed wetlands: a review. Crit Rev Environ Sci Technol 40:561–661

Heberer T (2002) Occurrence, fate, and removal of pharmaceutical residues in the aquatic environment: a review of recent research data. Toxicol Lett 131:5–17

Helt CD, Weber KP, Legge RL, Slawson RM (2012) Antibiotic resistance profiles of representative wetland bacteria and faecal indicators following ciprofloxacin exposure in lab-scale constructed mesocosms. Ecol Eng 39:113–122

Hijosa-Valsero M, Matamoros V, Martín-Villacorta J, Bécares E, Bayona JM (2010a) Assessment of full-scale natural systems for the removal of PPCPs from wastewater in small communities. Water Res 44:1429–1439

Hijosa-Valsero M, Matamoros V, Sidrach-Cardona R, Martín-Villacorta J, Bécares E, Bayona JM (2010b) Comprehensive assessment of the design configuration of constructed wetlands for the removal of pharmaceuticals and personal care products from urban wastewaters. Water Res 44:3669–3678

Kadlec R, Knight R, Vymazal J, Brix H, Cooper P, Haberl R (2000) Constructed wetlands for pollution control. IWA, London

Klamerth N, Rizzo L, Malato S, Maldonado MI, Agüera A, Fernández-Alba AR (2010) Degradation of fifteen emerging contaminants at μg L^{-1} initial concentrations by mild solar photo-fenton in MWTP effluents. Water Res 44:545–554

Kolpin DW, Furlong ET, Meyer MT, Thurman EM, Zaugg SD, Barber LB, Buxton HT (2002) Pharmaceuticals, hormones, and other organic wastewater contaminants in U.S. streams, 1999–2000: a national reconnaissance. Environ Sci Technol 36:1202–1211

Korkusuz EA (2005) Manual of practice on constructed wetlands for wastewater treatment and reuse in Mediterranean countries. Report, MED-REUNET II Support Programme (EC Project No: INCO-CT-2003–502453), AGBAR FOUNDATION, Spain

Lichtenthaler HK, Wellburn AR (1983) Determinations of total carotenoids and chlorophylls a and b of leaf extracts in different solvents. 603rd Meeting Held at the University of Liverpool, Liverpool, 11–13 April 1983

Lindström A, Buerge IJ, Poiger T, Bergqvist PA, Müller MD, Buser HR (2002) Occurrence and environmental behavior of the bactericide triclosan and its methyl derivative in surface waters and in wastewater. Environ Sci Technol 36:2322–2329

Loraine GA, Pettigrove ME (2006) Seasonal variations in concentrations of pharmaceuticals and personal care products in drinking water and reclaimed wastewater in southern California. Environ Sci Technol 40:687–695

Mackova M, Dowling D, Macek T (2006) Phytoremediation and rhizoremediation, vol 9, 1st edn. Springer, Dordrecht, pp 1–3

Magi E, Di Carro M, Scapolla C, Nguyen KTN (2012) Stir bar sorptive extraction and LC–MS/MS for trace analysis of UV filters in different water matrices. Chromatographia 75:973–982

Matamoros V, García J, Bayona JM (2005) Behavior of selected pharmaceuticals in subsurface flow constructed wetlands: a pilot-scale study. Environ Sci Technol 39:5449–5454

Matamoros V, Arias C, Brix H, Bayona JM (2007) Removal of pharmaceuticals and personal care products (PPCPs) from urban wastewater in a pilot vertical flow constructed wetland and a sand filter. Environ Sci Technol 41:8171–8177

Matamoros V, Caselles-Osorio A, Garcia J, Bayona JM (2008) Behaviour of pharmaceutical products and biodegradation intermediates in horizontal subsurface flow constructed wetland. A microcosm experiment. Sci Total Environ 394:171–176

Matamoros V, Nguyen LX, Arias CA, Salvadó V, Brix H (2012) Evaluation of aquatic plants for removing polar microcontaminants: a microcosm experiment. Chemosphere 88:1257–1264
McCully M (1999) Roots in soil: unearthing the complexities of roots and their rhizospheres. Annu Rev Plant Physiol Plant Mol Biol 50:695–718
McCutcheon SC, Schnoor JL (2003) Phytoremediation—transformation and control of contaminants, vol 9. Wiley, Hoboken, pp 1–298
Meagher RB (2000) Phytoremediation of toxic elemental and organic pollutants. Curr Opin Plant Biol 3:153–162
Newman LA, Reynolds CM (2004) Phytodegradation of organic compounds. Curr Opin Biotechnol 15:225–230
Onesios KM, Yu JT, Bouwer EJ (2009) Biodegradation and removal of pharmaceuticals and personal care products in treatment systems: a review. Biodegradation 20:441–466
Pereira P, Caçador I, Vale C, Caetano M, Costa AL (2007) Decomposition of belowground litter and metal dynamics in salt marshes (Tagus Estuary, Portugal). Sci Total Environ 380:93–101
Pilon-Smits E (2005) Phytoremediation. Annu Rev Plant Biol 56:15–39
Pubchem (2014) Available in: http://pubchem.ncbi.nlm.nih.gov
Punshon T, Gaines KF, Bertsch PM, Burger J (2003) Bioavailability of uranium and nickel to vegetation in a contaminated riparian ecosystem. Environ Toxicol Chem 22:1146–1154
Reboreda R, Caçador I (2007a) Copper, zinc and lead speciation in salt marsh sediments colonised by *Halimione portulacoides* and *Spartina maritima*. Chemosphere 69:1655–1661
Reboreda R, Caçador I (2007b) Halophyte vegetation influences in salt marsh retention capacity for heavy metals. Environ Pollut 146:147–154
Reinhold D, Vishwanathan S, Park JJ, Oh D, Michael Saunders F (2010) Assessment of plant-driven removal of emerging organic pollutants by duckweed. Chemosphere 80:687–692
Ruiz-Rueda O, Hallin S, Bañeras L (2009) Structure and function of denitrifying and nitrifying bacterial communities in relation to the plant species in a constructed wetland. FEMS Microbiol Ecol 67:308–319
Salt DE, Smith RD, Raskin I (1998) Phytoremediation. Annu Rev Plant Physiol Plant Mol Biol 68:49–643
Sigma-Aldrich (2014) Available in: https://www.sigmaaldrich.com/
Stevens KJ, Kim SY, Adhikari S, Vadapalli V, Venables BJ (2009) Effects of triclosan on seed germination and seedling development of three wetland plants: Sesbania herbacea, Eclipta prostrata, and Bidens frondosa. Environ Toxicol Chem 28(12):2598–2609
Susarla S, Medina VF, McCutcheon SC (2002) Phytoremediation: an ecological solution to organic chemical contamination. Ecol Eng 18:647–658
Ternes TA, Bonerz M, Schmidt T (2001) Determination of neutral pharmaceuticals in wastewater and rivers by liquid chromatography electrospray tandem mass spectrometry. J Chromatogr 398:175–185
Verlicchi P, Al Aukidy M, Zambello E (2012) Occurrence of pharmaceutical compounds in urban wastewater: removal, mass load and environmental risk after a secondary treatment-a review. Sci Total Environ 429:123–155
Vymazal J (2005) Horizontal sub-surface flow and hybrid constructed wetlands for wastewater treatment. Ecol Eng 25:478–490
WFD-UKTAG—Water Framework Directive-United Kingdom Technical Advisory Group (2009) Proposed EQS for water framework directive annex VIII substances: triclosan (For consultation). Scotland
Williams JB (2002) Phytoremediation in wetland ecosystems: progress, problems, and potential. Crit Rev Plant Sci 21:607–635
Ying GG, Kookana RS (2007) Triclosan in wastewaters and biosolids from Australian wastewater treatment plants. Environ Int 33:199–205
Zarate FM Jr, Schulwitz SE, Stevens KJ, Venables BJ (2012) Bioconcentration of triclosan, methyl-triclosan, and triclocarban in the plants and sediments of a constructed wetland. Chemosphere 88:323–329

Zhang D, Gersberg RM, Ng WJ, Tan SK (2013a) Removal of pharmaceuticals and personal care products in aquatic plant-based systems: a review. Environ Pollut 1–20
Zhang DQ, Hua T, Gersberg RM, Zhu J, Ng WJ, Tan SK (2013b) Fate of caffeine in mesocosms wetland planted with *Scirpus validus*. Chemosphere 90:1568–1572
Zhang DQ, Gersberg RM, Hua T, Zhu JF, Goyal MK, Ng WJ, Tan SK (2013c) Fate of pharmaceutical compounds in hydroponic mesocosms planted with *Scirpus validus*. Environ Pollut 181:98–106

Chapter 19
Phytoremediation of Inorganic Compounds

Bruno Barbosa, Jorge Costa, Sara Boléo, Maria Paula Duarte, and Ana Luisa Fernando

19.1 Introduction

Inorganic compounds occur as natural elements in the earth crust and both natural geological processes and human activities contribute to their dissemination in the environment (Benjamin and Honeyman 1992). In the absence of human activities, inorganic elements as heavy metals are leached in terrestrial and aquatic environments corresponding to natural rates for chemical and mechanic erosion (Alloway 1995; Benjamin and Honeyman 1992; McIntyre 2003). The use of municipal waste, pesticides, fertilizers, emissions from waste incinerators, car emissions, steel, metallurgical and petrochemical industries, the operation of nuclear facilities and the fallout from nuclear weapons, mining, as well as construction are examples of anthropogenic activities that induced large scale changes in natural environments, altering the rate of release of inorganic compounds within the ecosphere, contributing significantly to the disturbance, degradation, contamination, and pollution of the air, water systems, and worldwide soils, where ultimately these compounds tend to deposit (Alloway 1995; Fergusson 1991; Garbisu and Alkorta 2003; He et al. 2005; Kabata-Pendias 2011; Wuana and Okieimen 2011; Zhu and Shaw 2000). These factors and activities, sole or combined induce marginality to soils (Dauber et al. 2012) reducing crop yields and quality of agricultural products, posing an imminent risk to humans, animals, and the environment, via bioaccumulation and transfer among trophic levels. Phytoremediation, the use of vegetation with potential to remove or immobilize inorganic compounds, may

B. Barbosa • J. Costa • S. Boléo • M.P. Duarte
MEtRiCS, Departamento de Ciências e Tecnologia da Biomassa, Faculdade de Ciências e Tecnologia Universidade Nova de Lisboa, 2829-516 Caparica, Portugal

A.L. Fernando (✉)
CENSE, Departamento de Ciências e Engenharia do Ambiente, Faculdade de Ciências e Tecnologia, Universidade Nova de Lisboa, 2829-516 Caparica, Portugal
e-mail: ala@fct.unl.pt

© Springer International Publishing Switzerland 2016

A.B. Ribeiro et al. (eds.), *Electrokinetics Across Disciplines and Continents*,
DOI 10.1007/978-3-319-20179-5_19

provide a cost-effective option over physicochemical soil remediation techniques (Lasat 2000; McIntyre 2003; Pilon-Smits 2005). Hyperaccumulator plants can be used on phytoremediation as these plants are able to uptake contaminants at high rates. However, yields are usually low, and the resulting biomass is not economically interesting to be explored. Therefore, the use of industrial crops (non-food crops), tolerant to contamination, can be a viable approach to effectively clean contaminated environments, as polluted soils and sediments, or polluted waters, once significant yields can be obtained and the biomass being produced can be economically valorized. This chapter tries to identify and describe the main contaminated sites localized in Europe, with more focus on contaminated soils, the main activities that are releasing inorganic compounds to environment and related risks to humans, animals, and environment, and the possibility to associate their remediation with the generation of economic value by using plants phytoremediation mechanisms. The main remediation technologies, and, in particular, the application of phytoremediation technology and their mechanisms as well as the advantages and limitations of this approach are also discussed.

19.2 Inorganic Contamination, Remediation Technologies, and Market for Phytoremediation Technologies

Contaminated sites inventory for inorganic contaminants as heavy metals and radionuclides worldwide is still an ongoing process, being difficult to study market opportunities for the application of phytoremediation projects. In Europe, for example, many countries do not have national inventories and many of the methodological approaches and remediation standards for soil contamination are inexistent (Lasat 2000; McIntyre 2003). Nevertheless, many efforts are being made in order to overtake the lack of information. Recently, the elaboration of the "Geochemical Atlas of Europe" that includes a database of about 3000 samples for topsoil, floodplain sediments, stream sediments, and humus in addition with many maps for inorganic elements in soils of Europe can provide a perspective on the present status of pollution of the European soils (Lado et al. 2008), serving as a basis for future phytoremediation projects. Yet, the use of phytoremediation technologies is limited in Europe as compared with the USA or Canada because: (a) there is limited knowledge and poor dissemination; (b) current regulations do not clearly consider phytoremediation as an applicable technology; (c) unfavorable competition with standard clean-up methods; (d) lack of investments to encourage private initiative; and (e) property rights, which hinder the development of these kind of projects, with proved success in countries like the USA (Sharma and Pandey 2014). For all polluted soils, world remediation market values grew from 15–18 billion USD in 1998 to 50 billion USD in 2002 and specifically in the USA, markets could generate 4.5–6 million from the phytoremediation of metals in soils and 0.5–1 million from phytoremediation of radionuclides (Lasat 2000; McIntyre 2003). European market was estimated to be tenfold smaller than the US market (Lasat 2000). Lewandowski et al. (2006) studied the economic value of the phytoremediation process, taking into account the remediation results of willow (*Salix* spp.) in Cd contaminated soils. The conclusions point to a positive economic value for both producers and authorities when willows are used in phytoremediation. That value is dependent on many factors, namely: (a) the methodology used in the calculation; (b) the products obtained from the produced biomass by phytoremediation; (c) soil and climatic conditions; (d) local financial situation, the dynamics of local markets and the possibility of producing crops and their products with high economic value; (e) the depreciation period, that means the value that might be obtained until the soil remediation is complete; (f) the time required to remediate the soil; (g) initial investment foreseen, driven by producers in contaminated areas; (h) costs of phytoremediation, which depend on crop production costs, with additional costs associated with the treatment of heavy metals contained in the biomass and the income derived from the sale of biomass and co-products.

Many research studies conducted by universities and research centers, national bodies, etc., are being made in many countries, in order to provide information about contaminated sites, type of contamination and appropriate approaches to their remediation, as well as in order to overtake the referred limitations on the application of phytoremediation technologies. Examples of areas with high occurrence of local contamination with inorganic elements included the North-West Europe from Nord-Pas de Calais, one of the most polluted canals in northern France, where total metallic concentrations of cadmium, zinc, lead, or copper exceed several times French standards (Sabra et al. 2011), the Rhein-Ruhr region in Germany (Ginneken et al. 2007; Hüffmeyer et al. 2009), as well Belgium and the Netherlands. It is estimated that in Belgium, 1.4 % of the total area (28,000 ha) is contaminated with heavy metals (Meers et al. 2005), while in Germany, contamination with these compounds reach 10,000 ha of farm land (Lewandowski et al. 2006). Other studies suggest higher contamination in the Netherlands, Germany, and Poland with several elements as Cr, Cu, Fe, Ni, Pb, V, and Zn (Herpin et al. 1996). Other contaminated areas comprise the Saar region in Germany (Ginneken et al. 2007), northern Italy (Bini et al. 2012; Giuseppe et al. 2014) and Iberian Peninsula (Clemente et al. 2005; Costa and Jesus-Rydin 2001; Pratas et al. 2013), southern Poland, Czech Republic and Slovakia (Ginneken et al. 2007) as well as Ukraine (Shcheglov et al. 2014; Vystavna et al. 2014) and Bulgaria (Yordanova et al. 2014). In many regions of Greece, there is a high level of soil contamination with multiple metals including Cd and Ni resulting from intense mining and metallurgical activities which take place on site for over 2700 years (Papazoglou et al. 2005). According to official emission data, total anthropogenic emissions of lead in European countries in 2004 were 5580 Mg and in the same year total deposition of cadmium was estimated as 181 Mg (Ginneken et al. 2007). The main point of these citations, made only in the European context, but with similar pattern worldwide, highlights the fact that the main hot spots for soil contamination with inorganic compounds as heavy metals and radionuclides are placed in the nearby areas of all major urban agglomerations in Europe, and for sure, soil contamination by heavy industries is not a marginal phenomenon.

The presence of inorganic compounds in soils and in other media is carried out by several human activities that: (a) change its generation rate via more faster and larger manmade cycles; (b) change their original and well-known location in ore mines to random environments and ecosystems where higher potential of direct exposure may occur; (c) increase metal concentrations in the receiving ecosystems; (d) may change the chemical form of the inorganic compound to other more bioavailable (Wuana and Okieimen 2011). Metal mine tailings, municipal waste, disposal of high metal wastes in landfills, emissions from waste incinerators, car emissions, steel, metallurgical and petrochemical industries, construction, the application of fertilizers, pesticides, animal manures, biosolids and compost to soils, fossil fuel combustion residues, petrochemicals, atmospheric deposition, the operation of nuclear facilities and the fallout from nuclear weapons, constitute main activities that induced large scale changes in natural environments, changing the rate of release of nutrients, heavy metals, and radionuclides within the ecosphere, contributing significantly to the disturbance, degradation, contamination, and pollution of the air, water systems, and worldwide soils, where ultimately these compounds tend to deposit (Alloway 1995; Fergusson 1991; Garbisu and Alkorta 2003; He et al. 2005; Kabata-Pendias 2011; Wuana and Okieimen 2011; Zhu and Shaw 2000).

The presence of inorganic compounds in soils, as well as in sediments, water bodies, and in the atmosphere involve several risks, and there are also innumerous scientific studies reporting causality relations between, e.g., heavy metal contamination worldwide and the most diverse effects in human and animal health, as well as environmental dysfunctions: DNA damage in animals and humans, teratogenic effects, implications for the food chain and ecosystem functioning, soil erosion and desertification, crop yield loss and fodder crop contamination, quality loss of human semen, food contamination, surface and groundwater contamination, various forms of cancers and Alzheimer, etc. (Barbosa et al. 2010; Giaccio et al. 2011; He et al. 2005; Khlifi and Chaffai 2010; Lewis et al 2011; McClintock et al. 2012; Pan and Wang 2012; Radwan and Salama 2006; Videa et al. 2009; Zoche et al. 2010). The exposure to radionuclides may pose several threats to humans, animals, plant species, and the environment. Many studies report threats in teeth and bones of humans and animals (Almayahi et al. 2014; Hong et al. 2011; Marshall et al. 2010), influences in biological parameters important for health and reproduction of many plant populations (Kin et al. 2013) and in the soil environment (Charro et al. 2013; Molchanova et al. 2014). Radioactive contamination may also affect the abundance of living beings through the radiation and chemical toxic effects of radionuclides inducing effects of mutation accumulation over time (Møller et al. 2013). Apart from the problems associated with heavy metals and radionuclides, excess of nitrates and other inorganics in waters is also a matter of concern, among other issues related with the contamination with inorganics. All these risks should be avoided and it is urgent the removal, immobilization, and decontamination of soils and other media contaminated with inorganic contaminants.

19.2.1 Soil and Water Remediation Technologies

Given the widespread of inorganic compounds in all the world's ecosystems and the risks they may pose to public health as well as for the normal functioning of the environment, and in the case of agricultural ecosystems where the area of contaminated soil is no longer available for the production of food crops, it is necessary to find solutions that promote decontamination and its normal function, if possible in a cost-effective way. Regarding heavy metal contaminated soils, several types of remediation technologies have been used. The physicochemical techniques are among the most vulgarized and could be generally classified as approaches that promote: (a) isolation; (b) immobilization; (c) toxicity reduction; (d) physical separation; and (e) extraction, of inorganic contaminants from soils (Wuana and Okieimen 2011). The isolation/contention approach promotes leaching minimization of the pollutants by the installation of subsurface barriers: clay layers, organo-clay layers, and plastic cover. In the excavation approach, contaminated soil is removed and disposed of elsewhere (landfills, for example). It is a technique that comprises high costs and may involve groundwater contamination (Mulligan et al. 2001; Kabata-Pendias 2011). The application of the removal technology is by exposing the contaminated soil to chemical extraction and/or thermal treatment to remove volatile elements or compounds; involves also leaching processes and immobilization as well as high application costs (Kabata-Pendias 2011). Encapsulation approach covers small places on soil with a layer of low permeability material (e.g., clay) to prevent infiltration of water and dust (Kabata-Pendias 2011; Wuana and Okieimen 2011). The immobilization of heavy metals could be promoted by Fe/Mn oxides that adsorb or occlude various heavy metals or by phosphorus compounds that decrease their mobility by using other substances such as clay minerals, Ca-carbonates, and zeolites (Kabata-Pendias 2011; Mulligan et al. 2001). The technique of solidification/stabilization promotes changes in soil using materials with high binding capacity with fractions with lower mobility of metals and/or by immobilization, keeping the pH of soil neutral. For the required effectiveness of some particular locations it is a very expensive technology (Kabata-Pendias 2011). With the application of vitrification technique, the pollutants are immobilized with electric current. It is rarely used and involves high costs (Kabata-Pendias 2011; Wuana and Okieimen 2011). The soil could be also washed with water or with a surfactant. It is a widely applied technology, but limited by high water needs. The results are very dependent on the physical parameters of the soil. In addition to the technologies previously mentioned, chemical oxidation and reduction technologies are also used in the decontamination of soils containing many different types of contaminants (Mulligan et al. 2001). Other approach consists in the application of biodegradation technologies by which substances containing heavy metals are decomposed due to microbial action and the metals are mobilized and washed. It is a technique that shows long term effects being practiced in specific locations (Kabata-Pendias 2011). Other technology is electroremediation, a technique for removal of contaminants from polluted sites under the influence of an applied current (Ribeiro and Mexia 1997). Nevertheless, the technologies described above are expensive and may not be the most attractive option, especially when there are no obvious risks for the population (Fernando 2005). Besides, these techniques involve in some cases the removal of soil, ignoring the possibility of preserving some of its properties and characteristics, as well as various biotic and abiotic relations. This chapter gives a special emphasis on the application of the phytoremediation technology, an environmentally sustainable and economically attractive technology (Mulligan et al. 2001; Raskin et al. 1994), less expensive than physicochemical approaches that uses vegetation in situ and associated microorganisms in their rhizosphere to remove pollutants from contaminated soils (Fernando and Oliveira 2004). Phytoremediation techniques can also be applied to polluted waters as sewage and municipal wastewaters (for example, for phosphate, ammonium, and nitrate ions removal), as well as for agricultural runoff and drainage waters (phytoextraction/phytodepuration of metals and selenium, for example) (Pilon-Smits 2005). Phytodepuration could be merged with fiber and energy production providing simultaneously tertiary treatment for wastewaters, which serve as a source of water and nutrients (instead of mineral fertilizers) for the plants irrigation (Barbosa et al. 2014). This strategy could be applied as an approach to mitigate and reverse desertification. Phytoremediation technology can also be coupled with other remediation technologies.

19.2.2 Phytoremediation Technology: Characteristics of Hyperaccumulator Plants

Many plants show the ability to absorb, translocate, and accumulate metals and other inorganic compounds as radionuclides in their belowground and aerial components and for that reason could be used in their removal from contaminated media, by a technology known as phytoremediation (Baker 1981; Bañuelos et al. 2000; Cunningham and Ow 1996). It is a field of research that has been received increased attention from many researchers in the environmental area, because it presents itself as an environmentally sustainable, safe, and cost-effective technology.

The phytoremediation technology shows also an additional interest, related to the possibility of metal recovery from the biomass after harvest (Mirza et al. 2011). It applies to media such as contaminated soil, sludge, sediment, groundwater, surface water, wastewater, and air (Pivetz 2001). When the contamination is related with heavy metals, this approach is linked with the use of hypertolerant and hyperaccumulator plants. These plants will act on the bioavailable fraction of heavy metals in the soil, which is dependent on various factors that affect their mobility, such as: (a) the reaction of the soil; (b) clay mineral content and iron aluminum and manganese compounds; (c) the cation exchange capacity; (d) the organic matter content; (e) the presence of other elements; (f) the redox potential; and (g) methylation by microorganisms (Fergusson 1991; Ghosh and Singh 2005; Prasad 2004; Varennes 2003).

Many researches have focused on the physiological processes involved in hyperaccumulation of heavy metals by different plants. The plants may be protected externally against the metal, or else tolerate high concentrations of metals in their tissues, through specific physiological mechanisms that lead to its phytotoxicity minimization (Baker 1987). By this way these plants prevent or minimize extraction of metals at the level of the plasma membrane on root cells, or tolerate metal concentrations by accumulation on symplasm (Quaghebeur and Rengel 2005). Many hypotheses have been proposed to give an explanation to the possible role of a high concentration of metals at leaves level in hyperaccumulator plant species (Boyd 2007): (a) tolerance to metals; (b) drought resistance; (c) interference with plants in the neighborhood; and (d) defense against natural enemies. The hypothesis that attracts more attention suggests that elevated concentrations of heavy metals in aerial tissues can function as a self-defense strategy against herbivores and pathogens. Heavy metals can provide protection against a variety of enemies.

Currently research is directed either to the discovery of species with these characteristics, as for genetic manipulation of hyperaccumulator plants. Especially in the second case, it is essential to determine the molecular mechanisms of metal accumulation, since most hyperaccumulator plants identified to date show root systems which penetrate at low depths, displaying reduced yields. Hyperaccumulator plants from natural populations are plants that typically contain more than 1000 $mg \cdot kg^{-1}$ (0.1 % of dry weight) of Co, Cu, Cr, Pb, and Ni; more than 10,000 $mg \cdot kg^{-1}$ (1 % of dry weight) of Mn or Zn (Prasad 2004) in its biomass. The strategies used in the development of a phytoremediation plan are: (a) the research of hyperaccumulator plants with the required characteristics for phytoremediation; (b) breeding of plants; (c) development and improvement of hyperaccumulator plants using genetic engineering tools (Jabeen et al. 2009). Heavy metal hyperaccumulation by higher plants is a complex phenomenon that involves a series of steps, such as: (a) the transport of metals throughout the plasma membrane of the cells of the root; (b) its intake and translocation within the xylem; (c) the rehab and sequestration of metals at the level of the plant, for certain plant components, or at the cellular level (Lombi et al. 2001).

Generally speaking, the ability of a plant to accumulate a particular metal is determined by its ability to extract this element to its biomass and by the intracellular transport within the plant. The general processes that influence the rates of accumulation of metals in plants are: (a) bioactivation in the rhizosphere, an interaction between root and microorganisms; (b) absorption and compartmentalization at the root level, involving molecular transporters, channels, and chelating agents in the cytoplasm—the phytochelatins; (c) transport in xylem—simplasma loading and ion exchange; (d) distribution and sequestration to cell wall, sequestration for the vacuole and cytoplasmic chelation (Yang et al. 2005). The grade of extraction and accumulation for a given metal, and for a particular species of plant, is dependent on the phytoextraction coefficient (Kumar et al. 1995): a ratio between the mass of extracted metal by the dry biomass of the plant used.

The ideal plants for phytoremediation should display: (a) fast growth; (b) high biomass yield; (c) extensive and deep root system; (d) easiness to harvest;

(e) tolerance and ability to accumulate several heavy metals in their aerial components; and (f) known agronomic techniques (Yang et al. 2005). Thus, given the slow growth rates and lower biomass yields associated to metal hyperaccumulator plants we may conclude that these plants show low exploitation potential (Chaney et al. 1997). Ideally, therefore, should be to find out and apply into contaminated matrices, fast-growing species with high yields, with deep and well-developed root systems and from which there is some agronomic knowledge. In addition, these plant species should possess if possible, high efficiencies in the use of resources such as water and nutrients, high tolerance to pests, high adaptive capacity to soil and climatic conditions, as well as low ecological requirements. The produced biomass should be inserted into an integrated management of the available resources and its processing should return economic viable products for fiber, energy, and other by-product's markets when grown under contaminated soils and other marginal soils (Dauber et al. 2012). This concept fits in the logic of biorefinery, which seeks the full use of biomass, in a sustainable way for the production of biofuels, energy, materials, and chemicals with high value embedded. Industrial (energy and fiber) crops seem to display the required properties for phytoremediation purposes. Their main mechanisms and case studies of their applications, as well as their main advantages and disadvantages of its use are discussed in the next sub-chapters.

19.3 The Use of Industrial Crops for Phytoremediation Purposes and Main Mechanisms

Plants have the ability to serve as filters, lowering the contaminant level of inorganic compounds in polluted soils, wastewaters, or landfill leachates. Several non-food crops, such as energy crops and fiber crops, have been thoroughly documented as apt remediators of contaminated soils, landfill leachates, wastewaters, or sewage sludge amended soil.

Perennial trees, such as willow and poplar have been documented as efficient landfill caps treating its leachates (Börjesson 1999; Duggan 2005). Willow plantations have been irrigated with wastewater and sewage sludge (Hansson et al. 1999; Heller et al. 2003; Rosenqvist and Dawson 2005). Poplar was tested with success for remediation of soil amended with non-hazardous levels of industrial waste (Giachetti and Sebastiani 2006). Guo et al. (2002) reported the irrigation of Eucalyptus plantations with effluent from meat industries.

Perennial herbaceous crops, such as reed canary grass, *Miscanthus* and switchgrass are considered suitable for disposal of sewage sludge in soils (Bullard and Metcalfe 2001; Fernando 2005). Börjesson (1999) reports reed canary grass appropriate for treatment of landfill leachate as well. Irrigation with wastewater from municipal and/or industrial sources is reported a cultivation practice alternative for reed canary grass (USDA 2006) and giant reed (Mavrogianopoulos et al. 2002).

The latter is further documented to have high tolerance to metals in the soils treated with sewage sludge (Papazoglou 2007). Liquid manure application from pig farms as nitrogen substitute is an added value strategy for cardoon cultivation but also for the sugar beet cultivation, an annual crop (Luger 2003; Draycott 2006).

Concerning annual crops, rape seed is documented for phytoextraction of heavy metals (Rossi et al. 2002; Sheng et al. 2008), although Marchiol et al. (2004) reported low phytoextraction potential. Batchelor et al. (1995) indicate that sewage sludge and animal excreta can be used as fertilizers on rape seed plantations. Niu et al. (2007) successfully used the oilseed crops sunflower and Ethiopian mustard for phytoextraction of metals from sewage sludge. Bioremediation capabilities have also been suggested for hemp (Linger et al. 2002), flax (Bjelková et al. 2001; Grabowska and Baraniecki 1997), and sweet sorghum (Epelde et al. 2009).

19.3.1 Synergism Possibilities Between Industrial Crops Production and Phytoremediation to Remove Inorganic Contaminants from Soils and Water

The growing demand of biomass for bioenergy and biomaterials production, and in particular the production of energy and fiber crops, has been generating varied conflicts by the use of the land. Such conflicts can be solved by means of spatial segregation of the area of production of crops for energy and biomaterials, for the so-called surplus land, which encompass various types of marginal lands, among which heavy metal contaminated soils (Dauber et al. 2012). Fiber yielding crops are one of the major groups of plants with more economic importance, being used to produce textiles, papers, mats, ropes, and cordage material for various uses (Pandey and Gupta 2003), as also in composite applications for the automotive, construction, sports, leisure, and other mass production industries (Akil et al. 2011; Ardente et al. 2008). In many cases, and because a particular crop could be used for both fiber and bioenergy purposes (multi-purpose crops) fiber crops could be also produced under surplus land, mainly because over the past few decades, the demand for natural fibers increased sharply (Akil et al. 2011; Faruk et al. 2012). Notwithstanding its production sustains employment and income in many regions, its intensification, namely by intensive use of land, may involve soil nutrient, water and mineral resources depletion, along with soil and water pollution and quality degradation. Crops such as jute, hemp, kenaf, sisal, safflower, *A. donax* L., *Miscanthus*, and sunflower are examples of crops with potential for fiber (Faruk et al. 2012) and bioenergy purposes that could be used for phytoremediation of contaminated soils with inorganic compounds as heavy metals, metalloids, and radionuclides and for phytodepuration of many types of wastewaters. Energy and fiber crops can contribute for the recovery and remediation of contaminated soils through the production of various products and because these plants hold several mechanisms that may contribute for the degradation, accumulation, dissipation, or immobilization of the contaminant (EPA 2000; Mirza et al. 2011; Pivetz 2001). Each crop and associated mechanisms are applied according to the nature of the contaminated media, metallic element and extent of local contamination, local soil and weather conditions as well as according to socioeconomic local logics.

Industrial crops show many particular features from species to species, but in general, its introduction in contaminated land has several common advantages. Industrial crops when grown on marginal or degraded lands, can improve soil fertility and soil structure, increasing the organic matter content, helping in the control of erosion and increasing biological and landscape diversity (Fazio and Monti 2011; Fernando et al. 2010; Finnan and Styles 2013; Zegada-Lizarazu and Monti 2011). Generally speaking, annuals show low costs of implementation and can be grown in crop rotation systems with other types of crops (obviously not on contaminated land, when in rotation with food crops). Their insertion in a crop rotation system can increase the yield and profitability of soil over time; promote disease and pest control on site by increasing biological and landscape diversity, as well as providing a stable source of biomass for fiber, bioenergy, and other by-products. These plants can also be integrated into a waste management strategy and introduced as a mitigation strategy for the reversal of the desertification process, once they show low water and nutrients use, show commercial value for a given region, and because its production involves less problems in natural ecosystems since there is no competition with fodder crops production (Araus 2004; Kassam et al. 2012). For example, these crops can be implemented in a deserted region and its irrigation and fertilization provided by domestic wastewaters (Barbosa et al. 2014). Compared with annual crops, perennials offer additional ecological advantages, providing a wider range of ecological services such as greater vegetation cover and permanence in the soil, erosion control, show lower susceptibility to diseases, reduced need for pesticides, and, due to its extensive and deep root systems, possess high resource utilization efficiencies such as for water and nutrients, being also able to provide minimization of the contamination by leaching (Fernando 2005; Fernando et al. 2010; Zhang et al. 2011).

19.3.2 Industrial Crops Phytoremediation Mechanisms

Phytoremediation may lead to the accumulation, immobilization, dissipation, and degradation of the contaminants. The most effective phytoremediation mechanisms for inorganic compounds treatment approach are those that promote their accumulation in aerial tissues or their immobilization on root biomass. These mechanisms are utilized by hyperaccumulator plant species but also by industrial crops. The different mechanisms are depicted in Fig. 19.1.

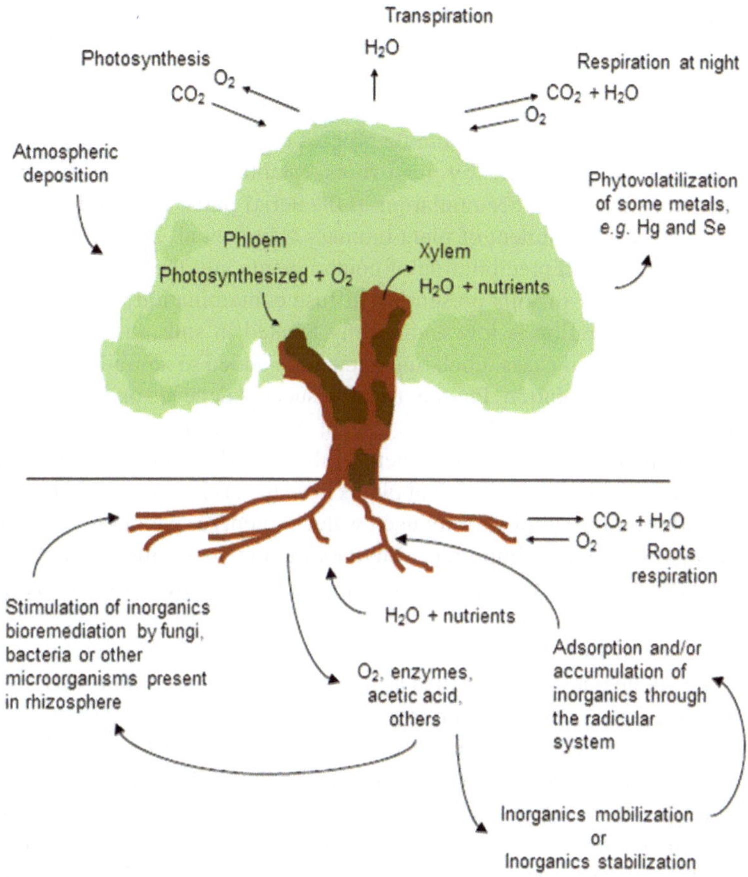

Fig. 19.1 Phytoremediation mechanisms of inorganic contaminants (adapted from Fernando 2005, Courtesy of AL Fernando)

19.3.2.1 Accumulation of Inorganic Compounds in Plant Biomass

Accumulation consists in the contention in plant material of contaminants (e.g., metals, nutrients, radionuclides) removed from the soil or other matrices, involving the mechanisms of phytoextraction and/or rhizofiltration (Pivetz 2001). Rhizofiltration refers to the use of plants in hydroponics setup for filtering polluted water. Because it is a mechanism that applies to water bodies, we will focus mainly on the description of phytoextraction mechanism, nevertheless it is an application that uses plants as Indian mustard, sunflower, tobacco, rye, spinach, and corn, for examplc, to accumulate in their root system, Ni, Zn, and/or Cr from polluted aqueous sources (Ghosh and Singh 2005); as well as in the removing of Pb and U (Dushenkov et al. 1995).

Phytoextraction mechanism consists in the use of plants to clean up pollutants from soils and other media via accumulation in harvestable tissues (Pilon-Smits 2005). Phytoextraction is the best approach to remove contamination from soil and then isolate it, via biomass harvest, without destroying soil structure and fertility (Ghosh and Singh 2005). It is a common process of phytoremediation that involves the extraction of contaminants by plant roots, established by common agronomic practices, with subsequent accumulation in the aerial parts, followed generally by harvest and disposal/treatment of plant biomass (Mirza et al. 2010, 2011). Absorption, concentration, and precipitation of contaminants into the biomass, is a viable strategy to be applied in soils presenting diffuse contamination as well as in sites where their concentration is low and mainly located in soils surface (Ghosh and Singh 2005). In certain cases, chelating agents are added to solubilize metals with low solubility in soil solution. Phytoextraction success depends on the ability of the plant to accumulate and store large amounts of the extracted metals in their tissues, translocating them to aerial components, along with a fast-growing biomass presenting high yields. The ability that plants have to survive in marginal conditions (i.e., low soil pH, high salinity, or reduced water content) as well as to produce deep root systems, are equally important. Phytoextraction is usually limited by roots deep and because of that, regions closer to the surface are those where greater extraction occur.

The phytoextraction mechanism is one of the most studied for energy and fiber crops. Many authors tested crops such as flax, hemp, *Cynara cardunculus*, *Jatropha curcas*, castor bean, soybean, sunflower, corn, poplar, wheat, *Brassica juncea*, sorghum, rapeseed, and safflower in the phytoextraction of Cd, Ni, Cu, Fe, Pb, Mn, Zn, and Au obtaining successful results (Baker 1987; Kumar et al. 2008; Papazoglou 2011; Zayed et al. 1998). Many industrial crops display this mechanism, showing good results in its application. Hammer et al. (2003) tested the heavy metals (Cd and Zn) phytoextraction in two types of contaminated soils, a limestone (during 5 years) and other acid (2 years) with *Salix viminalis*. The plant showed best results in acidic soil, presenting a higher biomass production and higher concentrations of metals in its aerial parts. The addition of elemental sulphur to soil caused no enhancement in plant yield, but the application of Fe in the chelate form, increased the production of biomass. Cd and Zn concentrations were significantly higher in leaves than stems. In both soils, the concentration in the aerial plant components decreased over time. Meers et al. (2007) tested the phytoextraction with five *Salix* clones, in Cd, Cr, Cu, Ni, Pb, and Zn contaminated soils. The clones exhibited a higher accumulation in Zn and Cd, and an estimation of the amount that could potentially be extracted, resulted in 5–27 $kg{\cdot}ha^{-1}$ $year^{-1}$ for Zn and 0.25–0.65 $kg{\cdot}ha^{-1}$ $year^{-1}$ for Cd. Arduini et al. (2006b) tested the phytoextraction of Cr (50–200 $mg.dm^{-3}$) during the period of higher growth of *Miscanthus sinensis* l. var. Giganteus. The production of aerial biomass was more affected than the radicular biomass; however, the addition of 100 $mg.dm^{-3}$ [Cr $(NO_3)_3$] or higher induced a profound change in root morphology. At 150 $mg.dm^{-3}$ Cr, production of lateral roots was inhibited, and its total length and biomass were reduced, while the average diameter of the roots or root volume increased significantly. Stems and green leaves reached higher concentrations of Cr at 200 $mg.dm^{-3}$ Cr exposure, while on roots and rhizomes such happened at 150 $mg.dm^{-3}$, and then decreased. This observation can be explained by a higher Cr translocation to aerial organs in an active process operated by plants under the presence of extreme toxic conditions. Also Arduini et al. (2004, 2006a) described several physiological processes of the plant in response to Cd. Plants were exposed to 0–3 $mg \cdot L^{-1}$ cadmium for 36 days, and all cadmium levels showed to be toxic to *Miscanthus*. At 0.75 $mg \cdot L^{-1}$ Cd, toxicity caused a biomass decrease by about 50 %, whereas between 2.25 and 3 $mg \cdot L^{-1}$ Cd it completely inhibited growth and disrupted the mechanisms that restricted Cd translocation to the shoot. The Cd concentration in the biomass markedly increased with all Cd levels, being higher in the radicular system by opposition to culms and leaves. Fernando (2005) under individual contamination with Cd, Cu, Hg, Ni, Pb, Zn, and Cr and, in other tests with sludge contaminated with the same metals, and also with *Miscanthus*, found that the species is able to accumulate in its biomass heavy metals from the growing medium in an efficient phytoextraction perspective. The largest proportion of contamination, however, remained in the rhizomes and roots. Borghi et al. (2007) investigated the tolerance and phytoextraction of Cu in poplars (*Populus* × *euramericana* Adda clone). At 20 μM Cu there has been no reduction in the growth or in foliar areas, showing high chlorophyll content and high photosynthetic potential. The plant accumulated more in roots and at higher concentrations of Cu (500–1000 μM), the clone Adda showed symptoms of leaf chlorosis and decreased photosynthetic efficiency. Willow also show phytoextraction capacity for radiocaesium from soil (Vandenhove 2013) and many other energy crops are successfully used in the phytoextraction of radionuclides. The plant species *Brassica juncea*, pennycress, alyssum, sunflowers, and hybrid poplars are being used in laboratory, pilot, and field applications in the phytoextraction of Ag, Cd, Co, Cr, Cu, Hg, Mn, Mo, Ni, Pb, Zn, and the radionuclides ^{90}Sr, ^{137}Cs, ^{239}Pu, and $^{238,\ 234}U$ (EPA 2000).

19.3.2.2 Immobilization of Inorganic Compounds

Immobilization of contaminants involves the containment of the contaminant and includes the mechanism of phytostabilization (Pivetz 2001). Phytostabilization is mostly used for the remediation of soil, sediment, and sludge and occur through sorption, precipitation, complexation, or contaminant valence change (Ghosh and Singh 2005). Plants that provide the immobilization of the contaminants by using the phytostabilization mechanism, transform toxic forms present in the soil, into other less toxic, by adsorption on root system, immobilizing/stabilizing the contaminants, or by the release of compounds (exudates) to the soil by the root system that may form complexes or precipitate thus stabilizing/immobilizing the contaminants (Fernando 2005). Plant transpiration and root growth mechanisms accounts for the chemical or physical stabilization of the contaminants. A dense root system: (a) stabilizes the soil and prevents water and wind erosion, providing hydraulic control/vertical migration (leaching) of contaminants to aquifers and groundwater; (b) creates an aerobic environment in the rhizosphere by the releasing of exudates (organic compounds) that promote microbial activity accelerating the degradation of organic contaminants (sometimes bounded to inorganic compounds, e.g., organometalics) to nontoxic forms or less toxic forms (phytostimulation); (c) provide retention of the contaminant in the rhizosphere by humification or lignification (Mirza et al. 2011). Examples of phytostabilization are the ability of some plants with deep root systems to reduce the Cr (VI) form highly toxic to Cr (III) which is much less soluble and less bioavailable (James 2001). Reinstallation of plant species also contributes to increase the ability of ecological restoration (EPA 2000). The ability of immobilization of toxic contaminants, such as metals, can be optimized by operations on soils that lead to the increase of pH and soil organic matter content or by bonding contaminants with certain constituents (e.g., carbonates and phosphates). Soil operations are more efficient in soils that exhibit fine texture, with high organic matter content, and could be induced to treat a variety of contaminated surfaces: soils, sediments, and sludge. Soil operations could lead to a more quickly fixation of metals, followed by its incorporation into root biomass, while chemical changes should be permanent. Most important operations that could be induced on soils are the incorporation of: (a) phosphate fertilizers; (b) soil organic matter or biosolids; (c) iron and manganese minerals; (d) natural or artificial clay; and (e) combination of these operations. It is a very effective technique when rapid immobilization is needed to preserve ground and surface water. Nevertheless, the major disadvantage when that approach is applied is that the contaminant remains in soil and therefore regular monitoring is also required (Ghosh and Singh 2005). Industrial crops could be used successfully in the immobilization of inorganic compounds. Ho et al. (2008) tested kenaf (*Hibiscus cannabinus* L.) on the phytoextraction and phytostabilization of Pb. Kenaf root system accumulated up to 20 mg Pb (per plant), but its aerial components accumulated less than 1 mg Pb (per plant). Most of the Pb was retained by the radicular system, both absorbed and adsorbed, phytoimmobilizing the metal. The already cited studies of Arduini et al. (2004), Arduini et al. (2006a), and Fernando (2005) showed also that *Miscanthus* is able to phytostabilize heavy metals in the radicular system. Plants are able to use one or more of the mechanisms associated with phytoremediation, simultaneously. Indian mustard, hybrid poplars, and herbaceous plants have been used with success in the phytostabilization of As, Cd, Cr, Cu, Hg, Pb, and Zn from field applications in soils, sediments, and sludge (EPA 2000). Vanek et al. (2006) tested the phytostabilization of radioactive elements (I, U and Ra) with sunflower (*Helianthus annuus* L.). Results indicate that uranium remained mostly in the root system (removal efficiency of 24 %), but 86 % of the Ra extracted from soil (42 %) translocated to the aerial part, thus showing a high removal efficiency. These crops are useful, combining restoration of degraded sites and biomass production, especially in tropical regions.

19.3.2.3 Dissipation of Inorganic Compounds

Phytovolatilization mechanism involves the use of plants to take up contaminants from the soil converting them into volatile forms or other less toxic compounds and transpiring them into the atmosphere (Ghosh and Singh 2005). The use of plants able to absorb metal contaminants and convert into less toxic chemical species, through transpiration is called phytovolatilization (dissipation mechanism). There are being used some natural strains of certain species of plants, or other genetically manipulated that show the ability to absorb heavy metals, to convert into volatile compounds within its tissues, and release them into the atmosphere. It was primarily used in the removal of mercury, where the metal, in an organometallic form, was transformed into the elementary form Hg, less toxic (Ghosh and Singh 2005). Some plants, such as *Arabidopsis thaliana* and *Brassica juncea* can grow under Se rich media and produce volatile compounds such as the dimethyl selenide (Bañuelos et al. 2000), a compound that shows toxicities 500–600 times smaller than its inorganic forms, which can be found in the soil. The phytovolatilization can be used with success in removing tritium (3H) (Dushenkov 2003), mercury (Hg), and arsenic (Mirza et al. 2010, 2011).

19.3.2.4 Stimulation of the Remediation Process and Contaminants Degradation

Phytoremediation can be stimulated by several mechanisms. One option is to stimulate the microbial community in the rhizosphere. Rajkumar and Freitas (2008) conducted the inoculation of *Pseudomonas* SP. PsM6 and *Pseudomonas jessenii* PjM15 in *Ricinus communis* causing an increment in its aerial and root biomass, under heavy metal contaminated soils as well as in other soils without any type of contamination. The stimulation of phytoremediation could be also operated by agronomic practices and the addition of complexing agents as EDTA that can increase metals bioavailability and subsequent phytoextraction/immobilization in plant tissues. That was tested for *Sorghum bicolor* and sunflower (*Helianthus annuus* L.) (Marchiol et al. 2004; Zhuang et al. 2009). Mleczek et al. (2009) and Adler et al. (2008) refer also to the addition of EDTA, contributing significantly to increase the bioavailability of Cd, Co, Cr, Cu, Ni, Pb, and Zn in soils, increasing also the accumulation of these metals by willow. Other solutions can be applied to increase the bioavailability of contaminants, such as citric acid (Fernando 2005).

Contaminant degradation refers to the destruction or alteration of contaminants. Degradation can only be applied to organic contaminants but is referred here because many contaminated soils show both organic and inorganic contamination, and degradation of some contaminants, such as organometallic complexes, releases inorganic contaminants, allowing the treatment by phytoextraction or phytostabilization mechanisms. Biodegradation is a process operated mainly by the microbial activity in the rhizosphere (Pilon-Smits 2005). The mechanism refers to the microbial presence in the plants rhizosphere and to the consequent pollutants biodegradation; the growth of this microbial community is stimulated by root exudates and/or supply of plant tissues. The phytodegradation or phytotransformation involves the conversion of the pollutant within the plants on its surface, and through catabolism or anabolism is converted to less toxic forms (Pivetz 2001). Usually the process occurs inside tissues and refers to the breakdown of pollutants by enzymes (Pilon-Smits 2005). In phytodegradation mechanism, the plants degrade organic pollutants through metabolic processes, and use their associations in rhizosphere. Enzymes are released in the rhizosphere playing active roles in the transformation of contaminants as ammunition, chlorinated solvents, herbicides, and insecticides, but also inorganic nutrients (Schnoor et al. 1995).

It should be noted that the mechanisms of phytoremediation presented executed by many plant species occur simultaneously, and for that reason, all the factors described above should be taken into account when a project of phytoremediation is implemented. From the analysis, it may be noted that the main mechanisms of phytoremediation of inorganic compounds by industrial crops are the phytoextraction, and the phytostabilization and their performances in phytoremediation are dependent on various factors of several natures. There is a huge potential for exploitation of matrices contaminated with inorganics using this type of crops.

19.3.3 Phytodepuration of Inorganic Compounds from Wastewaters

Energy and fiber crops as *Hibiscus cannabinus* L., hemp (*Cannabis sativa* L.), and nettle (*Urtica dioica* L.) could ally high growth and yield responses with higher phosphate, nitrate, and ammonium removal from polluted water bodies (Adler et al. 2008; Fernando 2013) The option of cultivating these crops by using wastewater in the irrigation have been reported as important in water-scarce regions and wastewater reuse considered an alternative water and nutrient supply. Perennial crops as *Arundo donax* L., *Miscanthus*, *Bamboo* sp., *Juncus effusus* L., and *Cyperus papyrus* L. could also be used successfully with the same purposes but at the same time with additional benefits derived from its perennial character, namely regarding erodibility and biodiversity, due to lower input needs and land disturbance, higher yields and higher permanence periods (Barbosa et al. 2014; Fernando et al. 2010). Literature indicates that industrial crops show potential simultaneously to deliver high yields, restore soil properties, and promote water quality improvement. Their production in water-scarce regions could provide environmental benefits and social and economic opportunities, safeguarding freshwater resources (Barbosa et al. 2014).

19.3.4 Utilization of Phytoremediation By-products

Producing industrial crops in contaminated land may also limit its use due to the quality of the biomass. At the moment, this issue represents the bottle neck of the valorization of the biomass produced by phytoremediation. But several approaches can already be applied in order to reduce biomass volume, recover heavy metals and other inorganic contaminants, as well as obtaining a product with economic value. Only by itself the accumulation of inorganic contaminants in biomass offers the chance to reduce significantly the volume of waste material for disposal; but the question here is: is it possible to recover the contaminant in order to recycle it? During the process of its recovery is it possible to produce other by-products with economic value? The main constituents of phytoremediation biomass material are lignin, hemicellulose, cellulose, mineral matter, and ash (Ghosh and Singh 2005). Composting may reduce the volume of the biomass. Furthermore, leaching tests for composted biomass may enhance metal solubility by creating soluble organic compounds. Also compaction may reduce biomass volume, but as well as for composting it is needed to collect and treat leachates (Ghosh and Singh 2005). Thermochemical conversion processes constitute the main potential routes for converting phytoremediation biomass in an integrated manner with generation of electrical and thermal energy. Combustion could be applied under controlled conditions in order to avoid gas and particles released into the atmosphere; biomass volume may be reduced to 2–5 % and the resulting ash should be disposed properly. By gasification, biomass material is subjected to a series of chemical changes in order to produce clean and combustive gas with high thermal efficiencies. The process involves drying, heating, thermal decomposition (pyrolysis) and gasification, and combustion chemical reactions, occurring simultaneously (Ghosh and Singh 2005). Gasification involves organic matter destruction and the release of metals as oxides. Liberated metals remain in the slag and techniques, such as modern flue gas clearing, assures effective capture of the metal containing dust. The effects of pyrolysis temperature on the composition and evolution of liquid and gaseous fractions of the process also have been studied in order to valorize heavy metals contained in biomass (Lievens et al. 2008). Pyrolysis decomposes material under anaerobic conditions without emissions into the air. The final products are pyrolytic fluid oil and coke, the pyrolysis residue where heavy metals will remain (Ghosh and Singh 2005). Yet, producing industrial crops in contaminated matrices may result in a biomass with higher ash content, higher metals content (Barbosa et al. 2013), and higher nitrogen content (Costa et al. 2013), which is problematic either technologically or environmentally. Higher ash and metals content may increase costs for disposal, and costs associated with the slagging and corrosion processes in combustion furnaces, and higher nitrogen in the biomass limits its use for combustion once, N will emit as NO_x, which has an acidifying potential (Dauber et al. 2012).

19.4 Advantages and Limitations of Phytoremediation with Industrial Crops

The application of the phytoremediation technology includes several advantages of different natures and the application of energy and fiber crops instead of hyperaccumulator plant species, in the phytoremediation of contaminated media, increases the spectrum of advantages of this technology. Transversal to all types of plants used, phytoremediation can be applied in situ but plants could be also used ex situ. The main advantage of the in situ approach, and in particular for soils contaminated with inorganic compounds, refers to the fact that it is not needed the removal of the soil, meaning that soils properties as fertility, structure, porosity, and texture could be preserved and in some cases improved at the same time that contamination is remediated. Phytoremediation should be seen as an approach that could restore all the ecosystem services of the soil and not only as a way to remediate contamination. This is a main advantage in relation to the traditional physicochemical technologies. This approach is passive and driven by solar energy. It is considered also a "green" technology when compared with physicochemical technologies because many of the impacts of the soil removal could be avoided. But when the phytoremediation plant species is an industrial crop this concept could enlarge, since the use of these crops to produce biomass for bioenergy or biomaterials could provide an economic benefit to farmers. Moreover, the use of industrial crops for different uses, including energy and biomaterials, provide an opportunity to reduce the dependence on nonrenewable resources due to their renewable and sustainable character, and its associated environmental impacts (e.g., greenhouse gases emissions, disposal at the end of the technical life).

This technology could be applied to one or to a mixture of contaminants and because of that it is possible to apply in sites not ready to be remediated by other methods (McIntyre 2003). However, that is also dependent on the concentration of the contaminants and the physical characteristics of the site. Other benefits rely on the fact that the knowledge applied to food crops as well in silvicultural practices can be applied to industrial crops in soils contaminated with inorganic compounds.

Phytoremediation offers the advantage of eliminating secondary air or water borne wastes, wastewaters, coal pile runoff, landfill leachate, mine drainage, and groundwater plumes (Pilon-Smits 2005). Plants could also provide ground cover that prevents wind and water erosion-particulate matter that can become suspended in the air or water will be deposited in site, reducing also its exposure to human, animals, and other ecosystems; its roots may provide stabilization of the contaminant (immobilization), reducing its potential spread. In sloping terrains deep and extensive root systems may also stabilize soil/terrain structure (McIntyre 2003). By reducing dust and other particles spread, vegetation cover is reducing also exposure pathways of certain diseases. One of the advantages more obvious of the application of plants in the phytoextraction of heavy metals and other inorganic compounds is the possibility to recover the metal content accumulated in the biomass (McIntyre 2003). The biomass enriched in heavy metals or other inorganic compounds consist by itself a way to reduce the volume of disposal material, when the contaminant moves from the contaminated matrices to the plant. By techniques like composting and compaction it is possible to reduce harvested biomass and further volume reduction of the disposal materials is achieved. Furthermore, if phytoextraction is combined with biomass generation and commercial utilization as an energy source, also the remaining ash content could be used as bio-ore (Ghosh and Singh 2005), and further reduction in the volume of disposal remaining materials is achieved. It is a cheaper technology when compared with traditional physicochemical remediation technologies. Phytoremediation technology also has the advantage of converting the contaminated sites into a more aesthetically appealing landscape, a fact that could find easier public interest and support (McIntyre 2003). Perennial energy/fiber crops show lower input costs associated when compared with annual energy/fiber crops (Fazio and Monti 2011), to be used in phytoremediation. The associated costs to maintain these crops in the soil during all the necessary cycles to achieve the possible remediation of the site are significantly lower than the associated costs of applying physicochemical technologies in the remediation of contaminated soils. Perennial crops offer additional environmental advantages over energy/fiber annual crops and provide a wider range of ecosystem services such as higher ground cover, longer permanence in the soil, erosion limitation, lower susceptibility to diseases, reduced pesticides needs, and due to their extensive rooting system they have high nutrient and water efficiencies, thus minimizing nutrient and contaminant leaching (Fernando 2005; Fernando et al. 2010; Zhang et al. 2011).

Many hyperaccumulator plants can only accumulate a specific element and its application is limited in sites with mixed contaminants (organic, inorganic, or in combination) (McIntyre 2003). Nevertheless, perennial crops, such as *A. donax* L. or *Miscanthus* spp., can tolerate contaminated soils with one or mixed contaminants, showing interesting yields and a biomass with good quality for fiber or bioenergy purposes (Fernando 2005). One of the disadvantages on the use of energy and fiber crops in the phytoremediation of inorganic compounds is the amount that can be annually extracted from soil. Comparison with hyperaccumulator plants show that industrial crops may present less contaminants concentration in the biomass. Yet, the higher yields obtained with industrial crops, permitting its economical valorization, may allow extracting from contaminated sites higher amounts of contaminants. Nevertheless, the use of wild or non-indigenous plants in the phytoremediation process could provoke its spread as well induce several risks for ecosystems functions nearby to phytotreatment sites (McIntyre 2003). In fact, the attributes required for optimal phytoremediation by a crop correspond to the typical weeds and invasive plants herbs (Low et al. 2011): (a) rapid growth; (b) low fertilizer inputs; (c) high water and nutrient use efficiency; and (d) absence of pests and diseases. Because of that, the implementation of this type of crops requires containment plans for their potential spread. Other disadvantage associated with potential invasive plant crops is the existing limited information, required to perform an appropriate risk assessment of the species (McIntyre 2003). Other advantages of the application of industrial crops over hyperaccumulator plants is that they can reach huge population numbers, they can have more than one growth cycle in 1 year, and they present higher adaptability to different environments. That could be useful in cases where it is needed a different genetic background to approach the remediation in a particular site (for example, inexistence of diseases and plagues or other trophic relations); sometimes hyperaccumulator occurs in metal-rich soils and show a genetic background that allows them to remediate those sites but not with different inorganic contaminants and with similar features. The growth of some hyperaccumulator plants could be hampered by specific germination requirements and pollination mechanisms (McIntyre 2003). Nevertheless, the use of perennial energy/fiber crops as *A. donax* L. which propagation depends on rhizomes cuttings with higher success/efficiency of establishment guaranteed overtake this advantages for hyperaccumulator phytoremediation plant species. Most hyperaccumulators show slow growth rates and produce small amounts of biomass. The introduction of industrial crops offers the chance to introduce phytoremediation plants with fast growth rates, with high amounts of biomass associated, resistant to diseases and plagues, not being the food basis of any herbivore, especially if not native. This fact overtakes also the problem of some hyperaccumulator plants that may induce the transfer of inorganic contaminants within the food chain. Thus, metal immobilization in energy and fiber crops represents a way of metal containment in biomass but also containment/immobilization within the food chain. Many of these crops show naturally higher physiological characteristics promoting them for both purposes: to clean contaminated media and to obtain high quality biomass; that fact could economize money invested in breeding programs focusing hyperaccumulator plants. Nevertheless, genetic engineering could also be applied to improve natural features of these crops.

One disadvantage associated with phytoremediation is that this technology requires several growing seasons to clean a contaminated site. Furthermore, this technology is not appropriate when contaminant presents an imminent danger to environment or humans (McIntyre 2003) posed by huge amounts of the contaminant that plants could not tolerate or by the time required to eliminate that risk. Owners of contaminated lands may not wish to wait several growing seasons, when hyperaccumulator plants are used to clean the sites. Energy/fiber crops by offering the possibility to generate a commercial value for the biomass produced in contaminated soils overtake this problem, even if the number of growing seasons needed to clean the place is also high. Microclimates on site may be inhospitable resulting in lower growth, establishment, yields, or lower quality of the biomass. Normally, phytoremediation with hyperaccumulator plants applies to shallow contaminated sites (McIntyre 2003), but the introduction of industrial crops with extensive and deep root systems could increase phytoextraction and allow higher phytostabilization of the contaminants. In sloped terrains its production may affect its efficacy and cost (McIntyre 2003). Deep and extensive root systems could also provide slope support in contaminated terrains. In this type of soils, as well as in tight sites the application of machinery and other traditional agricultural cropping techniques may not be performed (McIntyre 2003). The application of energy and fiber crops in phytoremediation opens the possibility to recycle all constituents contained in phytoremediation biomass. Understanding the main pathways and main techniques to achieve that objective is fundamental to project phytoremediation in the remediation market as well as for commercialization of its by-products (McIntyre 2003).

19.5 Concluding Remarks

The use of plants, especially industrial (non-food) crops, and associated microorganisms in a phytoremediation action could be a sustainable, cost-effective option with dual goals, pollution reduction and production of marketable biomass. According to several authors, fiber and energy crops have demonstrated the capacity to be tolerant to inorganic contamination, being able to uptake these pollutants into the root system, alleviating the contamination and restoring ecosystem services. At the same time, they represent a promising option for the bioeconomy, notably the biorefinery and bioenergy industries. Production of industrial crops in contaminated land also avoids changes in arable land, helping to mitigate the food vs. biofuel debate.

However, sustainability of industrial crops production in contaminated matrices depends on yields, on the aptitude of the crop to restore value to the contaminated matrices based on the uptake and accumulation/degradation rates for the substance of interest, and on the quality of the biomass being produced.

An evaluation of several works presented in the literature suggests that the production of non-food crops associated with inorganic contaminants have both positive and less positive aspects. The productivity loss due to the toxicity of the contamination diminishes the energy balance and the carbon sequestration by the biomass, reducing the greenhouse savings. Uptake of contaminants by biomass, if on one hand desirable to enhance the phytotreatment technology, reducing the time needed for decontamination, on the other hand, the increased accumulation of contaminants in the biomass can be detrimental for its use and economic valorization. Another drawback of this technique is the time needed to accomplish phytoremediation of soils, once several growing cycles are required to accomplish it, by comparison with the traditional physical–chemical techniques. But the presence of vegetation may contribute to improve the quality of soil and waters and the biological and landscape diversity. Moreover, the prospect of the valorization of the biomass, for bioenergy or bio-products production purposes, could lessen the financial costs of contaminants remediation, compared to the traditional physical–chemical processes, with the associated revenue of environmental benefits.

As part of active planning, bench-scale treatability studies should be conducted prior to field implementation. These studies represent a cost-effective tool for simultaneously evaluating multiple variables, optimizing performance and ultimately reducing environmental, social, and economic costs. In order to promote the sustainable production of biomass for the phytoremediation of inorganic contaminants, further research is needed, factoring in issues such as yields, inputs, and costs, as well as potential environmental and socioeconomic impacts.

References

Adler A, Dimitriou I, Aronsson P, Verwijst T, Weih M (2008) Wood fuel quality of two *Salix viminalis* stands fertilized with sludge, ash and sludge-ash mixtures. Biomass Bioenerg 32:914–925

Akil HM, Omar MF, Mazuki AA, Safiee S, Ishak ZAM, Abu-Bakar A (2011) Kenaf fiber reinforced composites: a review. Mater Design 32:4107–4121

Alloway B (1995) Heavy metals in soils. Blackie Academic, London

Almayahi BA, Tajuddin AA, Jaafar MS (2014) Measurements of natural radionuclides in human teeth and animal bones as markers of radiation exposure from soil in the Northern Malaysian Peninsula. Radiat Phys Chem 97:56–67

Araus JL (2004) The problems of sustainable water use in the Mediterranean and research requirements for agriculture. Ann Appl Biol 144:259–272

Ardente F, Beccali M, Cellura M, Mistretta M (2008) Building energy performance: a LCA case study of kenaf-fibres insulation board. Energ Buildings 40:1–10

Arduini I, Masoni A, Mariotti M, Ercoli L (2004) Low cadmium application increase *Miscanthus* growth and cadmium translocation. Environ Exp Bot 52:89–100

Arduini I, Ercoli L, Mariotti M, Masoni A (2006a) Response of *Miscanthus* to toxic cadmium applications during the period of maximum growth. Environ Exp Bot 55:29–40

Arduini I, Masoni A, Ercoli L (2006b) Effects of high chromium applications on *Miscanthus* during the period of maximum growth. Environ Exp Bot 58:234–243

Baker A (1981) Accumulators and excluders: strategies in the response of plants to heavy metals. J Plant Nutr 3:643–654

Baker A (1987) Metal tolerance. New Phytol 106:93–111

Bañuelos G, Zambrzuski S, Mackey B (2000) Phytoextraction of Se from soils irrigated with selenium-laden effluent. Plant and Soil 224:251–258

Barbosa J, Cabral T, Ferreira D, Lima L, Medeiros S (2010) Genotoxicity assessment in aquatic environment impacted by the presence of heavy metals. Ecotox Environ Safe 73:320–325

Barbosa B, Fernando AL, Lino J, Costa J, Sidella S, Boléo S, Bandarra V, Duarte MP, Mendes B (2013) Phytoremediation response of *Arundodonax* L. in soils contaminated with Zinc and Chromium. In: Eldrup A, Baxter D, Grassi A, Helm P (eds) Proceedings of the 21st European Biomass Conference and Exhibition, Setting the course for a Biobased Economy, Copenhagen, Denmark, 3–7 June 2013. ETA-Renewable Energies and WIP-Renewable Energies, pp 315–318

Barbosa B, Costa J, Fernando AL, Papazoglou EG (2014) Wastewater reuse for fiber crops cultivation as a strategy to mitigate desertification. Ind Crop Prod 68:17–23. doi:10.1016/j.indcrop.2014.07.007

Batchelor SE, Booth EJ, Walker KC (1995) Energy analysis of rape methyl ester (RME) production from winter oilseed rape. Ind Crop Prod 4:193–202

Benjamin M, Honeyman B (1992) Trace metals. In: Butcher S, Charlson R, Orians G, Wolfe G (eds) Global biogeochemical cycles. Academic, San Diego

Bini C, Wahsha M, Fontana S, Maleci L (2012) Effects of heavy metals on morphological characteristics of *Taraxacumofficinale* web growing on mine soils in NE Italy. J Geochem Explor 123:101–108

Bjelková M, Tejklová E, Griga M, Zajíková I, Genurová V (2001) Flax, linseed and hemp in phytoremediation, natural fibres (Poznan)—special edition: Proceedings 2nd Global Workshop Bast Plants in the New Millennium, Borovets, Bulgaria, p 285

Borghi M, Tognetti R, Monteforti G, Sebastiani L (2007) Responses of *Populus* × *euramericana* (*P. deltoids* × *P. nigra*) clone Adda to increasing copper concentrations. Environ Exp Bot 61:66–73

Börjesson P (1999) Environmental effects of energy crop cultivation in Sweden—I: identification and quantification. Biomass Bioenerg 16:137–154

Boyd R (2007) The defense hypothesis of elemental hyperaccumulation: status, challenges and new directions. Plant and Soil 293:153–176

Bullard M, Metcalfe P (2001) Estimating the energy requirements and CO_2 emissions from production of the perennial grasses *Miscanthus*, Switchgrass and Reed canary grass. ETSU/ ADAS, London

Chaney R, Malik M, Li Y, Brown S, Brewer E, Angle J, Baker A (1997) Phytoremediation of soil metals. Curr Opin Biotech 8:279–284

Charro E, Pardo R, Peña V (2013) Statistical analysis of the spatial distribution of radionuclides in soils around a coal-fired power plant in Spain. J Environ Radioact 124:84–92

Clemente R, Walker DJ, Bernal MP (2005) Uptake of heavy metals and As by *Brassica juncea* grown in a contaminated soil in Aznalcóllar (Spain): the effect of soil amendments. Environ Pollut 138:46–58

Costa C, Jesus-Rydin C (2001) Site investigation on heavy metals contaminated ground in Estarreja—Portugal. Eng Geol 60:39–47

Costa J, Fernando AL, Coutinho M, Barbosa B, Sidella S, Boléo S, Bandarra V, Duarte MP, Mendes B (2013) Growth, productivity and biomass quality of *Arundo* irrigated with Zn and Cu contaminated wastewaters. In: Eldrup A, Baxter D, Grassi A, Helm P (eds), Proceedings of the 21st European Biomass Conference and Exhibition, Setting the course for a Biobased Economy, Copenhagen, Denmark, 3–7 June 2013, ETA-Renewable Energies and WIP-Renewable Energies, pp 308–310

Cunningham S, Ow D (1996) Promises and prospects of phytoremediation. Plant Physiol 110:715–719

Dauber J, Brown C, Fernando AL, Finnan J, Krasuska E, Ponitka J, Styles D, Thrän D, Van Groeningen KJ, Weih M, Zah R (2012) Bioenergy from "surplus" land: environmental and socio-economic implications. BioRisk 7:5–50

Draycott AP (2006) Sugar beet. Blackwell, London

Duggan J (2005) The potential for landfill leachate treatment using willows in the UK—a critical review. Resour Conserv Recy 45:97–113

Dushenkov D (2003) Trends in phytoremediation of radionuclides. Plant and Soil 249:167–175

Dushenkov V, Kumar N, Motto H, Raskin I (1995) Rhizofiltration: the use of plants to remove heavy metals from aqueous streams. Environ Sci Technol 29:1239–1245

EPA (2000) Introduction to phytoremediation. Cincinnati. EPA/600/R-99/107

Epelde L, Mijangos I, Becerril JM, Garbisu C (2009) Soil microbial community as bioindicator of the recovery of soil functioning derived from metal phytoextraction with sorghum. Soil Biol Biochem 41:1788–1794

Faruk O, Bledzki AK, Fink H, Sain M (2012) Biocomposites reinforced with natural fibers: 2000–2010. Prog Polym Sci 37:1552–1596

Fazio S, Monti A (2011) Life cycle assessment of different bioenergy production systems including perennial and annual crops. Biomass Bioenerg 35:4868–4878

Fergusson J (1991) The heavy elements: chemistry, environmental impact and health effects. Pergamon, Oxford

Fernando ALAC (2005) Fitorremediação por *Miscanthus* × *giganteus* de solos contaminadoscom metais pesados, Ph.D. thesis, Faculdade de Ciencias e Tecnologia, Universidade Nova de Lisboa, Portugal (in Portuguese)

Fernando AL (2013) Environmental aspects of kenaf production and use. In: Monti A, Alexopoulou E (eds) Kenaf: a multi-purpose crop for several industrial applications, vol 7, Green energy and technology. Springer, London, pp 83–104

Fernando A, Oliveira J (2004) Fitorremediação de solos contaminados com metais pesados—Mecanismos, vantagens e limitações. Biologia Vegetal Agro-Industrial 1:103–114

Fernando AL, Duarte MP, Almeida J, Boléo S, Mendes B (2010) Environmental impact assessment (EIA) of energy crops production in Europe. Biofuels Bioprod Bioref 4:594–604

Finnan J, Styles D (2013) Hemp: a more sustainable annual energy crop for climate and energy policy. Energ Policy 58:152–162

Garbisu C, Alkorta I (2003) Basic concepts on heavy metal soil bioremediation. Eur J Miner Process Environ Prot 3:58–66

Ghosh M, Singh S (2005) A review on phytoremediation of heavy metals and utilization of its byproducts. Appl Ecol Environ Res 3:1–18

Giaccio L, Cicchella D, Vivo B, Lombardi G, Rosa M (2011) Does heavy metals pollution affects semen quality in men? A case of study in the metropolitan area of Naples (Italy). J Geochem Explor 112:218–225

Giachetti G, Sebastiani L (2006) Metal accumulation in poplar plant grown with industrial wastes. Chemosphere 64:446–454

Ginneken LV, Meers E, Guisson R, Ruttens A, Elst K, Tack FMG, Vangronsveld J, Diels L, Dejonghe W (2007) Phytoremediation for heavy metal-contaminated soils combined with bioenergy production. J Environ Eng Landsc 15:227–236

Giuseppe DD, Antisari LV, Ferronato C, Branchini G (2014) New insights on mobility and bioavailability of heavy metals in soils of the Padanian alluvial plain (Ferrara Province, Northern Italy). ChemErde–Geochem 74:615–623

Grabowska L, Baraniecki P (1997) Three year results on utilization soil polluted by copper-producing industry. Proc. of the Flax and other Bast Plants Symp. Natural Fibres, Spec. Ed. INF Poznan, pp 123–131

Guo LB, Sims REH, Horne DJ (2002) Biomass production and nutrient cycling in Eucalyptus short rotation energy forests in New Zealand. I: biomass and nutrient accumulation. Biores Technol 85:273–283

Hammer D, Kayser A, Keller C (2003) Phytoextraction of Cd and Zn with *Salix viminalis* in field trials. Soil Use Manage 3:187–192

Hansson PA, Svensson SE, Hallefält F, Diedrichs H (1999) Nutrient and cost optimization of fertilizing strategies for *Salix* including use of organic waste products. Biomass Bioenerg 17:377–387

He Z, Yang X, Stoffella P (2005) Trace elements in agroecosystems and impacts on the environment. J Trace Elem Med Biol 19:125–140

Heller MC, Keoleian GA, Volk TA (2003) Life cycle assessment of a willow bioenergy cropping system. Biomass Bioenerg 25:147–165

Herpin U, Berlekamp J, Markert B, Wolterbeek B, Grodzinska K, Siewers U, Lieth H, Weckert V (1996) The distribution of heavy metals in a transect of the three states the Netherlands, Germany and Poland, determined with the aid of moss monitoring. Sci Total Environ 187:185–198

Ho W, Ang L, Lee D (2008) Assessment of Pb uptake, translocation and immobilization in kenaf (*Hibiscus cannabinus* L.) for phytoremediation of sand tailings. J Environ Sci 20:1341–1347

Hong GH, Baskaran M, Molaroni SM, Lee H, Burger J (2011) Anthropogenic and natural radionuclides in caribou and muskoxen in the Western Alaskan Arctic and marine fish in the Aleutian Islands in the first half of 2000s. Sci Total Environ 409:3638–3648

Hüffmeyer N, Klasmeier J, Matthies M (2009) Geo-referenced modeling of zinc concentrations in the Ruhr river basin (Germany) using the model GREAT-ER. Sci Total Environ 407:2296–2305

Jabeen R, Ahmad A, Iqbal M (2009) Phytoremediation of heavy metals: physiological and molecular mechanisms. Bot Rev 75:339–364

James B (2001) Remediation—by—reduction strategies for chromate-contaminated soils. Environ Geochem Health 23:175–179

Kabata-Pendias A (2011) Trace elements in soils and plants, 4th edn. CRC, Boca Raton

Kassam A, Friedrich T, Derpsch R, Lahmar R, Mrabet R, Basch G, González-Sánchez E, Serraj R (2012) Conservation agriculture in the dry Mediterranean climate. Field Crop Res 132:7–17

Khlifi R, Chaffai A (2010) Head and neck cancer due to heavy metal exposure via tobacco smoking and professional exposure: a review. Toxicol Appl Pharm 248:71–88

Kin SG, Evseeva T, Oudalova A (2013) Effects of long-term chronic exposure to radionuclides in plant populations. J Environ Radioact 121:22–32

Kumar P, Dushenkov V, Motto H, Raskin I (1995) Phytoextraction: the use of plants to remove heavy metals from soils. Environ Sci Technol 29:1232–1238

Kumar G, Yadav S, Thawale P, Singh S, Juwarkar A (2008) Growth of *Jatropha curcas* on heavy metal contaminated soil amended with industrial wastes and Azotobacter—a greenhouse study. Biores Technol 99:2078–2082

Lado LR, Hengl T, Reuter H (2008) Heavy metals in European soils: a geostatistical analysis of the FOREGS geochemical database. Geoderma 148:189–199

Lasat MM (2000) Phytoextraction of metals from contaminated soil: a review of plant/soil/metal interaction and assessment of pertinent agronomic issues. J Hazard Subst Res 2:5–1–5–25

Lewandowski I, Schmidt U, Londo M, Faaij A (2006) The economic value of the phytoremediation function—assessed by the example of cadmium remediation by willow (*Salix*ssp). Agr Syst 89:68–89

Lewis M, Pryor R, Wilking L (2011) Fate and effects of anthropogenic chemicals in mangrove ecosystems: a review. Environ Pollut 159:2328–2346

Lievens C, Yperman J, Cornelissen T, Carleer R (2008) Study of the potential valorization of heavy metal contaminated biomass via phytoremediation by fast pyrolysis: part II. Characterization of the liquid and gaseous fraction as a function of the temperature. Fuel 87:1906–1916

Linger P, Müssig J, Fischer H, Kobert J (2002) Industrial hemp (*Cannabis sativa* L.) growing on heavy metal contaminated soil: fibre quality and phytoremediation potential. Ind Crop Prod 16:33–42

Lombi E, Zhao F, Dunham S, McGrath S (2001) Phytoremediation of heavy metal contaminated soils: natural hyperacumulation versus chemically enhanced phytoextraction. J Environ Qual 30:1919–1926

Low T, Booth C, Sheppard A (2011) Weedy biofuels: what can be done? Curr Opin Environ Sustain 3:55–59

Luger E (2003) Cardoon: introduction as an energy crop. BLT. http://blt.josephinum.at/. Accessed Nov 2014

Marchiol L, Assolari S, Sacco P, Zerbi G (2004) Phytoextraction of heavy metals by canola (*Brassica napus*) and radish (*Raphanussativus*) grown on multicontaminated soil. Environ Pollut 132:21–27

Marshall K, Watson S, McDonald P, Copplestone D, Watts SJ (2010) Exposure of birds to radionuclides and other contaminants in Special Protection Areas (SPAs) in North-West England. Sci Total Environ 408:2567–2575

Mavrogianopoulos G, Vogli V, Kyritsis S (2002) Use of wastewater as a nutrient solution in a closed gravel hydroponic culture of giant reed (*Arundodonax*). Bioresour Technol 82:103–107

McClintock T, Chen Y, Bundschun J, Oliver J, Navoni J, Olmos V, Lepori E, Ahsan H, Parvez F (2012) Arsenic exposure in Latin America: biomarkers, risk assessments and related health effects. Sci Total Environ 429:76–91

McIntyre T (2003) Phytoremediation of heavy metals from soils. Adv Biochem Eng/Biotechnol 78:97–123

Meers E, Ruttens A, Hopgood M, Lesage E, Tack F (2005) Potential of *Brassica rapa*, *Cannabis sativa*, *Helianthus annuus* and *Zea mays* for phytoextraction of heavy metals from calcareous dredged sediment derived soils. Chemosphere 61:561–572

Meers E, Vandecasteele B, Ruttens A, Vangronsveld J, Tack F (2007) Potential of five willow species (*Salix* spp.) for phytoremediation of heavy metals. Environ Exp Bot 60:57–68

Mirza N, Mahmood Q, Pervez A, Ahmad R, Farooq R, Shah MM, Azim MR (2010) Phytoremediation potencial of *Arundodonax*in arsenic-contaminated synthetic wastewater. Biores Technol 101:5815–5819

Mirza N, Pervez A, Mahmood Q, Shah M, Shafqat M (2011) Ecological restoration of arsenic contaminated soil by *Arundodonax* L. Ecol Eng 37:1949–1956

Mleczek M, Łukaszewski M, Kaczmarek Z, Rissmann I, Golinski P (2009) Efficiency of selected heavy metals accumulation by *Salix viminalis* roots. Environ Exp Bot 65:48–53

Molchanova I, Mikhailovskaya L, Antonov K, Pozolotina V, Antonova E (2014) Current assessment of integrated content of long-lived radionuclides in soils of the head part of the East Ural Radioactive Trace. J Environ Radioact 138:238–248
Møller AP, Nishiumi I, Suzuki H, Ueda K, Mousseau TA (2013) Differences in effects of radiation on abundance of animals in Fukushima and Chernobyl. Ecol Indic 24:75–81
Mulligan CN, Yong RN, Gibbs BF (2001) Remediation technologies for metal-contaminated soils and groundwater: an evaluation. Eng Geol 60:193–207
Niu Z, Sun L, Sun T, Li Y, Wang H (2007) Evaluation of phytoextracting cadmium and lead by sunflower, ricinus, alfalfa and mustard in hydroponic culture. J Environ Sci 19:961–967
Pan K, Wang W (2012) Trace metal contamination in estuarine and coastal environments in China. Sci Total Environ 421–422:3–16
Pandey A, Gupta R (2003) Fiber yielding plants of India, Genetic resources, perspective for collection and utilization. Nat Prod Rad 2:194–204
Papazoglou EG (2007) *Arundodonax* L. stress tolerance under irrigation with heavy metal aqueous solutions. Desalination 211:304–313
Papazoglou E (2011) Responses of *Cynaracardunculus* L. to single and combined cadmium and nickel treatment conditions. Ecotox Environ Safe 74:195–202
Papazoglou E, Karantounias G, Vemmos S, Bouranis D (2005) Photosynthesis and growth responses of giant reed (*Arundodonax* L.) to the heavy metals Cd and Ni. Environ Int 31:243–249
Pilon-Smits E (2005) Phytoremediation. Annu Rev Plant Physiol Plant Mol Biol 56:15–39
Pivetz B (2001) Ground water issue, phytoremediation of contaminated soil and ground water at hazardous waste sites, EPA/540/S-01/500
Prasad M (2004) Heavy metal stress in plants, from biomolecules to ecosystems, 2nd edn. Springer, Hyderabad
Pratas J, Favas PJC, D'Souza R, Varun M, Paul MS (2013) Phytoremedial assessment of flora tolerant to heavy metals in the contaminated soils of an abandoned Pb mine in Central Portugal. Chemosphere 90:2216–2225
Quaghebeur M, Rengel Z (2005) Review: arsenic speciation governs arsenic uptake and transport in terrestrial plants. Microchim Acta 151:141–152
Radwan M, Salama A (2006) Market basket survey for some heavy metals in Egyptian fruits and vegetables. Food Chem Toxicol 44:1273–1278
Rajkumar M, Freitas H (2008) Influence of metal resistant-plant growth-promoting bacteria on the growth of *Ricinuscommunis* in soil contaminated with heavy metals. Chemosphere 71:834–842
Raskin I, Kumar PBAN, Dushenkov S, Salt DE (1994) Bioconcentration of heavy metals by plants. Curr Opin Biotechnol 5:285–290
Ribeiro AB, Mexia JT (1997) A dynamic model for the electrokinetic removal of copper from a polluted soil. J Hazard Mater 56(3):257–271
Rosenqvist H, Dawson M (2005) Economics of using wastewater irrigation of willow in Northern Ireland. Biomass Bioenerg 29:83–92
Rossi G, Figliolia A, Socciarelli S, Pennelli B (2002) Capability of *Brassica napus* to accumulate Cadmium, Zinc and Copper from Soil. Acta Biotechnol 22:133–140
Sabra N, Dubourguier H, Hamieh T (2011) Sequential extraction and particle size analysis of heavy metals in sediments dredged from the Deûle Canal, France. The Open Environ Eng J 4:11–17
Schnoor L, Light S, McCutcheon N, Wolfe L, Carreira L (1995) Phytoremediation of organic and nutrient contaminants. Environ Sci Technol 29:318–323
Sharma P, Pandey S (2014) Status of phytoremediation in world scenario. Int J Environ Biorem Biodeg 4:178–191
Shcheglov AI, Olga B, Tsvetnova OB, Klyashtorin A (2014) The fate of Cs-137 in forest soils of Russian Federation and Ukraine contaminated due to the Chernobyl accident. J Geochem Explor 142:75–81

Sheng XF, Xia JJ, Jiang CY, He LY, Qian M (2008) Characterization of heavy metal-resistant endophytic bacteria from rape (*Brassica napus*) roots and their potential in promoting the growth and lead accumulation of rape. Environ Pollut 156:1164–1170

USDA (2006) Plant guide: reed canary grass. http://plants.usda.gov. Accessed Nov 2014

Vandenhove H (2013) Phytoremediation options for radioactively contaminated sites evaluated. Annu Nucl Energy 62:596–606

Vanek T, Soudek P, Tykva R, Vankova R (2006) Accumulation of radioiodine from aqueous solution by hydroponically cultivated sunflower (*Helianthus annuus* L.). Environ Exp Bot 57:220–225

Varennes A (2003) Produtividade de solos e ambiente. Escolar Editora, Lisboa (in Portuguese)

Videa J, Lopez M, Narayan M, Saupe G, Torresdey J (2009) The biochemistry of environmental heavy metal uptake by plants: implications for the food chain. Int J Biochem Cell B 41:1665–1677

Vystavna Y, Rushenko L, Diadin D, Klymenko O, Klymenko M (2014) Trace metals in wine and vineyard environment in southern Ukraine. Food Chem 146:339–344

Wuana RA, Okieimen FE (2011) Heavy metals in contaminated soils: a review of sources, chemistry, risks and best available strategies for remediation. ISRN Ecology. doi:10.5402/2011/402647

Yang X, Feng Y, He Z, Stoffella P (2005) Molecular mechanisms of heavy metal hyperaccumulation and phytoremediation. J Trace Elem Med Biol 18:339–353

Yordanova I, Staneva D, Misheva L, Bineva T, Banov M (2014) Technogenic radionuclides in undisturbed Bulgarian soils. J Geochem Explor 142:69–74

Zayed A, Gowthaman S, Terry N (1998) Phytoaccumulation of trace elements by wetland plants: I. Duckweed. J Environ Q 27:715–721

Zegada-Lizarazu W, Monti A (2011) Energy crops in rotation. A review. Biomass Bioenerg 35:12–25

Zhang Y, Li Y, Jiang L, Tian C, Li J, Xiao Z (2011) Potential of perennial crop on environmental sustainability of agriculture. Proc Environ Sci 10:1141–1147

Zhu YG, Shaw G (2000) Soil contamination with radionuclides and potential remediation. Chemosphere 41:121–128

Zhuang P, Shu W, Li Z, Liao B, Li J, Shao J (2009) Removal of metals by sorghum plants from contaminated land. J Environ Sci 21:1432–1437

Zoche J, Leffa D, Damiani A, Carvalho F, Mendonça R, Santos C, Boufleur L, Dias J, Andrade V (2010) Heavy metals and DNA damage in blood cells of insectivore bats in coal mining areas of Catarinense coal basin, Brazil. Environ Res 110:684–691

Sheng X, Xia [illegible] (2008) Characterization of heavy metal-resistant endophytic bacteria [illegible]

[illegible]

Chapter 20
Sensing of Component Traces in Complex Systems

Maria Raposo, Paulo A. Ribeiro, Nezha El Bari, and Benachir Bouchikhi

20.1 Introduction

Water pollution is a contemporary issue of mankind (WHO 2009, 2011). Conventional pollutants are pesticides and industrial intermediates. Over the last decade, it was found that beyond the conventional water pollutants, toxic/carcinogenic pesticides, and industrial intermediates, which are displaying persistence in the environment, a new class of pollutants is being introduced in the environment, namely, in the aquatic environment as complex mixtures via a number of routes but primarily by both untreated and treated sewage. This new class of pollutants consists of pharmaceuticals and active ingredients in personal care products (PPCPs), both human and veterinary, including not just prescription drugs and biologic products, but also diagnostic agents, "nutraceuticals," soaps, fragrances, sun-screen agents, and many others (Daughton and Ternes 1999). As example, one of these compounds is the triclosan, 5-chloro-2(2,4-dichlorophenoxy)-phenol, which is an antibacterial agent widely used in soaps, toothpastes, first-aid products, fabrics, and plastic goods. This compound is stable and lipophilic and its large consumption is a great deal of concern over its environmental fate since it has been

M. Raposo (✉) • P.A. Ribeiro
CEFITEC, Departamento de Física, Faculdade de Ciências e Tecnologia,
Universidade Nova de Lisboa, Caparica 2829-516, Portugal
e-mail: mfr@fct.unl.pt

N. El Bari
Biotechnology Agroalimentary and Biomedical Analysis Group, Department of Biology,
Faculty of Sciences, Moulay Ismaïl University, B.P. 11201, Zitoune, Meknes, Morocco

B. Bouchikhi
Sensor Electronic & Instrumentation Group, Department of Physics, Faculty of Sciences,
Moulay Ismaïl University, B.P. 11201, Zitoune, Meknes, Morocco

© Springer International Publishing Switzerland 2016

A.B. Ribeiro et al. (eds.), *Electrokinetics Across Disciplines and Continents*,
DOI 10.1007/978-3-319-20179-5_20

widely found in river water, lake water, sediments, fish, and in human milk (Ying and Kookana 2007; Lindstrom et al. 2002; Singer et al. 2002; Adolfsson-Erici et al. 2002). Recent studies have shown that triclosan has several biological activities that are unrelated to its antibacterial action, for example, affecting thyroxine homeostasis in weanling rats (Zorrilla et al. 2009) and exhibiting estrogenic and androgenic activity in breast cancer cells (Gee et al. 2008). More recently, James et al. (2010) make aware of that triclosan might endanger the pregnancy by reducing total placental estrogen secretion and thereby reducing estrogen action in target tissues critical for pregnancy maintenance, since this compound is known to be a potent inhibitor of estradiol and estrone sulfonation in sheep placenta. Unfortunately, this compound is not the only one to have a strong effect in the human health and, although several contaminants lists have been published (Daughton and Ternes 1999), the use of these chemicals and biologically active substances is in expansion. Increased amounts of PPCPs have been detected in wastewaters, worldwide surface waters, and other environmental matrices meaning that the development of sensing units for environmental monitoring is a main issue. Nevertheless, no commercial sensors for monitoring and detecting these pollutants have been made available so far, though for water treatment companies it is important to constantly monitor their presence in the distributed water and avoid further costumer complaints. This is mainly because the target pollutants molecules are in an aqueous medium together with countless spurious molecules and microscopic life, forming a rather complex matrix. However, some attempts have been developed with the desire to get sensors of these pollutant molecules. From the point of view of detection, the sensing devices described in the literature generally use a physical (electrical, optical, magnetic, etc.) property as detection principle together with the adoption of sensitive layers on the electrodes to make the sensor more selective to a particular molecule or molecular group. This type of sensing enables the determination of individual compounds or subgroups of related pollutants in complex matrix water samples (Albareda-Sirvent et al. 2001; Nunes et al. 1998). This means that the development of pollutant sensors should be directed for the concept of classification type device with quantifying capabilities, which can be attained by using a set or array of sensors is used and the response data is treated by statistical methods.

The concept of sensor arrays appears in the 1960s with the necessity of providing reliable data in the ambitious of aerospace activity (Fehr and McGahan 1967), but it was in the 1990s when a marked increase in the development of sensors of molecules in an aqueous medium using the sensor arrays concept. By 1995, Meyer et al. (1995) developed a monolithic array of 400 individually addressable microelectrodes in a modified CMOS wafer and demonstrated that is possible to characterize oxygen, hydrogen peroxide, and glucose distributions via the amperometric imaging. This work also warned of that: (1) sensor arrays can be used for pattern recognition analysis by using membranes with different sensitivities or selectivities or by applying different working potentials to the individual electrodes of a single chip; (2) the use of neural networks or fuzzy logic is a powerful tool for data processing. Following this work, Sangodkar et al. (1996) developed a glucose, urea, and triglycerides sensor based on polyaniline microsensor arrays demonstrating the strategy is quite general and can be extended to other enzyme-substrate systems leading to the so-called electronic tongue concept—a parallel concept of the already defined "electronic nose" (Gardner and Barlett 1994; Winquist et al. 1993). Therefore, within this concept and parallel definitions, several authors promptly applied the electronic tongue concept to various systems for classifying various samples as fruit juices, still drinks, milk, tea, coffee, beer, wine, etc. (Winquist et al. 1997; Legin et al. 1997; Toko 1998). It has also been applied to water and polluted water (Di Natale et al. 1997) through electrochemical detection methods. In 2002, Riul et al. used electric impedance as detection probe to develop electronic tongues (Riul et al. 2002, 2003a, b, c). Complete reviews of the advances in electronic tongues and noses can be found in Riul et al. (2010) and Askim et al. (2013).

This chapter provides a review of the literature and a forwarding of how the concepts of electronic tongue or nose can be applied to detection of pollutants molecules on aqueous environment. For this, methods of detection, preparation of sensitive layers, and data analysis with emphasis on information visualization approaches are described and examples in literature associated to water recognition are reviewed to provide a general conclusion on the reliability of this approach to detect a substance in a complex matrix, using the appropriated measurement methods, an array of sensor layers, and pattern recognition methods.

20.1.1 Concepts of Electronic Nose and Electronic Tongue

The concept of electronic nose has been well defined by Winquist et al. (1993) as "an array of gas sensors with different selectivity patterns, a signal collecting unit and pattern recognition software." Electronic nose is now commonly mentioned in the literature to refer to this type of instrument, although other terms have also been used, such as artificial nose, odor-sensing system, or electronic olfactometry. Also, Gardner and Bartlett (1994) defined the electronic nose as an instrument which comprises an array of chemical sensors with partial specificity, combined with an appropriate pattern recognition system for recognizing simple or complex odors. Measurements of single sensor transients provide an alternate method for vapor detection and identification to the conventional electronic nose systems based on sensor arrays functionalized for a broad range of chemical selectivity (Yadava 2012; Vilanova et al. 1996; Llobet et al. 1997; Hines et al. 1999; Osuna et al. 1999; Hoyt et al. 2002). Detection of volatile organic compounds (VOCs) is important for several applications like explosives detection, environment monitoring, chemical hazard detection, breath analysis for disease biomarkers, food quality monitoring, etc. (Llobet et al. 1997; Phaisangittisagul and Nagle 2008; Pearce et al. 2003; Francesco et al. 2005; Tothil 2003). Therefore, the conventional electronic nose operation is analogous to that of human nose, where olfactory receptor neurons generate response patterns to vapor inhalation which are brain processed for identification (Llobet et al. 1997).

The electronic tongues are one of the most promising tendencies to develop a fast, cheap, and objective method to evaluate food taste. Since the development of the first prototype in 1990 by K. Hayashi (Hayashi et al. 1990) a certain number of research groups have focused their efforts on the improvement of those systems though the use of diverse strategies and measuring techniques. There are several types of sensors that can be used in electronic tongues. The most common systems developed so far are based on electrochemical techniques such as potentiometry and voltammetry, as revealed by the numerous papers where these systems have found numerous applications (Arrieta et al. 2010a, b; Mimendia et al. 2010; Parra et al. 2006; Wei and Wang 2011; Chen et al. 2008; Woertz et al. 2011). However, other types of transduction, such as optical or gravimetric sensors, have also been addressed (Sun et al. 2008; Leonte et al. 2006; Veríssimo et al. 2010). Thus, an electronic tongue is a multisensor system, formed by an array of low-selective sensors, combined with advanced mathematical procedures for signal processing based on Pattern Recognition and/or Multivariate data analysis (Vlasov et al. 2005).

Over the past two decades, electronic nose and tongue have found a wide range of applications, for example, new applications, their use in beverage and food industries is rising steadily. Concerning to the electronic nose concept use, it has not found to be sensitive enough or be perturbed by the majority of compounds present in the headspace which are not relevant for the aroma to be detected. Under these circumstances, the electronic tongue could be addressed, for example, bitterness evaluation (Apetrei et al. 2004, 2007), polyphenolic content (Rodríguez-Méndez et al. 2008), discrimination of edible oils (Oliveri et al. 2009). Up until now, one of the most important drawbacks of the electronic tongue was their build-up based on the same type of sensors, i.e., potentiometric, voltammetric, or interdigitated electrodes (IDEs) (Moreno et al. 2006). These techniques restricted the amount of data containing useful information that could be attained from within these analyses. Emerging strategies have recently been proved to efficiently overcome these problems, namely, multi-sensor data fusion techniques. This has been applied to numerous fields and new applications are constantly being explored. For example, Gutiérrez et al. has reported the classification between beer samples, where two electronic tongues were combined using a data fusion method (Gutierrez et al. 2013). Di Natale et al. succeeded in classifying urine and milk by integrating electronic nose and electronic tongue and merging the data obtained from each of the sensor arrays (Di Natale et al. 2000). Also, Apetrei et al. (2010) characterized olive oils having different degrees of bitterness and found superior results by combining the three analytical instruments, an electronic nose, tongue, and eye.

20.2 Detection of Pollutants

20.2.1 Electrochemical Methods

Electrochemical methods, namely, potentiometry, amperometry, or cyclic voltammetry have been often used for detection of water pollutants.

Potentiometric measurements are made at equilibrium using a high-impedance voltmeter with currents that approach to zero. Only a two-electrode setup is required since the current is negligible. When the analyte solution and reference solution are not the same, a potential difference is observed. Since the electric potential of the reference electrode is constant, changes in the potential are due to the indicator (working) electrode. For the membrane potentiometric method, the observed potential is dominated by the Donnan potential that develops across a membrane (Ivarsson et al. 2001a).

Amperometry requires more complicated and expensive specialized equipment and qualified personnel for handling. However, this technique is the preferred mode to monitor a time-dependent change with a wide dynamic range (Wang et al. 2015). Amperometric current comes from the oxidation or reduction of electroactive compounds at a working electrode while a constant potential is applied; the measured current, generally of the order of μA, is a direct measurement of the electrochemical reactions rates taking place at the electrodes (Buratti et al. 2004).

In voltammetric measurements a current is measured between a metal working electrode and a counter-electrode when a voltage pulse is applied over the working electrode and the reference electrode. A set of pulses can be put together to form a pulse train in order to extract as much information as possible from the solution. When the potential is applied, electro-active compounds that react will be reduced or oxidized at a given potential value and a current, which is measured, will arise. Generally, one can remark that the voltammetric method can operate with complex liquids as long as compounds can be actively oxidized or reduced onto the work electrode while potentiometric method can operate in terms of the system net charge being a disadvantageous method in nonelectrolytes media (Legin et al. 1999; Ivarsson et al. 2001b).

20.2.2 Impedance Spectroscopy

Impedance spectroscopy has become a mature and well-understood technique to acquire, validate, and quantitatively interpret the experimental impedances (Lasia 1999). This technique provides an accurate method for in situ capacitance measurements of organic coatings and has been widely employed for the characterization of organic barrier coatings for many years (Hu et al. 2003). This technique is based on the electrode perturbation caused by an external signal of small magnitude, thus allowing to perform measurements in equilibrium or in stationary state.

The experimental values of impedance are usually represented by Nyquist and Bode diagrams. In the Nyquist diagrams, the imaginary part of the electrical impedance is represented versus the real part. The Nyquist diagrams, which form either simple arcs or multiple arcs in the complex plane, are very useful for obtaining parameters from the impedance spectra (Labrador et al. 2010).

By 1987, Taylor and Macdonald (1987) demonstrated that impedance spectroscopy can be used to characterize aqueous environments, by analyzing the electrical impedance as a function of frequency signal applied to nanostructures adsorbed onto solid substrates, with previously deposited interdigitated (IDT) electrodes and immersed in a water sample to be characterized. The electrical response analysis of the sensorial units was carried out by modelling the response to an equivalent circuit representative of the changes taking place at the double layer formed on the film/solution interface, in the electrolyte, and in the nanostructure adsorbed onto the IDT electrodes. This response will contain the fingerprint that will allow to identify the solution, and to determine the concentration of target molecules. This technique was subsequently and successfully applied to an artificial taste sensor based on Langmuir-Blodgett films of conducting polymers and a ruthenium complex and of self-assembled films of an azobenzene-containing polymer (Riul et al. 2002).

20.2.3 Surface Acoustic Wave Technique

The surface acoustic wave (SAW) technique consists of two IDTs deposited on a piezoelectric substrate such as quartz, see Fig. 20.1. Piezoelectric materials are anisotropic media, which yields different material properties versus the direction of signal propagation. A thin chemically sensitive layer is placed between the IDTs on the top surface of the piezoelectric substrate and in one of the IDTs (transmitting IDT) is connected to a high frequency signal and when the target chemical is adsorbed, the mass of the film increases, causing a change in the velocity and phase of the propagating acoustic wave signal, which results in a change in the amplitude and frequency of the output voltage measured at the load impedance, receiving IDT. The change is expected to be proportional to the concentration of the

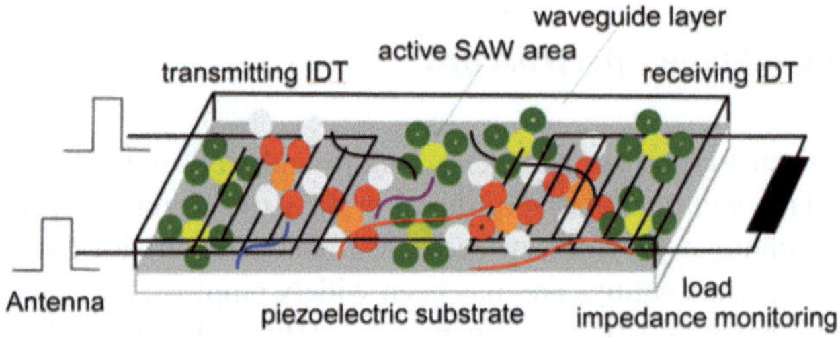

Fig. 20.1 SAW sensors based on IDT electrodes lying under the bound analyte layer on the bottom

target molecule. SAW sensors are extremely sensitive in detecting the properties of solid or fluid materials in contact with their surfaces, including the surface mass change, liquid density, liquid viscosity, and electrical conductivity (Shiokawa and Kondoh 2004). SAW sensor systems are quite stable against temperature fluctuation and external noise.

20.2.4 Sensors Integrated on FETs

Ion Sensitive Field-Effect Transistor (ISFET) biosensors can be also used to detect the pollutants in water, because of their high performance among these, their potentiality of batch production and low output impedance (Jaffrezic-Renault et al. 1999). The specific technology used for the fabrication of these ISFETs allows perform a differential measurement between both FETs, the output signals being measured against the common silicon substrate (Elbhiri et al. 1999). Also, the responses of the grafted ISFETs obtained with different leaving groups on the silylated molecules are analyzed by the modified site-binding model which allows determining the density of grafted sites (D'Amico et al. 2005).

20.2.5 Spectrophotometric Methods

In 1987, Worsfold and Clinch (1987) showed that spectrophotometry can be used for the determination of phosphate in natural water. The basic principle is that each compound absorbs or transmits light over a certain range of wavelength, which can also be used to measure the amount of a given chemical substance. Ultraviolet–visible (UV–Vis) spectroscopy is one of the most useful methods in analytical chemistry for the quantitative determination of different analytes. This technique uses light in the visible and adjacent (near-UV and near-infrared (NIR)) ranges. The absorption or reflectance in the visible range directly affects the perceived color of the chemicals involved (Hoffman 2010).

Fluorescence is the optical emission from molecules that have been excited to higher energy levels by absorption of electromagnetic radiation. Broad-band excitation light from a lamp passes through a monochromator, which passes only a selected wavelength. The fluorescence is dispersed by another monochromator and detected by a photomultiplier tube. By scanning the excitation monochromator through a given wavelength range the excitation spectrum can be obtained and scanning the fluorescence monochromator gives the fluorescence spectrum (Eugen 2013). In 1973, Keizer and Gordon Jr demonstrated that fluorescence spectroscopy can be a valuable technique for estimating petroleum residual concentrations in seawaters when a large numbers of samples are analyzed (Keizer and Gordon 1973). This method appears to offer a fast and sensitive analytical technique for the estimation of petroleum residues in seawater.

20.2.6 Other Methods

The High Performance Liquid Chromatography also called high-pressure liquid chromatography (HPLC) is one of the major techniques of modern analytical chemistry. It provides high efficiency and high resolution for a wide range of organic compounds and it has been extensively used in environmental analysis. This technique is used to separate the components in a liquid mixture and to identify and quantify each component. This approach has been used by several researchers and led to satisfactory results provided that some general rules (Realini 1981; Gawdzik et al. 1990). Realini (1981) used this technique for the analysis of waste water and drinking water. According to his paper, HPLC has shown to be an accurate, precise, and sensitive method, to quantitative and qualitative determination of ng/L levels of phenols in water. The use of ion-pair extraction in basic solution gave good recoveries, and the use of reversed-phase chromatography gave excellent separation of the priority pollutant phenols (Butler and Guthertz 2001).

Also, gas chromatography is nowadays the most important analytical method in organic chemical analysis for the determination of individual substances in complex mixtures. Mass spectrometry is a detection method which gives the meaningful data, as it arises from the direct determination of the substance, molecule, or of fragments. The results of mass spectrometry are therefore used as a reference for other indirect methods, namely for confirmation. The complete integration of mass spectrometry and gas chromatography into a single GC/MS system has shown itself to be synergistic at all levels (Hübschmann 2009). The area of application of GC and GC/MS is limited to substances which are volatile enough to be analyzed by gas chromatography. For this reason, the trend of chromatographic analysis in the environmental field is the development of accurate, easy to automate, and sensitive methods that reduce sample handling. The classical methodologies for the analysis of volatile compounds in water demonstrate this trend (León et al. 2006). According to Sánchez-Avila et al. (2010), the GC/MS method has proven to be useful in screening persistent organic pollutants as nonylphenol with a ng/L sensitivity level. Its high reliability, reproducibility, and robustness ensure the viability of its use for the routine multi-residual analysis of toxic substances in seawater following recent European Union legislation standards and US-EPA recommendations for seawater contaminants (Sahil et al. 2011).

20.3 Sensing Layers

20.3.1 Methods for Preparation of Sensing Layers

Several methods can be used to prepare sensing layers based on organic molecules, namely, casting, spin coating, sol–gel, etc. (Oliveira et al. 2001). However, the sensor should be immersed in aqueous solutions and it is advisable that the sensing layer does not leave the support. Therefore methods which enable the organic molecules become deposited on the support after immersion in water are most appropriate. Following this rule, the functionalization of the sensor surface can be based on the adsorption phenomena of organic and/or biological molecules onto solid electrodes also called supports or substrates, lying on either solid-support substrates or compliant thin substrates. Commonly, this technique is designated by self-assembly (SA) but both chemical and physical adsorption processes onto solid supports can be implemented (Oliveira et al. 2001). When the adsorption of layers of molecules onto substrates is based on chemical adsorption, the molecules should have at least two functional groups: one to be adsorbed onto the substrate and the other to be placed on the layer on the top, allowing the chemisorption of the same or/and other molecules. On the other hand, when the adsorption of layers is based on physical adsorption processes, the technique is usually designated of layer-by-layer (LbL), and consists of alternating cationic and anionic polyelectrolyte layers onto a solid substrate (Decher 1997), as schematized in Fig. 20.2. The adsorption of each layer is dependent on the polyelectrolyte solution parameters (molecular type, concentration, etc.), substrate, and the adsorption time. The LbL technique can also be used to patterning the sensor base, so that molecules are adsorbed onto a defined spatial region or a pattern.

Thin layers of organic molecules can be also deposited by electrochemical deposition or electropolymerization (Paunovic and Schlesinger 2006) and by Langmuir-Blodgett (LB) technique (Oliveira et al. 2001).

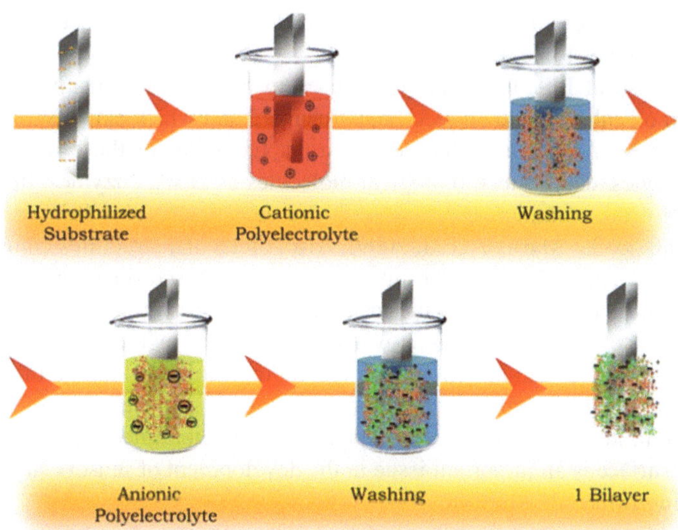

Fig. 20.2 Scheme of the layer-by-layer (LbL) technique for preparation of molecular layers

20.3.2 Materials

The materials to be used as sensor layers should be chosen in accordance with the sensor transducing principles and also with layers preparation technique.

In the case of electrochemical sensors, different metals as Copper (Cu), Glassy Carbon (GC), Gold (Au), Nickel (Ni), Palladium (Pd), Platinum (Pt), and Silver (Ag) have been used for detection. However, the deposition of adequate molecules on the electrodes, functionalization of the sensor surface, is an important issue in the development of sensors. Following the example of triclosan, one of the most PPCPs's pollutants investigated, sensors have been developed using several materials and methods for the last years. A strand of carbon fibers have been designed and characterized for use as voltammetric detector for high-performance liquid chromatography to detect triclosan in rabbit serum and urine (Wang and Chu 2004). Electropolymerizing *o*-phenylenediamine (*o*-PD) on glassy carbon electrodes has been used to create amperometric triclosan sensors working over a linear range of 2.0×10^{-7}–3.0×10^{-6} mol/L, with a detection limit as low as 8.0×10^{-8} mol/L (Liu et al. 2009). Multiwall carbon nanotube films were also developed for the fast detection of triclosan as electrochemical sensor, having a linear range from 50 μg L^{-1} to 1.75 mg L^{-1} and a limit of detection of 16.5 μg L^{-1} (about 57 nM) (Yang et al. 2009). Therefore, the choices of different materials which can be deposited on the electrodes are of extreme relevance in the detection of a determined molecule and each case should be analyzed carefully.

For the case of chemical SA, the molecules should have a chemical group adequate to be chemically bonded to the solid support and also to have another specific group which allows the chemical bond between the molecules of the next layer or the adsorbent. In addition, the adsorbed molecules are expected to interact with the molecules to be detected.

For the case of LbL technique to be used to prepare the sensitive layers, the adsorbed molecules should be electrically charged and which extends the range of different types of molecules that can be used. Examples of these molecules are polyelectrolytes, dendrimers, biological molecules as DNA, enzymes, lipids, gene and/or anti-gene, ceramics materials and also small molecules (Liu et al. 2009; Duarte et al. 2013a; Marletta et al. 2009). Also, molecular structures composed by groups of molecules as the case of liposomes can be adsorbed on surfaces (Duarte et al. 2013b). As a remark, liposomes can play an important role in the detection of molecules since most of pernicious water contaminants as they are lipid soluble. Moreover, liposomes can incorporate biological molecules and maintain its biological activities during months due to the presence of water in the inside of liposomes (Moraes et al. 2010). Therefore, the deposition of a determined type of molecules on a solid surface is an important issue in the development of sensors since it is the conditioning factor.

20.4 Pattern Recognition Methods

Electronic sensing gives rise to large amounts of data from a large number of sensors in the array that can be analyzed in a relatively short time. This leads to the availability of multivariate data matrices that require the use of mathematical and statistical procedures, to efficiently extract the maximum useful information from data.

20.4.1 Feature Extraction and Data Pre-processing

The purpose of feature extraction from data is to attain a low-dimensional mapping that preserves most of the information in its original feature vector. The features used for data analysis are directly extracted from the responses of sensors arrays, in order to fully exploit the maximum information existing in the response (Haddi et al. 2011). Pre-processing of electronic sensing data consists of extracting the most significant features from the sensor response curves. With respect to devices as electronic nose or tongue, pattern recognition methods are decisive factors in obtaining a versatile instrument able to reliably recognize a wide variety of odors and/or flavors.

20.4.2 Data Analysis

Raw data obtained separately from the electronic nose and/or the electronic tongue, combined or not, are analyzed by supervised and unsupervised pattern recognition techniques such as Principal Component Analysis (PCA), Support Vector Machines (SVMs), and Clustering Analysis (CA).

20.4.2.1 Principal Component Analysis

PCA is a well-known unsupervised method, often employed to analyze data from several kinds of gas and taste sensor arrays (Castro et al. 2011; Chen et al. 2008). The main objective of PCA (Gardner 1991; Llobet et al. 1997) consists of expressing information contained in a dataset by a lower number of variables, called principal components. These are linear combinations of the original response vectors and are chosen to contain the maximum data variance and to be orthogonal. Hence, PCA allows the reduction of multidimensional data to a lower dimensional approximation, while simplifying the data interpretation by the first two or three principal components (PC1, PC2, and PC3) in two or three dimensions, preserving most of the variance in the data.

20.4.2.2 Support Vector Machines

SVMs, which are based on Statistical Learning Theory (SLT), have been recently introduced as a new technique for solving a variety of learning classification and prediction problems (Cristianini and Shawe-Taylor 2000). SVMs have been successfully applied to a number of problems ranging from face identification and text categorization, to bioinformatics and data mining. The statistical classification method was proposed by Vapnik (1995), and the main idea of SVMs is to separate the classes with a particular hyperplane, which maximizes a quantity called margin. The margin is the distance from a hyperplane separating the classes to the nearest point in the dataset. SVMs were originally designed for binary classification, and there are currently two types of approaches for multi-class SVMs. One is the construction and combination of several binary classifiers "one against one or one against all methods," while the other is the direct consideration of all data in one optimization formulation (Hsu and Lin 2002). The direct approach consists of the construction of a classifier that recognizes the set of all the classes: the determination of the hyperplane between these different classes permits the choice of class when a new input is presented.

20.4.2.3 Clustering Analysis

CA has been employed to examine the electronic nose and tongue data by testing the relationships of various compounds. CA technique attempts to separate data into specific groups (Everitt 1974), based on a measure of similarity. Each data point was initially assumed to be a lone cluster and then the threshold is incrementally lowered. As a result, more and more samples were linked together and aggregated into larger and larger clusters of increasingly dissimilar elements. Finally, all samples were joined together. The results of hierarchical clustering methods are often displayed as a dendrogram connection. There are many types of similarity linkage, the three most common of these are: single linkage (nearest neighbor), where the distance between two clusters is determined by the distance between the two closest objects in the clusters; complete linkage (furthest neighbor), where the distances between clusters are determined by the greatest distance between any two objects in the different clusters; and group average, where the distances between clusters are determined by the mean distance of objects in different clusters.

20.4.2.4 Other Methods

(a) Data Fusion Approach

Amongst the three approaches offered by data fusion technique, namely low-level of abstraction, mid-level of abstraction, and high level of abstraction, the first approach is the most often used and generally provides good results. Several investigations have reported on low-level of abstraction (Di Natale et al. 2000; Rudnitskaya et al. 2006) providing evidence of its beneficial effects on merging data from different instruments, while also demonstrating its wide potential for applications in multi-sensor data fusion. Typically, in low-level of abstraction, the data from all sources are simply concatenated before model construction. The resulting data matrix has the number of rows equal to the number of samples, and the number of columns equal to the total number of signals from all sources. This approach was elaborated by several researchers and led to satisfactory results when following some general rules (Di Natale et al. 2001; Khaleghi et al. 2013). The merging of measurements from two sources could potentially provide more redundant information and this can grievously affect the ability of identification. For this reason, it will be suitable to couple low-level of abstraction to features selection technique in order to overcome the problem of dimensionality (Boilot et al. 2003). Another precaution to take into account is the dataset size of each source, which must be comparable in order to the largest one not "dominate" the other (Di Natale et al. 2000; Söderström et al. 2005). In mid-level fusion, feature extraction is applied to each data source before the extracted features are combined. Mid-level fusion is popular owing to its adequate performance and easy adaptability with well-established feature extraction methods. The third method of data fusion, called high-level of abstraction, assumes that data from each source are analyzed and afterwards a model is constructed for each data source separately. The results from all the models are then merged (Liu and Brown 2004).

(b) Fuzzy ARTMAP Artificial Neural Network

The Fuzzy ARTMAP neural network is a nonlinear supervised classifier based on fuzzy Adaptive Resonance Theory (ART). It is a promising method since fuzzy ARTMAP is able to carry out online learning without forgetting previously learnt patterns (stable learning). It can also recode previously learnt categories and is self-organizing (Llobet et al. 1999). In its most general form, Fuzzy ARTMAP includes two fuzzy ART modules (ART_a and ART_b), interconnected by an associative memory and some internal control structures that regulate learning and information flow (Carpenter et al. 1992). During supervised learning, ART_a receives a stream of input patterns $\{a^M\}$ and ART_b also receives a stream of patterns $\{b^M\}$, where b^M is the correct prediction given a^M. When a prediction by ART_a is not confirmed by ART_b, inhibition of the inter-ART associative memory activates a match tracking process. This increases ART_a vigilance by the minimum amount required for the system either to activate an ART_a category that matches the ART_b category or to learn a new ART_a category. Fuzzy ARTMAP generally shows superior performance in training when compared with the Multi-Layer Perceptron (MLP) (Carpenter et al. 1995; El Bari et al. 2007).

20.5 Sensor Arrays Applied to Sensing Pollutants Molecules in Water: A State of Art

20.5.1 Water Pollution Recognition by Electronic Noses

The electronic nose KAMINA was tested on the headspace water pollution recognition. The headspace of water samples polluted with chloroform or ammonia was investigated simulating stagnant and flowing waters (Goschnick et al. 2005). The KAMINA is based on the use of metal oxide gas sensing. This investigation demonstrated that the electronic nose KAMINA is able to discriminate between ammonia or chloroform contaminated water and clean water by analyzing the headspace of the corresponding water samples. For discrimination of the pollutants, the supervised linear discriminant analysis was employed. Experimental results demonstrated that a good discrimination and recognition was obtained of clean tap water or contaminated water containing either ammonia or chloroform can be achieved. The detection limits for the contaminants were below 1 mppm (ppm by mass).

More recently Sohn et al. (2008) used an electronic nose to continuously understand the complex odor-generating mechanisms within poultry housing as well as to identify strategies to reduce the impact of odor emissions on local communities. These actions aim to reduce negative public attitudes towards poultry farming (Jiang and Sands 2000; PAE 2003). Recently, a hybrid system based on electronic nose coupled with an electronic tongue and combined with multivariate analysis was applied to the different kinds of Moroccan waters (Haddi et al. 2014). This system allowed discriminate amongst the potable and non-potable water samples with a total variance of 86.98 %. This study also confirmed the usefulness of the hybrid systems.

20.5.2 Water Pollution Recognition by Electronic Tongues

20.5.2.1 Drinking Water Quality

Amongst the applications of electronic tongue concept, discrimination between different types of waters has been attempted from the beginning. In a first approach, experiments were conducted using an electronic tongue to virtually monitor the drinking water quality, measured from the raw water in the river to the tap water of the consumer. It was demonstrated by using signal analysis and statistical multivariate methods that the proposed multi-electrode virtual sensor system is able to detect water quality changes and consequently to estimate the water quality (Lindquist and Wide 2001). Also, Scozzari et al. (2007) used voltammetric techniques for the qualitative analysis of water, focusing the work on the signal analysis approach and its evaluation in terms of discrimination capability. The fundamental idea was to investigate how an adequate signal processing approach applied to a mature and affordable sensor technique (voltammetry) can address the issue of extracting aggregate chemical information, useful to characterize the liquid under analysis. In the proposed approach, dimensionality reduction was performed in a transformed domain via discrete cosine transform with an appropriate selection of a low dimensionality subset of the transformed coefficients.

An electronic tongue system, inductively coupled to a plasma atomic emission spectroscopy and an ion chromatography system, was also used as a sensory evaluation of mineral, spring, and tap water samples of different geographical origins (seven classes). Samples from the same geographical origin were correctly classified by both chemical analysis and the electronic tongue system (100 %). However, it was found that only 80 % classification rate can be achieved by sensory evaluation. Different water brands (different brand names) from the same geographical origin did not show definite differences, as expected. Forward stepwise algorithm selected three chemical parameters, namely, chloride (Cl), sulphate (SO_4^{2-}), and magnesium (Mg) contents and two electronic tongue sensor signals to discriminate according to the geographical origins (Sipos et al. 2012). Recently, a flow-cell electronic tongue composed by a sensor array comprised of six interdigitated microelectrodes coated with nanofiber films of poly(lactic acid)/multiwalled carbon nanotube (MWCNT) composites, and coupled with PCA, was used to discriminate potable water samples from non-potable water contaminated with metals or traces of pesticides (Oliveira et al. 2013). To analyze the sensory profiles of mineral water have also been used to investigate the similarities between the sensitivity of a trained human panel and an electronic tongue device. In this study, the trained sensory panel evaluated six flavored mineral water samples. The samples consisted of three different brands, each with two flavors (pear-lemon grass and jostaberry). The applied sensory method was profile analysis, and the sensory attributes obtained from electronic tongue measurements were implemented by using the multivariate statistical method Partial Least Squares regression (PLS). The results showed that the products manufactured under different brand names having the same aromas presented very similar sensory profiles (Sipos et al. 2013).

20.5.2.2 Detection of Metal Ions

Di Natale et al. (1997) used a sensor array of ion sensitive electrodes to measure the concentrations of a number of chemical species in solutions and the sensor array output data was analyzed by chemometrics, nonlinear least squares and neural networks. The best results have been achieved by the introduction of modular models which make use, at the same time, of both qualitative and quantitative information. Two years later, Legin et al. (1999) developed a multisensor system made of a set of 29 different chemical sensors based on all-solid-state crystalline and vitreous materials with enhanced electronic conductivity and redox and ionic cross-sensitivity. This system was used to the determination of low contents of uranium(VI), uranium(IV), iron(II), and iron(III) in complex aqueous media.

Multidimensional data, processed by pattern recognition methods such as artificial neural networks and partial least squares, have been shown that Fe(II) and Fe(III) contents in the range from 10^{-7} to 10^{-4} mol L^{-1} of total iron concentration can be determined with the average precision of about 25 %. U(VI) and U(IV) contents could be determined with the average precision of 10–40 % depending on the concentration. A similar approach has been applied to recognition and evaluation of mineral water samples solutions containing mixtures of ions of different nature and concentration (Colilla et al. 2002). More recently, a detection system based on the measurement of impedance of two modified electrodes, each one containing a chelating agent (pyrocatechol violet and a nitrilotriacetic derivative) combined with principal component data analysis procedure was used to detect Al^{3+}, Fe^{3+}, Cd^{2+}, Pb^{2+}, Hg^{2+}, Cu^{2+}, Ca^{2+}, and Ag^{+} metal ions at micromolar levels in ultrapure water. This work also demonstrated that selecting the appropriate working frequencies and sensors, the array can also be applied to different aqueous systems such as bottled mineral water or concentrated NaCl (27 %) yielding similar results (Yánez Heras et al. 2010).

20.5.2.3 Detection of Water Toxicity Due Organisms

Lower-cost and easy-to-read instrumentation would be very promising also in regarding to the detection of sub-products from algae decomposition or from other organisms which depend on their concentration, can be toxic or, otherwise, can give unpleasant taste and odor to water. The detection of sub-products from algae decomposition as 2-methylisoborneol (MIB) and geosmin (GSM) in water samples has been also evaluated by an electronic tongue system based on nonspecific polymeric sensors and impedance measurements. PCA, applied to the generated data matrix, indicated that this electronic tongue was capable to perform, with remarkable reproducibility, the discrimination of these two contaminants in either distilled or tap water, in concentrations as low as 25 ng L^{-1} (Braga et al. 2012). Another example of the application of the electronic tongue to estimate water toxicity is the potentiometric multisensory system conjugated with projection on latent structures (PLS) regression. This approach was used to test waters of a bioassay with three living test organisms: *Daphnia magna*, *Chlorella vulgaris*, and *Paramecium caudatum*. In this case, the prediction of water toxicity with relative errors 15–26 % (depending on the test object) was achieved (Kirsanov et al. 2013).

20.5.2.4 Detection of Pollutants

There are several examples of developed systems for detection of water pollutants worth to be mentioned. In 2006, Carvalho et al. (2006) developed a sensor system adapted to detect chloroform, a potentially carcinogenic compound which is often used in water disinfection. This system consisted of interdigitated gold-coated microelectrodes covered with various conducting polymers through the LbL technique. Analysis of impedance data obtained from these sensors, over a range of frequencies from 1 Hz to 1 MHz, revealed to able to detect chloroform in water and to distinguish different concentrations, the detection limit being of the order of 0.01 mg L^{-1}. One year later, the same team developed a similar system, using also impedance spectroscopy as measurement technique, to detect small amounts of the carcinogenic trihalomethanes (THM) as bromoform, bromodichloromethane, and dibromochloromethane, compounds which may be generated as by-products of water-treatment processes in public water supply systems. The detection limits were found as low as 0.02 mg L^{-1} (Carvalho et al. 2007). More recently the same team demonstrated that modified nylon nanofibers with LbL films of conductive polypyrrole (PPy) and poly(o-ethoxyaniline) (POEA) assembled onto graphite interdigitated polyethylene terephthalate substrates are suitable to detect paraoxon pesticide in water by discriminating the pesticide in the water of corn washing in postharvest treatment, compared to water of other sources (Oliveira et al. 2012). Meanwhile, a new protocol has been developed and evaluated for the determination of trihalomethanes (THMs) at the submicromolar concentration level in water (Peverly and Peters 2012). The protocol was based on a three-step stripping analysis that uses a single electrochemical cell and that entails (a) direct electrochemical reduction of a trihalomethane at a silver cathode to form halide ions in an aqueous sample containing tetraethylammonium benzoate, (b) capture of the released halide ions as a silver halide film on the surface of a silver gauze anode, and (c) cathodic reduction and quantitation of the silver halide film by means of differential pulse voltammetry. With this method the THMs as bromoform and chloroform were successfully quantitated in about 30 min, with a detection limit of 3.0 μg L^{-1} (12 nM) and 6.0 μg L^{-1} (50 nM), respectively, as well as the total trihalomethane content in a prepared water sample at a level commensurate with the maximum allowable annual average of 80 μg L^{-1}, as mandated by the United States Environmental Protection Agency.

Aqueous solutions of petroleum hydrocarbons as cyclohexane, naphthalene, benzene, toluene, ethylbenzene, and the three isomers of xylene (BTEX analytes) were analyzed an array of chemiresistor sensors, made from thin film assemblies of single-wall carbon nanotubes, multiwall carbon nanotubes, reduced graphene oxide nanosheets (RGON), and gold nanoparticles (AuNP). The carbon nanotube and RGON chemiresistors functionalized with octadecyl-1-amine and the AuNP chemiresistors were functionalized with 1-hexanethiol (Cooper et al. 2013). This study demonstrated that the AuNP chemiresistor was the most sensitive to all the analytes with limits of detection ranging 0.2 and 0.6 ppm in water. In contrast, the multiwall carbon nanotube chemiresistor was the less sensitive to the analytes with limits of detection between 20 and 200 ppm. These results reveal that these nanomaterials have great potential, with further optimization, of being incorporated into devices that would detect hydrocarbons in water at concentrations according with the regulations of the US Environmental Protection Agency. Mixtures of benzene, toluene, ethylbenzene, *p*-xylene, and naphthalene dissolved in water have also been probed with an array of partially selective gold nanoparticle chemiresistor sensors (Cooper et al. 2014). The overall root mean square error between the predicted and measured concentrations (residuals) was of 0.2–1.5 mg/L, for mixtures with a nominal component concentration of 10 mg/L. The accuracy of the random forests predictions was not unduly affected by increasing mixture complexity. It should be referred that random forests analysis is a statistical technique suitable for quantifying the relationship between responses of partially selective sensors to the concentration of different hydrocarbons in water.

The presence of ammonium nitrate in water was detected by an electronic tongue based on pulse voltammetry by using of an array of eight metallic working electrodes encapsulated in a stainless steel cylinder (Campos et al. 2013). In a first step the electrochemical response of the different electrodes was investigated in the presence of ammonium nitrate in water in order to further design the wave form used in the voltammetric tongue. The response of the electronic tongue was then tested in the presence of a set of 15 common inorganic salts; i.e., NH_4NO_3; $MgSO_4$; NH_4Cl; NaCl; Na_2CO_3; $(NH_4)_2SO_4$; $MgCl_2$; Na_3PO_4; K_2SO_4; K_2CO_3; $CaCl_2$; NaH_2PO_4; KCl; $NaNO_3$; K_2HPO_4. A PCA plot showed a fairly good discrimination between ammonium nitrate and the remaining salts studied. In addition Fuzzy Art map analyses determined that the best classification was obtained using the Pt; Co; Cu; and Ni electrodes. Moreover, PLS regression allowed the creation of a model to correlate the voltammetric response of the electrodes with concentrations of ammonium nitrate in the presence of potential interferents such as ammonium chloride and sodium nitrate. A voltammetric electronic tongue device have been used for evaluation of ammonia and orthophosphate concentrations from influent and effluent wastewater since the quantification of these compounds in wastewater treatment plants is important due to their implication in the eutrophication process (Campos et al. 2014). The performed electrochemical study of the response to the presence of ammonium and orthophosphate ions was carried out in order to design a suitable waveform for each electrode.

Concerning the detection of PPCPs with electronic tongue, excluding the cases of pharmaceutical applications, are practically inexistent in literature. Recently, Pimentel developed an electronic tongue based in LbL and impedance spectroscopy measurements to determine the concentration, between nano and picoMolar, of ibuprofen in water (Pimentel 2014).

20.6 Conclusions

From the different systems of electronic nose and electronic tongue discussed in this review, one can conclude that complex solutions can be accurately analyzed by using a set of dedicated sensors together with adequate data treatment. The extent of the analysis covers both qualitative and quantitative aspects, with solution classification close to 100 % of success in certain cases and nanomolar level detection capabilities.

In all prototypes considered in this review, the process of building such devices involves the following parts: sensor unit development, selection of transducing mechanism-measuring technique, and finally data analysis to be undertaken. These parts can be summarized as:

- The sensor units are mostly made of sensitive molecule layers deposited onto solid supports which, for the cases of electrical transducing, usually present either a single or IDEs systems. The preparation of molecular layers can be achieved through of several techniques as casting, spin-coating, sol–gel, Langmuir-Blodgett, electrodeposition, self-assembly, layer-by-layer. Provided that these layers should be stable in the support when immersed in the solution, self-assembly and layer-by-layer techniques are promising. These techniques also allow accurate layer thickness control, film architecture shaping, and a wide range of molecular systems to be assembled. The last two features are crucial towards the improvement of sensor selectivity capabilities. Films of nanostructures have proved to be the most promising as revealed by both the large sensitivities and discrimination capabilities achieved, as a result of their featured response. It should be referred that the knowledge about of the production of thin films deposited on surfaces is a crucial for the development of these systems.
- Concerning the transduction mechanisms, several sensor probes can be implemented, namely based on electrochemistry, impedance, SAWs, semiconductors, spectrophotometry among other. The simplest and often used for detection of water pollutants are through the electrochemical and impedance analysis. In addition, prototypes based on assemble of different types of measuring methods can also be a way to succeed to detect of traces of a given component, as supported from data obtained with both electronic nose and tongue integrated in the same instrument.
- Concerning the data analysis, particularly in what concerns to component classification it is fundamental the use of supervised and unsupervised well-established pattern recognition techniques. From these one finds the PCA the most often used method mainly because of its easy implementation. However, to go deep in data analysis, to get the best featured information, more sophisticated data treatment has to be addressed.

As a final remark, one believes this approach is adequate for in situ monitoring main water pollutants.

Acknowledgements Financial support for this work was provided by Fundação para a Ciência e Tecnologia, Grant No. PEst-OE/FIS/UI0068/2011, (FCT-Portugal) and the Centre National pour la Recherche Scientifique et Technique (CNRST—Maroc) and under project N° Chimie/01 13/14, and by Moulay Ismaïl University through project "Appui à la recherche." This work also resulted from taking part in bilateral collaboration projects under the programs Morocco-Portugal.

References

Adolfsson-Erici M, Pettersson M, Parkkonen J, Sturve J (2002) Triclosan, a commonly used bactericide found in human milk and in the aquatic environment in Sweden. Chemosphere 46:1485–1489

Albareda-Sirvent M, Merkoçi A, Alegret S (2001) Pesticide determination in tap water and juice samples using disposable amperometric biosensors made using thick- film technology. Anal Chim Acta 442:35–44

Apetrei C, Rodríguez-Méndez ML, Parra V, Gutierrez F, de Saja JA (2004) Array of voltammetric sensors for the discrimination of bitter solutions. Sens Actuators B 103:145–152

Apetrei C, Gutierez F, Rodriguez-Mendez ML, de Saja JA (2007) Novel method based on carbon paste electrodes for the evaluation of bitterness in extra virgin olive oils. Sens Actuators B 121:567–575

Apetrei C, Apetrei IM, Villanueva S, de Saja JA, Gutierrez-Rosalesc F, Rodriguez-Mendez ML (2010) Combination of an e-nose, an e-tongue and an e-eye for the characterisation of olive oils with different degree of bitterness. Anal Chim Acta 663:91–97

Arrieta AA, Rodríguez Méndez ML, De Saja JA, Blanco CA (2010a) Prediction of bitterness and alcoholic strength in beer using an electronic tongue. Food Chem 123:642–646

Arrieta AA, Rodríguez Méndez ML, De Saja JA (2010b) Aplicación de una lengua electrónica voltamétrica para la clasificación de vinos y estudio de correlación con la caracterización química y sensorial. Quimica Nova 33:787

Askim JR, Mahmoudi M, Suslick KS (2013) Optical sensor arrays for chemical sensing: the optoelectronic nose. Chem Soc Rev 42:8649–8682

Boilot P, Hines EL, Gongora MA, Folland RS (2003) Electronic noses intercomparison, data fusion and sensor selection in discrimination of standard fruit solutions. Sens Actuators B 88:80–88

Braga GS, Paterno LG, Fonseca FJ (2012) Performance of an electronic tongue during monitoring 2-methylisoborneol and geosmin in water samples. Sens Actuators B 171:181–189

Buratti S, Benedetti S, Scampicchio M, Pangerod EC (2004) Characterization and classification of Italian Barbera wines by using an electronic nose and an amperometric electronic tongue. Anal Chim Acta 525:133–139

Butler WR, Guthertz LS (2001) Mycolic acid analysis by high-performance liquid chromatography for identification of mycobacterium species. Clin Microbiol Rev 14:704–726

Campos I, Pascual L, Soto J, Gil-Sánchez L, Martínez-Máez R (2013) An electronic tongue designed to detect ammonium nitrate in aqueous solutions. Sensors 13(10):14064–14078

Campos I, Sangrador A, Bataller R, Aguado D, Barat R, Soto J, Martinez-Manez R (2014) Ammonium and phosphate quantification in wastewater by using a voltammetric electronic tongue. Electroanalysis 26(3):588–595

Carpenter G, Grossberg S, Markuzon N, Reynolds J, Rosen D (1992) Fuzzy ARTMAP: a neural network architecture for incremental supervised learning of analog multidimensional maps. IEEE Trans Neural Network 3:698–713

Carpenter G, Grossberg S, Reynolds J (1995) A fuzzy ARTMAP non parametric probability estimator for non-stationary pattern recognition problems. IEEE Trans Neural Network 6:1330–1336

Carvalho ER, Filho NC, Firmino A, Oliveira ON Jr, Mattoso LH, Martin-Neto L (2006) Sensorial system to detect chloroform in water. Sensors Lett 4(2):129–134

Carvalho ER, Filho NC, Venancio EC, Oliveira ON Jr, Mattoso LHC, Martin-Neto L (2007) Detection of brominated by-products using a sensor array based on nanostructured thin films of conducting polymers. Sensors 7(12):3258–3271

Castro M, Kumar B, Feller JF, Haddi Z, Amari A, Bouchikhi B (2011) Novel e-nose for the discrimination of volatile organic biomarkers with an array of carbon nanotubes (CNT) conductive polymer nanocomposites (CPC) sensors. Sens Actuators B 159:213–219

Chen Q, Zhao J, Vittayapadung S (2008) Identification of the green tea grade level using electronic tongue and pattern recognition. Food Res Int 41:500–504

Colilla M, Fernández CJ, Ruiz-Hitzky E (2002) Case-based reasoning (CBR) for multicomponent analysis using sensor arrays: application to water quality evaluation. Analyst 127:1580–1582

Cooper JS, Myers M, Chow E, Hubble LJ, Cairney JM, Pejcic B, Mueller K-H, Wieczorek L, Raguse B (2013) Performance of graphene, carbon nanotube, and gold nanoparticle chemi resistor sensors for the detection of petroleum hydrocarbons in water. J Nanopart Res 16 (1):2173

Cooper JS, Kiiveri E, Chow E, Hubble LJ, Webster MS, Müller K-H, Raguse B, Wieczorek L (2014) Quantifying mixtures of hydrocarbons dissolved in water with a partially selective sensor array using random forests analysis. Sens Actuators B 202:279–285

Cristianini N, Shawe-Taylor J (2000) An introduction to support vector machines, 1st edn. Cambridge University Press, New York

D'Amico A, Di Natale C, Martinelli E, Sandro L, Baccarani G (2005) Sensors small and numerous: always a winning strategy. Sens Actuators B 106:144–152

Daughton CG, Ternes TA (1999) Pharmaceuticals and personal care products in the environment: agents of subtle change? Environ Health Perspect 107(Suppl 6):907–938

Decher G (1997) Fuzzy nanoassemblies: toward layered polymeric multicomposites. Science 277:1232

Di Natale C, Macagnano A, Davide F, D'Amico A, Legin A, Vlasov Y, Rudnitskaya A, Selezenev B (1997) Multicomponent analysis on polluted waters by means of an electronic tongue. Sens Actuators B 44:423–428

Di Natale C, Paolesse R, Macagnano A, Mantini A, D'Amico A, Legin A, Lvova L, Rudnitskaya A, Vlasov Y (2000) Electronic nose and electronic tongue integration for improved classification of clinical and food samples. Sens Actuators B 64:15–21

Di Natale C, Macagnano A, Nardis S, Paolesse R, Falconi C, Proietti E, Siciliano P, Rella R, Taurino A, D'Amico A (2001) Comparison and integration of arrays of quartz resonators and metal-oxide semiconductor chemoresistors in the quality evaluation of olive oils. Sens Actuators B 78(1–3):303–309

Duarte AA, Gomes PJ, Ribeiro JHF, Ribeiro PA, Hoffmann SV, Mason NJ, Oliveira ON Jr, Raposo M (2013a) Characterization of PAH/DPPG layer-by-layer films by VUV spectroscopy. Eur Phys J E 36(9):9912

Duarte AA, Filipe SL, Abegão LMG, Gomes PG, Ribeiro PA, Raposo M (2013b) Adsorption kinetics of DPPG liposome layers: a quantitative analysis of surface roughness. Microsc Microanal 19:867–887

El Bari N, Amari A, Vinaixa M, Bouchikhi B, Correig X, Llobet E (2007) Building of a metal oxide gas sensor-based electronic nose to assess the freshness of sardines under cold storage. Sens Actuators B 128:235–244

Elbhiri Z, Chovelon JM, Renault NJ, Chevalier Y (1999) Chemically grafted field effect transistors for the detection of potassium ions. Sens Actuators B 58:491–496

Eugen CA (2013) Spectrophotometry: principle and applications. Workshop. Environment 7:94–100

Everitt B (1974) Cluster analysis. Heinemann Educational, London

Fehr U, McGahan LC (1967) Instrumentation for an array of infrasonic- and hydromagnetic-wave sensors. J Acoust Soc Am 41:587

Francesco FD, Fuoco R, Trivella MG, Ceccarini A (2005) Breath analysis: trends in techniques and clinical applications. Microchem J 79:405–410

Gardner JW (1991) Detection of vapours and odours from a multisensory array using pattern recognition: Part 1. Principal component and cluster analysis. Sens Actuators B 4:109–115

Gardner JW, Barlett PN (1994) A brief history of electronic noses. Sens Actuators B 18–19:211–220

Gawdzik B, Gawdzik J, Czerwlińska-Bil U (1990) Use of polymeric sorbents for the off-line preconcentration of priority pollutant phenols from water for high-performance liquid chromatographic analysis. J Chromatogr 509:135–140

Gee RH, Charles A, Taylor N, Darbre PD (2008) Oestrogenic and androgenic activity of triclosanin breast cancer cells. J Appl Toxicol 28:78–91

Goschnick J, Koronczi I, Frietsch M, Kiselev I (2005) Water pollution recognition with the electronic nose KAMINA. Sens Actuators B 106:182–186

Gutierrez JM, Haddi Z, Amari A, Bouchikhi B, Mimendia A, Ceto X, del Valle M (2013) Hybrid electronic tongue based on multisensor data fusion for discrimination of beers. Sens Actuators B 177:989–996

Haddi Z, Amari A, Alami H, El Bari N, Llobet E, Bouchikhi B (2011) A portable electronic nose system for the identification of cannabis-based drugs. Sens Actuators B 155:456–463

Haddi Z, Bougrini M, Tahri K, Braham Y, Souiri M, El Bari N, Othmane A, Jaffrezic-Renault N, Bouchikhi B (2014) A hybrid system based on an electronic nose coupled with an electronic tongue for characterization of Moroccan waters. Sensors Trans 27:190–197

Hayashi K, Yamanaka M, Toko K, Yamafuji K (1990) Multichannel taste sensor using lipid-membranes. Sens Actuators B 2(3):205–213

Hines EL, Llobet E, Gardner JW (1999) Electronic noses: a review of signal processing techniques. IEEE Proc Circuits Dev Syst 156:297–309

Hoffman A (2010) Spectroscopic techniques: I spectrophotometric techniques. In: Wilson K, Walker JM (eds) Principles and techniques of biochemistry and molecular biology, 7th edn. Cambridge University Press, Cambridge, pp 477–521

Hoyt S, McKennoch S, Wilson DM (2002) Transient response chemical discrimination module. Proc IEEE Sensors 1:376–381

Hsu CW, Lin CJ (2002) A comparison of methods for multi-class support vector machines. IEEE Trans Neural Network 13:415–425

Hu JM, Zhang JQ, Cao CN (2003) Determination of water uptake and diffusion of Cl^- ion in epoxy primer on aluminum alloys in NaCl solution by electrochemical impedance spectroscopy. Prog Org Coat 46:273–279

Hübschmann HJ (2009) Handbook of GC/MS: fundamentals and applications, 2nd edn. Wiley, Weinheim

Ivarsson P, Kikkawa Y, Winquist F, Krantz-Rulcker C, Hôjer NE, Hayashi K, Toko K, Lundstrôm I (2001a) Comparison of a voltammetric electronic tongue and a lipid membrane taste sensor. Anal Chim Acta 449:59–68

Ivarsson P, Holmin S, Hôjer NE, Rûlcker CK, Winquist F (2001b) Discrimination of tea by means of a voltammetric electronic tongue and different applied waveforms. Sens Actuators B 76:449–454

Jaffrezic-Renault N, Senillou A, Martelet C, Wan K, Chovelon JM (1999) ISFET microsensors for the detection of pollutants in liquid media. Sens Actuators B 59:154–164

James MO, Li W, Summerlot DP, Rowland-Faux L, Wood CE (2010) Triclosan is a potent inhibitor of estradiol and estronesulfonation in sheep placenta. Environ Int 36:942–949

Jiang JK, Sands JR (2000) Odour and ammonia emission from broiler farms, Rural Industries Research and Development Corporation, Publication No. 00/2

Keizer PD, Gordon DC Jr (1973) Detection of trace amounts of oil in sea water by fluorescence spectroscopy. J Fisheries Res Board of Canada 30(8):1039–1046

Khaleghi B, Khamis A, Karray FO, Razavi SN (2013) Multisensor data fusion: a review of the state-of-the-art. Inform Fusion 14:28–44

Kirsanov D, Zadorozhnaya O, Krasheninnikov A, Komarova N, Popov A, Legin A (2013) Water toxicity evaluation in terms of bioassay with an electronic tongue. Sens Actuators B 179:282–286

Labrador RH, Masot R, Alcañiz M, Baigts D, Soto J, Martínez-Mañez R, García-Breijo E, Gil L, Barat JM (2010) Prediction of NaCl, nitrate and nitrite contents in minced meat by using a voltammetric electronic tongue and an impedimetric sensor. Food Chem 122:864–870

Lasia A (1999) Electrochemical impedance spectroscopy and its applications. Mod Aspect Elect 32:143–248

Legin A, Rudnitskaya A, Vlasov Y, Di Natale C, Davide F, D'Amico A (1997) Tasting of beverages using an electronic tongue. Sens Actuators B 44:291–296

Legin AV, Seleznev BL, Rudnitskaya AM, Vlasov YG, Tverdokhlebov SV, Mack B, Abraham A, Arnold T, Baraniak L, Nitsche H (1999) Multisensor system for determination of iron(ii), iron (iii), uranium(vi) and uranium(iv) in complex solutions. Czech J Phys 49:679–685

León VM, Llorca-Pórcel J, Álvarez B, Cobollo MA, Munõz S, Valor I (2006) Analysis of 35 priority semivolatile compounds in water by stir bar sorptive extraction–thermal desorption–gas chromatography–mass spectrometry part II: method validation. Anal Chim Acta 558:261–266

Leonte II, Sehra G, Cole M, Hesketh P, Gardner JW (2006) Taste sensors utilizing high-frequency SHSAW devices. Sens Actuators B 118:349

Lindquist M, Wide P (2001) Virtual water quality tests with an electronic tongue. In: IMTC/2001: Proceedings of the 18th IEEE Instrum Meas Technol Conference: Rediscovering Measurement in the Age of Informatics 1–3, pp 1320–1324

Lindstrom A, Buerge IJ, Poiger T, Bergvist P, Muller MD, Buser H (2002) Occurrence and environmental behavior of the bactericide triclosan and its methyl derivative in surface waters and in wastewater. Environ Sci Technol 36:2322–2329

Liu Y, Brown SD (2004) Wavelet multiscale regression from the perspective of data fusion: new conceptual approaches. Anal Bioanal Chem 380:445–452

Liu Y, Song Q-J, Wang L (2009) Development and characterization of an amperometric sensor for triclosan detection based on electropolymerized molecularly imprinted polymer. Microchem J 91(2):222–226

Llobet E, Brezmes J, Vilanova X, Sueiras JE, Correig X (1997) Qualitative and quantitative analysis of volatile organic compounds using transient and steady state responses of a thick film tin oxide gas sensor array. Sens Actuators B 41:13–21

Llobet E, Hines EL, Gardner JW, Bartlett PN, Mottram TT (1999) Fuzzy ARTMAP based electronic nose data analysis. Sens Actuators B 61:183–190

Marletta A, Silva RA, Ribeiro PA, Raposo M, Gonçalves D (2009) Modeling adsorption processes of poly-p-phenylenevinylene precursor and sodium acid dodecylbenzenesulfonate onto layer-by-layer films using a Langmuir-type meta stable equilibrium model. Langmuir 25 (4):2166–2171

Meyer H, Drewer H, Ortindig B, Cammann K (1995) Two-dimensional imaging of O_2, H_2O_2, and glucose distributions by an array of 400 individually addressable microelectrodes. Anal Chem 67:1164–1170

Mimendia A, Legin A, Merkoçi A, Del Valle M (2010) Use of sequential injection analysis to construct a potentiometric electronic tongue: application to the multidetermination of heavy metals. Sens Actuators B 146:420–426

Moraes ML, Gomes PJ, Ribeiro PA, Vieira P, Freitas AA, Köhler R, Oliveira ON Jr, Raposo M (2010) Polymeric scaffolds for enhanced stability of melanin in incorporated in liposomes. J Colloid Interf Sci 350:268–274

Moreno L, Merlos A, Abramova N, Jimenez C, Bratov A (2006) Multi-sensor array used as an "electronictongue" for mineral water analysis. Sens Actuators B 116:130–134

Nunes GS, Skládal P, Yamanaka H, Barceló D (1998) Determination of carbamate residues in crop samples by cholinesterase-based biosensors and chromatographic techniques. Anal Chim Acta 362:59–68

Oliveira ON, Raposo M, Dhanabalan A (2001) Langmuir-Blodgett and self-assembled polymeric films. In: Nalwa HS (ed) Handbook of surfaces and interfaces of materials, vol 4. Academic, New York, pp 1–63

Oliveira JE, Scagion VP, Grassi V, Correa DS, Mattoso LHC (2012) Modification of electro spun nylon nanofibers using layer-by-layer films for application in flow injection electronic tongue: Detection of paraoxon pesticide in corn crop. Sens Actuators B 171–172:249–255

Oliveira JE, Grassi V, Scagion VP, Mattoso LHC, Glenn GM, Medeiros ES (2013) Sensor array for water analysis based on interdigitated electrodes modified with fiber films of poly(lactic acid)/multiwalled carbon nanotubes. IEEE Sensors J 13(2)

Oliveri P, Baldo MA, Daniele S, Forina M (2009) Development of a voltammetric electronic tongue for discrimination of edible oils. Anal Bioanal Chem 395:1135–1143

Osuna RG, Nagle HT, Schiffman SS (1999) Transient response analysis of an electronic nose using multi-exponential models. Sens Actuators B 61:170–182

PAE (2003) Odour impacts and separation distances for chicken meat farms in Queensland, Report prepared for Qld Chicken Meat Council, June 2003

Parra V, Arrieta AA, FernándezEscudero JA, Rodríguez Méndez ML, De Saja JA (2006) Electronic tongue based on chemically modified electrodes and voltammetry for the detection of adulterations in wines. Sens Actuators B 118:448–453

Paunovic M, Schlesinger M (2006) Fundamentals of electrochemical deposition, 2nd edn. Wiley, New York

Pearce TC, Schiffman SS, Nagle HT, Gardner JW (2003) Handbook of machine olfaction. Wiley, Weinheim

Peverly AA, Peters DG (2012) Electrochemical determination of trihalomethanes in water by means of stripping analysis. Anal Chem 84(14):6110–6115

Phaisangittisagul E, Nagle HT (2008) Sensor selection for machine olfaction based on transient feature extraction. IEEE Trans Instrum Meas 57:369–378

Pimentel R (2014) Desenvolvimento de um Sensor de Ibuprofeno em Meio Aquoso. Master Degree Thesis, Faculdade de Ciencias e Tecnologia, Universidade Nova de Lisboa, Portugal (in Portuguese)

Realini PA (1981) Determination of priority pollutant phenols in water by HPLC. J Chromatogr Sci 19:124–129

Riul A Jr, dos Santos Jr DS, Wohnrath K, Di Tommazo R, CarvalhoACPLF FFJ, Oliveira ON Jr, Taylor DM, Mattoso LHC (2002) Artificial taste sensor: efficient combination of sensors made from Langmuir-Blodgett films of conducting polymers and a ruthenium complex and self-assembled films of an azobenzene-containing polymer. Langmuir 18:239

Riul A Jr, Soto AMG, Mello SV, Bone S, Taylor DM, Mattoso LHC (2003a) An electronic tongue using polypyrrole and polyaniline. Synth Met 132:109–116

Riul A Jr, Malmegrim RR, Fonseca FJ, Mattoso LHC (2003b) Nano-assembled films for taste sensor application. Artif Organs 27:469–472

Riul A Jr, Malmegrim RR, Fonseca FJ, Mattoso LHC (2003c) An artificial taste sensor based on conducting polymers. Biosens Bioelectron 18:1365–1369

Riul A Jr, Dantas CAR, Miyazakic CM, Oliveira ON Jr (2010) Recent advances in electronic tongues. Analyst 135:2481–2495

Rodríguez-Méndez ML, Apetrei C, de Saja JA (2008) Evaluation of the polyphenolic content of extra virgin olive oils using an array of voltammetric sensors. Electrochim Acta 53:5867–5872

Rudnitskaya A, Kirsanov D, Legin A, Beullens K, Lammertyn J, Nicolaï BM, Irudayaraj J (2006) Analysis of apples varieties—comparison of electronic tongue with different analytical techniques. Sens Actuators B 116:23–28

Sahil K, Prashant B, Akanksha M, Premjeet S, Devashish R (2011) Gas chromatography-mass spectrometry: applications. Int J Pharm Biol 12:1544–1560

Sánchez-Avila J, Quintana J, Ventura F, Tauler R, Duarte CM, Lacorte S (2010) Stir bar sorptive extraction-thermal desorption-gas chromatography–mass spectrometry: an effective tool for determining persistent organic pollutants and nonylphenol in coastal waters in compliance with existing directives. Mar Pollut Bull 60:103–112

Sangodkar H, Sukeerthi S, Srinivasa RS, Lal R, Contractor AQ (1996) A biosensor array based on polyaniline. Anal Chem 68:779–783

Scozzari A, Acito N, Corsini G (2007) A novel method based on voltammetry for the qualitative analysis of water. IEEE Trans Instrum Meas 56(6):2688–2697

Shiokawa S, Kondoh J (2004) Surface acoustic wave sensors. Jpn J Appl Phys 43:2799–2802
Singer H, Muller S, Tixier C, Pillonel L (2002) Triclosan: occurrence and fate of a widely used biocide in the aquatic environment: field measurements in wastewater treatment plants, surface waters, and lake sediments. Environ Sci Technol 36:4998–5004
Sipos L, Kovács Z, Sági-Kiss V, Csiki T, Kókai Z, Fekete A, Héberger K (2012) Discrimination of mineral waters by electronic tongue, sensory evaluation and chemical analysis. Food Chem 135:2947–2953
Sipos L, Gere A, Szollosi D, Kovacs Z, Kokai Z, Fekete A (2013) Sensory evaluation and electronic tongue for sensing flavored mineral water taste attributes. J Food Sci 78(10): S1602–S1608
Söderström C, Rudnitskaya A, Legin A, Krantz-Rülcker C (2005) Differentiation of four Aspergillus species and one Zygosaccharomyces with two electronic tongues based on different measurement techniques. J Biotechnol 119:300–308
Sohn JH, Hudson N, Gallagher E, Dunlop M, Zeller L, Atzeni M (2008) Implementation of an electronic nose for continuous odour monitoring in a poultry shed. Sens Actuators B 133:60–69
Sun H, Mo ZH, Choy JTS, Zhu DR, Fung YS (2008) Piezoelectric quartz crystal sensor for sensing taste-causing compounds in food. Sens Actuators B 131:148–158
Taylor DM, Macdonald AG (1987) AC admittance of the metal insulator electrolyte interface. J Phys D Appl Phys 20:1277–1283
Toko K (1998) Electronic tongue. Biosens Bioelectron 13(6):701–709
Tothil IE (2003) Rapid and on-line instrumentation for food quality assurance. CRC, Boca Raton
Vapnik V (1995) The nature of statistical learning theory, 1st edn. Springer, New York
Veríssimo MIS, Oliveira JABP, Gomes MTSR (2010) Contribution of compressional waves to the identification and quantification of a water contaminant. Sens Actuators B 151:21–25
Vilanova X, Llobet E, Alcubilla R, Sueiras JE, Correig X (1996) Analysis of the conductance transient in thick-film tin oxide gas sensors. Sens Actuators B 31:175–180
Vlasov Y, Legin A, Rudnitskaya A, Di Natale C, D'Amico A (2005) Nonspecific sensor arrays ("electronic tongue") for chemical analysis of liquids (IUPAC Technical Report). Pure Appl Chem 77:1965–1983
Wang L-H, Chu S-C (2004) Voltammetric detector for liquid chromatography: determination of triclosan in rabbit urine and serum. Chromatographia 60(7–8):385–390
Wang H, Ohnuki H, Endo H, Izumi M (2015) Impedimetric and amperometric bifunctional glucose biosensor based on hybrid organic-inorganic thin films. Bioelectrochemistry 101:1–7
Wei Z, Wang J (2011) Detection of antibiotic residues in bovine milk by a voltammetric electronic tongue system. Anal Chim Acta 694:46–56
Winquist F, HoÈrnsten EG, Sundgren H, Lundström I (1993) Performance of an electronic nose for quality estimation of ground meat. Meas Sci Technol 4:1493–1500
Winquist F, Wide P, Lundström I (1997) An electronic tongue based on voltammetry. Anal Chim Acta 357:21–31
Woertz K, Tissen C, Kleinebudde P, Breitkreutz J (2011) A comparative study on two electronic tongues for pharmaceutical formulation development. J Pharmaceut Biomed Anal 55:272–281
World Health Organization (WHO) (2009) Report: "Global health risks: mortality and burden of disease attributable to selected major risks"
World Health Organization (WHO) (2011) Guidelines for drinking water 4th edn
Worsfold PJ, Clinch JR (1987) Spectrophotometric field monitor for water quality parameters the determination of phosphate. Anal Chim Acta 197:43–50
Yadava RDS (2012) Modeling, simulation and information processing for development of a polymeric electronic nose system. In: Korotcenkov G (ed) Chemical sensors—simulation and modelling. Momentum Press, New York, pp 411–502, Ch. 10
Yánez Heras J, Rodriguez SD, Negri RM, Battaglini F (2010) Chelating electrodes as taste sensor for the trace assessment of metal ions. Sens Actuators B 145:726–733

Yang J, Wang P, Zhang X, Wu K (2009) Electrochemical sensor for rapid detection of triclosan using a multiwall carbon nanotube film. J Agric Food Chem 57(20):9403–9407

Ying G-G, Kookana RS (2007) Triclosan in wastewaters and biosolids from Australian wastewater treatment plants. Environ Int 33:199–205

Zorrilla LM, Gibson EK, Jeffay SC, Crofton KM, Setzer WR, Cooper RL (2009) The effects of triclosan on puberty and thyroid hormones in male Wistar rats. Toxicol Sci 107(1):56–64

Chapter 21
Analysis of Endocrine Disrupting Chemicals in Food Samples

Miriany A. Moreira, Leiliane C. André, Marco D.R. Gomes da Silva, and Zenilda L. Cardeal

21.1 Introduction

21.1.1 Endocrine Disrupting Chemicals

The synthesis of new chemical substances is increasingly providing better quality of life. Several synthesized chemical compounds are applied in packaging, cosmetics, and toiletries, among others. For many years, organic compounds have been used interchangeably in the manufacture of packaging and product formulation, but in recent decades studies have shown that some of these substances may be harmful to man. Some organic compounds in plastic containers and household items are suspected to cause harm, both to human's and animal's health. Several studies have been conducted to show the association between exposure to these compounds and the effects on the endocrine system.

Chemicals that can damage the endocrine system are called as endocrine disrupting chemicals—EDCs. According to the World Health Organization (WHO), "an endocrine disrupter is an exogenous substance or mixture that alters one or more functions of the endocrine system and consequently causes adverse

M.A. Moreira • Z.L. Cardeal (✉)
Departamento de Química, Instituto de Ciências Exatas (ICEx) Universidade Federal de Minas Gerais, Av. Antônio Carlos, 6627, Belo Horizonte, MG 31270-901, Brazil
e-mail: zenilda@ufmg.br

L.C. André
Departamento de Análises Clínicas e Toxicológicas, Faculdade de Farmácia (FaFar), Universidade Federal de Minas Gerais, Av. Antônio Carlos, 6627, Belo Horizonte, MG 31270-901, Brazil

M.D.R. Gomes da Silva
LAQV, REQUIMTE, Departamento de Química, Faculdade de Ciências e Tecnologia, Universidade Nova de Lisboa, Caparica 2829-516, Portugal

© Springer International Publishing Switzerland 2016

A.B. Ribeiro et al. (eds.), *Electrokinetics Across Disciplines and Continents*,
DOI 10.1007/978-3-319-20179-5_21

health effects in the intact organism, or its progeny or sub-populations" (IPCS/WHO 2002). Some pesticides, surfactants, plastics in general are classified as EDC. In 2000, the European Union has established a list of 562 suspected compounds which cause dysfunction in the endocrine system (EC 2001). These compounds can be divided into different classes such as alkylphenols and plasticizers. These two classes of compounds are widely studied, considering the concentrations they are found in food, environmental, and biological samples and are therefore subject of this chapter.

Alkylphenols such as 4-nonylphenol (NP-4) and 4-octylphenol (OP-4) are by-products of degradation of the ethoxylatedalkylphenol, nonionic surfactant used in domestic and industrial applications (Schroder 2001). An example of use of surfactants is their application in the manufacture of detergents, which are commonly used for cleaning and sanitizing in various industries, including the food industry. Ethoxylatenonylphenol is present in the formulation of various pesticides and their major metabolites. Nonylphenols are considered as presenting low degradability (Brix et al. 2001; IPCS/WHO 2002). Alkylphenols such as 4-tert-butylphenol (4-t-BP), 4-OP, and 4-NP are also used as monomers in the manufacture of phenolic resins (Ozaki and Baba 2003). Nonylphenol may also be used in the form of tris(nonylphenol)-phosphite, a substance used as antioxidant for plastic materials, commonly used as packaging in the food storage (Guenther et al. 2002).

Some phthalates, as benzylbutyl-phthalate (BBP) and di (ethylhexyl)-phthalate (DEHP), among others, are also cited as possible causes of disturbances in the endocrine system (EC 2000). Phthalates have multiple applications and are used in industry primarily as plasticizers to increase the flexibility of plastics applied in the manufacture of toys, household items, and packaging, among others. Other applications include the use of some phthalates, such as diethyl-phthalate (DEP), in cosmetics and pesticides (USEPA 2001). Phthalates are not chemically bound to the polymer matrix, and therefore the migration to the food can easily occur (Fromme et al. 2002). This is the reason why they are found as contaminants in many food products.

Several studies have demonstrated the association of these organic compounds with endocrine dysfunction (human and animal). Their majority is related to chronic exposure and concern the action of some compounds suspected of causing endocrine dysfunction. Studies in animals show that phthalates may cause changes in kidneys, liver, and fetal malformation (Xia et al. 2011). In humans, the possible toxic effects of these compounds are described in some studies. Hauser et al. showed a relationship between the concentration of DEHP metabolites in semen samples and sperm DNA damage (Hauser et al. 2007). A study by Meeker and Ferguson determined a positive relationship between the concentration of phthalates and bisphenol A in urine and changes in thyroid hormones (Meeker and Ferguson 2011). In vivo studies show that some phthalates can cause the increase in uterine leiomyoma cells and changes in the secondary structure of the protein albumin (Shen et al. 2014; Xie et al. 2013). Ferguson et al. showed that maternal exposure to certain phthalates during pregnancy is related to systemic inflammation and systemic oxidative stress (Ferguson et al. 2014).

21.2 Alkylphenols and Phthalates in Food Samples

The sources of phthalates and alkylphenols exposure are varied, and the absorption can occur by oral, pulmonary, or skin contact (WHO/UNEP 2012). Studies report the identification of these compounds in air samples, drinking water, river water, and cosmetics. However, one of the major sources of exposure is from food products (Huang et al. 2011; Romero-Franco et al. 2011; Moreira et al. 2009). The food contamination occurs primarily through the contact of the food matrices with the packaging material or during the food manufacturing process. The European Food Safety Authority (EFSA), based on toxicological studies, established a tolerable daily intake (TDI) for phthalates, 0.01 mg kg^{-1} of body weight per day for the BPD, 0.5 mg kg^{-1} of body weight per day for BBP, and 0.05 mg kg^{-1} of body weight per day for DEHP (EFSA 2005a, b, c).

The alkylphenols and phthalates have been reported in food samples in several studies. Wu et al. showed that the concentrations of phthalates in soft drinks ranged from 0.015 to 0.159 mg L^{-1} (Wu et al. 2014). Ostrovsky et al. observed phthalate concentrations of 1.5–12.5 μg g^{-1} in fatty food products (Ostrovsky et al. 2011). Alkylphenols and plasticizers are also found in other matrices. Nonylphenol was determined in concentrations ranging from 0.3 to 72.9 ng L^{-1} in surface waters (Klosterhaus et al. 2013). In cosmetics, phthalates were measured in concentrations ranging from 0.41 to 38.34 mg L^{-1} (Feng and Jiang 2012). This persistent presence described in environmental and food matrices shows that the human being is regularly ingesting these compounds which can be toxic even at low concentrations.

Studies related to the use of plastic films and plastic packaging for food contact show that these packages may also be possible sources of food contamination. A study by Cirillo et al. (2011), in which the concentration of phthalates was determined in cereals, legumes, and vegetables stored in plastic containers, presents levels of 22.8–270.3 ng g^{-1} and 10.2–142.8 ng g^{-1} for DEHP and DBP, respectively (Cirillo et al. 2011). The amount of alkylphenols and plasticizers found is related to the characteristics of foods and food compounds that they are in contact with. Compounds such as DEHP, dioctyl-phthalate (DOP), and diisononyl-phthalate (DINP) have high K_{ow}, a measure of the lipophilicity of the compound, and thus have affinity for fatty foods such as oils and mayonnaise (Nanni et al. 2011; Ostrovsky et al. 2011).

Phthalates are also found in many food products such as olive oil, milk, and wine (Feng et al. 2005; Cavaliere et al. 2008; del Carlo et al. 2008). The knowledge of the migration magnitude of phthalates present in packaging into food is of main importance. Therefore, many studies have been conducted to determine phthalates in food samples. DBP and DOP were founded in a concentration range between 0.023 and 0.664 μg L^{-1}, respectively, in food simulants when they were put in contact with plastic packaging, in a study described by Gonzalez-Castro et al. (Gonzalez-Castro et al. 2011). In another work, Kueseng et al. determined concentrations of 0.61 μg g^{-1} of DEHP, which migrated from condiment packaging (Curry) (Kueseng et al. 2007). Cacho et al. determined concentrations ranging from 0.42 to 2.3 ng g^{-1} for octylphenol and 46 to 50 ng g^{-1} for nonylphenol in plant samples (Cacho et al. 2012).

21.3 Determination of Alkylphenols and Phthalates in Food Matrices

21.3.1 *Sample Preparation Methods: Less Solvent and Higher Sensitivity*

Since EDCs are present in food matrices at a very low concentration level, the development of highly sensitive analytical methods and sample preparation procedures are required. In addition to the analysis of food samples, prior treatment of the sample is necessary for proper isolation/extraction and concentration of the compounds. For the extraction step, techniques may be used as solid phase extraction (SPE) (Ibrahim et al. 2014), solid phase microextraction (SPME) (Wu et al. 2012), single drop microextraction (SDME) (Batlle and Nerín 2004), dispersive liquid–liquid microextraction (DLLME) (Farajzadeh and Mogaddam 2012; Yilmaz et al. 2014), assisted air liquid–liquid microextraction (AALLME) (Farajzadeh and Mogaddam 2012), and liquid phase microextraction (LPME) (Villar-Navarro et al. 2013), among others. Techniques such as liquid–liquid extraction for liquid food samples and classic solvent extraction for solid food matrices can also be used. However, these methods have the disadvantage of using large quantities of organic solvents. Due to this drawback, other extraction techniques using low amounts of organic solvents have been developed and applied in food analysis. SPE and SPME have been used in several studies to extract phthalates and alkylphenols from food samples. SPME is a technique that combines the simultaneous isolation and extraction of the target compounds together with their concentration in a fiber coated with a suitable polymeric film. This fiber is placed in direct or indirect contact with the analyte or analytes in the sample. Analytes are adsorbed/absorbed in the fiber according to its chemical nature (Ribeiro et al. 2008). The fiber can be exposed by direct immersion in solution (Fig. 21.1a) or in the sample headspace. In this latter case, the fiber is exposed to the headspace of the sample which is present in the void volume of the vial used (Fig. 21.1b). Various types of materials are used as fiber coatings, such as polydimethylsiloxane, polyacrylate, and divinylbenzene, among several others. The fiber is subsequently inserted into the injector port of the gas chromatograph, to thermally desorb the analyte to the analytical column. SPME has been successfully applied in the analysis of liquid foods. Feng et al. obtained detection limits of 0.02 and 0.23 μg kg^{-1} for DBP and BBP, respectively, in a phthalate analysis study in which cow's milk and SPME-GC/MS were used for extraction and analysis of compounds (Feng et al. 2005). Li et al. obtained detection limits in the range of 0.006–0.03 ng mL^{-1} for DEHP, DBP, and DEP in a study in which phthalates were analyzed in samples of alcoholic beverages using SPME and GC/MS.

As the process of absorption/adsorption of analytes on the fiber is an exothermic process, cooling of the fiber enables increased sensitivity. Moreira et al. (2014) used the system shown in Fig. 21.2 for the analysis of phthalates. Their migration to food simulants in plastic containers heated in microwave was investigated. DBP and

Fig. 21.1 SPME system by immersion (**a**) and by headspace (**b**)

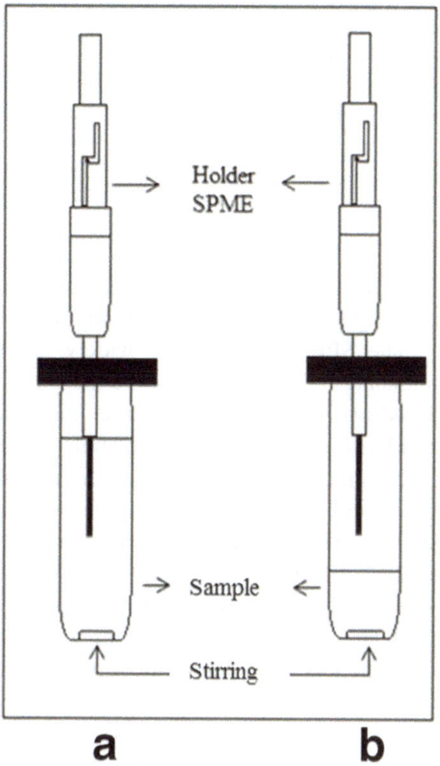

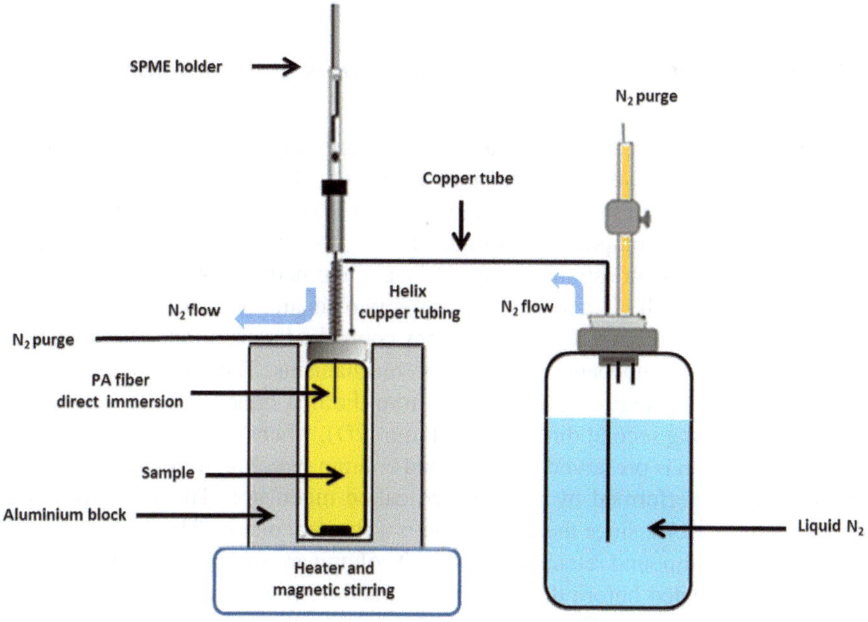

Fig. 21.2 Cold SPME system used to analyze the phthalate migration by GC/MS (Menezes and Cardeal 2011)

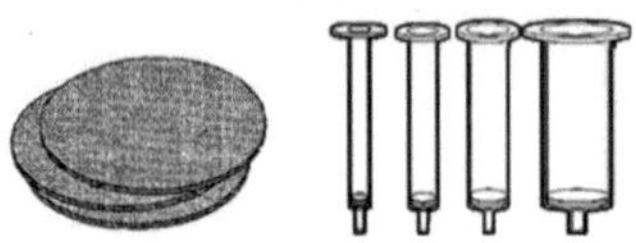

Fig. 21.3 Cartridges and disks used in SPE according to USEPA (2015)

BBP were analyzed by GC/MS, yielding limit of quantitation (LOQ) of 0.20 and 0.30 mg kg^{-1}, respectively (Moreira et al. 2014).

The SPE has also been widely applied to the analysis of alkylphenols and phthalates. The SPE consists of the use of a cartridge or disk (Fig. 21.3) containing an adsorbent which will be in contact with the sample. The extraction process is similar to a liquid chromatographic separation, where the separation of the target compounds from the matrix occurs through polarity similarities between the solid phase and liquid phase where the sample is dissolved (Petrovic et al. 2002). In a study by Gu et al. (2014), detection limits 0.015–2.2 ng g^{-1} were obtained in the analysis of alkylphenols and phthalates in seafood samples. SPE and HPLC/MS were used in the alkylphenols analysis, and SPE GC/MS analysis was used for phthalate determinations (Gu et al. 2014). In another study by Li et al., where alkylphenols were determined in soft drinks samples, the detection limit obtained for OP and NP was 0.03 μgL^{-1} when SPE and GC/MS were used for extraction and analysis of the target compounds (Li et al. 2013). Del Carlo et al. (2008) obtained detection limits of 0.015–0.018 mg L^{-1} for analysis of phthalates in wine samples, using SPE and GC/MS (Del Carlo et al. 2008).

21.3.2 GC × GC Analysis of Alkylphenols and Phthalates

Gas chromatography coupled to mass spectrometry has often been applied in the analysis of alkylphenols and phthalates. However, other techniques may also be used for the analysis of these compounds in order to obtain a greater separation and detectability of the compounds. Comprehensive two-dimensional gas chromatography (GC × GC) demonstrates to be an elegant alternative in the analysis of these compounds. GC × GC is a system that uses two separation processes (two chromatographic columns) in order to perform compound separation. The analytical columns used have independent separation mechanisms. The term *comprehensive* is used because the entire sample eluted from the first dimension (1D) column is introduced into the second dimension column (2D), where the separation occurred in the first column is preserved in the second column (von Muhlen et al. 2007). This transference is performed by an interface called modulator. The 1D column has classical dimensions, since the 2D column is a shorter column. This is necessary in order that the compound retentions in the 2D column are sufficiently short and their elution is completed before the next eluate fraction from the 1D column is introduced in the 2D column. Normally, 1D column is a capillary column containing an apolar stationary phase, while in the 2D column a high polarity medium stationary phase is employed (classical set of columns). The difference in the chemical selectivity between the two columns allows compounds which are not adequately separated in the 1D column to be fully resolved in the 2D column (Bruckner et al. 1998). The modulator is used to fractionate the eluate of 1D column, refocusing it and subsequently transferring it to 2D column (Dallüge et al. 2003; Vendeuvre et al. 2004). The modulators may be of cryogenic type, controlled by valves or by heat. In a valve modulator, a valve is inserted between the columns for fraction transfer control of the eluate from the 1D to the 2D column (Bruckner et al. 1998; Pedroso et al. 2009). Thermal and cryogenic modulators make the transference control of the 1D eluate to the 2D column using temperature changes. In a thermal modulator, 1D eluate is trapped on a sorbent and subsequently by thermal desorption sent to the 2D column. The cryogenic modulators use a cryogenic medium to perform the eluate trapping from the 1D column. The subsequent eluate release to the 2D column is performed by an external heat source. With this cryogenic trap, the eluate is reconcentrated (focused) and sent to 2D column (Marriott and Shellie 2002; Pedroso et al. 2009).

In many complex sample analyses, where one-dimensional gas chromatography is used, the presence of interferents co-eluting with target compounds jeopardizing the entire analysis. In GC × GC, using two columns allows an increase on the separation of target compounds, eliminating the influence of sample and chromatographic system contaminants. Another advantage resulting from the use of GC × GC is the increased detectability (sensitivity) achieved. The modulation process compresses the band from the first chromatographic column eluate, and consequently each analyte elutes as a number of narrow and intense peaks which are more easily detectable (Marriott and Shellie 2002).

The GC × GC technique has been applied to the analysis of many organic compounds from several different matrices (Murray 2012; Mateus et al. 2008, 2010; Cardeal et al. 2006, 2008). Djokicet al. showed the efficiency of separation and analysis of organic compounds in biodiesel applying GC × GC-FID (Djokic et al. 2012). Other studies report the use of GC × GC in biological matrices. In a study by Amorim et al. (2009), in which polycyclic aromatic hydrocarbons (PAHs) in urine were analyzed, detection limits were obtained in a range between 0.03 and 0.18 $\mu g\ L^{-1}$ (Amorim et al. 2009). GC × GC has also proven to be effective in the analysis and quantitation of illicit drugs in blood (Andrews and Paterson 2012). Some studies have reported the use of GC × GC for qualitative analysis of alkylphenols and plasticizers in different matrices. A study by Vallejo et al. showed the possibility of separation of the different OP and NP isomers when GC × GC coupled with a flame ionization detector (FID) and mass spectrometry detector (MSD) were used (Guenther et al. 2006; Vallejo et al. 2011). The use of GC × GC coupled with mass spectrometry (GC × GC/MS) allowed the separation and analysis of different isomers of nonylphenol in sewage samples in a study by Zhang et al. (2012). Studies by Vallejo et al. and Zhag et al. using GC × GC, exemplify the application possibilities for alkylphenols analysis. However, the application of GC × GC-FID for quantitative analysis of alkylphenols and phthalates still little reported for food analysis.

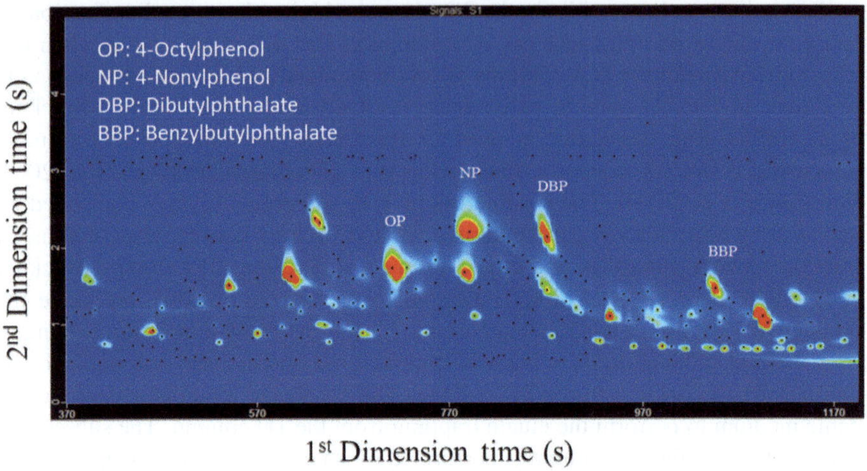

Fig. 21.4 Bubble diagram obtained by GC × GC-FID of alkylphenols and phthalates in aqueous solution at a concentration of 5.0 μg L^{-1}, using a polyacrylate fiber for SPME extraction (65 °C, 30 min, by fiber immersion); carrier gas: H_2 flow at 1.5 mL min^{-1}; 5 s modulation period; Splitless mode analysis for 2 min. Chromatographic conditions according to Moreira et al. (2015)

The efficient separation of analytes made possible by GC × GC allows its use with low-cost universal detectors such as flame ionization detector (FID). Therefore, it is an effective alternative for the quantitation of alkylphenols and plasticizers. The contour diagram obtained in the analysis and quantitation of phthalates and alkylphenols (Moreira et al. 2015) shows an efficient separation of some target compounds (Fig. 21.4), even when present in a low concentration range, μg L^{-1}.

In general, food matrices are complex samples and the use of GC × GC reduces the possibility of false positive or negative results, since the clear separation of compounds is obtained. GC × GC enables to reduce or even eliminate matrix interference in the analysis process of alkylphenols and phthalates. Furthermore, an increase in detectability is obtained when this chromatographic technique is applied. In this work, LOD ranging from 0.07 to 0.32 μg L^{-1} and LOQ ranging from 0.17 to 0.55 μg L^{-1} were obtained for some EDCs.

Acknowledgment Financial support was provided by FP7-PEOPLE-2010-IRSES-269289-*ELECTROACROSS—Electrokinetics across disciplines and continents: an integrated approach to finding new strategies for sustainable development* and the Brazilian foundations CNPq.

References

Amorim LCA, Dimandja J-M, Cardeal ZL (2009) Analysis of hydroxylated polycyclic aromatic hydrocarbons in urine using comprehensive two-dimensional gas chromatography with a flame ionization detector. J Chromatogr A 1216(14):2900–2904

Andrews R, Paterson S (2012) A validated method for the analysis of cannabinoids in post-mortem blood using liquid–liquid extraction and two-dimensional gas chromatography-mass spectrometry. Forensic Sci Int 222(1–3):111–117

Batlle R, Nerín C (2004) Application of single-drop microextraction to the determination of dialkyl phthalate esters in food simulants. J Chromatogr A 1045(1–2):29–35

Brix RS, Hvidt S, Carlsen L (2001) Solubility of nonylphenol and nonylphenol ethoxylates. On the possible role of micelles. Chemosphere 44(4):759–763

Bruckner CA, Prazen BJ, Synovec RE (1998) Comprehensive two dimensional high-speed gas chromatography with chemometric analysis. Anal Chem 70(14):2796–2804

Cacho JI, Campillo N, Viñas P, Hernández-Córdoba M (2012) Determination of alkylphenols and phthalate esters in vegetables and migration studies from their packages by means of stir bar sorptive extraction coupled to gas chromatography-mass spectrometry. J Chromatogr A 1241:21–27

Cardeal ZL, Gomes da Silva MD, Marriott PJ (2006) Comprehensive two-dimensional gas chromatography/mass spectrometric analysis of pepper volatiles. Rapid Commun Mass Spectrom 20:2823–2836

Cardeal ZL, Patricio P, Gomes da Silva MDR, Marriott PJ (2008) Comprehensive two-dimensional gas chromatography for fingerprint pattern recognition in cachaça production. Talanta 74:793–799

Cavaliere B, Macchione B, Sindona G, Tagarelli A (2008) Tandem mass spectrometry in food safety assessment: the determination of phthalates in olive oil. J Chromatogr A 1205 (1–2):137–143

Cirillo T, Fasano E, Castaldi E, Montuori P, Cocchieri RA (2011) Children's exposure to di (2-ethylhexyl)phthalate and dibutylphthalate plasticizers from school meals. J Agric Food Chem 59(19):10532–10538

Dallüge J, Beens J, Brinkman UAT (2003) Comprehensive two-dimensional gas chromatography: a powerful and versatile analytical tool. J Chromatogr A 1000(1–2):69–108

Del Carlo M, Pepe A, Sacchetti G, Compagnone D, Mastrocola D, Cichelli A (2008) Determination of phthalate esters in wine using solid-phase extraction and gas chromatography-mass spectrometry. Food Chem 111(3):771–777

Djokic MR, Dijkmans T, Yildiz G, Prins W, Van Geem KM (2012) Quantitative analysis of crude and stabilized bio-oils by comprehensive two-dimensional gas-chromatography. J Chromatogr A 1257:131–140

EC (2000) European Commission DG ENV-towards the establishment of a priority list of substances for further evaluation of their role in endocrine disruption. http://ec.europa.eu/environment/archives/docum/pdf/bkh_main.pdf. Accessed November 2013

EC (2001) European Commission Environment—communication from the commission to the council and the European parliament. http://ec.europa.eu/environment/archives/docum/01262_en.htm#bkh. Accessed November 2013

EFSA (2005a) Opinion of the scientific panel on food additives, flavouring, processing aids and material in contact with food on a request from the Commission related to bis(2-ethylhexyl) phthalate (DEHP) for use in food contact materials—European Food Safety Authority. EFSA J 243:1–20

EFSA (2005b) Opinion of the scientific panel on food additives, flavouring, processing aids and material in contact with food on a request from the commission related to butylbenzyl phthalate (BBP) for use in food contact materials—European Food Safety Authority. EFSA J 241:1–14

EFSA (2005c) Opinion of the scientific panel on food additives, flavouring, processing aids and material in contact with food on a request from the Commission related to di-butyl phthalate (DBP) for use in food contact materials—European Food Safety Authority. EFSA J 242:1–17

Farajzadeh MA, Mogaddam MRA (2012) Air-assisted liquid–liquid microextraction method as a novel microextraction technique; Application in extraction and preconcentration of phthalate esters in aqueous sample followed by gas chromatography-flame ionization detection. Anal Chim Acta 728:31–38

Feng CH, Jiang SR (2012) Micro-scale quantitation of ten phthalate esters in water samples and cosmetics using capillary liquid chromatography coupled to UV detection: effective strategies to reduce the production of organic waste. Microchim Acta 177(1–2):167–175

Feng YL, Zhu JP, Sensenstein R (2005) Development of a headspace solid-phase microextraction method combined with gas chromatography mass spectrometry for the determination of phthalate esters in cow milk. Anal Chim Acta 538(1–2):41–48

Ferguson KK, Cantonwine DE, Rivera-Gonzalez LO, Loch-Caruso R, Mukherjee B, Del Toro LVA, Jimenez-Velez B, Calafat AM, Ye X, Alshawabkeh AN, Cordero JF, Meeker JD (2014) Urinary phthalate metabolite associations with biomarkers of inflammation and oxidative stress across pregnancy in Puerto Rico. Environ Sci Technol 48:7018–7025

Fromme H, Kuchler T, Otto T, Pilz K, Müller J, Wenzel A (2002) Occurrence of phthalates and bisphenol A and F in the environment. Water Res 36(6):1429–1438

Gonzalez-Castro MI, Olea-Serrano MF, Rivas-Velasco AM, Medina-Riviero E, Ordoñez-Acevedo LG, De León-Rodríguez A (2011) Phthalates and bisphenols migration in mexican food cans and plastic food containers. Bull Environ Contam Toxicol 86(6):627–631

Gu YY, Yu XJ, Peng JF, Chen SB, Zhong YY, Yin DQ, Hu XL (2014) Simultaneous solid phase extraction coupled with liquid chromatography tandem mass spectrometry and gas chromatography tandem mass spectrometry for the highly sensitive determination of 15 endocrine disrupting chemicals in seafood. J Chromatogr B 965:164–172

Guenther K, Heinke V, Thiele B, Kleist E, Prast H, Raecker T (2002) Endocrine disrupting nonylphenols are ubiquitous in food. Environ Sci Technol 36(8):1676–1680

Guenther K, Kleist E, Thiele B (2006) Estrogen-active nonylphenols from an isomer-specific viewpoint: a systematic numbering system and future trends. Anal Bioanal Chem 384 (2):542–546

Hauser R, Meeker JD, Singh NP, Silva MJ, Ryan L, Duty S, Calafat AM (2007) DNA damage in human sperm is related to urinary levels of phthalate monoester and oxidative metabolites. Hum Reprod 22(3):688–695

Huang L-P, Lee C-C, Hsu PC, Shih TS (2011) The association between semen quality in workers and the concentration of di(2-ethylhexyl) phthalate in polyvinyl chloride pellet plant air. Fertil Steril 96(1):90–94

Ibrahim N, Osman R, Abdullah A, Saim N (2014) Determination of phthalate plasticisers in palm oil using online solid phase extraction-liquid chromatography (SPE-LC). J Chem. Article ID 682975, 9 pp., doi:10.1155/2014/682975

IPCS/WHO (2002) International Programme on Chemical Safety-Global assessment of the state of the science of endocrine disruptors. http://www.who.int/ipcs/publications/new_issues/endocrine_disruptors/en/

Klosterhaus SL, Grace R, Hamilton MC, Yee D (2013) Method validation and reconnaissance of pharmaceuticals, personal care products, and alkylphenols in surface waters, sediments, and mussels in an urban estuary. Environ Int 54:92–99

Kueseng P, Thavarungkul P, Kanatharana P (2007) Trace phthalate and adipate esters contaminated in packaged food. J Environ Sci Health B 42(5):569–576

Li Y, Zhang S, Song C, You J (2013) Determination of bisphenol a and alkylphenols in soft drinks by high-performance liquid chromatography with fluorescence detection. Food Anal Methods 6(5):1284–1290

Marriott P, Shellie R (2002) Principles and applications of comprehensive two-dimensional gas chromatography. TrAC Trends Anal Chem 21(9–10):573–583

Mateus EP, Gomes da Silva MDR, Ribeiro AB, Marriott PJ (2008) Qualitative mass spectrometric analysis of the volatile fraction of creosote treated railway wood sleepers by using comprehensive two-dimensional gas chromatography. J Chromatogr A 1178:215–222

Mateus EP, Barata RC, Zrostlíková J, Gomes da Silva MDR, Paiva MR (2010) Characterization of the volatile fraction emitted by Pinus spp by one and two dimensional chromatographic techniques with mass spectrometric detection. J Chromatogr A 1217:1845–1855

Meeker JD, Ferguson KK (2011) Relationship between urinary phthalate and bisphenol A concentrations and serum thyroid measures in US Adults and Adolescents from the National Health and Nutrition Examination Survey (NHANES) 2007–2008. Environ Health Perspect 119(10):1396–1402

Menezes HC, Cardeal ZL (2011) Determination of polycyclic aromatic hydrocarbons from ambient air particulate matter using a cold fiber solid phase microextraction gas chromatography–mass spectrometry method. J Chromatogr A 1218:3300–3305

Moreira DS, Aquino SF, Afonso RJ, Santos EP, de Pádua VL (2009) Occurrence of endocrine disrupting compounds in water sources of Belo Horizonte Metropolitan Area, Brazil. Environ Technol 30(10):1041–1049

Moreira MA, André LC, Cardeal ZL (2014) Analysis of phthalate migration to food simulants in plastic containers during microwave operations. Int J Environ Res Public Health 11 (1):507–526

Moreira MA, André LC, Ribeiro AB, Gomes da Silva MDR, Cardeal ZL (2015) Quantitative analysis of endocrine disruptors by comprehensive two-dimensional gas chromatography. J Braz Chem Soc 26(3):531–536

Murray JA (2012) Qualitative and quantitative approaches in comprehensive two-dimensional gas chromatography. J Chromatogr A 1261:58–68

Nanni N, Fiselier K, Grob K, Di Pasquale M, Fabrizi L, Aureli P, Coni E (2011) Contamination of vegetable oils marketed in Italy by phthalic acid esters. Food Control 22(2):209–214

Ostrovsky I, Cabala R, Kubinec R, Górová R, Blaško J, Kubincová J, Řimnáčová L, Lorenz W (2011) Determination of phthalate sum in fatty food by gas chromatography. Food Chem 124 (1):392–395

Ozaki A, Baba T (2003) Alkylphenol and bisphenol A levels in rubber products. Food Addit Contam 20(1):92–98

Pedroso MP, de Godoy LAF, Fidélis CHV, Ferreira EC, Poppi RJ, Augusto F (2009) Comprehensive two-dimensional gas chromatography (GC × GC). Quim Nova 32(2):421–430

Petrovic M, Eljarrat E, de Alda MJL, Barceló D (2002) Recent advances in the mass spectrometric analysis related to endocrine disrupting compounds in aquatic environmental samples. J Chromatogr A 974(1–2):23–51

Ribeiro LH, Costa Freitas AM, Gomes da Silva MDR (2008) The use of headspace solid phase microextraction for characterization of volatile compounds in olive oil matrices. Talanta 77:110–117

Romero-Franco M, Hernandez-Ramirez RU, Calafat AM, Cebrián ME, Needham LL, Teitelbaum S, Wolff MS, López-Carrillo L (2011) Personal care product use and urinary levels of phthalate metabolites in Mexican women. Environ Int 37(5):867–871

Schroder HF (2001) Tracing of surfactants in the biological wastewater treatment process and the identification of their metabolites by flow injection-mass spectrometry and liquid chromatography-mass spectrometry and -tandem mass spectrometry. J Chromatogr A 926 (1):127–150

Shen Y, Ren ML, Feng X, Gao YX, Xu Q, Cai YL (2014) Does nonylphenol promote the growth of uterine fibroids? Eur J Obstet Gynecol Reprod Biol 178:134–137

USEPA (2001) U.S. Environmental Protection Agency. Report No. EPA/625/R-00/015, Washington, DC

USEPA (2015) United States-Environmental Protection Agency—National Exposure Research Laboratory Environmental Sciences. http://www.epa.gov/esd/chemistry/org-anal/faq.htm. Accessed February 2015

Vallejo A, Olivares M, Fernández LA, Etxebarria N, Arrasate S, Anakabe E, Usobiaga A, Zuloaga O (2011) Optimization of comprehensive two dimensional gas chromatography-flame ionization detection-quadrupole mass spectrometry for the separation of octyl- and nonylphenol isomers. J Chromatogr A 1218(20):3064–3069

Vendeuvre C, Bertoncini F, Duval L, Duplan JL, Thiébaut D, Hennion MC (2004) Comparison of conventional gas chromatography and comprehensive two-dimensional gas chromatography for the detailed analysis of petrochemical samples. J Chromatogr A 1056(1–2):155–162

Villar-Navarro M, Ramos-Payan M, Fernández-Torres R, Callejón-Mochón M, Bello-López MA (2013) A novel application of three phase hollow fiber based liquid phase microextraction (HF-LPME) for the HPLC determination of two endocrine disrupting compounds (EDCs), n-octylphenol and n-nonylphenol, in environmental waters. Sci Total Environ 443:1–6

von Muhlen C, Zini CA, Caramão EB, Marriott PJ (2007) Nomenclature for comprehensive multidimensional chromatography in Portuguese language. Quim Nova 30(3):682–687

WHO/UNEP (2012) World Health Organization—United Nations Environment Programme. State of the science of endocrine disrupting chemicals-2012. http://unep.org/pdf/9789241505031_eng.pdf. Accessed 10 December 2013

Wu YP, Wang YC, Ding WH (2012) Rapid determination of alkylphenols in aqueous samples by in situ acetylation and microwave-assisted headspace solid-phase microextraction coupled with gas chromatography-mass spectrometry. J Sep Sci 35(16):2122–2130

Wu P-G, Pan X-D, Ma B-J, Wang L-Y, Zhang J (2014) Determination of phthalate esters in non-alcoholic beverages by GC-MS and optimization of the extraction conditions. Eur Food Res Technol 238(4):607–612

Xia H, Chi Y, Qi X, Su M, Cao Y, Song P, Li X, Chen T, Zhao A, Zhang Y, Cao Y, Ma X, Jia W (2011) Metabolomic evaluation of di-n-butyl phthalate-induced teratogenesis in mice. Metabolomics 7(4):559–571

Xie X, Lü W, Chen X (2013) Binding of the endocrine disruptors 4-tert-octylphenol and 4-nonylphenol to human serum albumin. J Hazard Mater 248–249:347–354

Yilmaz PK, Ertas A, Kolak U (2014) Simultaneous determination of seven phthalic acid esters in beverages using ultrasound and vortex-assisted dispersive liquid–liquid microextraction followed by high-performance liquid chromatography. J Sep Sci 37(16):2111–2117

Zhang C, Eganhouse RP, Pontolillo J, Cozzarelli IM, Wang Y (2012) Determination of nonylphenol isomers in landfill leachate and municipal wastewater using steam distillation extraction coupled with comprehensive two-dimensional gas chromatography/time-of-flight mass spectrometry. J Chromatogr A 1230:110–116

Chapter 22
Multidimensional Chromatographic Techniques for Monitoring and Characterization of Environmental Samples

Eduardo P. Mateus, Marco D.R. Gomes da Silva, Alexandra B. Ribeiro, and Philip Marriott

22.1 Introduction

The electrokinetic process (EK) has been applied for removal of inorganic and organic contaminants in polluted soils (Lageman et al. 1989; Pamukcu and Wittle 1992; Probstein and Hicks 1993; Ribeiro and Mexia 1997; Ribeiro 1998; Ribeiro et al. 1998; Virkutyte et al. 2002), Cu-Cr-As impregnated wood waste (Ribeiro et al. 2000), fly ash from straw combustion (Hansen et al. 2001), or municipal solid waste incinerators fly ash (Pedersen et al. 2001). Research on the application of the EK has mainly been performed for heavy metals remediation. EK research aiming organic pollutants remediation can be found on the removal of pesticide/herbicides (Ribeiro and Mateus 2009; Polcaro et al. 2007), pharmaceutical and personal care compounds (Guedes et al. 2014), chlorinated solvents (Rohrs et al. 2002; Rabbi et al. 2002), petroleum hydrocarbons (Murillo-Rivera et al. 2009; Park et al. 2005), phenol (Acar et al. 1995), polycyclic aromatic hydrocarbons (Alcantara et al. 2008; Niqui-Arroyo et al. 2006; Maini et al. 2000), or polychlorinated biphenyls (Gomes et al. 2015). Those studies, conducted with the emphasis on the movement of target contaminants in the EK system, in order to assess whether the method can be applied to remove them from polluted matrices, have been performed by spiking the matrices with the analytes at higher concentrations far from those present in real

E.P. Mateus (✉) • A.B. Ribeiro
CENSE, Departamento de Ciências e Engenharia do Ambiente, Faculdade de Ciências e Tecnologia, Universidade Nova de Lisboa, Caparica 2829-516, Portugal
e-mail: epm@fct.unl.pt

M.D.R. Gomes da Silva
LAQV, REQUIMTE, Departamento de Química, Faculdade de Ciências e Tecnologia, Universidade Nova de Lisboa, Caparica 2829-516, Portugal

P. Marriott
Australian Centre for Research on Separation Science, School of Chemistry, Monash University, Wellington Road, Clayton, VIC 3800, Australia

© Springer International Publishing Switzerland 2016

A.B. Ribeiro et al. (eds.), *Electrokinetics Across Disciplines and Continents*,
DOI 10.1007/978-3-319-20179-5_22

samples or by means of model matrices (e.g., kaolinite), which are much less complex than real environmental matrices.

The setup of a remediation strategy requires the previous characterization of the polluted matrices in terms of its chemical composition, and the success of the remediation process improves with the level of the characterization and contaminants monitorization.

However, the elucidation of the behavior of organic contaminants present in real environmental matrices, when submitted to the EK remediation, is a hard and challenging task. Environmental matrices produce complex samples, which are composed by a wide array of components, through volatiles and semi-volatiles compounds to more heavy compounds, comprising several polarity and concentration ranges. Thus, the qualitative and quantitative analysis of pollutants in complex environmental matrices requires a technique able to separate the analytes from the other components of the matrix. In the past decades, for environmental characterization, analysis and identification of environmental contaminants one-dimensional gas chromatography (1D-GC), often coupled to a mass spectrometry (GC/MS) as a specific detection method, has been the analytical method of choice. Nevertheless, in spite of its resolution power and the continuous development of equipment and analytical methodologies and techniques, the components from environmental samples may remain difficult or unachievable to separate by 1D-GC. This is specially the case if additional sample fractionation methodologies prior to the final analysis are not used, thus promoting coelutions and difficult or ambiguous compound identification and monitorization (Meyer et al. 1999). This drawback is due to the complexity of the environmental samples with their large number of potential components that are usually present in a wider range of concentrations. Consequently, the trace level analytes, that sometimes are the toxically active components in the matrix under study, may never be detected, if they are co-eluting with high concentration compounds. Another potential problematic issue is related to the existence of possible isomeric configurations of analytes that, due to its structural similarity, promote almost identical mass spectra and retention times. This reality, when using GC/MS, jeopardizes analytes identifications and demands the acquisition of pure mass spectra, a task that makes the identification process very difficult and sometimes even impossible without the full separation of peaks in order to assure clean mass spectra.

Consequently, the 1D-GC approach may not always achieve satisfactory results, although valid, for the chemical composition of environmental matrices, resulting in a considerable amount of information that remains unexploited or hidden.

The development of new analytical techniques, in order to maximize analyte separations, has always been a target that is historically highlighted by the progression of packed column to capillary column chromatography and by the upgrade of one-dimensional (1D) to multidimensional (MD) chromatography systems. Such advances aim to reach a higher chromatographic capacity, in order to achieve the separation for all sample analytes (David and Sandra 1987; Bertsch 1999).

22.2 Some Theoretical Considerations

The pursuit for multidimensional systems is supported by the limitations of the achievable maximum number of theoretical plates for a single column in 1D systems due to inherent physical and statistical constraints (Grushka 1970; Chaves das Neves and Freitas 1996; Bartle 2002).

The physical limitations may be exemplified, using equation (22.1) (Grushka 1970), to calculate the chromatographic capacity of a capillary column with 50 m × 0.25 mm and $d_f = 0.25$ μm, assuming $R = 1$ and t_2/t_1 as 10:

$$n = \frac{\sqrt{N}}{4R} ln\left(\frac{t_2}{t_1}\right) + 1 \qquad (22.1)$$

where:

n = peak capacity of a single column chromatographic system
N = number of theoretical plates of a single column
t_1 and t_2 = retention time window, t_1 and t_2 are retention times
R = resolution for the separation of two compounds with retention times t_1 and t_2.

The estimated value will tell us that theoretically the column will be able to separate 260 analytes under ideal conditions (Grob et al. 1978, 1981; Bartle 2002), a number that will be insufficient for some contaminated environmental matrices.

However, when the statistical theory of overlap (STO) is applied (Davis and Giddings 1983; Martin et al. 1986; Bertsch 1999; Bartle 2002) for the same column, the result will give us an even more limited picture of separation power in 1D systems. The maximum number of analytes that the chromatographic column can theoretically separate is, in equation I, expressed by the capacity factor (n). Statistically, this value decreases because the analytes will be on reality randomly and not discretely distributed through the chromatographic separation. The STO points out (22.2) that in a chromatographic analysis the number of separated peaks (S) is related to the capacity factor of the column (n) and with the total number of analytes (m) present in the sample (Davis and Giddings 1983; Bertsch 1999; Bartle 2002).

From (22.2), the probability (P) of observing a peak, consisting of one single component, will be estimated by (22.3).

$$S = m \cdot \exp\left(-\frac{2m}{n}\right) \qquad (22.2)$$

$$P = \exp\left(-\frac{2m}{n}\right) \qquad (22.3)$$

where:

S = number of separated single peaks
n = peak capacity of a single column chromatographic system
m = number of sample components

The application of equation (22.3) leads to the theoretical conclusion that for a sample with 100 components, a column with a capacity for 290 analytes will be needed in order to guaranty the separation of half of the components (Bartle 2002). This shows the limitation of 1D-GC, in spite of its high separation power, even under ideal separation conditions without the variables, peak tailing, and/or the wide ranges of analytes concentrations that are not considered by the STO (Bertsch 1999).

Due to the potential high complexity of some of the environmental samples, which easily reach 100 components, in a wide range of concentrations, the occurrence of co-elutions became inevitable, even in most efficient columns in 1D-GC.

22.3 Multidimensional Systems

22.3.1 Multidimensional Gas Chromatography with a Flow Switching Device: GC–GC

The potential high complexity of the chromatograms that result from environmental samples forwarded the analysts to new ways of chromatography, such as multidimensional systems, where the analytes are submitted to two or more independent separation steps, on independent columns, in order to achieve separation efficiency.

The classical example came from the "heart-cut" systems using flow switching devices, interfacing two columns (GC–GC), such as the Deans switch (Fig. 22.1),

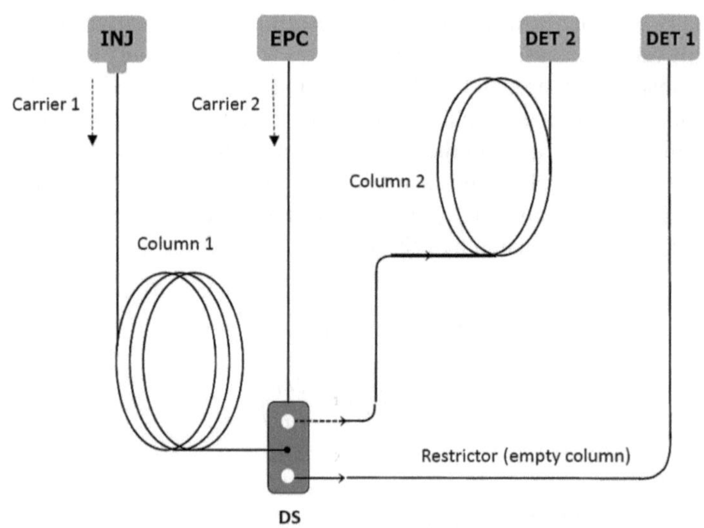

Fig. 22.1 Schematic representation of a multidimensional GC system with a heart-cut configuration, using a Dean switch device (DS), *INJ* injector, *DET* detector, *EPC* electronic pressure control

based on pneumatic pressure balancing. This configuration allows the isolation of target peaks, or packet of compounds, by partial selective transference (online heart-cut), from a primary column to a second column with different selectivity. On the second column/dimension, the transferred target analytes or the components of a retention time window (packet) are submitted to improved chromatographic separation, avoiding the potential co-elution with the transferred interfering peaks and with the non-transferred compounds that will resume on the first dimension (Schomburg et al. 1984; David and Sandra 1987; Bertsch 1999; Poole 2003).

Figure 22.2 shows an example of GC–GC, with flame ionization detection (FID), performed with a Polycyclic Aromatic Hydrocarbons (PAHs) standard mix sample, using the heart-cut system through a Deans switch device, to target the peak of Benzo[a]pyrene (B[a]P) that is eluting on a nonpolar column (5 % phenyl in polydimethylsiloxane) and transferring it to a medium polar column (50 % phenyl in polydimethylsiloxane) for secondary analysis. Figures 22.3 and 22.4 demonstrate the applicability of the same system, using the same columns to isolate B[a]P, on complex chromatograms. In Fig. 22.3, the B[a]P peak is targeted on an atmospheric particulate matter (PM_{10}) extract and sent to the second dimension where it is

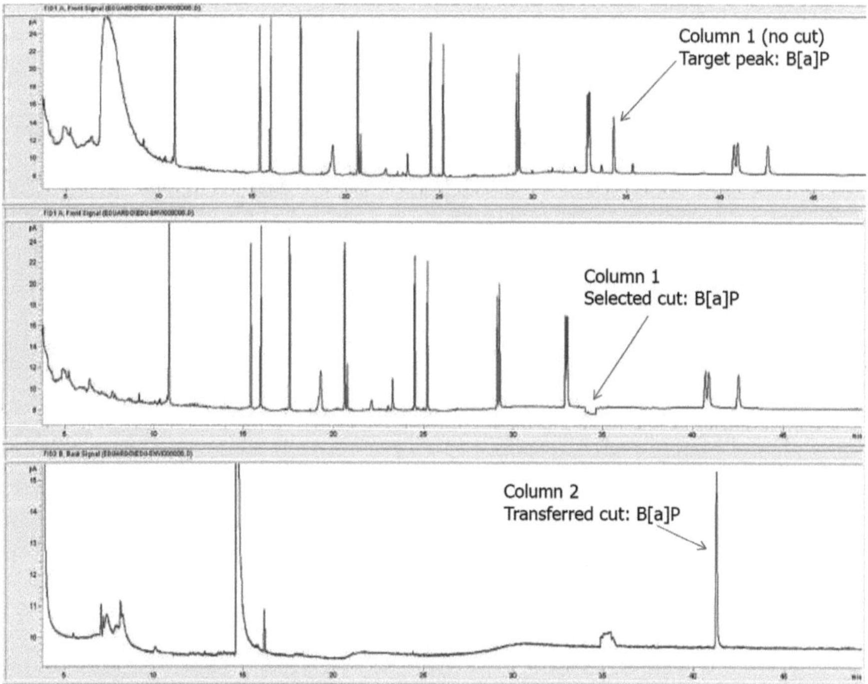

Fig. 22.2 Representation of multidimensional GC–GC analysis from a PAHs mix test sample, using a heart-cut system to transfer Benzo[a]pyrene. The first top chromatograms show the primary separation and the bottom ones the resulting heart-cut chromatograms on both dimensions (Detector: FID, Injection: 1 μL splitless)

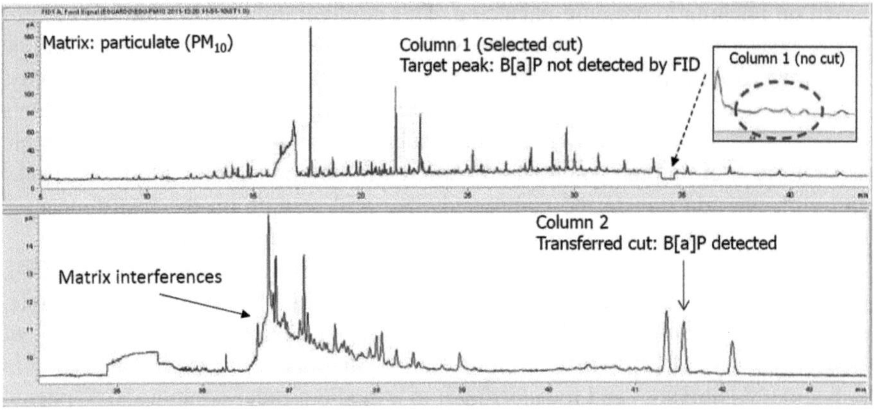

Fig. 22.3 Representation of multidimensional GC–GC analysis from atmospheric particulate matter (PM10), using a heart-cut system, targeting Benzo[a]pyrene. The top chromatogram shows the primary separation, and the bottom one the resulting heart-cut chromatograms on second dimension (Detector: FID, Injection: 1 μL splitless)

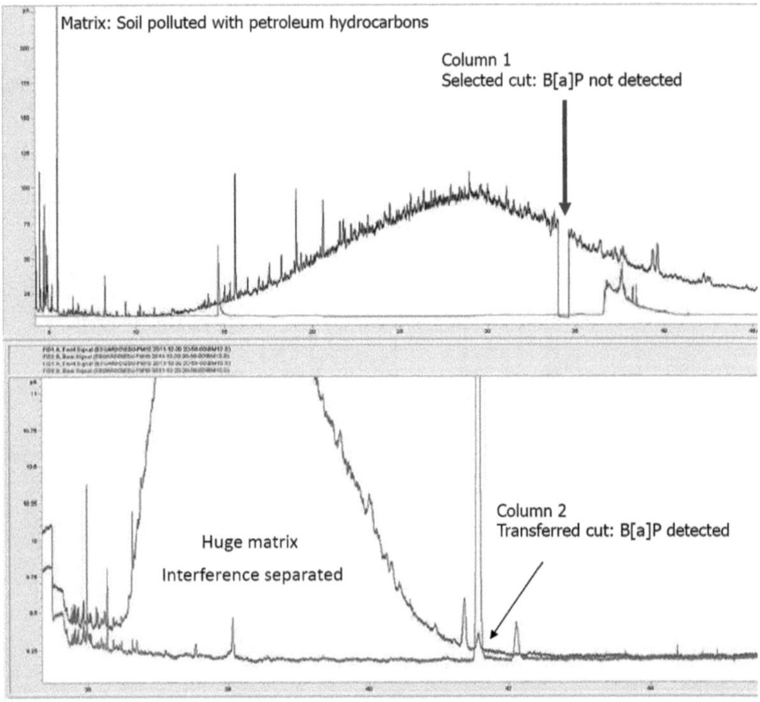

Fig. 22.4 Representation of multidimensional GC–GC analysis from a soil polluted with petroleum hydrocarbons, using a heart-cut system, targeting Benzo[a]pyrene. The top chromatogram shows the primary separation, and the bottom one the resulting heart-cut chromatograms on second dimension. The major peak overlaying B[a]P transferred peak is its standard used for retention time confirmation (Detector: FID, Injection: 1 μL splitless)

separated from the interferences that co-elute with it on the first dimension. Note that on the first dimension the baseline hides the peaks. A more dramatic example is shown in Fig. 22.4, where the B[a]P peak is targeted on a chromatogram from an extract of a soil polluted with petroleum hydrocarbons. The chromatogram is very complex, being impossible to detect B[a]P. After the cut, a large gap is easily observed on the first dimension, a sign of the amount of compounds that are co-eluting. In spite of its trace amount, the B[a]P peak is detected on the second dimension and isolated from other peaks, including a significant amount of matrix interference components. These examples clearly show the power of the technique for target analysis, even using an "almost universal" detector such as the FID, which do not perform structural identifications.

In spite of its efficiency for target analysis, the MDGC is a time-consuming technique, with long analysis times, which may not fit with the demands of routine analysis. Additionally, it may be technically impractical to analyze, on the fly, all the target compounds that elute from the first dimension or carry out sequential transfers in a narrow window of retentions times. The probability of new co-elutions on the second column (Poole 2003) will increase, leading to system inefficiency if large amount of analytes need to be monitored on the same chromatographic run. However, this apparent disadvantage may be compensated by the accessible information content of the data that is processed in the same way as in 1D-GC and by its operational simplicity.

22.3.2 Comprehensive Two-Dimensional Gas Chromatography: GC × GC

In 1991, Liu and Phillips (1991) introduced the comprehensive two-dimensional gas chromatography (GC × GC). The GC × GC system consists of two columns with different selectivities that are serially connected through a suitable interface, which usually is a thermal modulator (Fig. 22.5) (Phillips and Beens 1999; Marriott and Shellie 2002; Dimandja 2003). At the GC × GC technique, the entire sample separated on the first column is transferred to the second one, resulting in an enhanced chromatographic resolution into two independent dimensions, where the analytes are separated by two independent mechanisms (orthogonal separation) (Schoenmakers et al. 2003).

By definition, a chromatographic method is considered comprehensive if (1) the sample transfer from the first to the second column is qualitatively and quantitatively complete; (2) the orthogonality principle is respected, meaning that both separation mechanisms are independent; and (3) the separation on the first column is preserved on the second one (Bertsch 2000; Dallüge et al. 2003; Schoenmakers et al. 2003).

Operationally, the most important component in the GC × GC system is the interface. The common interface comprises usually a thermal modulator, which either performs heating (to accelerate solute into a narrow band in the second column) or cooling (to retard analyte and cause on-column trapping or cryofocusing of the bands) or both, depending on the design.

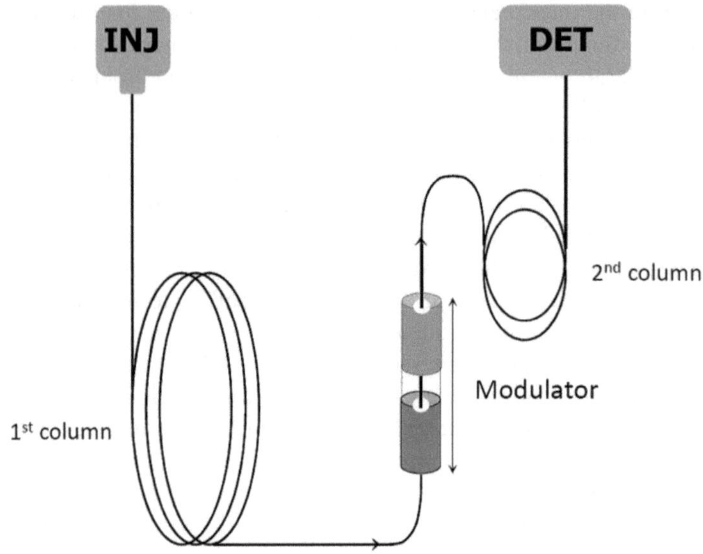

Fig. 22.5 Schematic representation of a GC × GC system (modulator: longitudinally modulated cryogenic system—LMCS)

The modulator collects and cuts the effluent from the first column into small portions, which are refocused and sampled onto the second column, in the form of narrow pulses. The modulation is thus the sequential liberation of the solute from the first column onto the second column, preserving and further fractionating the separation obtained on the first column. Additionally, since the modulated zones of a peak analyte are thermally focused before the separation on the second column, in a mass conservative process, the resulting segments (peaks) of the modulation are now more intense, with higher *S/N* ratios, and much narrower than in conventional GC (Lee et al. 2001; Dallüge et al. 2002b), improving the detection of trace analytes and the chromatographic resolution (Fig. 22.6). An example of the modulation effect, with the promoted increase of analytes *S/N* ratio, and the resulting increase in sensibility, can be observed on Fig. 22.6 for an extract from a soil polluted with hydrocarbons.

In theory, the total peak capacity n in GC × GC is the product of the peak capacities of the two individual columns n_1 (column 1) and n_2 (column 2), which results in separation potential $n_t = n_1 \times n_2$, that theoretically is much higher than in any other chromatographic arrangements (Fig. 22.7) (Venkatramani et al. 1996; Bertsch 2000).

In order to maximize the column peak capacities and the separation power of the GC × GC system, the separation mechanisms in both columns should be based on different and independent physical–chemical interactions (orthogonality) (Venkatramani et al. 1996; Phillips and Beens 1999; Marriott and Shellie 2002; Dallüge et al. 2003; Dimandja 2003; Schoenmakers et al. 2003), meaning that columns with stationary phases with maximum different selectivity should be selected for both dimensions. The first dimension columns have usually nonpolar

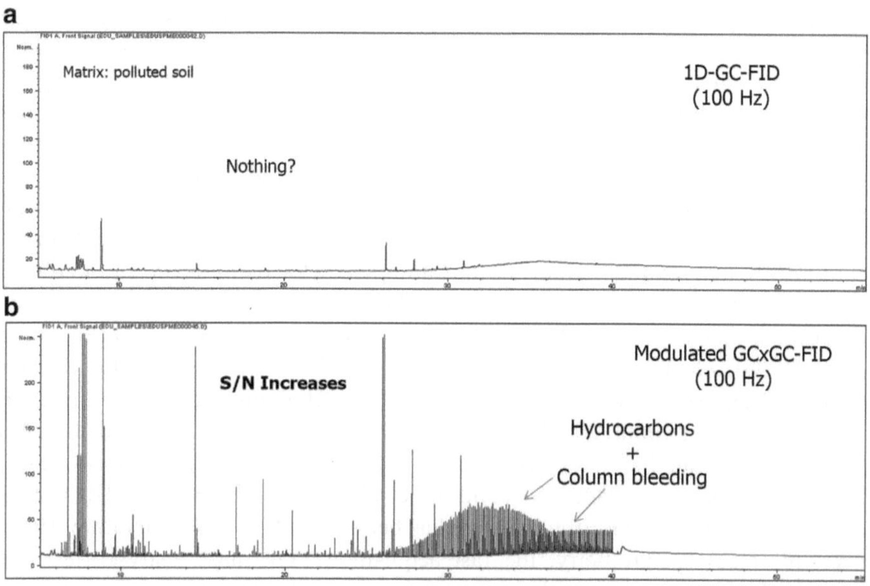

Fig. 22.6 Chromatograms from the same extract of a polluted soil sample obtained by conventional 1D-GC-FID (**a**), and by modulated GC × GC-FID (**b**), showing the peak signal increment (Injection: 1 μL in splitless)

Fig. 22.7 Schematic representation of peak capacity for two individual columns A (column 1) and B (column 2) in 1D-GC and coupled for multidimensional GC using heart cut and GC × GC configurations

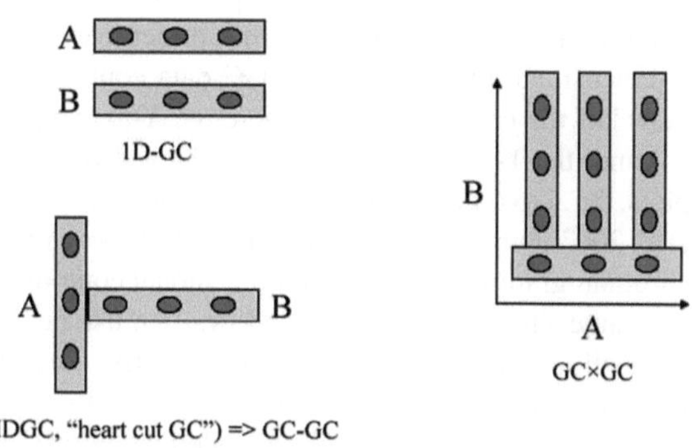

stationary phases and promote separation mainly due to the volatility of sample components (boiling point separation) (Phillips and Beens 1999; Marriott and Shellie 2002; Dallüge et al. 2003). The first column is usually of standard dimensions (e.g., 30 m × 0.25 mm; $d_f = 0.25$ μm), and the time scale of the first dimension separation corresponds to a normal GC separation, resulting in peak widths of several seconds. The second-column stationary phase is usually polar, or mid-polar (Beens et al. 2000; de Geus et al. 2001; Dallüge et al. 2002a). Fast GC can be performed in the second dimension column if a short narrow bore column (e.g., 1 m × 0.1 mm × 0.1 μm) is used. This configuration allows that the total time of the chromatographic run on the second column never exceeds more than a few seconds. This means that the second-column separation is performed essentially under isothermal conditions (Beens et al. 1998) and, therefore, analyte separation is based only on their polarity or on another property independent from the first dimension. Additionally, using the short narrow bore column the actual total run time of both 1D-GC and GC × GC analysis, for the same sample, will be approximately the same (Marriott and Shellie 2002; Dallüge et al. 2003).

The very fast separation in the second column results in very narrow peaks with widths between 0.1 and 0.6 s (Beens et al. 1998; Lee et al. 2000; Marriott et al. 2000), that require high data acquisition rate detectors (50–100 Hz) to obtain sufficient number of data points over a chromatographic peak for its accurate description (Zrostlíková et al. 2003). The use of detectors, such as FID (Frysinger et al. 1999), electron capture detector (ECD) (de Geus et al. 2000; Korytár et al. 2002), nitrogen phosphorous detector (NPD) (Khummueng et al. 2006; Ochiai et al. 2007; Mateus et al. 2008), and mass spectrometer detectors with quadrupoles (qMS) (Song et al. 2004; Adahchour et al. 2005; Mateus et al. 2008) and time-of-flight mass analysers (TOFMS) (Dallüge et al. 2002c; Ma et al. 2007; Mateus et al. 2008), on GC × GC applications, has been described in the literature. However, for mass spectrometry, the fast acquisition TOF mass analysers are the suitable detectors for this technique and have considerably driven and enlarged the application potential of GC × GC.

After data acquisition, suitable software is used to generate a reconstructed two-dimensional chromatogram, or a 3D plot; which is the representation of the linear modulated chromatograms projected on the second dimension for each modulation. The independent second dimension chromatograms are aligned in a bidimensional plane, the GC × GC data contour plots representing the "bird's-eye view" of the chromatogram, where the *X*-axis represents the separation on the first column, the *Y*-axis the separation achieved on the second column and, for the 3D plots, the *Z*-axis the intensity of detector response (Marriott and Shellie 2002; Dallüge et al. 2003; Dimandja 2003). Figure 22.8 shows a representation of a one-dimensional chromatogram versus the two-dimensional contour plot GC × GC chromatogram for the same chromatographic window. It must be pointed that any peaks that are vertically aligned are co-eluting on the first column, and thus preventing 1D-GC-FID to be used for their monitorization purposes, a situation that is overtaken for GC × GC-FID since the peaks are adequately resolved on the 2D plot.

In Fig. 22.9, one can observe the separation obtained for a creosote sample, where the compounds show a very good resolution between themselves and from the matrix interferences.

Due to the orthogonal separation occurring in both columns, the chromatograms resulting from GC × GC are ordered, producing structured chromatograms, where the analytes have their spatial location, in the contour plot, based on their structures and thus physical–chemical nature (Phillips and Beens 1999). In the reconstructed 2D contour plots, characteristic patterns are obtained, in which the members of homological series with different volatilities are ordered along the first dimension axis, and the compounds are scattered along the second dimension axis according to their polarity (for a nonpolar–polar column set). This cluster representation of

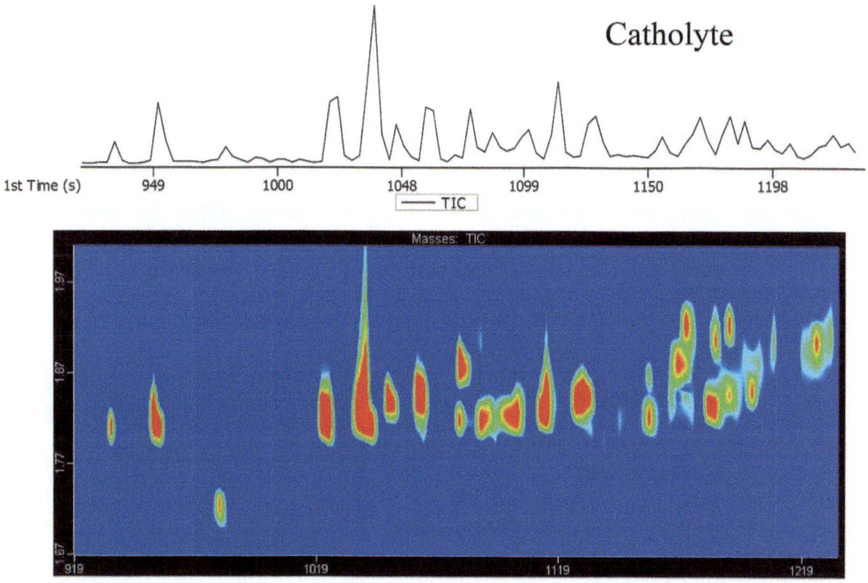

Fig. 22.8 Representation of the process of a GC × GC chromatogram: (*Top*) raw linear chromatogram showing the elution profile in 1D-GC; (*Bottom*) the software processed GC × GC chromatogram visualized as a two-dimensional contour plot (Electrokinetic remediation of wood treated with creosote: Catholyte sample)

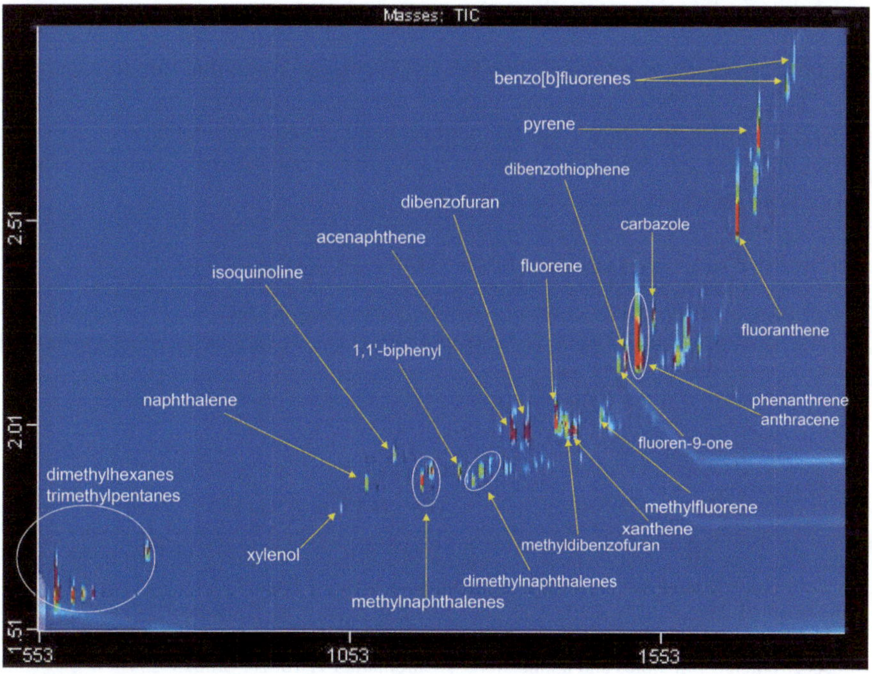

Fig. 22.9 Comprehensive two-dimensional separation space for a creosote sample GC × GC analysis

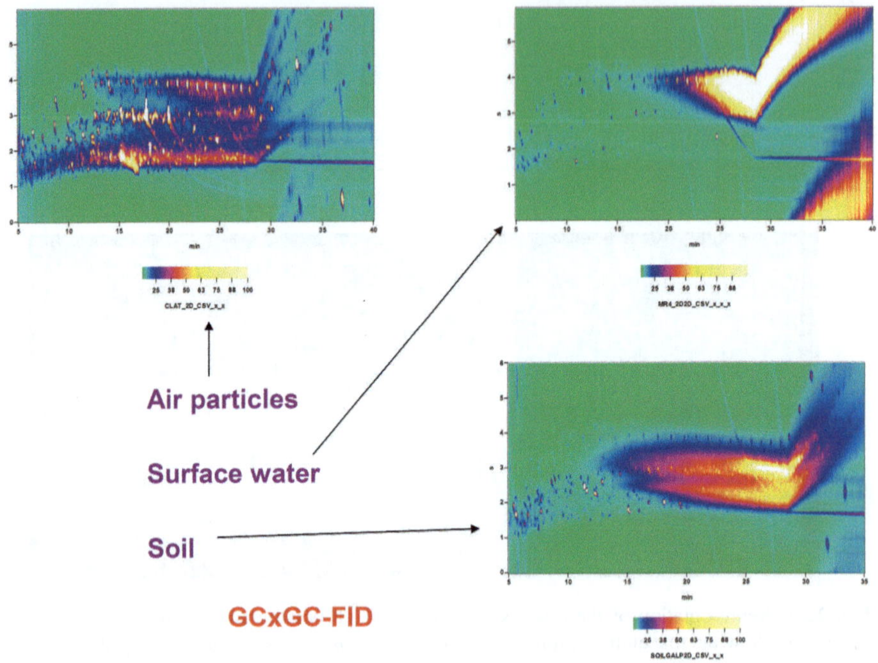

Fig. 22.10 GC × GC-FID contour plots for three different environmental matrices: atmospheric particulate matter (PM10), surface water, and polluted soil. The column set was mid-polar × nonpolar (Injection: 1 μL in splitless)

various subgroups of analytes in the GC × GC contour plots, that turn in some way the 2D space in chromatographic maps of chemical properties, may be used as a tool for compound class analysis, tentative identification/detection of analytes (Frysinger et al. 1999; Korytár et al. 2002), and matrix characterization (Fig. 22.10) with the use of any detector.

Acknowledgments Financial support for the work was provided by project FP7-PEOPLE-2010-IRSES-269289-*ELECTROACROSS—Electrokinetics across disciplines and continents*: *an integrated approach to finding new strategies for sustainable development.*

Very special thanks are due to Dr. Blagoj Mitrevski for his valuable technical support and advice.

References

Acar YB, Gale RJ, Alshawabkeh AN, Marks RE, Puppula S, Bricka M, Parker RJ (1995) Electrokinetic remediation: basics and technology status. J Hazard Mater 40(2):117–137

Adahchour M, Brandt M, Baier H-U, Vreuls RJJ, Batenburg AM, Brinkman UAT (2005) Comprehensive two-dimensional gas chromatography coupled to a rapid-scanning quadrupole mass spectrometer: principles and applications. J Chromatogr A 1067(1–2):245–254

Alcantara T, Pazos M, Cameselle C, Sanroman MA (2008) Electrochemical remediation of phenanthrene from contaminated kaolinite. Environ Geochem Health 30(2):89–94
Bartle KD (2002) Introduction. In: Mondello L, Lewis AC, Bartle KD (eds) Multidimensional chromatography. Wiley, Chichester, pp 3–15
Beens J, Boelens H, Tijssen R, Blomberg J (1998) Quantitative aspects of comprehensive two-dimensional gas chromatography (GC × GC). J High Resolut Chromatogr 21:47–54
Beens J, Blomberg J, Schoenmakers PJ (2000) Proper tuning of comprehensive two-dimensional gas chromatography (GC × GC) to optimize the separation of complex oil fractions. J High Resolut Chromatogr 23:182–188
Bertsch WJ (1999) Two-Dimensional gas chromatography. Concepts, instrumentation, and applications—part 1: fundamentals, conventional two-dimensional gas chromatography, selected applications. J High Resolut Chromatogr 22(12):647–665
Bertsch WJ (2000) Two-dimensional gas chromatography. Concepts, instrumentation and applications—part 2: comprehensive two-dimensional gas chromatography. J High Resolut Chromatogr 23:167–181
Dallüge J, van Rijn M, Beens J, Vreuls RJJ, Brinkmann UAT (2002a) Comprehensive two-dimensional gas chromatography with time-of-flight mass spectrometric detection applied to the determination of pesticides in food extracts. J Chromatogr A 965:207–217
Dallüge J, Vreuls RJJ, Beens J, Brinkmann UAT (2002b) Optimisation and characterisation of comprehensive two-dimensional gas chromatography with time-of-flight mass spectrometric detection. J Sep Sci 25:201–214
Dallüge J, van Stee LLP, Xu X, Williams J, Beens J, Vreuls RJJ, Brinkman UAT (2002c) Unravelling the composition of very complex samples by comprehensive gas chromatography coupled to time-of-flight mass spectrometry: cigarette smoke. J Chromatogr A 974(1–2):169–184
Dallüge J, Beens J, Brinkman UAT (2003) Comprehensive two-dimensional gas chromatography: a powerful and versatile analytical tool. J Chromatogr A 1000(1–2):69–108
David F, Sandra P (1987) In: Sandra P, Bicchi C (eds) Capillary gas chromatography in essential oil analysis. Huething, Heidelberg, p 387
Davis JM, Giddings JC (1983) Statistical theory of component overlap in multicomponent chromatograms. Anal Chem 55:418–424
de Geus H-J, Schelvis A, de Boehr J, Brinkman UAT (2000) Comprehensive two-dimensional gas chromatography with a rotating thermal desorption modulator and independently temperature programmable columns. J High Resolut Chromatogr 23:189–196
de Geus H-J, Aidos I, de Boer J, Luten JB, Brinkman UAT (2001) Characterisation of fatty acids in biological oil samples using comprehensive multidimensional gas chromatography. J Chromatogr A 910:95–103
Dimandja JMD (2003) A new tool for the optimized analysis of complex volatile mixtures: comprehensive two-dimensional gas chromatography/time-of-flight mass spectrometry. Am Lab 25(3):42–53
Frysinger GS, Gaines RB, Ledford EB Jr (1999) Quantitative determination of BTEX and total aromatic compounds by comprehensive two-dimensional gas chromatography. J High Resolut Chromatogr 22:195–200
Gomes HI, Ottosen LM, Ribeiro AB, Dias-Ferreira C (2015) Treatment of a suspension of PCB contaminated soil using iron nanoparticles and electric current. J Environ Manage 151:550–555. doi:10.1016/j.jenvman.2015.01.015
Grob K Jr, Grob G, Grob K (1978) Comprehensive, standardized quality test for glass capillary columns. J Chromatogr A 156(1):1–20
Grob K, Grob G, Grob K Jr (1981) Testing capillary gas chromatographic columns. J Chromatogr A 219(1):13–20
Grushka E (1970) Chromatographic peak capacity and the factors influencing it. Anal Chem 42 (11):1142–1147
Guedes P, Mateus EP, Couto N, Rodríguez Y, Ribeiro AB (2014) Electrokinetic remediation of six emerging organic contaminants from soil. Chemosphere 117:124–131

Hansen HK, Ottosen LM, Pedersen AJ, Villumsen A (2001) Speciation and mobility of cadmium in straw and wood combustion fly ash. Chemosphere 45:123–128

Khummueng W, Trenerry C, Rose G, Marriott PJ (2006) Application of comprehensive two-dimensional gas chromatography with nitrogen-selective detection for the analysis of fungicide residues in vegetable samples. J Chromatogr A 1131(1–2):203–214

Korytár P, Leonards PEG, de Boer J, Brinkman UAT (2002) High-resolution separation of polychlorinated biphenyls by comprehensive two-dimensional gas chromatography. J Chromatogr A 958:203–218

Lageman R, Pool W, Seffinga G (1989) Electro-reclamation: theory and practice. Chem Ind 18:585–590

Lee AL, Lewis AC, Bartle KD, McQuaid JB, Marriott PJ (2000) A comparison of modulating interface technologies in comprehensive two-dimensional gas chromatography (GC × GC). J Microcolumn Sep 12:187–193

Lee AL, Bartle KL, Lewis AC (2001) A model of peak amplitude enhancement in orthogonal two dimensional gas chromatography. Anal Chem 73:1330–1335

Liu Z, Phillips JB (1991) Comprehensive two-dimensional gas chromatography using an on-column thermal modulator interface. J Chromatogr Sci 29:227–231

Ma C, Wang H, Lu X, Li H, Liu B, Xu G (2007) Analysis of Artemisia annua L. volatile oil by comprehensive two-dimensional gas chromatography time-of-flight mass spectrometry. J Chromatogr A 1150(1–2):50–53

Maini G, Sharman AK, Knowles CJ, Sunderland G, Jackman SA (2000) Electrokinetic remediation of metals and organics from historically contaminated soil. J Chem Technol Biotechnol 75(8):657–664

Marriott PJ, Ong RCY, Kinghorn RM, Morrison PD (2000) Time-resolved cryogenic modulation for targeted multidimensional capillary gas chromatography analysis. J Chromatogr A 892 (1–2):15–28

Marriott P, Shellie R (2002) Principles and applications of comprehensive two-dimensional gas chromatography. Trends Anal Chem 21(9–10):573–583

Martin M, Herman DP, Guiochon G (1986) Probability distributions of the number of chromatographically resolved peaks and resolvable components in mixtures. Anal Chem 58:2200–2207

Mateus EP, Gomes da Silva MDR, Ribeiro AB, Marriott PJ (2008) Qualitative mass spectrometric analysis of the volatile fraction of creosote-treated railway wood sleepers by using comprehensive two-dimensional gas chromatography. J Chromatogr A 1178(1–2):215–222

Meyer S, Cartellieri S, Steinhart H (1999) Simultaneous determination of PAHs, hetero-PAHs (N, S, O), and their degradation products in creosote-contaminated soils. Method development, validation, and application to hazardous waste sites. Anal Chem 71(18):4023–4029

Murillo-Rivera LI, Barrón J, Oropeza-Guzman MT, González I, Teutli-Leon MMM (2009) Influence of anolyte and catholyte composition on TPHs removal from low permeability soil by electrokinetic reclamation. Electrochim Acta 54(7):2119–2124

Chaves das Neves HJ, Freitas AMC (1996) Introdução à Cromatografia Gás-Líquido de Alta Resolução. Dias de Sousa, Ltd

Niqui-Arroyo JL, Bueno-Montes M, Posada-Baquero R, Ortega-Calvo JJ (2006) Electrokinetic enhancement of phenanthrene biodegradation in creosote-polluted clay soil. Environ Pollut 142(2):326–332

Ochiai N, Ieda T, Sasamoto K, Fushimi A, Hasegawa S, Tanabe K, Kobayashi S (2007) Comprehensive two-dimensional gas chromatography coupled to high-resolution time-of-flight mass spectrometry and simultaneous nitrogen phosphorous and mass spectrometric detection for characterization of nanoparticles in roadside atmosphere. J Chromatogr A 1150(1–2):13–20

Pamukcu S, Wittle JK (1992) Electrokinetic removal of selected heavy metals from soil. Environ Prog 11(3):241–250

Park JY, Kim SJ, Lee YJ, Baek K, Yang JW (2005) EK-Fenton process for removal of phenanthrene in a two-dimensional soil system. Eng Geol 77(3–4):217–224

Pedersen AJ, Ottosen LM, Villumsen A (2001) Electrodialytic removal of heavy metals from MSWI fly ashes. In: Proceedings of the 3rd symposium and status report on electrokinetic remediation, Karlsruhe
Phillips JB, Beens J (1999) Comprehensive two-dimensional gas chromatography: a hyphenated method with strong coupling between the two dimensions. J Chromatogr A 856:331–347
Polcaro AM, Vacca A, Mascia M, Palmas S (2007) Electrokinetic removal of 2,6-dichlorophenol and diuron from kaolinite and humic acid-clay system. J Hazard Mater 148(3):505–512
Poole CF (2003) The essence of chromatography. Elsevier, Boston
Probstein RF, Hicks R (1993) Removal of contaminants from soils by electric fields. Science 260:498–503
Rabbi MF, Clark B, Galeb RJ, Ozsu-Acar E, Pardue J, Jackson A (2002) In situ TCE bioremediation study using electrokinetic cometabolite injection. Waste Manag 20(4):279–286
Ribeiro AB (1998) Use of electrodialytic remediation technique for removal of selected heavy metals and metalloids from soils. Ph.D. Thesis, Technical University of Denmark, Denmark
Ribeiro AB, Mateus EP (2009) In: Krishna RR, Cameselle C (eds) Electrochemical remediation technologies for polluted soils, sediments and groundwater, 1st edn. Wiley, New York
Ribeiro AB, Mexia JT (1997) A dynamic model for the electrokinetic removal of copper from a polluted soil. J Hazard Mater 56(3):257–271
Ribeiro AB, Villumsen A, Bech-Nielsen G, Réffega A, Vieira e Silva JM (1998) Electrodialytic remediation of a soil from a wood preservation industry polluted by CCA. In: Proceedings of the 4th international symposium the challenge safety and environment in wood preservation. Cannes-Mandelieu
Ribeiro AB, Mateus EP, Ottosen LM, Bech-Nielsen G (2000) Electrodialytic removal of Cu, Cr, and As from chromated copper arsenate-treated timber waste. Environ Sci Technol 34 (5):784–788
Rohrs J, Ludwig G, Rahner D (2002) Electrochemically induced reactions in soils—a new approach to the in-situ remediation of contaminated soils? Part 2: remediation experiments with a natural soil containing highly chlorinated hydrocarbons. Electrochim Acta 47 (9):1405–1414
Schoenmakers P, Marriott P, Beens J (2003) Nomenclature and conventions in comprehensive multidimensional chromatography. LCGC Europe 16:335–339
Schomburg G, Husmann H, Hübinger E, König W (1984) Multidimensional capillary gas chromatography-enantiomeric separations of selected cuts using a chiral second column. J High Resolut Chromatogr 7:404–410
Song SM, Marriott P, Wynne P (2004) Comprehensive two-dimensional gas chromatography-quadrupole mass spectrometric analysis of drugs. J Chromatogr A 1058(1–2):223–232
Venkatramani CJ, Xu J, Phillips JB (1996) Separation orthogonality in temperature-programmed comprehensive two-dimensional gas chromatography. Anal Chem 68:1486–1492
Virkutyte J, Sillanpaa M, Latostenmaa P (2002) Electrokinetic soil remediation—critical overview. Sci Total Environ 289(1–3):97–121
Zrostlíková J, Hajslová J, Cajka T (2003) Evaluation of two-dimensional gas chromatography–time-of-flight mass spectrometry for the determination of multiple pesticide residues in fruit. J Chromatogr A 1019(1–2):173–186

Index

A
Acar, Y.B., 5, 8
Acid-enhanced EKR treatment, 254–255
Adler, A., 387
Adsorption diffusion
external mass transfer
coefficients, 98–100
model, 87–88
intra-particle
diffusion coefficient, 101
mass transfer model, 88–89
rate constant, 100, 101
kinetics analysis, liquid-solid interface, 98–101
Adsorption dynamic models
conservation balance, 103
distributed parameter systems, 103
heterogeneous model
for batch experiment, 104
for dynamic mode, 104
kinetics at liquid-solid interface, 105
lumped parameter systems, 103
systems dynamics, 103
Adsorption efficiency, 93
Adsorption processes
activated carbons
parameters for, 106–107
surface groups, 84
textural and surface characteristics, 84
adsorbate/adsorbent surfaces interaction, 82
batch laboratory technique, 86
chemical posttreatment, 84–85
Cr(III) adsorption on activated carbon
capacity *vs.* time, 92–93
efficiency *vs.* time, 92
diffusion coefficients, 98–101
diffusion models, 83
external mass transfer, 87–88
intra-particle mass transfer, 88–89
dynamic models, 102–105
energy of activation, 101–102
gas–solid interface, 82
mass action process, 86
pore size, 105–106
reaction models, 83
chemisorptions kinetics model, 91–92
pseudo-first-order rate model, 89–90
pseudo-second-order rate model, 90–91
stages, 87
uptake of metal ions on carbon
chemisorptions Elovich model, 97, 100
pseudo-first-order model, 93–95
pseudo-second-order model, 95–97
Adsorption reaction models, 83
chemisorptions kinetics model, 91–92
pseudo-first-order rate model, 89–90
pseudo-second-order rate model, 90–91
AEM. *See* Anion-exchange membranes (AEM)
Air pollution control residues
applications, 231–232
characteristics, 229–230
elements, 228
management strategies, 227, 228
mineral compounds, 230
treatment with hydraulic binders, 227
Alkylphenols and phthalates
GC×GC analysis (*see* Two-dimensional gas chromatography)
sample preparation methods

© Springer International Publishing Switzerland 2016

A.B. Ribeiro et al. (eds.), *Electrokinetics Across Disciplines and Continents*,
DOI 10.1007/978-3-319-20179-5

Alkylphenols and phthalates (*cont.*)
cartridges and disks, solid phase extraction, 432
cold SPME system, phthalate migration, 430–431
solid phase microextraction, 430–431
sources, 429
Al-Najjar, M., 266
Alshawabkeh, A., 5
Amorim, L.C.A., 433
Anawar, H.M., 318
André, L.C., 427–434
Anion-exchange membranes (AEM), 62–63, 133
ANN. *See* Artificial neural networks (ANN)
Anthropogenic dispersion, 297
Apetrei, C., 404
Arduini, I., 384–386
Arsenic remediation, 323–324
Artificial neural networks (ANN), 167
cation exchange capacity values, 169
general scheme, 168
multilayer perceptrons, 168
tool for soil science, 167
training and prediction, 168
Askim, J.R., 403
Available energy, 58–59

B
Baek, K., 6
Barbosa, B., 373–393
Bartlett, P.N., 403
Batchelor, S.E., 381
BBP. *See* Benzylbutyl-phthalate (BBP)
Bentonite clay
column test, $CaCl_2$ concentration, 46–47
electroosmotic flow volume, 48–50
physicochemical properties, 45–46
volume expansion, free swelling tests, 47–48
Benzylbutyl-phthalate (BBP), 428
Bertolini, L., 212
Binder, 227–228, 231–232
Biodegradation, 190
Biogeochemical model, 191
Blair, B.D., 353
Blandón, L., 161–170
Boléo, S., 373–393
Borghi, M., 385
Börjesson, P., 380
Bouchikhi, B., 401–419
Buerge, I.J., 352
Burken, J.G., 365
Burland, J.B., 29

C
Cacador, I., 359
Cacho, J.I., 429
Caffeine, 352
concentrations of, 353
physico-chemical properties, 355, 356
Calcium concentration distribution, in clay, 52–54
Cameselle, C., 8, 183
Cang, L., 322
Cardeal, Z.L., 427–434
Cardoso, R., 19–40
Carvalho, E.R., 416
Carvalho, P.N., 362
Casagrande, L., 1
Cation-exchange membranes (CEM), 62–63, 133
CEM. *See* Cation-exchange membranes (CEM)
Chandler, A.J., 227
Characterization factor
adapted
for groundwater, 194–195
Monte Carlo simulation, 193–194
probability density functions, 194–195
scores and ranks calculation, 194
substances chemical properties, 193
proposed, 190
Chemical equilibrium
numerical model
chemical system, 255, 259
Newton-Raphson method, 261–262
pore electrolyte composition, 261, 263
theoretical model, 257–258
Chemisorptions
Elovich model, 97, 100
kinetics model, 91–92
Chen, W., 241
Chiang, T.-S., 319
Chien, S.H., 92
Chinthamreddy, S., 295
5-Chloro-2-(2,4-dichlorophenoxy)phenol. *See* Triclosan
Christensen, I.V., 211
Chronopotentiometry, 69–70
Cirillo, T., 429
Clay
electrokinetic process
calcium concentration distribution, 52–54

electroosmotic flow, 48–50
pH distribution, 49–52
reactor, 45–46
water content, 50
zeta potential, 50–52
electroosmotic dewatering, 44
electroosmotic flow test specimens, 48
swelling behavior, bentonite, 46–48
Clayey soils
hydraulic *vs.* electroosmotic conductivity, 27
kaolin, 29
mechanisms, 27–28
pores size, 26
types of pores and structures, 27–28
Clayton, W.R., 92
Clinch, J.R., 407
Clustering analysis, 412
Complex matrix, 402
Compost
definition, 165
discharge current, 166–167
normalized current, 166
Comprehensive two-dimensional gas chromatography
environmental matrices, 450
flame ionization detection, 446–447
heart cut configurations, 446–447
vs. one-dimensional chromatogram, 448–449
schematic representation, 445–446
separation space for creosote sample, 448–449
thermal modulator, 445–446
Concentration polarization, 67
Conditioning, 19
Consolidation, 19
Constructed wetlands, 359–360
Contaminated area, Almaden mining district, 297
Contaminated soils
energy and fiber crops, 381
heavy metal, 377
phytoextraction, 384
phytoremediation, 390
and sediments
classification of remediation technologies, 335
electrokinetics with nanoremediation, 333–343
electroremediation, 336–337
nanoremediation, 337–338
Copper mine tailings
analytical and tailings preparation, 146–147
EKR cells, 145–146
electric resistance, 147–148
high frequency electric fields, 155–157
low frequency electric fields, 153–155
mineral composition, 144–145
pulsed electric DC fields, 148–151
sinusoidal AC/DC electric field, 152–153
Copper removal. *See* Copper mine tailings
Corrugated membranes, 73–74
Costa, J., 373–393
Coupled model
algorithm, 253
chemical equilibrium model, 257–262
reactive transport model, 262–269
Couto, M.N.P.F.S., 351–368
Couto, N., 113–124, 129–139, 313–325
Crespo, J.G., 57–76
Current reversal, 152
Cyanotoxins
production, 114–115
removal
membrane filtration, 117
permeate and concentrate streams, 118
photodegradation, 116
photolysis, 116
TiO_2 photocatalysis, 117
ultrafiltration and nanofiltration, 117–118

D

Data fusion approach, 412–413
DC. *See* Direct current (DC)
Dean, J., 192
Decontamination, 19
DEHP. *See* Di (ethylhexyl)-phthalate (DEHP)
Del Carlo, M., 432
DEP. *See* Diethyl-phthalate (DEP)
De-swelling, in clay. *See* Clay
Dettenmaier, E.M., 354
Dewatering process, 19
electrokinetic processes in soils, 22–24
and hydrodynamic consolidation of soils, 24–26
soil structure and effects on hydraulic behavior, 26–29
Diamond, L., 174
Dias-Ferreira, C., 173–196, 279–291, 331–345
Diethyl-phthalate (DEP), 428
Diffusion, 182

Di Natale, C., 404, 415
Di (ethylhexyl)-phthalate (DEHP), 428
Direct current (DC), 1, 5, 10
Distributed parameter systems, 103
  heterogeneous model
    for batch experiment, 104
    for dynamic mode, 104
Drogui, P., 117
Duarte, M.P., 373–393

E

Earth's water cycle, 59
Ebbers, B., 135
EDC. *See* Endocrine disrupting chemicals (EDC)
ED process. *See* Electrodialytic (ED) process
EDS. *See* Electrodialytic separation (EDS)
EDTA. *See* Ethylenediaminetetraacetic acid (EDTA)
Efflorescence, 205
EKG. *See* Electrokinetic geosynthetics (EKG)
EK process. *See* Electrokinetic (EK) process
EKR. *See* Electrokinetic remediation (EKR)
El Bari, N., 401–419
Electrical resistance and capacitance, 162–163
Electric fields
  electric resistance, 147–148
  high frequency, 155–157
  low frequency, 153–155
  pulsed electric DC fields, 148–151
  sinusoidal AC/DC, 152–153
Electrochemical methods, pollutants detection, 405
Electrochemical process
  algal blooms and cyanotoxins production, 114–115
  analytical methodologies, phosphorus recovery, 120
  arsenic removal from soil
    case study, 323–324
    coupling EK process with permeable reactive barrier, 319
    electrokinetic-enhanced phytoremediation, 321–322
    electrokinetic remediation, 316–318
    phytotechnologies, 320–321
    soil arsenic contamination, 314–316
  cyanotoxins removal
    membrane filtration, 117
    permeate and concentrate streams, 118
    photodegradation, 116
    photolysis, 116
    TiO_2 photocatalysis, 117
    ultrafiltration and nanofiltration, 117–118
  electrodialytic cell designs, phosphorus recovery, 119–120
  electrodialytic process, 118–119
    characteristics and duration, 122
    electrodegradation, 124
    phosphorus migration, 123
    photodegradation, 124
  electrokinetic P recovery, 133–138
  electrokinetic process, 118–119
  eutrophization, 113
  membrane concentrate production, 119–122
  phosphorus percentage, 122
  phosphorus recovery potential
    ashes, 132
    biosolids, 131
    concentrations, 132
    effluent of WWTP, 131
    liquid phase, 131
    sewage sludge ashes, 133
  wastewater treatment plants, 131
    factors, 130
    human excreta, 130
    sources, 129
  water treatment plant, 119
Electrodes, 182–185, 216–217
Electro-desalination
  definition, 206
  efflorescence, 205
  electrode placements, 216–217
  electromigration, 208
  electroosmosis and electromigration *vs.* diffusion and advection, 210–211
  material characteristics
    fired-clay bricks, 211–212
    historic Portuguese tiles, 213–214
    natural stones, 212–213
  phase changes for soluble salts in pores, 206, 207
  pH neutralization, 208–210
  pilot scale test, 217–221
  poulticing, 207
  principle of, 207–208
  removal rate, 215
  sub-efflorescence, 206
  transference numbers, 215–216
Electrodialytic cell designs, 119–120
Electrodialytic pre-treatment, MSWI, 234–238
  compressive strength, 234, 239, 240
  leaching behavior, 241–243
  workability, 239

Electrodialytic (ED) process, 118–119
  characteristics and duration, 122
  electrodegradation, 124
  municipal solid waste incineration
    cell design, 232–233
    removal of heavy metals, 233–234
  phosphorus migration, 123
  photodegradation, 124
Electrodialytic separation (EDS), 134–135
Electrodialytic soil remediation (EDR), 336
Electro-kinetically injection, 48, 52
Electrokinetic-enhanced phytoremediation
  DC/AC field, 322
  phytoextraction, 322
  schematic representation, 321
Electrokinetic geosynthetics (EKG), 20
Electrokinetic P recovery
  anion-exchange membranes, 133
  cation-exchange membranes, 133
  SSA (*see* Sewage sludge ashes (SSA))
Electrokinetic (EK) process, 118–119. *See also* Electrokinetic remediation (EKR)
  multidimensional chromatographic techniques
    comprehensive two-dimensional GC, 445–450
    environmental matrices, 440
    with flow switching device:, 442–445
    limitations, 441–442
    organic pollutants, 439–440
    statistical theory of overlap, 441
  in soils
    double layer in clays and, 22–23
    electroosmotic efficiency, 22
    Helmholtz-Smoluchowski theory, 23
    zeta potential, 23–24
Electrokinetic remediation (EKR), 143
  advantages, 5, 183
  algorithm for coupled model, 253
  arsenic removal from soil, 316–318
  cells
    anode and cathode zones, 146
    with ion-exchange membranes, 145
    with nylon mesh and fillter paper, 145–146
  characteristics, 252
  chemical equilibrium
    numerical model, 259–262
    theoretical model, 257–258
  commercial establishments, 5
  vs. copper mine tailings characteristics
    analytical and tailings preparation, 146–147
    EKR cells, 145–146
    electric field enhancements, 143
    electric resistance, 147–148
    high frequency electric fields, 155–157
    low frequency electric fields, 153–155
    mine tailings, 144–145
    pulsed electric DC fields, 148–151
    sinusoidal AC/DC electric field, 152–153
  definitions, 181–182
  diffusion, 11, 182
  ELECTROACROSS, 8–9
  electrochemical reactions, 182–183
  electrodes, 184–185
  electrolysis, 10
  electromigration, 10, 182
  electroosmosis, 3, 182
  electrophoresis, 5, 10–11
  evaluation, 186
  finite element method, 254
  full-scale applications, 183–184
  geochemical conditions, 185
  groups, 6–7
  Lasagna ™, 185
  LCA (*see* Life Cycle Assessment (LCA))
  mercury-polluted soil
    advantages, 295
    Almaden mining district, 296–299
    BCR distribution, 300, 302
    diagram of experimental system, 299–300
    distribution of Hg, 302
    mathematical model, 303–307
    reducible fraction, 300–301
    sequential extraction procedures, 296
    weak acid soluble fraction, 300–301
  model-based research processes, 253
  modelled system
    acid-enhanced EKR treatment, 254–255
    chemical system, 254–255
    pH and minerals concentration, 256
  organic contaminants, 183
  phases, 186
  PHREEQC, 253
  pilot-and full-scale applications, 4
  porous characteristics, 15
  porous material, 11
  practical application *vs.* academical research
    EREM symposia special editions, 14
    number of scientific publications, 12, 13
    number of US patents, 12
    origin of publications, 12, 13

Electrokinetic remediation (EKR) (*cont.*)
principle of, 9
processes in civil engineering, 14
reactive transport
numerical model, 265–269
theoretical model, 262–265
schematic representation, 182
scientific textbooks and handbooks, 8
simulation
dissolution of minerals, 270–271
pH, 270
sustainability assessment, 186
symposium, 8
waste products and environmental problems, 15
WATEQ4F, 253
workshop, 7
zero valent iron nanoparticles, 281, 282
Electrokinetics
application of artificial neural networks, 167–169
compost, 165–167
definition, 336
electrical resistance and capacitance, 162–163
electrolysis current *vs*. treatment time, 163
electromigration, 161
electroosmosis, 161
electroremediation laboratory cell, 162
for PCB removal from soil (*see* Polychlorinated biphenyls (PCB))
reverse electrodialysis (*see* Rese electrodialysis (RED))
soils with thermal shock, 164–165
technology, 162
zero valent iron nanoparticles, in porous media
chemical equilibrium, 287
coupling, 281, 282
electroosmotic permeability and mobility, 285–286
electrophoretic cell, experimental setup, 281, 283–284
Nernst–Planck equations, 285
operations, 285
transport of iron nanoparticles, 287–290
Electrokinetic treatment (EKT)
artificial neural networks application
cation exchange capacity values, 169
general scheme, 168
multilayer perceptrons, 168
tool for soil science, 167
training and prediction, 168
coefficients of hydraulic and electroosmotic permeability, 20–21
compacted specimens characteristics, 31–32
electrical conductivity, molding water content, 32–33
electroosmotic conductivity, 37, 39
electroosmotic permeability, 37–38
geosynthetics, 20
kaolin
consistency limits for salt concentrations, 31
destructured kaolin, 29–30
hydraulic and electroosmotic conductivities, 30
pore size distributions, 29–30
zeta potential, 30
microporosity, 39
oedometer, 35–36
saturated hydraulic conductivity, 33
silver electrodes corrosion, 35, 37
in soils
double layer in clays and, 22–23
electroosmotic efficiency, 22
Helmholtz-Smoluchowski theory, 23
properties, 20
zeta potential, 23–24
void ratios, 31–32, 35
volume of water collection, 37–398
Electrolysis, 10, 163
Electromigration, 10, 161, 182, 183, 208
Electronic nose, 403–404
Electronic tongue, 403–404
Electroosmosis, 161, 182
advection, 290
definition, 3
permeability and mobility, 285–286
remediation (*see* Electrokinetic remediation (EKR))
transport, 281
Electroosmotic flow
bentonite content test, 49
clay, 48–50
dewatering, 44
in hydrated clay specimens, 45
reactive transport theoretical model, 264–265
test specimen preparation, 48
transient variation of cumulative volume, 48–50
Electrophoresis, 5, 10–11
Electrophoretic transport, 281

Electroreclamation. *See* Electrokinetic remediation (EKR)
Electroremediation. *See also* Electrokinetic remediation (EKR)
  cells, 162
  definition, 336
  electrodialytic soil remediation, 336
  heavy metals removal, 336
  PCB-contaminated soils with iron nanoparticles, 343, 344
  spiked kaolin, 337
  surfactant, 337
Elimelech, M., 60
Elutriate, 362
EN-206-1, 243
Endocrine disrupting chemicals (EDC)
  alkylphenols and phthalates
    GC×GC analysis, 432–434
    sample preparation methods, 430–432
    sources of, 429
  definition, 427
Energy of activation, 101–102
Environmental matrices, 440, 450
Environmental risk assessment (ERA), 174
Equivalent circuit, electrokinetic system, 162–163
ERA. *See* Environmental risk assessment (ERA)
Ethylenediaminetetraacetic acid (EDTA), 317–318
Exergy, 58–59
External diffusion, 87
External mass transfer diffusion model, 87–88

F
Fan, G., 331–345
Feijoo, J., 209, 213
FEM. *See* Finite element method (FEM)
Ferguson, J.F., 7
Fernando, A.L., 373–393
Ferreira, A.R., 351–368
FID. *See* Flame ionization detection (FID)
Figueiredo, A.M.S., 81–107
Finite element method (FEM), 254
Fired-clay bricks, 211–212
First-order biodegradation model, 191
Flame ionization detection (FID)
  bubble diagram, two-dimensional GC, 434
  comprehensive two-dimensional GC, 446–447
  multidimensional GC with flow switching device, 443–444
Flow switching device
  benzo[a]pyrene, 443–445
  flame ionization detection, 443–444
  heart-cut configuration, 442
Fly ash
  applications, 231–232
  chemical and physical characteristics, 228
  definition, 227
  management strategies, 227
  treatment with hydraulic binders, 227
Fonseca, I.M., 81–107
Fouling, 75
Fraga, L.E., 117
Freitas, H., 387
Frutuoso, A.R., 234
Fuzzy ARTMAP artificial neural network, 413

G
García-Herruzo, F., 295–307
García-Rubio, A., 251–276, 295–308
Gardner, J.W., 403
Gas chromatography (GC)
  alkylphenols and phthalates (*see* Two-dimensional gas chromatography)
  comprehensive two-dimensional
    environmental matrices, 450
    flame ionization detection, 446–447
    heart cut configurations, 446–447
    vs. one-dimensional chromatogram, 448–449
    schematic representation, 445–446
    separation space for creosote sample, 448–449
    thermal modulator, 445–446
  with flame ionization detection, 363
  with flow switching device, 442–445
    benzo[a]pyrene, 443–44
    flame ionization detection, 443–444
    schematic representation, heart-cut configuration, 442
  polycyclic aromatic hydrocarbons, 443
Gent, D.B., 296
Gingine, V., 19–40
Godin, J., 173
Gomes da Silva, M.D.R., 427–434, 439–450
Gomes, H.I., 173–196, 279–291, 331–345
Gómez-Lahoz, C., 251–276, 295–308
Gonzalez-Castro, M.I., 429
Gordon, D.C. Jr., 407
Gray, D.H., 7
Green remediation evaluation matrix (GREM), 186

Groundwater resources
approaches, 187
characterization factor, 188–189
contamination, 186
improved life cycle impact assessment methods, 187
proposed characterization factor, 190
ranking order of substances
biodegradation, 190
degradability classification, 191–192
groundwater ubiquity score, 192
mobility classification, 190–191
models, 191
naphthalene exposure risk index, 192–193
Groundwater ubiquity score (GUS), 192
Guedes, P., 113–124, 129–139, 313–325, 351–368
Guedes, P.R., 225–244
Guo, L.B., 380
GUS. *See* Groundwater ubiquity score (GUS)
Gutiérrez, C., 143–158
Gutierrez, J.M., 404
Gu, Y.Y., 432

H
Hall, S., 251–276
Hammer, D., 384
Hansen, H.K., 3–16, 143–158, 205–221
Hayashi, K., 404
Heavy metal, 227–228, 233–234
contamination, 296, 303
Henry constant, 355
Herinckx, S., 208
Heterogeneous EKR systems, 252
Hicks, R.E., 5
High frequency electric fields, 155–157
High performance liquid chromatography, 363, 408
High-pressure liquid chromatography (HPLC), 408
Higueras, P., 297
Hijosa-Valsero, M., 361
Historic Portuguese tiles, 213–214
HMB. *See* 2-Hydroxy-4-methoxybenzophenone (HMB)
Ho, W., 386
Hoyos, J.A., 161–170
Ho, Y.S, 90
HPLC. *See* High performance liquid chromatography (HPLC)
2-Hydroxy-4-methoxybenzophenone (HMB), 352
concentrations of, 353
physico-chemical properties, 355, 356
Hyperaccumulator plants, 391, 392
biomass, 380
factors, 378
heavy metals concentrations, 379

I
Impedance spectroscopy, pollutants detection, 405–406
Industrial crops
advantages and limitations, 390–393
by-products utilization, 389
energy and fiber crops, 381
perennial herbaceous crops, 380
phytodepuration of inorganic compounds, 388
phytoremediation mechanisms
accumulation in plant biomass, 383–385
contaminants degradation, 387–388
dissipation of inorganic compounds, 387
immobilization of inorganic compounds, 385–386
schematic representation, 382–383
rape seed, 381
Inorganic compounds. *See* Phytoremediation
Instantaneous reaction model, 191
Internal diffusion, 87
Intra-particle mass transfer diffusion, 88–89
Ion-exchange membranes
levelized cost of electricity, 64
permselectivity, 66
properties, 65–66
Ionic mobility, 215, 221
Ion sensitive field-effect transistor (ISFET) biosensors, 407
Ion transport
and concentration polarization, 67–68
electrokinetics, RED
concentration polarization, 67
limiting current density, 68
schematic representation, 67–68
Iron nanoparticles
coupling electrokinetics, 281, 282
transport of
concentration across electrophoretic cell, 287–288

oxidation-reduction potential, 288–289
pH variation, 289–290
Iskanda, I.K., 8
Isosaari, P., 318

J
Jacobs, L.C.V., 116
James, M.O., 402
Javadi, A., 266
Jensen, P.E., 143–158, 225–244
Jones, J.F.P.C., 20
Jury, W.A., 192

K
KAMINA, 414
Kamran, K., 208, 209, 211, 212
Kanel, S.R., 283
Kaolin
consistency limits for salt concentrations, 31
destructured kaolin, 29–30
hydraulic and electroosmotic conductivities, 30
pore size distributions, 29–30
zeta potential, 30
Kawamura, M., 243
Keizer, P.D., 407
Kirkelund, G.M., 225–244

L
Lageman, R., 5
Lagergren, S., 89–90
Layer-by-layer (LbL) technique, 409
LCA. *See* Life Cycle Assessment (LCA)
Leaching behavior, 241–243
LECA. *See* Light expanded clay aggregates
Legin, A.V., 415
Lemming, G., 186
Levelized cost of electricity (LCOE), 64
Lewandowski, I., 374
Lichtenthaler, H.K., 363
Life cycle assessment (LCA)
3D saturated/unsaturated FRAC3DVS software, 173–174
EASEWASTE model, 174
ecotoxicological impacts, 181
environmental risk assessment, 174
evaluation of soil remediation interventions, 174–180
impacts in groundwater resources (*see* Groundwater resources)
input categories, 174
output related categories, 181
pro memoria, 181
sources, 181
Light expanded clay aggregates, 360
Li, L.Y., 266
Lima, A.T., 229, 241
Lim, J., 266
Liquid-solid interface. *See* Adsorption processes
Liu, Z., 445
Li, Z.C., 191
Logan, B.E., 60
López-Vizcaíno, R., 296
Loureiro, J.M., 81–107
Low frequency electric fields, 153–155
Lu, B., 266
Lumped parameter systems, 103
Lygina, E.S., 81–107
Lyubchik, S.B., 81–107
Lyubchyk, A.I., 81–107

M
Macdonald, A.G., 406
Magro, C.C., 225–244
Marchiol, L., 381
Marriott, P., 439–450
Mass action process, 86
Mass transfer
external diffusion model, 87–88
intraparticle diffusion, 88–89
Mateus, E.P., 43–55, 113–124, 129–139, 351–368, 439–450
Matyscak, O., 216
MC-LR structure, 114, 115
Meeks, Y., 192
Meers, E., 384
Mercury-polluted soil
Almaden mining district, 296–299
electroremediation experiments
advantages, 295
BCR distribution, 300, 302
diagram of, 299–300
Hg distribution, 302
reducible fraction, 300–301
sequential extraction procedures, 296
weak acid soluble fraction, 300–301
mathematical model
chemical equilibria, 305–307
ionic flux, 303–304
transport equations, 304–305
Meriluoto, J., 120
Meyer, H., 402

Microcystins, 114, 115
Mine soil, 323
Mine tailings. *See* Copper mine tailings
Mitchell, J.K., 7
Mleczek, M., 387
MLP. *See* Multilayer perceptrons (MLP)
Model-based research processes, 253
Monod model, 191
Moreira, M.A., 427–434
Mortar, 234, 239–241
MSWI. *See* Municipal solid waste incineration (MSWI)
Multidimensional chromatographic techniques
 comprehensive two-dimensional GC
  environmental matrices, 450
  flame ionization detection, 446–447
  heart cut configurations, 446–447
  vs. one-dimensional chromatogram, 448–449
  schematic representation, 445–446
  separation space for creosote sample, 448–449
  thermal modulator, 445–446
 environmental matrices, 440
 with flow switching device
  benzo[a]pyrene, 443–445
  flame ionization detection, 443–444
  schematic representation, heart-cut configuration, 442
 limitations, 441–442
 organic pollutants, 439–440
 polycyclic aromatic hydrocarbons, 443
 statistical theory of overlap, 441
Multilayer perceptrons (MLP), 168
Multi-physical EKR systems, 252
Municipal solid waste, 226
Municipal solid waste incineration (MSWI)
 advantages and disadvantages, 227
 air pollution control residues and fly ash
  applications, 231–232
  characteristics, 229–230
  elements, 228
  management strategies, 227
  mineral compounds, 230
  treatment with hydraulic binders, 227
 electrodialytic pre-treatment, 234–238
  compressive strength, 234, 239, 240
  leaching behavior, 241–243
  workability, 239
 electrodialytic process
  cell design, 232–233
 management strategies, 227
 solid particles production, 227–229
 treatment with hydraulic binders, 227
 Waste Incineration Directive, 226–227

N

Nanofiltration, 117–118
Nanoremediation
 bimetallic nanoparticles, 338
 coupled to electrokinetics
  PCB concentrations, 338–339
  saponin *vs.* Tween 80, 341
  soil physical and chemical characteristics, 341–343
  surfactants, 338
  two-*vs.* three-compartment electrokinetic cylindrical cell, 338, 340
 dechlorination of PCB, 337–338
 soil organic matter, 338
Naphthalene exposure risk index (NERI), 192–193
Natural stones, 212–213
NERI. *See* Naphthalene exposure risk index (NERI)
Nernst–Planck equations, 285
Nernst–Planck–Poisson (NSS) system, 264
Net power density, 75–76
Newton-Raphson (NR) method, 261–262
Niu, Z., 381
Non-linear reactive-transport model, 254, 269
4-Nonylphenol, 428
Nord, A.G., 213
NR method. *See* Newton-Raphson (NR) method
Nunes, L.M., 173–196
nZVI. *See* Zero valent iron nanoparticles (nZVI)

O

Ockenden, W.A., 334
Octanol–water partition coefficient, 354
4-Octylphenol, 428
Olsen, H.W., 7
One-dimensional gas chromatography (1D-GC), 440, 448–449
Organic carbon partition coefficient, 354
Osmotic pressure, concentrated saline streams, 60–61
Ottosen, L.M., 3–16, 135, 143–158, 205–221, 225–244, 331–345

Owens, J.W., 174
Oxybenzone. *See* 2-Hydroxy-4-methoxybenzophenone (HMB)

P
Page, C.L., 266
Pamukcu, S., 5, 43–55, 279–291
Pattern recognition methods
  data analysis
    clustering analysis, 412
    principal component analysis, 411
    statistical learning theory, 412
    support vector machines, 412
  data fusion approach, 412–413
  data pre-processing, 411
  feature extraction, 411
  fuzzy ARTMAP artificial neural network, 413
Pattle, R.E., 62
Pawlowski, S., 57–76
Paz-Garcia, J.M., 209, 215, 251–276, 295–308
PCA. *See* Principal component analysis (PCA)
Pedersen, A.J., 233
Perennial herbaceous crops, 380
Permeable reactive barriers (PRB), 319
Persistent organic pollutants (POP). *See* Polychlorinated biphenyls (PCB)
pH
  distribution, 49–52
  and minerals concentration, 256
Pharmaceuticals
  concentrations, 367
  removal efficiency
    planted beds, 367
    unplanted beds, 366–367
Pharmaceuticals and personal care products (PPCP)
  contamination
    caffeine, 352, 353
    oxybenzone, 352, 353
    triclosan, 352
  experimental design, 362–363
  pharmaceuticals removal efficiency, 366–367
  physical and chemical characteristics
    Henry constant, 355, 356
    octanol–water partition coefficient, 354
    organic carbon partition coefficient, 354
  phytoremediation
    advantages, 355
    constructed wetlands, 359–360
    macrophytes plant species, 361–362
    process, 357–359
    SWOT analysis, 357
    wetlands, 359
  procedure, 363
  salt marsh plants, 363–365
  wastewater treatment plants, 351–353
Phillips, J.B., 445
Phosphorus recovery potential
  ashes, 132
  biosolids, 131
  concentrations, 132
  effluent of WWTP, 131
  liquid phase, 131
  sewage sludge ashes, 133 (*see also* Sewage sludge ashes (SSA))
PHREEQC, 253
Phytodepuration, 388
Phytoextraction, 384
Phytoremediation, 320–321
  advantages, 355
  constructed wetlands, 359–360
  contamination, 375
  factors, 375
  hyperaccumulator plants, 374
    biomass, 380
    factors, 378
    heavy metals concentrations, 379
  inorganic compounds in soils, 376
  macrophytes plant species, 361–362
  market values, 374
  process, 357–359
  radioactive contamination, 376
  soil and water remediation technologies, 377–378
  SWOT analysis, 357
  use of industrial crops
    advantages and limitations, 390–393
    by-products utilization, 389
    energy and fiber crops, 381
    mechanisms, 382–388
    perennial herbaceous crops, 380
    phytodepuration of inorganic compounds, 388
    rape seed, 381
  wetlands, 359
Phytotechnologies, 320–321
Phytovolatilization, 387
Pilon-Smits, E., 354
Pilot scale test, electro-desalination, 217–221
Pinto, M.M., 81–107

Pollutants, 174, 181, 184
electrochemical methods, 405
high performance liquid chromatography, 408
impedance spectroscopy, 405–406
sensors integrated on FETs, 407
spectrophotometric methods, 407
surface acoustic wave technique, 406–407
water recognition
by electronic noses, 414
by electronic tongues, 414–418
Polychlorinated biphenyls (PCB)
chemical structure, 333
contaminated soils and sediments
classification of remediation technologies, 335
electrokinetics with nanoremediation, 333–343
electroremediation, 336–337
nanoremediation, 337–338
industrial application, 333
soil, 332
toxic equivalency factor, 334
Polycyclic aromatic hydrocarbons, 443
POP. *See* Polychlorinated biphenyls (PCB)
Porous media. *See* Zero valent iron nanoparticles
Poulticing, 207
PPCP. *See* Pharmaceuticals and personal care products (PPCP)
Prasad, M.N.V., 8
Pressure drop, 74
Principal component analysis (PCA), 411
Probstein, R.F., 5, 7
Profiled membranes, 73–74. *See also* Corrugated membranes
Pseudo-first-order model application, 93–95
Pseudo-first-order rate model, 89–90
Pseudopersistent contaminants. *See* Pharmaceuticals and personal care products (PPCP)
Pseudo-second-order model application, 95–97
Pseudo-second-order rate model, 90–91
Pulsed electric DC fields
chemical mechanisms of dissolution, 149
electrodialytic remediation conditions, 150
normalized concentration of copper, 149–151
transport across cation-exchange membrane, 151
voltage *vs.* time, 148

Q
Quina, M.J., 232

R
Rajkumar, M., 387
Rao, P.S.C., 192
Raposo, M., 401–419
Reactive transport
numerical model
finite element discretization, 265–266
non-linear finite elements model, 269
theoretical model
electrochemically-induced transport, 264–265
Nernst–Planck–Poisson system, 264
representative volume element, 262–263
Realini, P.A., 407
Real soil, 296
Reboreda, R., 359
RED. *See* Rese electrodialysis (RED)
Reddy, K.R., 5, 8, 295
Redox couples, 72–73
Renaud, P.C., 7
Renewable energy, 58
Representative volume element, 262–263
Reuss, F.F., 1
Reverse electrodialysis (RED)
chronopotentiometry, 69–70
fouling, 75
ion-exchange membranes
levelized cost of electricity, 64
permselectivity, 66
properties, 65–66
ion transport
concentration polarization, 67
limiting current density, 68
schematic representation, 67–68
metrics, 63–64
monovalent *vs.* multivalent Ions, 70–72
net power density, 75–76
pressure drop, 74
principle, 62–63
profiled membranes, 73–74
redox couples, 72–73
Rhizofiltration, 383
Ribeiro, A.B., 3–16, 43–55, 81–107, 113–124, 129–139, 173–196, 225–244, 279–291, 313–325, 331–345, 351–368, 439–450

Ribeiro, P.A., 401–419
Ristinmaa, M., 251–276
Riul, A. Jr., 403
River–sea interface
  earth's water cycle, 59
  salinity gradient energy, potential of, 60
Rodríguez-Maroto, J.M., 6, 279–291, 295–308
Rojo, A., 143–158
Rörig-Dalgaard, I., 209, 212, 214
Ryu, B.-G., 323

S
Sabbas, T., 232
Salinity gradient energy (SGE)
  mechanism, 57–58
  osmotic pressure, concentrated saline streams, 60–61
  RED (*see* Rese electrodialysis (RED))
  renewable energy, 58
  river–sea interface
    Earth's water cycle, 59
    gradient potential, 60
  thermodynamic/available energy, 58–59
Salt weathering, 205–207
Sample preparation methods, alkylphenols and phthalates
  cartridges and disks, solid phase extraction, 432
  cold SPME system, phthalate migration, 430–431
  solid phase microextraction, 430–431
Sánchez-Avila, J., 408
Sangodkar, H., 402
Santos, J.N., 35
SAW technique. *See* Surface acoustic wave (SAW) technique
Scaling-up, 296, 299
Schnoor, J.L., 365
Scozzari, A., 414
SD. *See* Systems dynamics (SD)
Segall, B.A., 7
Sensors
  arrays applied to pollutants
    recognition by electronic noses, 414
    recognition by electronic tongues, 414–418
  detection of pollutants
    electrochemical methods, 405
    high performance liquid chromatography, 408
    impedance spectroscopy, 405–406
    sensors integrated on FETs, 407
    spectrophotometric methods, 407
    surface acoustic wave technique, 406–407
  electronic nose and tongue, 403–404
  layers
    materials, 410
    methods for preparation, 408–409
  pattern recognition methods
    data analysis, 411–413
    data pre-processing, 411
    feature extraction, 411
SEP. *See* Sequential extraction procedures (SEP)
Sequential extraction procedures (SEP), 296
Serna, A., 161–170
Sewage sludge ashes (SSA)
  acid extraction, 134
  electrodialytic separation, 134–135
  heavy metals percentage, 136–137
  organic contaminants percentage, 137–138
  pH, 133–134
  phosphorus percentage, 136
  sludge testing, 137
SGE. *See* Salinity gradient energy (SGE)
Shale rock, 44
Sharma, H.D., 8
Sheng, D., 266
Shrestha, R.A., 43–55
Sillanpää, M., 318
Sinusoidal AC/DC electric field, 152–153
SiteWise™, 186
Skibsted, G., 213, 215
SLT. *See* Statistical learning theory (SLT)
Smith, D.W., 266
Sohn, J.H., 414
Soils
  arsenic contamination, 314–316
  dewatering and hydrodynamic consolidation, 24–26
  EKR (*see* Electrokinetic remediation (EKR))
  electrokinetic processes
    double layer in clays and, 22–23
    electroosmotic efficiency, 22
    Helmholtz-Smoluchowski theory, 23
    zeta potential, 23–24
  electrokinetic treatment suitability, 20
  electrolysis charge and characterization parameters, 165
  LCA (*see* Life Cycle Assessment (LCA))
  structure and effects on hydraulic behavior, 26–29
  subjected to thermal shock, 164

Soils (*cont.*)
volcanic soil sample characteristics, 164
and water remediation technologies, 377–378
Solid-liquid interface, 82, 83, 86, 104
Solid phase extraction (SPE), 432
Solid phase microextraction (SPME), 430–431
SPE. *See* Solid phase extraction (SPE)
Spectrophotometric methods, pollutants detection, 407
SPME. *See* Solid phase microextraction (SPME)
Spoof, L., 120
SSA. *See* Sewage sludge ashes (SSA)
Statistical learning theory (SLT), 412
Stevens, K.J., 361
Strengths, weaknesses, opportunities and threats (SWOT) analysis, 357
Subefflorescence, 206
Sun, T.R., 341
Support vector machines, 412
Surface acoustic wave (SAW) technique, 406–407
Surface diffusion, 87
Sustainable power generation. *See* Salinity gradient energy (SGE)
Swelling behavior, bentonite clay, 46–48
SWOT analysis. *See* Strengths, weaknesses, opportunities and threats (SWOT) analysis
Systems dynamics (SD), 103

T
Tao, Y., 318
Taylor, D.M., 406
TCE. *See* Trichloroethylene (TCE)
TCS. *See* Triclosan
TEF. *See* Toxic equivalency factor (TEF)
Teixeira, M.R., 113–124, 173–196
4-Tert-butylphenol (4-t-BP), 428
Thermal modulator, 445
Thermal shock, 164–165
Thirumavalavan, M., 116
TiO_2 photocatalysis, 117
Toxic equivalency factor (TEF), 334
Tran, N., 117
Trichloroethylene (TCE), 193–196
Triclosan, 352
concentrations of, 353
physicochemical properties, 355, 356
Tronner, K., 213
Two-dimensional gas chromatography
advantages, 433
comprehensive, 432
flame ionization detector, bubble diagram, 434
modulator, 432–433

U
Ultrafiltration, 117–118

V
van der Sloot, H.A., 227
Vapnik, V., 412
Vázquez, M.V., 161–170
Velizarov, S., 57–76
Vereda-Alonso, C., 295–308
Villen-Guzman, M., 251–276, 295–308

W
Waste Incineration Directive, 226–227
Wastewaters, 378, 380–381, 388, 390
Wastewater treatment plants (WWTP), 131, 351
factors, 130
human excreta, 130
sources, 129
WATEQ4F, 253
Water content, in clay, 50
Water electrolysis, 10
Water pollution recognition
by electronic noses, 414
by electronic tongues
drinking water quality, 414–415
metal ions detection, 414–415
pollutants detection, 416–418
water toxicity detection, 416
Water treatment plant (WTP), 119
Wellburn, A.R., 363
Winquist, F., 403
Wise, D.L., 8
Worsfold, P.J., 407
WTP. *See* Water treatment plant (WTP)
WWTP. *See* Wastewater treatment plants (WWTP)

Y
Yan, S., 116
Yuan, C., 319

Z

Zelm, R.V., 187
Zero valent iron nanoparticles (nZVI), 332, 338–340, 342–343
  coupling electrokinetics, 281, 282
  electrophoretic cell, 281, 283–284
  model
    chemical equilibrium, 287
    electroosmotic permeability and mobility, 285–286
    Nernst–Planck equations, 285
    operations, 285
  transport of
    iron concentration across electrophoretic cell, 287–288
    oxidation-reduction potential, 288–289
    pH variation, 289–290
Zeta potential, 50–52
Zhang, A.P., 43–55
Zhang, C., 433
Zhang, D.Q., 354
Zhou, D.-M., 313–325

MIX
Papier aus verantwortungsvollen Quellen
Paper from responsible sources
FSC® C105338

If you have any concerns about our products,
you can contact us on
ProductSafety@springernature.com

In case Publisher is established outside the EU,
the EU authorized representative is:
Springer Nature Customer Service Center GmbH
Europaplatz 3, 69115 Heidelberg, Germany

Printed by Libri Plureos GmbH
in Hamburg, Germany